# THE 448TH BOMB GROUP (H)
## Liberators over Germany
### in World War II

# THE 448TH BOMB GROUP (H)
## Liberators over Germany
### in World War II

Jeffrey E. Brett

**Schiffer Military History**
Atglen, PA

**Dust jacket and aircraft profile artwork by S.W. Ferguson, Colorado Springs, CO.**

**COLOR OF THE DAY**
The dust jacket artwork depicts the late day return of the 448th BG Liberators on April 22, 1944, shortly before their interception by Me 410 fighter-bombers that had followed them home to their landing field at Seething, England. In the foreground is the Liberator piloted by 715th Sqdn. Lt. Marion Peek whose crew took grim consolation in gunning down one of the attackers in the onslaught. The flight is seen firing green identification flares – the color of the day – as their signal to the British coast watchers of their fateful homecoming.

Book design by Robert Biondi.

Copyright © 2002 by Jeffrey E. Brett.
Library of Congress Catalog Number: 2001091245.

All rights reserved. No part of this work may be reproduced or used in any forms or by any means – graphic, electronic or mechanical, including photocopying or information storage and retrieval systems – without written permission from the copyright holder.
"Schiffer," "Schiffer Publishing Ltd. & Design," and the "Design of pen and ink well" are registered trademarks of Schiffer Publishing, Ltd.

Printed in China.
ISBN: 0-7643-1464-5

We are always looking for people to write books on new and related subjects. If you have an idea for a book, please contact us at the address below.

| Published by Schiffer Publishing Ltd.<br>4880 Lower Valley Road<br>Atglen, PA 19310<br>Phone: (610) 593-1777<br>FAX: (610) 593-2002<br>E-mail: Schifferbk@aol.com.<br>Visit our web site at: www.schifferbooks.com<br>Please write for a free catalog.<br>This book may be purchased from the publisher.<br>Please include $3.95 postage.<br>Try your bookstore first. | In Europe, Schiffer books are distributed by:<br>Bushwood Books<br>6 Marksbury Ave.<br>Kew Gardens<br>Surrey TW9 4JF<br>England<br>Phone: 44 (0)208 392-8585<br>FAX: 44 (0)208 392-9876<br>E-mail: Bushwd@aol.com.<br>Free postage in the UK. Europe: air mail at cost.<br>Try your bookstore first. |
|---|---|

# Contents

Introduction ............................................................................................................ 6
Acknowledgments ................................................................................................. 8

| | | |
|---|---|---|
| *Chapter 1* | Origins: March-November 1943 ........................................................ | 10 |
| *Chapter 2* | Welcome "Home": November 1943 .................................................... | 20 |
| *Chapter 3* | Combat: December 1943 ..................................................................... | 32 |
| *Chapter 4* | The Enemy ... Weather: January 1944 ................................................ | 46 |
| *Chapter 5* | ... And More Weather: February 1944 ................................................ | 57 |
| *Chapter 6* | Combat Legs: March 1944 .................................................................. | 68 |
| *Chapter 7* | Luftwaffe over Seething: April 1944 ................................................... | 79 |
| *Chapter 8* | Buildup: May 1944 .............................................................................. | 99 |
| *Chapter 9* | Invasion: June 1944 ............................................................................. | 109 |
| *Chapter 10* | English Summer: July 1944 ................................................................ | 123 |
| *Chapter 11* | Home by Christmas?: August 1944 .................................................... | 133 |
| *Chapter 12* | Versatility: September 1944 ................................................................ | 147 |
| *Chapter 13* | Another Winter: October 1944 ............................................................ | 156 |
| *Chapter 14* | All Weather Air Force: November 1944 ............................................. | 166 |
| *Chapter 15* | Holiday Season: December 1944 ........................................................ | 175 |
| *Chapter 16* | Another Year: January 1945 ................................................................ | 182 |
| *Chapter 17* | How Much Longer?: February 1945 ................................................... | 192 |
| *Chapter 18* | Luftwaffe Resurgent: March 1945 ...................................................... | 202 |
| *Chapter 19* | Not Over Yet: April 1945 .................................................................... | 218 |
| *Chapter 20* | The End: May-June 1945 .................................................................... | 228 |

*Appendices*
*Appendix 1*: Honor Roll .................................................................................... 236
*Appendix 2*: Aircraft ......................................................................................... 246
*Appendix 3*: Aircrews ....................................................................................... 252
*Appendix 4*: Leadership .................................................................................... 291
*Appendix 5*: Missions ....................................................................................... 292
*Appendix 6*: Statistical Summary ..................................................................... 298

448th Bomb Group Aircraft in Profile ................................................................ 299
Index .................................................................................................................... 305

# Introduction

This is the simple story of the 448th Bombardment Group in World War II. Simple in the fact that it outlines the Group's involvement from its inception until its return to the U.S. following V-E Day. Complicated, however, are the personal stories of the men who spent years away from their civilian lives and families. Many lost their closest friends; numerous others suffered wounds that would never completely heal. Yet all of them persevered to provide the framework for the world we live in today.

It is the story of duty that is almost unheard of today. It is the story of young men whose childhoods were shaped by the greatest economic catastrophe our nation has ever endured yet as young adults they were called on once more. It is not a story of turning your back on responsibility when life is not fair. Life is not fair and these men are a perfect testament to the fact. Yet, they faced the adversity and horror of war with a resolve to win and responsibility to their fellow man.

This is their story. Little has been written about the 448th Bomb Group. No Medal of Honor recipients hailed from the Group. Although commended on several occasions, the Group did not receive as much attention or press as did some of their counterparts. Without the fanfare, the Group executed its mission to the utmost.

Like most unsung heroes of any war, the 448th consisted of ordinary men from all walks of life: mechanics, lawyers, farmers, doctors, students, fathers, and brothers. They arrived at recruiting stations from all corners of the United States. Some men even joined the Group from other countries. Unfortunate times thrust these men of diverse backgrounds into a situation which they had little control. There was plenty of complaining; after all, a soldier has a right to complain. But, when it came time to do the mission, they excelled.

From the inception of the 448th until the long-awaited trip home, many life-changing decisions were a normal everyday occurrence for hundreds of men. The war brutally claimed thousands of young, promising lives based on these decisions. For others, it graciously smiled upon them. No one knew "who it was going to be!"

These seemingly trivial decisions shaped the fate of thousands of families not just in America but throughout Europe. We should never allow ourselves to forget what these 'average' American men sacrificed to give us what we enjoy today. There is nothing average about their sacrifices. All sacrifices were minor compared to the eighty-five men killed in action, the 875 men who were missing in action, or the 119 men who were wounded or died from wounds received. Nevertheless, they were still sacrifices.

After the war, they continued to excel and directly contributed to the tremendous success we are privy to in this country. Men returned to their families and pieced their civilian careers together again. Several remained in the military and continued their faithful service. A future triple ace of the Korean War, Joseph McConnell, called the 448th home as did six future general officers.

These men set the standard. They made America great and it is our challenge to maintain it. Words can never express what thanks these men deserve. But to all of you, Thank You!

## BY PROXY

*I've often wished that I could go*
*To visit London Town*
*And other places round about*
*From "Cork to County Down."*
*To wander freely o'er the land*
*The miracle of God*
*Seeing scars that he has healed*
*And walk where heroes trod.*

*Many years have come and gone*
*Since all the anger stilled*
*The fields are green and peace pervades*
*The land where blood was spilled*
*The Drones, the Flak, the Shaking Roar*
*No more assault the shore*
*The trains now run on schedule*
*Unlike they did before.*

*The Past is not completely gone*
*Sure, it cannot be*
*Too many ghosts still walk the moors*
*And swim in from the sea*
*They gather at the Briefing Hut*
*For "Target for Tonight"*
*Then pause, shake hands, and shed a tear*
*And go out for the fight.*

*You may ask, and rightly so,*
*How can I know all this?*
*After all, I haven't been to the land*
*That God has kissed*
*I have not seen, nor have I felt*
*Her breath upon my face*
*But I have been, and lived it all*
*Through friends who took my place*

*- Anonymous*

# Acknowledgments

I have always enjoyed reading about World War II, but I never dreamed of writing about it. What started as a hobby grew into a passion that led to this book. When I started, my goal was to record the stories, especially the personal ones, for future generations to read, learn and appreciate. They are priceless. I have attempted to intertwine these first-hand personal accounts with the impersonal mission histories and other forms of "official" record keeping. I believe this provides the most accurate yet personal account of the events. I have tried to eliminate all interpretations and record just the facts. However, in some instances the facts are simply not recorded and all the evidence is purely circumstantial. In this case I documented what is absolute and what is speculation. I have diligently corroborated all information but errors are sure to happen. Hopefully, they are insignificant and rare. In one of his books, Mr. Ed Chu included a quote from the book "About Face" that best describes the process of compiling war stories. "The second problem with war stories is they have their genesis in the fog of war. In battle, your perception is often only as wide as your battle sights. Five participants in the same action, fighting side by side, will often tell entirely different stories of what happened, even within hours of the fight. The story each man tells might be virtually unrecognizable to the others, but that does not mean it any less true." I pray this book will factually represent the great sacrifices these men made.

Many people are responsible for this book. Without the tremendous and tireless efforts of Patricia Everson and her 448th Bomb Group Collection, this would have been impossible. Without a doubt, she is the single source for information on the 448th Bomb Group. She has graciously and patiently answered questions and provided the information that much of this book is based. Also a special thanks belongs to Bert LaPoint for the use of his outstanding photos. Working in the photo lab, he had unprecedented access to the photographic history of the Group and the foresight to preserve it. Another special thanks to Phyllis DuBois who proofread and corrected my attempts at writing. Finally a special thanks to all members of the 448th Veterans Association who have been so quick to accept me as a friend and let me ask the seemingly endless barrage of questions.

I used several books as reference during the writing of the book. The best books detailing the 8th Air Force are Roger Freeman's "The Mighty Eighth," "Mighty Eighth War Manual," and "Mighty Eighth War Diary." Other very good sources more specific to the 448th BG are James Hoseason's "The 1000 Day Battle," "Fields of Little America" by Martin Bowman, "Sporty Course" by Jack Swayze, and "Mission Failure and Survival" by Charles McBride. The official records of the Group were also accessed at both the USAF Historical Research Agency and the National Archives and Records Administration. The newsletters of the Station 146 Tower Association were also a great source of information concerning the airfield past and present.

The most fascinating and enjoyable part of the research involved the personal stories, diaries and correspondence with the veterans. The following individuals generously gave their time by sharing their stories and experiences. They are the reason for the book. They include: Carl Ahrendt, A.R. Albrecht, James Ames, Edward Anderson, Robert Ash, Stan Baldwin, Donald Beck, Barney Bernard, Ronald Berryhill, Dick Best, Chuck Blaney, William Blum, Don Bodiker, Ray Boll, Marjorie Bollschweiler, Chuck Bonner, Dale Bottoms, Tom Brittan, Ernest Brock, William Brown, Stephen Burzenski, Lawson Campbell, Charles Carn, Richard Casterline, Aubrey Cates, Edward Chu, Thaine Clark, Harold Closz, George Cooksey,

*Acknowledgments*

Lee Conner, George Copeland, Benedict Corsiglia, John Cushman, John Davis, Frank DeCola, Lois DiLorenzo, Miles Drawhorn, Ed Drouin, Reg Dunn, George Dupont, Paul Dwyer, Henry Ebeling, Peter Edgar, Carl Eggert, George Elkins, Leroy Engdahl, Herman Engel, Benjamin Everett, Wallace Forman, Merritt McGahan, William Gamble, Jack Garrett, Gene Gaskins, John Gedz, Noble Germany, George Glevanik, John Grunow, Chester Hackett, Robert Harper, Hugh Harries, Wade Hartley, Clyde Hatley, Hershel Hausman, J.A. Hey, Marvin Hicks, Arthur Hipkins, Paul Homan, Elmer Homelvig, Arthur Howell, Arthur Hunt, Matthew Hurley, Bill Jann, Allan Johnson, Ben Johnson, Stanley Johnson, Theresa Jones, Norman Kanwisher, Fred Kerniss, Robert Kessler, Richard Kimball, Jaap van der Kuylen, Chester Labus, Bob Lambertson, Ed Langton, William Lantz, Stephen Lawnicki, Cater Lee, Vincent Liedka, Elbert Lozes, Charles McBride, John McCune, Jack McDaniel, Thomas Miller, William Morris, Ralph Nicholas, Milton Nichols, Jack Parker, Pat Patterson, Jim Pegher, Ricardo Perez, Dean Peterson, Neal Pettit, Ray Pytel, Patrick Raspante, Julius Rebeles, Phillip Ray, Russel Reindal, John Richmond, John Rowe, Robert Sampson, Lewis Sarkovich, Paul Schauwecker, John Schlicher, Edward Schroeder, Elvin Sheffield, Vernon Siegel, Douglas Skaggs, Jack Smith, John Snider, Basil St. Dennis, John Stanford, Rocky Starek, Arthur Steele, James Straub, Roy Stroop, Pat Terranova, Morris Thomson, Len Thornton, Irv Toler, Jim Turner, Dale Van Blair, Robert Voight, Wayne Wanker, Ira Wells, Ralph Welsh, Nancy Westgate, Lucian Whipple, Dick Wickham, Emily Wilbur, Frel Youngblood, Stanley Zabrowski, Donald Zeldin, John Zima, Joseph Zonyk, and Mary Louise Zubialde. To all of you thank you for making this possible.

Jeff Brett

one

# ORIGINS:
# March-November 1943

America's entrance into World War II provided an opportunity for American airpower strategists to put their theory of precision bombing into practice. Many people believed air power alone could bring Germany and Japan to defeat. Britain's repulse of the German aerial onslaught during the Battle of Britain offered a contrasting view. Furthermore, British attempts at daylight bombing resulted in prohibitive losses. Allied fighters lacked the range to escort the bombers into Germany, leaving the bombers to the wrath of the Luftwaffe. The damage inflicted by the German fighters forced British bombers into the relative safety of the night. As a result, RAF leaders abandoned the idea of daylight precision bombing and warned their American counterparts of disaster if a policy of daylight bombing was adopted.

However, American airmen retained their belief in daylight precision bombing. Despite the German and British failures, they believed heavily defended bombers could carry the war to the doorsteps of Germany and do it in daylight. With the purpose of defeating Germany first, Allied strategists set out to develop a framework to accomplish this goal. In January 1943, at the Casablanca Conference, Allied leaders reached agreement on the objective of the strategic air campaign. Their goal was the "destruction and dislocation of the German military, industrial, and economic system and the undermining of the morale of the German people to the point where their capacity for armed resistance is fatally weakened."[1] This directive, the Casablanca Directive, set in motion planning by the RAF and USAAF to implement a strategic air plan.[2] The British, wary of continued high casualties, preferred attacking at night. The green and relatively inexperienced Americans insisted on daylight attacks. The two compromised and the resulting effort forced Germany to deal with Allied bombers both day and night.

In order to implement this plan, tacticians addressed nineteen targets. The German aircraft industry received the highest priority. Ball bearings and petroleum followed closely.[3] This huge undertaking called for an unprecedented build up of bomber and fighter forces. The first phase called for eight hundred heavy bombers by July 1943. Phase II, starting in October 1943, called for 1,192 heavy bombers. Phase III required 1,746 heavy bombers by January 1944 and 2,702 heavy bombers by June 1944 for Phase IV.[4] Phase III included the 448th Bombardment Group (Heavy) of the USAAF.

A restricted letter from the War Department dated 6 April 1943 issued instructions activating certain Army units, including the 448th Bombardment Group. Sixteen days later Special Order #58 from Headquarters Second Air Force at Fort George Wright, Washington, outlined the evolution of the 448th and its four squadrons. The 29th Bombardment Group at Gowen Field, Idaho, provided the bulk of the initial cadre for the Group Headquarters and the four flying squadrons, the 712th, 713th, 714th and 715th. Each squadron was assigned eleven officers and fifty-eight enlisted men, except the 713th BS which had twelve officers. Each squadron consisted of one "model" crew that would train new crews. A model crew consisted of a pilot, copilot, navigator, bombardier, operations officer, engineer-gunner, radio operator, line chief and two mechanics. These men provided the nucleus from which the squadrons grew.

Special Order #151 released the 448th cadre from the 29th Bomb Group for the specific purpose of attending the Army Air Forces School of Applied Tactics (AAFSAT) at Orlando, Florida. Established in 1942, AAFSAT consisted of two phases, one academic and one practical. The academic portion consisted of two weeks of lectures and conferences concerning operations, intelligence, and problems of command. The practical phase, conducted at an outlying AAFSAT base, provided crews with operational experience. Crews were scheduled to complete fifty hours of flying, simulating combat missions during the final two weeks. Following the thirty-day school, orders directed the group to Wendover Field, Utah. The 448th Bomb Group now had identity.

Four fully loaded B-24s departed Gowen Field headed south for AAFSAT. The luckier ones flew while the less fortunate individuals endured a five-day, four-night train ride to Florida. After the endless clatter of a train ride, the weary rail-bound members of the newly formed 448th Bomb Group disembarked in scenic Orlando. The anticipation of Florida quickly wore off. Sleeping assignments and a quick settling-in provided the only break before crews were instructed to report to the base theater. At the theater, the group received its first introduction to its Commanding Officer, Colonel James Thompson. Colonel Thompson, a thirty-seven year old regular army officer who received his commission and wings in 1930, quickly introduced his staff and then proceeded to get down to business. His previous experience working in the Inspector General Department taught him the importance of attention to details. He intended to impart this lesson to his new subordinates. He explained the 448th's purpose at the school was to prepare for combat and lay the foundation for an entire Heavy Bombardment Group. Each crewmember received class assignments according to his specialty and Col. Thompson dismissed the Group. The long road to combat and victory in Europe had begun.

The two weeks of academic study ended with the Group moving south to Pinecastle Army Air Base for the practical portion of the program. Life quickly changed. No permanent buildings existed at Pinecastle; the runways were the only completed structure. Gone were the nice barracks. The senior officers lived in tents with wooden floors. The remainder of the troops pitched tents over the bare ground. A trip to the far away latrine provided good training for future applications in England. They were there to fly, nothing else. B-24s were not the only thing flying at Pinecastle. Large mosquitoes roamed the night air making sleep impossible without the protection of mosquito netting. Adding to the misery, summer arrived in south Florida bringing heat and rain.

Life at Pinecastle revolved around twenty-four hour alerts. B-24s flew constantly. No notice launches kept crews scrambling at all hours of the day or night. The ground crews worked tirelessly keeping the planes airborne. Crews spent more time in the air at Pinecastle than they did on the ground. Simulated enemy fighter attacks, formation flying, and mock attacks on Tallahassee and Key West taught the new crews hard-learned lessons. In combat, the skies were less friendly and forgiving

Col. James Thompson was given the task of creating a combat-ready Heavy Bombardment Group from scratch. The thirty-seven year old regular army officer hand-picked his staff and led the 448th Bomb Group into combat less than nine months from its inception. (Gaskins)

Mechanic Ray Slocumb uses some free time on the way to Florida to catch some sleep. Once the men started AAFSAT, free time evaporated as days were spent learning the fundamentals of a heavy bomb group. (Skaggs)

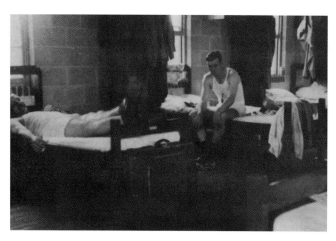

Col. James Thompson, left, and Lt. Maynard Ingalls, right, resting in the barracks at Orlando during the academic portion of AAFSAT. The conditions were plush compared to what the future held for the infant Group. (Skaggs)

Stan Filipowicz, radio operator, Andy Hau, bombardier, George Glevanik, engineer, line chief Paul R. "Frenchy" Riendeau (left to right) in front of 095 OLD FAITHFUL following a training flight in the summer of 1943. (Skaggs)

with American bomber formations taking regular poundings from German fighters and anti-aircraft batteries.

Life had its brighter moments at Pinecastle. After receiving a daylight only pass, one crew caught a rickety bus to San Alando recreation area in Orlando. On the way, the bus engine caught fire and the passengers were forced to help extinguish the blaze. Nonetheless, they reached the park. The large swimming hole and associated grounds provided a cool refreshing break as did the pleasant sights and civilian companionship. Good things never last and the guys returned to Pinecastle with fond memories of Florida but anticipation about the future. Wendover Field waited.

The 448th completed its thirty-day training on 30 June 1943 and left hot and humid Florida for the hot, parched salt flats of western Utah. One echelon left by rail, the other by air. On the same day, an advanced team from Gowen Field arrived at Wendover. Three days later the remaining personnel from Gowen arrived at Wendover Field and prepared for the arrival of the AAFSAT graduates. The flying echelon arrived on 6 July and the ground echelon arrived from Florida the next day, making the 448th an autonomous unit for the first time.

Wendover, Utah, located on the Utah/Nevada border, provided the perfect setting for combat preparation. Colonel Thompson remarked, "This is the place to train troops. No contamination with civilians!" He meant it, too. The base's purpose was to train B-24 crews. The tar-paper buildings provided shade from the sun and 120 degree heat, but little else. Trees, bushes, and grass did not exist. The olive drab airplanes were the only things green on base. A small base theater existed, but only occasionally operated. The Post Exchange, also very small, carried only the bare necessities. A single road led to the State Line Hotel, so named because the state line ran down the middle of the hotel. The town of Wendover consisted of "Spikes", "The A1 Cafe", a garage, and a gambling house two blocks from the base on the Nevada side of the line. It was spartan to say the least.

New crews arrived, swelling the ranks of the skeleton group. On 18 July, the first new crews arrived. Sixteen more crews arrived on 10 August. Sgt. George DuPont, a mechanic-propeller specialist, joined the Group at Wendover in July 1943. His long, hot truck ride from the 18th Replacement Wing at Salt Lake City, 130 miles away, ended at desolate, barren Wendover Field, Utah. As a bombardier in the 713th BS, Lt. Charles McBride arrived under different circumstances. He made the trip from Gowen Field to Wendover with his pregnant wife. The accommodations, or lack thereof, at Wendover forced him to leave her in Salt Lake City.

Lt. Lewis Sarkovich arrived at Wendover on 22 August and thought he had arrived at the end of the world. Several days later he wandered to the flight line to watch B-24s shooting landings. He watched in shock as a B-24, just after takeoff, slowly descended until it hit the ground in a fiery explosion. Making matters worse, this was the second such incident he had witnessed; the first occurred two weeks earlier at Gowen Field. It failed to instill confidence in the young pilot who had yet to fly the B-24. He got his chance the following day. Just after initial takeoff the aircraft started descending just like the ones he had watched crash. With his life flashing before his eyes, Sarkovich observed a life-saving technique. The instructor lowered full flaps allowing the aircraft to start a climb; he then slowly raised the flaps. Sarkovich never forgot this lesson.

The Group started their transformation into a combat-ready Heavy Bomb Group at Wendover. Days consisting of twelve-hour shifts; six to six, were the norm for ground personnel. Following breakfast, troops assembled for inspection and then classes on first aid, aircraft recognition, know your enemy, and many other combat related topics. After a break for lunch, the afternoon offered a rare chance to catch some sleep, provided there were no mock gas attacks. All too often, chemical warfare officers tossed smoke bombs simulating gas attacks to ensure gas masks were close at hand. Mail call at 1630 provided the bright spot of the day. It brought precious, sought-after news from home. After another trip to the mess hall, it was down to the flight line for a shift at work. Sgt. George DuPont catnapped on the ramp waiting for the only airplane of the squadron, ship #092, to return from its two-hour training mission. On 10 August, seven additional airplanes arrived and sleep disappeared. The amount of maintenance required to keep the old, over-used airplanes flying offered the perfect chance for the mechanics to become familiar with the planes. This hands-on experience paid dividends in the coming months.

Aircrews spent their time at Wendover in much the same fashion as the ground personnel. Incomplete crews received their full complement of individuals and new crews arrived almost daily. As the Group gained strength, training intensified. Formation flying, gunnery training, navigation and practice bombing runs continued to receive the most attention, day and night. Takeoffs and landings continued to receive emphasis on all missions. For the model crews that experienced the School of Applied Tactics in Orlando, Wendover mirrored the training they received but on a much larger scale. This time, however, they were the teachers.

An aerial gunnery school, nestled in the mountains just north of Wendover, taught gunners the specifics of a fifty caliber machine gun, the Emerson nose turret, left and right feed, headspace, holding paul, and other gunnery skills. An unmanned jeep carrying a mounted aircraft silhouette around a pre-planned course received many rounds from the novice gunners. Lt. Arthur Steele was one. As a bombardier, he also had to man a turret. The training sergeant ensured he and all the other gunners heeded the white-washed motto posted on the side of the mountain, "Kill or be killed."

Weekends granted a slight reprieve and a chance for some recreation. For Lt. Charles McBride, weekends at Wendover meant a round trip to Salt Lake City to visit his wife. On one of these trips, she went into labor. The baby was born on Sunday, just in time for him to make it back to Wendover for another week of training. Lt. A.D Skaggs used these weekends to visit Salt Lake City with his crew. Another favorite and more accessible recreation spot was located ten miles south of the base. This nice, freshwater lake located on the edge of the salt flats

Lt. Maynard Ingalls on his way to the showers during the applied portion of AAFSAT at Pinecastle Army Air Force Base. Life at Pinecastle revolved around flying operations twenty-four hours a day in austere conditions. (Skaggs)

offered a cool, soothing break from the Utah heat. Whenever time allowed, crews trekked the ten miles through choking dust. Despite the refreshing dip, the return trip left everyone caked in dust again.

The reality and seriousness of flying reared its ugly head for all to see at Wendover. Another group, the 399th Bomb Group, shared training facilities at Wendover with the 448th. They quickly earned the nickname "Dying 399th." The unfortunate group killed fifty-four men in five weeks. Sgt. George DuPont watched as a 399th B-24 flew overhead slow and low. The aircraft shuddered then stalled, crashing into a boxcar on a nearby railroad track. The crash killed all except for the pilot and a passenger. The pilot later died of injuries received in the crash. The black, charred remains of burned-out wrecks contrasted sharply with the white sand of the desert. To Lt. Arthur Steele they looked like a spider with the car and truck tracks for legs and the black spot on the white sand as the spider. Crashes were not uncommon at Wendover. But, to the young and immortal, it only happened to the other guy.

In the middle of August, the 448th was alerted for a move to Fairmount, Nebraska, but the order was promptly canceled. Instead, Group Headquarters authorized ten-day furloughs for everyone but only a small percentage of each squadron received furloughs at a time. The training schedule could not be affected and it continued at a fast and furious pace. Lt. Skaggs' broken-down Studebaker required repairs in order to make a trip to his home in Oklahoma. Sgt. George Glevanik, the engineer on Lt. Skaggs' crew and a superb mechanic, put ingenuity to work and made the parts for the car. He and Sgt. Ray Lee accompanied Lt. Skaggs to Oklahoma and from there they caught trains home. The rest of Skaggs' crew boarded buses and trains for home. Lt. Charles McBride used this chance to visit his wife, baby and family. This would be the last trip home for many men before shipping overseas.

After each crews' ten-day furlough, the intensity of training reached a new peak. Rumors of another move quickly spread throughout the ranks. Excitement grew as the Group concluded the second phase of its training with thirty-nine crews and eleven planes. In mid-September everyone packed up and prepared the Group for its next move. Destination: Sioux City, Iowa.

So far the group had avoided the training pitfalls associated with other Groups training at Wendover. Luckily, the 448th had no aircraft accidents. This all changed on 13 September as two 713th BS aircraft prepared to takeoff from Wendover bound for Sioux City. Lt. Chester Hackett, 713th BS commander, took off first followed immediately by another B-24E, tail number 42-66421, flown by Lt. Heber Thompson. As Lt. Hackett flew a 180-degree turn in order to fly back over the base, Lt. Thompson initiated his turn to join his leader. Lt. Cater Lee, riding in the nose of Lt. Thompson's aircraft, witnessed the low slow turn. The aircraft shadow on the salty flats slowly started rising to meet the aircraft. Recognizing the sink rate at such a low altitude, Lt. Thompson added full power as he pulled the nose up. The mushy controls could not counteract the sink rate and the airplane impacted the ground like a floundering fish.

The impact bounced Lt. Lee around the nose of the aircraft, bruising and cutting him. Bewildered, he climbed out a half-opened nose wheel door. In the bomb bay the impact sheared bolts holding bins full of the squadron's records. A "snowstorm of paper" resulted, adding to the confusion. Sgt. James Bricker was standing on the catwalk when the crash occurred. The loose bins pinned his foot to a bulkhead and, fearing the plane might explode, his shoes were cut off with a GI knife. In the haste tendons and muscles in his ankle were also cut. Fortunately, the only other injury was to Lt. James

On rare occasions day passes were handed out. One of the popular spots to visit was San Alando Park near Orlando. Front (left to right) Tommy Thompson, George Glevanik, Ray Lee. Back (left to right) A.D. Skaggs, Stanley Filipowicz, Andy Hau. (Skaggs)

Urban who broke his arm. Everyone else escaped with cuts and bruises. After one night at the Wendover Hospital, Lt. Hackett returned to fly everyone to Sioux City.

Unlike Wendover, Sioux City Army Air Base, located at Sergeants Bluff, Iowa, offered distractions. Five miles north were the city lights of Sioux City and plenty of things to do. So, in typical Army fashion, Group Headquarters restricted everyone to base. With combat just around the corner, the Group still required a lot of training. Colonel Thompson continued to stress the importance of teamwork and crew discipline. These traits would spell success in combat. Aircrews flew in larger and larger formations requiring increased reliance on one another. Confidence soared.

The Consolidated B-24 Liberator carried the men of the 448th to war and back. Although not as well known, more B-24s were produced during World War II than any other bomber. It served the 448th well flying 7,343 combat sorties with the Group alone. (Bailey)

Boosting confidence even more, new straight-from-the-factory B-24s arrived at Sioux City. Unlike the older, D and E models, these new H models offered the latest in technology and combat effectiveness. They showcased twin, fifty caliber machine guns in a new nose turret instead of the old flexible mounted gun in the nose. Lt. Albert DiLorenzo wrote his wife from Sioux City. "We got our 'fly-away' yesterday and we are now all set to go. A 'fly-away' is a new ship each crew gets before leaving the country. It is our own, our very own."

Lt. John McCune and the rest of his crew marveled at their new toy. "Air Corps number 42-7739. This was the only legend she bore the day she landed at Sioux City from the modification center in Birmingham, but to our eyes she seemed clothed in gold leaf. Every man on the crew fell in love with '739' when she was assigned to us on that 23rd day of October. A brand new B-24H with the new Emerson nose turret, she smelled new, she radiated newness. The plexiglass of her turrets all aglisten, even her drab war paint seemed to shine. We admired her for hours, then cudgeled our brains for a suitable name for her." They finally agreed on MAID OF ORLEANS in honor of Joan d'Arc of France.

Crews quickly learned these new planes were not perfect. On 2 October Sgt. George Glevanik flew with a ferry crew to pick up another "fly away" plane for the Group. They returned to Sioux City early on the morning of 4 October and turned in the appropriate documents before heading to the barracks for sleep. Unbeknownst to Sgt. Glevanik, that very plane had been assigned to his crew and was scheduled to takeoff early that same morning on a training mission. After only a few hours of sleep, someone shook him awake and informed him he was scheduled to fly. The crew was already at the plane and he was supposed to meet them there. When he arrived at the airplane, servicing personnel told Sgt. Glevanik the plane was serviced and ready for the mission. Accepting their word, the crew took to the air flying up the Missouri River in their new pride and joy. After several hours of navigation, bombing, and air to ground gunnery practice, the pilot, Lt. A.D. Skaggs, turned the aircraft for home

Upon reaching 1,500 feet, the number four engine started losing power. Just thirty minutes from the base, Sgt. Glevanik frantically searched for the problem. All the fuel sight gauges showed over a quarter tank of fuel remaining and no other problems seemed to exist. It did not take long to determine the gauges were inaccurate and the tanks were empty. Shortly thereafter another engine quit. Lt. Skaggs quickly called Lt. Don Todt, the navigator, for a heading to the nearest emergency field. Lt. Todt had already located a small grass airfield a few miles away just north of Vermillion, South Dakota. Lt. Skaggs

Men of the 448th enjoy one of the favorite pastimes at Wendover. The cool water of the nearby lake offered a relief to the otherwise arid and hot conditions. (Gaskins)

turned the plane to the heading and informed the crew to prepare for a crash landing. The steepness of the approach and the impending failure of the other two engines forced Lt. Skaggs to fly a higher than normal airspeed on final. As a result, after touchdown on the short field, the airplane quickly went off the end, crossed a ditch and tore through a fence into an adjoining hay field. Despite Lt. Skaggs frantic efforts, the aircraft's right wing struck a haystack and bounced through the top of it shearing off the wing tip. The left wing and landing gear struck another haystack, collapsing the gear and leaving the left wing tip scraping the ground. As the dust and straw settled, the plane came to rest with the remaining portion of the right wing pointing skyward and the left wing tip on the ground. The new, beautiful plane now sat like a helpless pile of junk in a farmer's field. However, due to superb piloting skills and excellent teamwork, the only major injuries were bruised egos.

Sgt. Glevanik, worried about the security of the airplane, immediately cordoned off the area. A telephone lineman, working on a nearby telephone pole, witnessed this incredible event. He relayed a message to the base for the crew explaining what had happened. While awaiting transportation back to the base, the unfortunate crew received royal treatment from the local townspeople. An elderly couple, who also witnessed the landing, hurried into town and returned with sandwiches and coffee for the grateful flyers. An hour later, the Army arrived with MPs to guard the plane and transport the crew back to base.

The following day, the Vermillion Newspaper headlines read, "Crew Member Prayed as Bomber is Forced Down at Vermillion." The article continued. "As at least one member prayed, a heavy bomber from Sioux City Air Base glided to a forced landing near here Monday afternoon, crashed through a fence, struck two haystacks and came to a rest damaged, its occupants uninjured. Lt. Austin W. Marshall, public relations officer at the Sioux City base who came here to investigate, praised the pilot and copilot for 'a beautiful job of landing under difficult circumstances.' 'I wasn't scared,' stated one crewmember. 'I didn't have time to be afraid. But I heard one of the men praying.'" Lt. Elbert Lozes, bombardier, was the praying crewmember, thinking of his wife and unborn child who was born fifteen days following the incident. This was the second and last training accident suffered by the Group. Miraculously, neither accident resulted in fatalities. It was an enviable record; would combat be so kind?

The training started paying dividends at Sioux City. The countless sorties and hours spent in the air increased proficiency and confidence in the airplanes. Sgt. George DuPont watched in wonder as Lt. E.P. Durley, a pilot in the 712th BS, struggled to land a B-24 with oil streaming out of the number four engine. Unable to feather the engine, the propeller spun like a runaway windmill. Despite all of the problems, Lt. Durley landed the crippled plane safely.

In sharp contrast to the intense training, men found ways to amuse themselves. Despite the restrictions to stay on base, Sgt. George DuPont often snuck off base to visit his wife. On one escapade he hid in a laundry bag to escape undetected. The truck driver never knew. Getting back on base proved to be a much more worthy challenge. Lt. Skaggs, Sgt. Glevanik, Sgt. Sheehan, and Sgt. Gaskins did not let their acquaintances in nearby Vermillion go untapped. The Saturday following their crash-landing, they planned a return trip to Vermillion. This time they went in search of pheasant. While waiting for the Army to pluck them from their crash site, they had noticed a large number of pheasants. They received permission to come back for some hunting, which they did. After several hours of hunting, the crew emerged with eight birds and no casualties. Sgt. Ray Lee arranged to have the birds prepared by a chef downtown. The delicious banquet, minus the lead from the birdshot, proved far better than typical Army chow.

While the 448th prepared for combat at Sioux City, advanced echelons of other units were preparing Seething, England, for operations. On 1 September Lt. Nikolas Gianakos and two enlisted men, Sgt. James Cleary and Cpl. Clarence Harrell, were transferred to Seething from the Headquarters unit at Station 114. Nine days later another officer and twenty enlisted men arrived to aid in establishing the Headquarters for the arriving Bomb Group. They were not the only ones arriving.

At the same time, troops of the 58th Station Complement Squadron, not yet associated with the 448th Bomb Group, boarded the USS General John Pope. They completed the eight-

Men congregate around the vehicles as they prepare to return to the main base after swimming. While the water was refreshing and cool, the ride back to the base was hot and dusty caking the men with a thick layer of dirt. (Gaskins)

day ocean crossing unescorted. The Squadron moved by rail to Seething where Lt. Col. V. B. Cagle, an officer from the 8th Air Force, welcomed them. He had assumed command of Seething on 14 September 1943 to oversee the construction of the base. The 200 enlisted men and six officers found the partially completed base in a state of hurried construction. The infrastructure of the base either did not exist or was not complete, including showers and mess facilities. The squadron, commanded by Capt. William Searles, set out to prepare the incomplete base for the arrival of the 448th.

Prior to the arrival of the 58th, Sgt. Reginald Dunn of the RAF arrived at Seething by jeep to prepare the control tower for the incoming 'Yanks.' Dunn and one colleague received keys to the control tower and little other instructions. With no vehicles, no timetable, and no clear duties, the two started preparing the control tower by setting up RAF communications equipment. Five days later the 58th arrived.

With the arrival of the 58th, the two RAF men fell under command of Lt. Wallace "Wally" Bollschweiller who was the Flight Control Officer. The Brits found his name difficult to say and spell, but enjoyed his leadership. The control tower personnel continued their preparations for the arrival of aircraft. Although the work on the runways continued, disabled aircraft started using the partially completed runways, much to the dismay of the Clerk of Works. Battle-damaged aircraft including three fighters, a Lancaster, and a Mosquito provided work for the tower crew even before workers completed the runways. Due to the proximity of the coast, Seething received this kind of traffic throughout the war.

Back stateside, the 448th prepared for the next move. Several of the men, including Lt. Col. Carl Elver, Capt. Ron Kramer, and Capt. Suel Arnold, departed early and reported to the Air Transport Command terminal in New York. Flying in a C-54, they arrived in England after stops at Goose Bay, Reykjavik, and Prestwick. After receiving briefings from the 8th Air Force leadership at High Wycombe, they arrived at Seething as the advanced echelon of the 448th and on 23 November 1943, Lt. Col. Elver assumed command from Lt. Col. Cagle. They worked closely with the men already at Seething to help prepare for the imminent arrival of the bombers.

The slow procession to combat picked up speed as orders arrived on 3 November temporarily assigning the 448th to the 21st Bombardment Wing at Herrington Field, Kansas for final preparations for movement overseas. Starting on 3 November, crews left in large formations bound for Kansas. Arriving at Herrington, they were greeted with blustery, cold weather. Winter had arrived and the overcast skies added gloom to the reality of leaving home for parts unknown. Immediately after landing, aircraft were towed to the hangars to undergo last minute combat modifications. Life rafts, emergency rations, flares, first aid kits, new radios, and IFF equipment were added to the airplanes. On 5 November crews started the myriad of details required to complete their overseas processing. They remained in the dark about their final destination but were told to prepare to leave 11 November 1943.

Final processing included auditing and processing of personnel records, physical fitness exams, clothing and equipment inspections, distribution of critical equipment items, communications instructions, assignment of crews and planes scheduled to fly overseas, briefings on routes to be traveled, and scheduling final departure for port of embarkation. All the crews received new equipment including jungle pack parachutes, 45 caliber pistols, ammunition, watches, Mae West life jackets, new style oxygen masks, clothing, and various other articles required for flight. Orders dated 6 November 1943 assigned crews to specific airplanes. For most of the crews the planes remained the same as the ones they flew in from Sioux City.

Lt. Thomas Apple's crew received their shiny new B-24 that they quickly christened BIG ASS BURD, complete with nose art. After the first flight, Maj. Hubert Judy, the Group Air Executive officer, and Lt. Apple discovered one of the wings on the new plane was warped. A replacement was quickly hurried to Herrington. Their new plane, a new top-of-the-line J model quickly named FASCINATING LADY, arrived just in time for the Group to depart. It was the first B-24 J in Europe.

The reality of leaving grew more apparent as time for departure neared. The bone-chilling cold added to the increasing gloom. Spare time consisted of sitting in the club or in the barracks. Since the club closed early, most of the evenings were spent trying to shake the cold in the barracks. The thin walls merely slowed the howling wind and the cold air filled the rooms. Letter writing or bull sessions occupied most of the time. Much of the evenings were spent speculating and trying to solve the riddle of the Group's final destination. Some thought the presence of tropical gear and mosquito netting on the airplanes ensured the Group was bound for North Africa or possibly even the Pacific. Others thought the heated flying suits pointed to Europe. Only time would tell.

Lt. Charles McBride found time to visit with his family one more time. His wife, new son, mother and father drove to Herrington to visit. This was the last time Lt. McBride saw his

Aircraft 41-29183 sits in a hayfield in Vermillion, South Dakota after running out of gas on its maiden flight. Fortunately no one was injured in this accident but the brand new aircraft was salvaged. (Gaskins)

father. His father died two months later. Lt. Henry Schroeder's crew used their time to name their ship and started to paint the nose art. The artwork on BOMB BOOGIE got as far as the outline, but the crew had to wait on the painting so final modifications could be completed.

Final processing rapidly drew to completion as did personal preparations. Lt. Arthur Steele witnessed the cash registers at the PX grow hot from the sales of silk stockings (for bargaining purposes), cigarettes, and candy. These personal supplies complimented the 45s in shoulder holsters, a few carbines, enough morphine vials to make a modern day drug dealer envious, and bomb bays full of assorted war goodies. Crews received final briefings, pilots signed receipts for their $350,000 airplanes, baggage was checked, and flight plans were filed. The route of flight would take the Group to Morrison Field, Florida, then to Marrakech, Morocco, via the southern route. Stops were planned in Puerto Rico, British Guyana, Brazil, across the Atlantic to Dakar, French West Africa, and finally Marrakech. The final destination remained classified. Crews went to bed on the 10th with knots in their stomachs. Training was complete. The airplanes, normally weighing 37,000 pounds, bulged with combat gear and fuel at 60,000 pounds. They were combat ready! Tomorrow would bring the Group one step closer to combat.

11 November 1943 dawned blustery and cold. Temperatures hovered near zero. Crews finalized flight planes, checked the weather, and prepared for departure. Lt. John McCune and crew, flying newly christened MAID OF ORLEANS, lifted off at 0630. The following day, Sgt. Brona Bottoms on SWEET SIOUX, flown by Lt. Thomas Keene, left for Morrison Field. One week later Lt. Elmer Hammer's crew left with Lt. Morris Thomson navigating. A total of sixty-four aircraft and crews of the 448th Bombardment Group left wind-swept Kansas bound for combat and parts unknown.

At 1800 on 8 November the ground echelon of the 448th, including ten air crews, was alerted for movement, but a raging snowstorm delayed the departure. Finally, the Group headed out on a secret military train on the Illinois Central Rail Line to Chicago. From there they continued via the Grand Trunk and Canadian National Railway to Port Huron, Michigan, and then on to Buffalo, New York. Finally, at 2230 on 11 November, the exhausted men arrived at Camp Shanks in Orangeburg, New York. Overall, it had been a pleasant trip with good food. However, upon arrival, it was back to the Army.

A one-mile march to their quarters awaited the men as they detrained. The following four days were spent in typical Army fashion, waiting in lines for in-processing and preparing for embarkation. Individuals received proper clothing, medi-

An unexpected benefit from this crash-landing was meeting the local farmers who gave their permission for the men to return and pheasant hunt. George Glevanik, Francis Sheehan, Gene Gaskins and Harold Whitaker (left to right) show of their bounty following their hunt. (Skaggs)

cal exams, security training, and lectures on issues ranging from insurance to censorship. Gas masks were distributed and instructions were given for their use. Everyone endured a five-mile hike in full combat gear, a drill at abandoning ship down a rope ladder and the standard Army obstacle course with a thirty-five pound pack. With relief, the men prepared to leave Camp Shanks. Orders were given to clean the barracks before departing. The officer in charge told the men in the 713th BS it was not necessary; after all, they were going into combat. The group left Camp Shanks at 1000 on Saturday, 21 November 1943 bound for the docks of New York City. Upon arrival, the 713th BS received word to stand aside; they would be the last to board. Due to the condition of their barracks, they would pull Kitchen Police, KP, for the entire trip. What happened to combat? By 1800 all were aboard the "Queen Elizabeth."

Bob Harper (left), the assistant intelligence officer and Andy Hau, the Group bombardier, pose in Sioux City, Iowa as the Group was undergoing its final preparations for combat. (Skaggs)

They left New York harbor and set sail for England at 1530 on 23 November 1943. The commandant of the ship, Captain A.K. Aspden, operated the ship under a joint command of the Royal Canadian Air Force and the Transportation Corps of the U.S. Army. The "Queen Elizabeth" traveled alone, relying on her speed to avoid German submarines. The common joke on the way over was "You don't have to worry about German torpedoes, only the water which would come in afterwards." This was no joke for Sgt. George DuPont who slept on R deck, two decks below the waterline. To everyone's chagrin, the ship's crew announced over the intercom that German radio had claimed the ship had been sunk.

The Group quickly realized the reality of war; comfort was a nicety of a past life. The former luxury liner overflowed with troops. Sgt. Stanley Zabrowski and twenty-three other men were assigned to a stateroom intended for two. Twelve bunks filled the tiny room. That meant you slept inside one night and the following night was spent topside on the open deck, usually in the rain. Tired troops quickly discovered they were not tired enough to sleep in the wet cold.

For some members of the 712th and 713th BS their misfortune of KP duty did contain a hidden blessing; they were the only ones who got enough to eat. Bert LaPoint paid the price however; he spent every day, from midnight until 0800, cutting breakfast bacon by the tons. Meals were served twice daily. Breakfast, served from 0600 to 1200, consisted of a hard-boiled egg, bacon, a slice of toast, and a cup of coffee. Cooks served dinner from 1300 until 1900. The round the clock meals barely allowed time for everyone to be fed.

Efforts were made to establish some routine to life on the ship. Squadrons established orderly rooms, programs and classes were held at regular intervals, and Church services were conducted in the Officers' Lounge on Sunday. Every day, the ship's central address system announced emergency drills and action station drills. Considering the large number of people involved, the drill proceeded smoothly. Troops spent Thanksgiving Day in the middle of the North Atlantic enjoying a meal of greasy pork chops. For Cpl. Bert LaPoint, Thanksgiving meant piles of pork chops instead of bacon.

Morale remained high in spite of the cramped quarters. Cpl. LaPoint listened as Captain Corsiglia, 713th BS adjutant, called out names of personnel receiving another stripe for going overseas. To his dismay, he failed to hear his name. Wondering why, he questioned Captain Corsiglia and was told he should not be on the "Queen Elizabeth" but on an airplane somewhere enroute. KP offered little consolation! Every man received a Red Cross package with a carton of cigarettes and each enlisted man received an additional package containing more cigarettes, reading material, cards, and other goodies. Some people passed the time playing dice games in the ship's stairways; Sgt. Sigifredo Perez, an aircraft mechanic, lost all of his money before getting off the boat in Scotland.

After six days of twisting and turning to avoid detection, the "Queen Elizabeth" anchored at the Mouth of the Clyde near Glasgow, Scotland, at 0500 on 29 November 1943. Sigifredo Perez watched the ship's great anchors splash into the sea. After cleaning the entire ship, the men finally disembarked at 1515 the following day. Much to everyone's delight, the American Red Cross served coffee, doughnuts, cigarettes and the typical English tea and crumpets despite the bitter cold and rain. After the long days at sea, the most popular sight was an American Red Cross "lassie." Following a two-hour Red Cross break, tired troops rode blacked-out trains to Ditchingham, located between Seething and Bungay. Three times air raids forced the train to stop. On the first occasion, Cpl. Bert LaPoint, being curious, opened the blackout curtain to observe the searchlights stabbing into the night sky and the bomb flashes off to the right. The conductor quickly admonished him for opening the curtain and pointedly explained the importance of keeping the curtains pulled. Arriving at midnight, Army trucks completed the trip by ferrying the ground echelon to their new home, Seething. In the pre-dawn hours of 1 December, the disheveled, tired and hungry ground echelon received bedding and underwent registration. Welcome Home!

NOTES:
[1] *The Army Air Force in World War II, Europe: Torch to Pointblank, August 1942 to December 1943.* Ed. Craven, W.F. and Cate, J.L. Chicago: University of Chicago Press, 1949. p. 305.

[2] Hansell, Haywood S., Jr. *The Strategic Airwar against Germany and Japan.* Washington D.C.: Office of Air Force History, 1986. p. 74.
[3] Ibid. p. 75.
[4] Ibid. p. 76.

two

# WELCOME "HOME":
# November 1943

In the skies over the U.S. during the next several days, sixty-four crews breathed a sigh of relief. They were now doing what they were trained to do: fly. No more sitting around speculating about their destination or trying to shake the bitter Kansas cold. Ironically, it was Armistice Day, 11 November 1943, when the first crews departed for combat. Lt. John McCune and crew left two days later. At 5,000 feet, the entire southeastern U.S. sprawled beneath the airplane. Their new Liberator with minimum cruise power set and the A-5 autopilot engaged offered a smooth platform to tour the countryside. They ate sandwiches and shot the bull while they admired the sights. Occasionally, another aircraft could be seen in the distance. Only position reports to Memphis, Atlanta, and Jacksonville interrupted the sightseeing. As the crews neared Orlando, air traffic increased from the nearby School of Applied Tactics. A mere six months earlier the fledgling Group, four crews strong, fought the mosquitoes of Florida during their stint. Now, combat ready and flying the latest B-24 Hs, they were bound for combat.

As crews brought their planes in to land at Morrison Field in sunny West Palm Beach, Florida, they immediately started shedding their flying gear. The eighty-degree weather was a far cry from the numbing cold of Kansas. Morrison Field was a garden spot. The beautifully landscaped and designed base provided a wonderful send off for the men going overseas. The large Officers' Club offered excellent food and the quarters were nestled among the shrubs, palms, and flowers overlooking the spacious lawn.

At Morrison Field, crews received detailed briefings on the routes to be flown, weather along the route and other pertinent information needed for the expected ten-day trip. Secrecy remained high during their stay. Everyone was restricted to base and kept under lock and key, no phone calls or contact with anyone outside the Group was allowed. Identification cards were pulled effectively restricting everyone to base. Some crews used this stop to christen their airplanes. A local artist emblazoned tail number 42-7754 with a freshly painted caricature of Betty Boop holding two bombs. HARMFUL LIL ARMFUL was now complete.

Each copilot received a large sum of money, one to two thousand dollars, from the base finance office. This money was to be used to cover emergencies enroute. For Lt. John McCune, this offered a perfect opportunity for a practical joke, he thought. While his copilot, Lt. Lloyd Morse, napped one afternoon, they deftly opened his shirt and unzipped the money belt without waking him. They removed the money, hid it in an envelope and left the room. After his nap, Lt. Morse wandered down the hall to watch a pinochle game involving the pranksters. He watched the game for awhile without making a comment, then walked to the phone and made a call. Fearing he was calling the Provost Marshall to tell of the theft, Lt. McCune and his accomplices, Lt. Maurice Hooks and Doc Joseph Kaiser, 715th BS flight surgeon, returned the money before his return. Much to their chagrin, he never missed the money. Instead of calling to report a theft, he had called the shoe store inquiring about shoes he had left for repair!

Crews left Florida on a staggered schedule to allow for the sparse accommodations at the enroute stops. In the humid Florida darkness of 14 November, a sergeant from operations switched on the light waking up Capt. William Blum, 712th BS Operations Officer, and the crew of FAT STUFF II. "Briefing at 0500 sir." "For many months, we had been working long, diligent hours in preparation for combat, and this morning, shrouded in a drizzling rain, would mark the first leg in the long route to that goal. The briefing room was crowded, cigarette smoke and strong lights presenting an air of intense activity. Naturally we were all conscious of the long flight before us, and unmistakable signs of nervousness were evident in spite of the horseplay and witticisms." The briefing, much like those of practice missions, outlined the route of flight and consisted of the standard roll call, order of takeoff, land marks, navigational aids, emergency procedures and finally a weather brief.

Capt. Blum, flying on FAT STUFF II, (Although Lt. Jack O'Brien was pilot of FAT STUFF II, Capt. Blum was the ranking officer of the formation) was the flight leader of the first six aircraft from the 712th BS to undertake the trip. With 2,500 gallons of gas, 2,800 pounds of baggage, and fourteen passengers, FAT STUFF II lifted into the early morning Florida sky. Exactly one hour to the minute after takeoff, Capt. Blum opened the sealed envelope containing their eventual destination. Taking the slip of paper from the envelope marked "SECRET", he read the message over the intercom. "Your destination – The British Isles. All crew members are warned against discussing this subject at any other time than in flight." To Capt. Blum, this was bad news. English weather meant hard flying and the Luftwaffe pilots were to be feared.

Squadron patches of the 448th Bomb Group. (Lee)

Lt. A.D. Skaggs checked his aircraft one last time, then started engines and his crew prepared for takeoff. The newly christened HARMFUL LIL ARMFUL lifted into the south Florida sky on 14 November bound for Puerto Rico. Once the designated coordinates were reached, the sealed envelope containing the final destination was ripped open.

The cockpit of a Consolidated B-24 Liberator. (USAF)

Mechanical problems with the auxiliary power unit and bad weather delayed Lt. John McCune and his crew by a day. Finally, MAID OF ORLEANS left U.S. soil at 0700 bound for Borinquen Field, Puerto Rico. Upon reaching their cruise altitude of 9,000 feet, Lt. McCune engaged the autopilot and broke open the orders. England! After the initial excitement subsided, the business of flying quickly took center stage. Dark, foreboding clouds appeared on the horizon. Following a quick consultation with the crew, it was decided to fly through the holes rather than try to climb over the clouds. After numerous turns and maneuvering, the MAID OF ORLEANS broke into clear skies. The navigator, Lt. Maurice Hooks quickly 'shot' the sun with a sextant and after some quick calculations determined they were on course. Confidence soared. Lunches were passed out and the crew relaxed by reading books or sleeping. Five and half-hours later the telltale Morse code dots and dashes from Borinquen echoed in the headset of the radio operator. The MAID settled to terra firma at 1300.

Borinquen Field, a permanent peacetime Army base, surpassed Morrison Field's beauty. Located on the northwest corner of the island near the town of Aguadilla, Borinquen was paradise. The tropical weather, warm and humid, provided the perfect environment for the lush growth. The airfield was carved from a mass of shrubs, palm trees, and flowers. All of the buildings were concrete and camouflaged. Everything was placed in a neat, orderly fashion. The officers were housed in the DeGink Hotel, affectionately called the "Biltmore." The Officers' club overlooked a wide sandy beach of the Atlantic Ocean from atop a large cliff. Col. Thompson ensured the Air Transport Command routed all crews through Puerto Rico; it would be the last "good deal" these crews would see for a long time.

Lt. Thomas Apple's crew experienced unwanted excitement as they attempted to leave Borinquen for the next leg of their journey. Just after takeoff, when the wheels of FASCINATING LADY were a mere two feet off the ground, the cover on the right side Tokyo tank rattled loose and the highly flammable fuel shot straight up before streaming back over the number three engine's exhaust. Lt. Apple instinctively reached to feather the engine, but Maj. Hubert Judy, much more experienced in the B-24 and the Group Air Executive Officer, stopped him before he shut down the engine. At such a low altitude, the results would have been catastrophic. After climbing to 500 feet, they feathered the engine and successfully circled to the field and landed. Lt. Cater Lee, a crewmember on FASCINATING LADY, watched an irate Maj. Judy commandeer a jeep and rush to the Base Commander's office. Oh, the poor private who failed to safety wire the cover.

Before any American GI reached Seething, Reginald Dunn of the RAF was already hard at work ensuring the still unfinished base was ready for its new tenants. Note the large T2 hangar across the field from the control tower. (Bollschweiler)

For Lt. McCune and crew, their stay at Borinquen lasted longer than most. The next morning, the crew found MAID OF ORLEANS sitting on her nose. Her plexiglass nose was shattered and the nose wheel was broken. It was discovered that crews at Morrison Field towed the plane with the brakes locked. The stress fractured the nose gear and, when brakes were applied during the preflight engine run, the gear broke. Fortunately, the gear held during the landing the previous day. They spent thirty-two days "enduring" double rooms, tile floors, and electric refrigerators. Reluctantly, Lt. McCune arranged for Doc Kaiser and Maj. Kenneth Squyres, 715th BS commander, to continue on with other crews. Major Squyres caught a ride with Lt. Skaggs and crew. Although they did not realize it, this was the last time Lt. McCune's crew would see their commander.

Lt. Skaggs and crew departed Borinquen for Atkinson Field near Georgetown, British Guyana, deep in the South American jungle, on 15 November. During the briefing, the briefing officer outlined the route of flight south-southeast to Trinidad where radio contact would be established with Waller Field. Over-flying Waller, the leg continued south to the Cuyuni River followed by a turn west. Contact could then be established with

Atkinson Field. Waller Field provided an excellent alternate in case of bad weather at Atkinson Field. In Lt. A.D. Skaggs' case, it was needed. Low clouds forced them to fly lower and lower; so low the wild orchids in the jungle were clearly visible. Finally, the poor weather forced them to divert to Waller Field, Trinidad.

Capt. William Blum and the crew of FAT STUFF II also visited Waller Field. Approaching the field the crew witnessed a tropical thunderstorm phenomenon. Rain pounded the aircraft and violent updrafts and downdrafts tossed the aircraft about in an uncontrolled dance. Scared stiff, the crew rode out the storm and broke out into the bright sunshine. At Waller they spent one night playing ping-pong at the Officers' club, watching a movie and stocking up on supplies from the PX. Another early morning briefing cut the evening short as everyone tried to rest for the next leg to Belem, Brazil.

Lt. Jack Swayze, pilot of aircraft 41-29235, safely completed the trip to the airfield in the jungle. After landing they taxied to their assigned parking spot and were quickly greeted with the customary spraying of DDT to rid the plane of any stowaway insects. Following the spraying, native young boys selling melons descended on the plane. Military Police quickly dispatched them. Accommodations were sparse. Open barracks with only a roof provided sleeping quarters. The following morning, men woke to a torrential, tropical downpour. However, within thirty minutes, the rain ceased and the sun baked the forest in a steam bath. The trip continued; next stop, Belem, Brazil.

Belem, located on the Amazon River, was also carved out of the jungle. Flight plans called for a direct flight across the endless jungle until reaching the Amazon River. A short jaunt up the river ended at Belem. Crews found Belem much more accommodating than Atkinson Field. The Post Exchange offered good shopping and, more importantly, individuals slept in individual rooms and not in open bay barracks.

Meanwhile, the entire Group stretched across the length of Air Transport Command's Southern Route. The only command structure existed within each crew. Group Headquarters relied on the ability of each pilot and crew to fly and navigate the perils found all along the route. Only after arriving in England would the Group re-emerge as a consolidated fighting force. Now, it was sixty-four individual crews against nature. So far the Group escaped unscathed.

Belem claimed the first casualty on 19 November. FAT STUFF II landed first followed shortly after by Lt. Jack Parker's HELLO NATURAL. As the crew of FAT STUFF II climbed out of the aircraft, a sudden tropical rain shower engulfed the field making it difficult for the following aircraft to line up

Tom and Maureen Apple, Sara and Cater Lee, and Susan and Rich Henderson (left to right) in front of their plane BIGASS BIRD. One of the Group's original planes, it never went overseas with the Group after it was determined to have a "warped" wing. (Lee)

with the runway. Third in the landing sequence, Lt. Carrol Key vainly peered through the rain searching for the field. Late in visually acquiring the airfield, he landed well down the runway and immediately applied brakes upon landing. The brakes locked on the wet runway and the aircraft started sliding sideways. The gear collapsed causing the number two propeller to shear off and enter the cockpit, severing Lt. Key's arm at the wrist. Running through the rain and mud to help, the crew of FAT STUFF II and the firefighters found the cockpit awash in blood. Although the aircraft, 42-52128, was destroyed, no other crewmember was injured. For everyone witnessing the event, it had a sobering effect!

HELL'S KITTEN 41-29236. One of the Group's original aircraft received at Herrington, Kansas, it was transferred to another Group soon after their arrival in the ETO. (Everson)

For the remaining sixty-three crews, the winding journey continued with the next stop Natal, Brazil. Natal would be the final stop before the ocean crossing. Natal resembled all the other South American bases – sparse, hot, and humid; it was another airfield reclaimed from the jungle. Monkeys and other wildlife could be heard all through the night as crews tried to sleep prior to their dangerous ocean crossing. For Sgt. Brona Bottoms, the town of Natal reminded him of Mexico, only with no cars. Another crew, Lt. Robert Ayrest's, spent fourteen days in the jungle while ground crews replaced a leaking fuel cell in LAKI-NUKI.

Lt. Jack O'Brien and Capt. William Blum landed FAT STUFF II at Natal on 22 November. FAT STUFF II underwent a complete overhaul in preparation for the ocean crossing giving the crew three days off. At the airfield, a mock dogfight featuring a British Mosquito and a P-38 offered brief entertainment but with the end of the 'battle', Lt. Arthur Steele and the remainder of the crew searched for another source of fun. The following day, they found their entertainment at the beach. A bare-footed Brazilian private led a horse to a Brazilian Army officer who mounted up and proceeded to put on a show. Ironically, the horse kicked up sand, showering several American nurses for whom the show was intended. Needless to say, they were not impressed. The crew spent some time shopping for bargaining items. Silk stockings, trinkets, inlaid book covers, stilettos, watches, mosquito boots and lipstick all found their way into the already heavily laden B-24.

Prior to leaving Belem, crews were reminded to keep accurate fuel-use records. The Belem-Natal leg offered the final chance for crews to tweak their airplanes in preparation for the Atlantic crossing. The 448th was one of the first Groups flying B-24 H models to Europe via the southern route. The previously necessary stop at Wideawake Field on Ascension Island would not be necessary. Extra fuel tanks in the bomb bay would extend the range of the new H models to enable them to reach Ekner Field, in Dakar, French West Africa. Tiny Ascension Island would provide an emergency divert base for the crews, but otherwise it would be a long lonesome trip across the Atlantic for the sixty-three crews.

Leaving Natal, Lt. Charles Knorr, flying CRUD WAGON, narrowly averted another disaster. Shortly after takeoff, an engine problem forced them to shut down the malfunctioning engine. Attempts to feather the propeller failed and the prop continued to windmill uncontrollably. The overweight airplane compounded the problem. Outstanding airmanship enabled Lt. Knorr to safely land the airplane back at Natal at night. Lt. A.D. Skaggs and crew logged six hours and forty minutes flying time, over four and a half at night, crossing the Atlantic. As always, Lt. Don Todt's navigation skills proved perfect and HARMFUL LIL ARMFUL's wheels touched African soil on 18 November. Africa was no different from previous stops. The first greeting was from the bug man who sprayed the plane with DDT. Lt. John Rhodes and his crew successfully completed the ocean crossing despite strong fuel odors in FINK'S JINX. They experienced these odors on previous legs but extensive checks after each leg failed to find a problem.

LADY LUCK 41-28578. Another of the Group's original aircraft, LADY LUCK survived the war and returned to the U.S. after flying numerous combat missions. (LaPoint)

Lt. Jack Swayze and crew took off from Natal and climbed to 8,000 feet for the ocean crossing. The weather showed the most favorable winds for the crossing at 14,000 feet. After a quick consultation, everyone agreed on 14,000 feet even though not everyone had an oxygen mask. No one used them anyway. After landing on the pierced steel runway at Dakar, fuel gauges showed 1,250 gallons remaining from the original 2,790!

Not all crossings were so easy. Lt. Knorr and crew reached the African continent late in the afternoon. Thick haze limited ground visibility and made landmark identification challenging. Dakar was not in sight. Lt. Stanley Baranofsky, the navigator, quickly determined their position to be south of Dakar. The radio compass confirmed the fact, and after ten hours of flying, with fuel reserves running low, their B-24 touched down on the pierced steel planking of Dakar.

Flying with Lt. Charles Billings and crew, Lt. John Grunow flew the trip on aircraft 41-29234. Their ocean crossing on Thanksgiving Day proved nearly fatal. Just as they passed the point of no return, the engineer reported a failure of the Tokyo tanks' transfer pump. Unable to pump precious fuel from the tanks, they had insufficient fuel to complete the trip to Africa. Since they had passed the point of no return, they also did not have enough fuel to return to Natal. They elected to turn around where at least air-sea rescue might reach them. The navigator, Lt. Everard Wandell, located a small island, San Fernando de Noronha, 150 miles off the coast of Brazil, where Navy Seabees were constructing an airfield. They landed on the unfinished airstrip, grateful to be alive!

Capt. Blum and the crew of FAT STUFF II reported to a large ATC hangar for the main briefing at 2100 on 25 November. The over 1600-mile ocean crossing to Dakar offered the most challenging phase of the trip so far. A stationary front 500 miles off the coast straddled the route of flight not to mention a maximum weight takeoff at night followed by six hours of night flying. Three minutes after midnight, FAT STUFF II rumbled down the runway followed by five more aircraft at three-minute intervals. The drone of the engines quickly put everyone to sleep except the navigator, engineer, and pilot. Just passing the 450-mile mark, the auxiliary tanks ran dry causing a vapor lock in the fuel lines and all four engines sputtered. Every sleeping person sat straight up in frozen terror. Using booster pumps, the engineer restarted the engines within ten seconds.

The rest of the trip proved uneventful except for a few moments of anxiety as they neared the coast. Capt. William Blum worried that German submarines may have compromised the radio beacon but the celestial and dead reckoning navigation of Lt. Seymour Ausfresser proved perfect. The nose gunner first sighted the coast of Africa at 0805. With an ample fuel supply, the crew started their letdown when a swarm of locusts greeted FAT STUFF II, covering the aircraft with ugly white mucus and blood. Lt. Elmer Hammer and his crew encountered similar conditions. Approaching the coast, a nasty sandstorm shrouded their plane making the coastline invisible. Only after crossing the coast did they ascertain their position.

As Lt. Jack O'Brien taxied FAT STUFF II to the parking ramp at Dakar, an African man wearing a large red cylindrical hat, brown shirt, brown shorts, heavy infantry shoes and carrying an antiquated rifle ran to greet them. After the crew piled out onto the dusty ramp, he saluted smartly. Lt. O'Brien offered him a cigarette that he gratefully accepted but he seemed stumped by a piece of lifesaver candy. After much deliberation, he put it in his mouth; his smile showed his liking for the candy. He guarded the plane for the duration of their stay in Dakar.

Exhausted crews sought the refuge of cots and mosquito netting. Mosquitoes rivaled the south Florida variety, except they carried malaria. Dakar offered little in the way of creature comforts. Quarters, housing eighteen men each, were adequate but the food was bad. Furthermore, no recreation facilities existed. It did not matter to the crews. The following day was another leg closer to combat.

Bad weather along the route to Marrakech created a backlog of crews. The planned one-day layover ballooned as they waited for favorable weather. For Lt. Albert DiLorenzo and Sgt. Roy Stroop, their stay narrowly missed becoming quite a bit longer. Lt. DiLorenzo wrote his wife on 8 December. "Our pilot has malaria and will be in the hospital three weeks. Probably won't be able to fly for several weeks after release from hospital ... we're still in French West Africa ... not a permanent base – no recreation facilities (coconuts and monkeys). Want to get to final destination so we can get some mail." Lt. Donald Coleman contracted malaria and was unable to continue so Lt. Marvin Onks, who was flying as an extra pilot, assumed command of the crew. Sgt. Roy Stroop kept his feverish symptoms to himself to avoid being left behind. Unfortunately, Lt. Coleman was left behind for medical care.

Crews took time off to sightsee and visit the docks, market place, French cafes, and entertainment sections of the city. Swimming in the Atlantic offered another enjoyable respite. The crews enjoyed their time off and found the culture inter-

William Blum (left) and Jack O'Brien sharing a cigarette with a guard in Dakar. Note how a censor carefully covered the squadron patch with ink. (Everson)

esting. Capt. Blum and the rest of the crew of FAT STUFF II enjoyed a Thanksgiving dinner of turkey, stuffing, potatoes, cranberry sauce and candy. After filling their stomachs, they retired to the barracks for a game of poker.

The route to Marrakech, French Morocco, took crews east into Algeria to avoid Spanish territory, then across the Sahara desert and the almost 14,000 foot high Atlas Mountains. Marrakech itself was situated in the plains just north of the mountains. Two weather-reporting stations operated from each side of the mountains and advised crews on the ever-changing weather situation. Tindouf, Algeria, in the Sahara, provided an emergency divert field and also offered an excellent update point for crews transiting the mountains.

Lt. Jack Swayze's crew left Dakar heading north and quickly encountered blowing sand and thick swirling dust storms. The storms topped out at over 14,000 feet making flying and navigation difficult. No one knew the effects of sand on the engines, but fortunately the engines withstood the sandstorms. The next hurdle, the Atlas Mountains, remained hidden in a thick cloudbank. After conferring with the navigator, Lt. Swayze estimated the tops of the mountains to be 3,500 to 4,000 feet high. Upon reaching the clouds, Lt. Swayze established a shallow descent to 5,000 feet in order to get below the weather. What a surprise when they broke out of the weather to find mountains rising above them on both sides. Fortunately, they had descended into a valley and after reviewing the charts, it was determined that the mountains were 4,000 meters not feet! Lt. Jack Parker, flying HELLO NATURAL, also narrowly averted disaster after losing an engine on takeoff. They spent a week in Dakar waiting on repairs.

Arrival procedures at Marrakech remained the same. First came the ritual bug spraying followed by tired crews searching for a place to rest. Accommodations also proved to be familiar. Some crews stayed in old Foreign Legion facilities furnished with a cot and one blanket per man. Others were less fortunate and stayed in olive drab tents. Neither proved to be a match for the winter cold.

Upon their arrival, a driver carried the crew of FAT STUFF II to a small adobe settlement. The retired Foreign Legion settlement was home for several days until they left for England. The crew spent their time playing basketball and sightseeing. The first evening in Marrakech, two native boys provided a 'special tour' of the town for an exorbitant price. The tour included dancing girls and all the weird sights not usually seen by Americans. The most interesting tour included the native quarter of Marrakech and the Sultan's palace. The walled area around the Sultan's palace consisted of clay and mud houses jam-packed with thousands of natives. The market reeked of

LAKI-NUKI burns after crash-landing at Predannack, Cornwall on 14 December 1943. The aircraft was severely damaged after straying over an enemy airfield in France on its way from North Africa to England. (USAF)

spoiled goat heads, fish and lamb. Vendors crowded the visitors, offering their goods for sale. Small donkeys trotted by with people perched precariously over the rear legs of the donkey with their feet dangling off the side. To Lt. Arthur Steele, it resembled scenes from "Ali Baba" or "The Arabian Nights."

Lt. Lewis Sarkovich, the copilot on BOMB BOOGIE, experienced radio problems at Dakar and did not depart as scheduled. Delayed until 4 December, they arrived at Marrakech to find they would be delayed another day. President Roosevelt was landing at Marrakech enroute to Tehran for the Tehran Conference with Stalin and Churchill. They stood down until the President's aircraft cleared the area.

Crews remained in North Africa for several days receiving briefings on procedures for the final leg to England. Takeoffs were scheduled over an eleven-day period to prevent major losses. On occasion, German fighters ventured out over the Atlantic intercepting airplanes bound for England. The Luftwaffe used false radio signals to lure unsuspecting bombers within reach of shore-based anti-aircraft batteries and fighters. All these tactics were discussed and briefed to crews preparing for the trip to England. The route of flight took crews in a northwesterly direction to remain outside the range of German fighters and the coast of Spain. Upon reaching the twelfth meridian, crews flew north until reaching the fifty-first parallel. A turn to the east carried crews to St. Mawgan near Newquay in Cornwall, England. Takeoffs were scheduled during the night to provide the cover of darkness for crews transiting these danger areas. Navigation errors could put the crews near neutral Spain and Portugal or, even worse, near German occupied

France. A late turn to the east could put the crews in neutral Ireland. There was little margin for error.

The first major tragedies experienced by the 448th occurred leaving Marrakech. Lt. Joseph Shank and crew left Marrakech in 42-52108 on the night of 22 November bound for England. Less than an hour into the flight, only seventy-five miles from the base, engine problems forced the crew to return to Marrakech. A navigation error caused the crew to miss the airfield on the return. Unknowingly, they flew into the Atlas Mountains where they hit a mountain killing all on board. The tragedy struck home for Lt. Arthur Steele. Hours before they left Marrakech, he had sat on the steps of base operations teaching Lt. Shank the words to the song "Whenever I walk through the roses ... " To make matters worse, he recalled Lt. Shank's wife and mother having breakfast with them the day they departed Herrington. The human side of war reared its ugly head and they were not even in England yet.

On 30 November an operations sergeant informed the crew of FAT STUFF II they were required to attend a 2200 briefing. The notice broke up a heated poker game as everyone scrambled to pack and prepare for the final leg of the trip. Lt. Jack O'Brien lifted FAT STUFF II into the air at 0030 on 1 December on the final leg. Several hundred miles of over-water flight lay between them and England. All the guns were manned due to the proximity of the enemy coast but after several hours of flight

Home! One of the many barracks areas at Seething, these were more permanent than the majority of the prefabricated huts. Both did little to shield the men from the cold. Note the bomb shelter in the foreground. (Bollschweiler)

the tension eased and everyone started to relax. Bad weather re-ignited the tension as a layer of rime ice covered the aircraft. For two hours the crew used the rubber de-icer boots on the wing and tail surfaces to break the ice. By daybreak, Lt. Seymour Ausfresser, "Aussie", navigated FAT STUFF II perfectly to a position so the code letters could be transmitted to the radio station at Land's End. After receiving permission to proceed, the station vectored them toward their destination. At 0935, FAT STUFF II touched down on English soil for the first time.

On 8 December 1943, Lt. John Rhodes and crew lifted off from Marrakech in their heavily laden B-24, FINK'S JINX. Just after takeoff an engine failed during the most critical phase of flight. Unable to keep control of the stricken plane, Lt. Rhodes crashed at the end of the runway killing all on board. Lt. John Grunow, Group Assistant Operations Officer, initiated his takeoff roll just moments before Lt. Rhodes' plane crashed. Insufficient runway to abort forced them to continue their takeoff and fly through the updraft from the smoke and flames of the burning plane. Fortunately, they gained sufficient altitude to avoid a similar fate. The cause of the engine loss was unknown, but speculation centered on the fuel odors the crew experienced on previous legs.

All did not go smoothly for Lt. Robert Ayrest and crew. Taking off from Marrakech at 0130 in the clear moonlit morning of 14 December, Lt. Ayrest headed LAKI-NUKI out to sea. Beautiful weather turned ugly as they turned north off the coast of Portugal. Concealed in the clouds they flew on instruments until 1100. To make navigation more difficult, the Germans

Unlike bases in the States, bases in England were dispersed to hinder enemy attacks. This aerial view of Seething shows how the living areas were scattered around the countryside away from the operations area at the airfield itself. (Gaskins)

jammed the radio frequencies. Trying to break out of the storm and icing, they descended to 6,500 feet and turned on the IFF (Identify Friend or Foe). Breaking into the clear they continued to descend to 1,000 feet while the radio operator attempted contact with their destination field. Upon reaching 1,000 ft, fog completely engulfed the aircraft, reducing visibility to near zero. Climbing back to 6,500 ft, they started circling in an effort to positively identify their position. After an hour, Lt. Ayrest turned to a 040 heading on what the navigator believed to be the correct heading. Overcast and undercast prevented the use of celestial navigation or drift. Descending, they finally broke out of the weather at 600 feet and crossed over an island, then a harbor and city. Seeing a blacktop airport with barrage balloons, they circled at 800 feet in preparation for landing. Recognizing swastikas on German Ju-88s and FW-200s on the field, they applied full power and started evasive action as they headed for the safety of the clouds. Heavy anti-aircraft fire bracketed the ship and machine gun fire reached them. Unable to reach the clouds, Lt. Ayrest descended and headed for the open sea.

Suddenly, LAKI-NUKI shuddered from a direct hit just forward of the copilot's rudder pedals. Bleeding profusely, the copilot, Lt. Irvin Litman, put out the fire but the burst destroyed the nose wheel, hydraulic system, radio, and all instruments. The engineer, Sgt. Frank Boula, carefully moved Lt. Litman to the radio compartment and applied first aid to his wounded leg while Lt. Ayrest struggled to outdistance the guns. Another burst blew the sides off the nose turret followed by another one exploding in the waist. Miraculously, flying shrapnel missed everyone else. Reaching the clouds, they breathed a sigh of relief. It only lasted for a second as the clouds thinned leaving the ship visible to ships just off the coast. They lit up from stem to stern as flak exploded around LAKI-NUKI. Ayrest frantically returned the ship to the haven of the clouds while Sgt. Edward Schroeder prepared the ditching gear and sent out a SOS. Heading in a westerly direction, the crew analyzed the damage. Two engines were severely shot up, all controls were gone with the exception of the ailerons, and the only instrument working was the liquid free compass. Staying in the clouds, LAKI-NUKI somehow maintained altitude.

After flying for an hour, Lt. Ayrest took the aircraft down to 100 feet and maintained the heading until the British mainland appeared. With only twenty minutes of fuel remaining, they flew inland searching for an airfield. Fortunately, it took minimal time to locate a field at Predannack, Cornwall and they proceeded to fire the colors of the day and received acknowledgment. After trying to extend the gear and flaps to no avail, they made several passes setting up for a gear up landing. Finally Lt. Ayrest brought the stricken plane in at a 150

Home for the enlisted men of two combat crews. (Everson)

miles per hour, cut all the switches and successfully crash-landed LAKI-NUKI. Upon striking the ground, blue flames engulfed the bomb bay. The crew scrambled to safety as flames quickly spread throughout the ship completely destroying the aircraft and all its contents despite RAF attempts to douse the flames with foam. Miraculously, the only injuries consisted of flak wounds to the copilot and a cut on the head of the ball turret gunner.

Lt. Albert DiLorenzo's letter dated 18 December explained his ill-fated trip to England. "I am in England – very cold and damp. We left our pilot in French West Africa with malaria. Our assistant operations officer flew us to North Africa and there fell ill himself with malaria. Our operations officer flew us here to England and today our crew chief is in the hospital with malaria." Unable to stand, Sgt. Roy Stroop finally succumbed to the fever-wreaking malaria. Medics later transferred him to a military hospital where he spent three months recovering.

All other aircraft completed the journey safely arriving at the airfield at St. Mawgan near Newquay. Most crews spent several days there preparing for the final leg to their new home.

Hut number 10, home to the officers of crew 64. These half circle buildings were the mainstays at bases in Europe. Segregated by officer and enlisted, each hut housed two crews. Although the same size, an officer's hut held eight men while an enlisted hut contained twelve. (Skaggs)

Early members of the Group pose at Seething soon after their arrival at Seething. Back row (left to right), two unknown men, Harvey Broxton, Chuck Billings, Ridd Solomon. Middle row (left to right), Cherry Pitts (between Broxton and Billings) Artie Pace, Albee Charette, "Moose" Tarrant. Front row (left to right), Dwight Covell, Ben Baer, Pierre Delcambre, John Merkling, Charles Wilder, "Duke." (Skaggs)

Once a sufficient force arrived, a mass formation departed and flew the 200 miles to Seething. Lt. A.D. Skaggs and crew arrived in England on 22 November. Three days later, the crew flew to their new home. Lt. Thomas Keene and crew landed at Seething on Wednesday 24 November. Lt. Charles Knorr and crew arrived on 1 December and spent three days waiting on other crews before reaching their new base. Lt. Elmer Hammer's crew landed at Seething on 4 December also. Slowly, crews straggled into Seething. Lt. Ayrest and crew, minus their personal gear, eventually caught a C-47 to Seething.

With the arrival of the ground echelon on 1 December, the 448th was now complete again. Airplanes and crews continued to trickle into Seething, but most crews arrived during the first week of December. The most notable exception was Lt. John McCune's crew. After enduring a month at Borinquen Field, they resorted to cannibalizing a crashed B-24 in an effort to get THE MAID OF ORLEANS airworthy. Finally, after over thirty days waiting, crew 63 was on their way again. They finally arrived at Seething on 4 January 1944, a fifty-one day trip!

Initial impressions of England were mixed. Although the countryside seemed green and lush from the air, crews sloshed through the mud and shivered from the cold on the ground. The hurried construction of the base at Seething churned the wet ground into a sea of mud. It seemed to always rain or snow increasing the morass of mud and adding a damp chill to the air that never went away.

Started in 1942 by John Laing & Sons, Ltd., the base at Seething was typical of most airbases built for the large influx of American air power. The airfield itself was laid out in a triangular pattern with three active runways. Runway 07/25, the 6,000-foot main runway, was intersected by two smaller 4,200-foot runways, 12/30 and 01/19. All three 150-foot wide runways were ringed by a fifty-foot wide perimeter track that stretched over three miles. Fifty-one hardstands were scattered along the perimeter track for aircraft parking. The most imposing features on the flight line were two large hangars, one on the north side of the airfield and one near the control tower on the south side. The small, two-story control tower was the center of the operations area and would soon be the focal point of most days.

Group Headquarters, located just down the road from the operations area, housed most of the administrative offices of the Group plus the briefing room and the chapel. The various living sites were dispersed around the headquarters site. Originally, there were two officer's clubs, one for the flyers and one for the non-flyers, but eventually the clubs were consolidated. The club consisted of three long, half cylindrical, Nissen huts connected together. One was the mess, another was the club, and the third was the kitchen. At one end of the club, an oil painting of Col. Thompson hung on the wall. Pvt. Tom Flannery, assigned to Headquarters and later transferred to Yank magazine in London as a cartoonist, and Lt. Bob Harper, Assistant Intelligence Officer, painted lively murals of aerial combat on the other walls. A large radio-phonograph, purchased with club money, provided the entertainment. The bar, staffed by RAF enlisted men, offered beer and gin to its customers; occasionally scotch was sold. A system of colored lights in the mess and club provided means of notifying crews of the Group's status. A white light indicated stand down, yellow for alert with no target yet, red for alert target known, purple for preparation complete and mission in progress.

Over 400 buildings, mostly Nissen and Quonset huts, were scattered across the countryside in several different living areas. Original plans provided accommodations for 460 officers and 2,660 enlisted men. Unlike bases in the U.S., Seething was scattered in thirteen different "sites" making it a harder target for attacking German aircraft. Scattered sites also put a premium on transportation; bicycles quickly became prized possessions. Enlisted men and officer air crewmembers lived in separate living sites. Two crews shared metal corrugated Nissen huts, eight officers to a hut and twelve enlisted men to a hut. The metal and concrete of the huts failed miserably at providing a warm living space. It was dry, but the bitter, damp cold of England seeped into every crack and crevice. A small stove fired by coke feebly tried to heat each hut. Coke, a poor grade of coal, burned slowly and produced very little heat. Furthermore, the weekly coke ration lasted only a few days. Guards kept vigil over the coke pile to keep it from disappearing completely. The men resorted to other methods. In Sgt. Joe Zonyk's barracks, they devised a contraption that dripped used aircraft oil onto the hot stove. After the heated oil turned the stove red-hot, they quickly decided this was more dangerous than flying missions. Other furnishings consisted of a steel cot per person with a three biscuit mattress, several wool Army blankets, a small footlocker, and space to hang up clothes. As time progressed, ingenuity with bomb crates produced more luxurious furniture.

Upon arrival, not all bath facilities were operational. To accommodate everyone, showers were rationed. Twice a week, between 1400 and 1600, men trekked to the bath sites for hot showers. Original crews recalled the similar treks to the bath facilities at Pinecastle, Florida six months earlier. Now, instead of the heat and humidity, everyone complained about the cold and the endless mud.

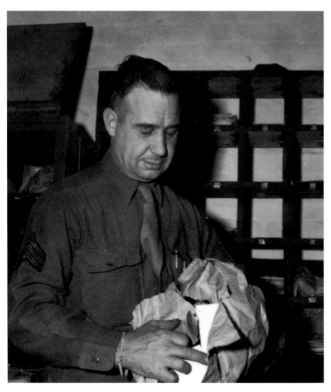

News from home and loved-ones provided an instant boost to morale. Despite the censorship, sending and receiving mail was an important activity for the men. (Bollschweiler)

Although some mail was waiting at Seething when the first men arrived, the majority arrived after the Group reached Seething. There was no such thing as too much mail. (USAF)

Work details knocked down the remaining walls of bombed-out buildings in nearby Norwich and trucked the debris to Seething. The crushed rubble provided the footings for new roads and sidewalks around the base. Sgt. Bert LaPoint worked on this detail until the middle of December, waiting for the completion of the photo lab. His introduction to British food occurred while on a work detail in Norwich. An elderly woman selling fresh milk and crumpets from a horse drawn wagon came by during a break. Sgt. LaPoint bought a quart of milk and sat on a curb enjoying his treat. An American major walked up to him and he stood up and saluted. The major promptly told him of the British practice of not testing or pasteurizing their milk. He quickly poured it out and never bought milk again.

The village of Seething, population 300 including those away on war duty, lay ten miles south of Norwich near the North Sea in a part of England known as East Anglia. The countryside was rural and several farms lay interspersed within the area occupied by the 448th. Many of these farms, undaunted by the huge influx of Americans disrupting their lives, were straight out of picture books. The thatched roofs and hedgerows captured the imagination and attention of the American flyers. The farmers continued their daily chores throughout the war. Many of these families opened their homes and provided a home away from home for many American GIs. Mrs. Mildred Davy, the enlisted men's Aero Club cashier, learned Sgt. George DuPont had attended school in Belgium, her home country. She immediately took a liking to him. She introduced him to her sons and provided him with keys to their house in nearby Loddon. She gave him a photo with the inscription "To my American son from his English mother" an expression reflecting many relationships between the locals and the airmen.

While the Group was still spread across the western hemisphere, the men already at Seething sent an invitation to the local school inviting the teacher and students to the base for Thanksgiving. Eighty-five school kids and two evacuee children accepted the offer. Patricia Knights, a nine-year old school girl from Seething, enjoyed her trip. "I still remember the excitement as the trucks drew up outside the school, with the USAAF men lifting us into the back and putting up the tailboard. Then on to the airfield where one of them carried me over the mud into a large Nissen building and we had to leave our coats on pegs. This caused me great concern as my mother had instructed me to look after my five-year old brother, Reggie's, coat. It was the only one he had and there were no more coupons or money left to replace it. We sat at long tables and some airmen were playing music. I had wanted to ask so many questions, about America, their families, and what they thought of our country, but as I was the only child at this table and they

STAR SPANGLED HELL 42-7767 (also known as SHACK RABBIT). Another "original" that was transferred. When aircraft arrived in England, they were immediately taken to depots to be made combat-ready. Due to the tremendous losses suffered by Groups in the fall and winter of 1943, many aircraft at the depots were sent to under-strength Groups after modifications were completed and never returned to their "original" Group. (Everson)

SHOO SHOO BABY 41-29208. A popular name for aircraft named after a hit song of the time, this original 448th aircraft was transferred shortly after arrival in Europe. (Everson)

were asking so many things that I was far too shy to say more than my name and either yes please or no thank you. We had turkey, (the first I had eaten!) and what I thought was jam, but was our introduction to cranberry sauce. I must have been such a disappointment to the men, neither talking or eating much, but they gave me their candies and wrapped them in paper so I could take them home."

The warm hospitality offered by the local population and the return gestures by the new airmen far surpassed the chill of winter. It even outshined the seemingly endless morass of mud. It was the start of many life-long friendships that united the two countries.

# three

# COMBAT:
# December 1943

Before leaving the U.S., Col. Thompson assembled a very capable but young staff to assist him. His deputy and Air Executive officer, Lt. Col. Hubert Stonewall Judy, Jr., was an experienced B-24 pilot and engineer who was called to active duty from the Reserve Officer Training Corps in the wake of Pearl Harbor. The operations officer, Capt. Ronald Kramer, was only twenty-three years old but already an experienced test pilot and B-17 instructor pilot. Lt. John Grunow, also twenty-three years old, served as the assistant operations officer. Like Capt. Kramer, he was an experienced four-engine pilot and test pilot. The ground executive officer was Lt. Col. Karl Elver and the base adjutant was Capt. Ken Parkinson. The oldest man on the staff was the intelligence officer, Capt. Seul Arnold, who was a lawyer in civilian life. This was his second war as he served in France during the First World War.

Other units joined the 448th to make Seething a combat-ready airfield. They all called Seething home and without their vitally important jobs the 448th could not complete its mission. They included the 862nd Chemical Company Air Operations commanded by 1Lt. David Taylor, 123rd Quartermaster Company, 58th Station Complement Squadron commanded by Capt. William Searles, 262nd Medical Dispensary (Aviation), 459th Sub Depot commanded by Maj. Frank Cruikshank, 2102nd Engineer Aviation Fire Fighting Platoon, 1193rd Military Police Company commanded by 1Lt. Samuel Gaines, and the 1596th Ordnance Supply and Maintenance Company commanded by 1Lt. P. J. Stokes.

With the Bomb Group back together again, the damage of the trans-oceanic trip was evaluated. The Group left the U.S. with sixty-four aircraft and seventy-four crews. Four aircraft were lost enroute as were two of the crews and another pilot. Lt. John McCune's aircraft and crew arrived late and another pilot remained hospitalized in Africa with malaria. Despite these setbacks, the Group was officially assigned to the 20th Combat Bomb Wing, 2nd Bomb Division of the VIII Bomber Command. The VIII Bomber Command, in conjunction with the VIII Fighter Command, made up the famed Eighth Air Force. With their new assignment, much work remained to be done before the first combat mission.

As everyone settled into their new surroundings, attention centered on preparations for combat. Upon arrival at Seething, all of the planes were ferried to the 3rd Strategic Air Depot at Watton for final combat modifications and inspections. Lt. A.D. Skaggs ferried five aircraft to Watton during the first two weeks of December. The most noticeable modification was armor plating bolted under the pilot's and copilot's windows for added protection. Mechanics overhauled engines as needed and all inspections were completed before the planes returned to Seething. One ferry trip to Watton was costly. On 26 November, MERRY MAX, flown by Lt. William Ferguson, ran into poor visibility and was unable to establish radio contact with anyone. The crew attempted to land at the 359th FG base at East Wretham, near Thetford. Landing into the sun, their visibility was reduced further making the landing extremely hazardous. The frozen ground made stopping nearly impossible and the B-24 skidded off the end of the runway, crashing into a tree and operations shack. Fortunately no one was injured but the aircraft was lost.

The planes were not the only thing needing attention. The base required much work to prepare for the great orchestra of events necessary to execute the war orders issued by Headquarters. The bomb dump existed in name only. In reality, bombs lay scattered where trucks unloaded them. No organization existed and the constant mud exacerbated the situation.

During the day, everyone worked to bring order to the bomb dump. At night, aircraft identified for a practice or real mission required loading. In the States, practice centered on loading one aircraft. Now six to eight aircraft per squadron needed loading. Benjamin Everett belonged to the 715th Ordnance section. "We would work all day in the bomb dump and then get called out about 2300 to load aircraft. Back to bed about 0600 and maybe you could sleep until 1200 then back to the dump again or else to unload the aircraft as the mission was scrubbed. I think we unloaded the same bombs on the aircraft about eight or ten times before the planes flew their first mission."

Everyone learned to cope with the British weather as best as they could. Before leaving the States, everyone turned in their overshoes and heavy work coats. As a result, ill-prepared troops suffered in the cold wet weather. Many men reported to sick call and spent time in the hospital. One man, Pvt. Sullivan, fell ill and was transferred to a general hospital and later home. The 1596th Ordnance Company diary recorded the woes. "Most of the men were having trouble with colds and flu due to a change in weather never quite experienced before."

Crews started flying practice missions on 6 December in preparation for combat. Practice sorties were flown daily with emphasis on formation flying. Experienced combat crews from the 93rd BG provided first hand accounts, combat lectures and survival tips for the green aviators. Other topics included flying control, prisoner of war and escape, medical procedures, aircraft recognition, and air/sea rescue including dinghy drills. Orientation flights familiarized crews with the local area procedures and intricacies of flying in wartime England. One of these procedures, known as "Darkey", would become quite familiar for many aircrews. The phrase "hello darkey" transmitted over a specific radio frequency three time signaled a lost aircraft. Ground stations provided an approximate fix and directions to the nearest airfield for all friendly planes. Many crews owed their lives to these operators who performed an invaluable service.

Another term crews familiarized themselves with was "squeakers." This name referred to the barrage balloons scattered around the country to discourage low flying aircraft, especially German. Low powered transmitters emitted air raid type sounds on a certain frequency to warn a plane when it was within ten miles of the balloon. Everyone learned to keep eyes and ears open.

Bunchers and splashers also entered the vocabulary of the flyers. Bunchers were medium frequency radio beacons for navigation that were generally located at the airfields. Splashers were more powerful, permanent, cousins of the Buncher. Navigators used radio beacons to identify these stations and safely

When the Group arrived at Seething, construction was still in progress. The wet English winter produced a quagmire of mud that engulfed the entire base. Slowly, the construction ended and thousands of tons of concrete alleviated some of the problem but not before everyone cursed the endless treks through the muck. (Skaggs)

Huts arranged in their typical fashion in one of the dispersed living sites at Seething. Note the bicycles leaning on the huts. Bicycles were the most popular means of transportation to work, chow, and nearby pubs. (Lozes)

Group Headquarters was located just down the road from the control tower and airfield. The chapel was located in the snow covered hut on the left. (LaPoint)

navigate the crowded airspace of East Anglia. Initially, crews flew several of these training sorties prior to the first mission. They quickly tired of these practice missions and confidently awaited a "real" mission.

Tragedy did not wait for combat. On 10 December, Sgt. Bert LaPoint was installing a camera in aircraft 099 when he heard what sounded like machine gun fire. He jumped out of the plane and what he saw was worse. The noise was not gunfire but an aircraft propeller striking the canvas top of a 6x6 truck carrying men to work on the flight line. The driver of the truck attempted to pull off the perimeter track at the first hardstand past the control tower to allow the taxiing aircraft to pass. Once off the concrete, the truck settled in the soft mud unable to move. The spinning propeller from aircraft 42-52116 ripped through the soft-top of the truck. Sgt. LaPoint heard a lot of screaming as he ran to the truck. What he saw made him sick but he retained enough presence of mind to run to the nearby control tower for an ambulance. Four men were taken to the base hospital. Cpl. Edward Haney, from the 712th BS, and Cpl. Arthur Lazarus of the 58th Complement Squadron, were killed. Pvt. Arnold Kapnick and Pvt. Martin Janson survived but they suffered serious injuries. Pvt. Kapnick spent over two months in various hospitals before returning to Seething as a crew chief.

Waldo Balzer, the chaplain's assistant with the 446th BG at Bungay received an urgent message for Chaplain Theodore Runyan stating two men from Seething were killed in an accident. Since the 448th did not have a chaplain (the Group's chaplain, Capt. William Reid, remained in the States because of medical reasons), they were to prepare the memorial service. Ironically, less than a week later, Chaplain Runyan and Waldo Balzer received orders transferring them to Seething permanently.

Crews were first alerted on the evening of 13 December for a mission the following morning. Ground crews worked tirelessly during the night loading bombs, fuel, and ammunition. Aircrews went to bed and spent a restless night anticipating the uncertainty of combat. CQs woke the crews at 0330. They stumbled out of bed into the chilled English night, dressed and found their way to the mess hall. Crews tasted their first combat breakfast consisting of eggs, greasy bacon or spam, toast and coffee. After breakfast crews went to the briefing room. Military police checked names as the men filed into the large, well-lit room. Everyone found a seat on benches and folding chairs facing a platform with a speaker's podium. A large map of Europe covered by a large motion picture screen hung behind the platform. Left of the platform a chalkboard contained all pertinent mission data. Still further left were tables and chairs for the senior officers. To the right of the platform was a board with all tail insignias of participating units.

At 0445, a sharp command of "Room, Ten-hut" brought the aircrews to attention. The Briefing Officer called the roll in order of formation position. Following the roll call, the chaplain read from the scriptures and said a quick prayer. The Group navigator provided a time hack to synchronize watches. Everyone waited in anticipation as the large map at the front of

The unrelenting weather granted no mercy to the new arrivals at Seething. Here Roland "Peanuts" Hallinger (left) and Elbert Lozes (right) brave the elements to visit the showers. (Lozes)

A.D. Skaggs and Elbert Lozes pose in the doorway of their hut, #10, soon after their arrival in England. (Skaggs)

the room was covered, hiding the route of flight and the target. After the Group commander spoke about the significance of the mission, the map was unveiled showing the target and route. Group operations provided details of the target, the Initial Point, Rally Point, fighter escorts, and position within the Wing and Division. Time for stations, engine start, taxi, takeoff, Group assembly, and Wing Assembly were briefed. The mission was discussed from start to finish. Intelligence discussed the route in more detail giving fighter escort specifics, flak threats, and fighter threats. Two large photographs of the target area were projected on the screen for all to see. Formation plans were distributed showing each plane's position in the formation. On the back of the formation plan was a blank form that pilots could use to jot down pertinent information. This was the same information that appeared on the chalkboard at the front of the briefing room. The weather officer briefed the forecast and current conditions applicable to the mission, such as freezing level and contrails. The "Old Man", Col. Thompson, spoke last. He offered words of encouragement and advice for the crews. Following the brief, chaplains heard confessions and offered communion to crewmembers.

After the briefing, everyone set about doing his job. Navigators moved to a separate room for their own briefing and to make flight plans and prepare the necessary charts. Bombardiers received specific briefings on the bomb loads to be carried. Radio operators received the codes of the day and the frequencies to be used. Copilots collected escape kits from intelligence and received the candy ration. Pilots stayed behind to receive a more detailed brief.

Crews collected their gear – flying coveralls, sheepskin pants and jackets, silk inserts and gloves, boots, scarves, parachutes, and Mae West life preservers. Everyone turned in personal effects such as wallets at a special counter. Then they caught trucks for the cold ride around the perimeter track to their appropriate hardstand and aircraft.

Once at the airplane, quiet crewmembers donned their flying clothes and checked their equipment in the frosty predawn

A.D. Skaggs and Elbert Lozes stand in the snow near their hut with Group Headquarters in the background. While the snow was initially a spectacle for some of the men, it soon became a curse to all, especially the men who worked outside. For them, there was inadequate gear and protection from the bitter cold. (Skaggs)

light. The steady rumbling of the auxiliary generator offered the only sound. Occasionally a crew chief checked something on the aircraft or made last minute repairs. Once everyone reassembled at the plane, pilots issued last minute instructions. Training took over and crews quickly set about accomplishing their specific preflight duties. At the designated check-in time, ten crewmembers answered the pilot's call over the intercom. A thunderous roar erupted in the quiet English countryside as almost one hundred radial engines started. Engine checks and final pre-flight checks were completed. Again, at the briefed time, the lead aircraft taxied out of its hardstand toward the

The interior of the huts were spartan to say the least. Men improvised and made their homes as comfortable as possible. A.D. Skaggs passes the time by playing checkers on a table and chairs constructed from bomb crates. (Skaggs)

Elbert Lozes plays checkers from the same spot. As the cold winter progressed and the coke shortage grew, most wood quickly disappeared as firewood for the inadequate stoves. Some ingenious yet dangerous methods were used involving burnt oil dripping onto a hot stove to increase the heat output. (Skaggs)

active runway. Pilots watched as aircraft taxied by on the perimeter track, searching for the plane they were to follow in the slow processional. Eventually, all the planes were lined up, poised for takeoff. Everyone kept an eye on the control tower watching for the flare signaling takeoff. Once the tower fired the flare, controllers posted at the approach end of the runway signaled each aircraft for takeoff by flashing a green aldus lamp.

At thirty second intervals, the fully loaded B-24s of the 448th rolled down the runway and slowly climbed into the early morning sky. Once all of the planes were airborne, crews formed into their respective formations at the assigned altitudes. The group formation then formed itself into its appropriate position in the wing formation. After almost two hours of flying, the formation left the English coast at 0730. On 14 December the Group flew within sight of the Dutch coast before receiving a recall message. The Group returned to Seething after four hours of flying without seeing combat.

The Group flew on 16 December as part of a mission to Bremen, Germany. Eighteen B-24s took off from Seething, but again the 448th received the recall message before entering enemy territory. The same thing occurred on 20 December. This time the Group participated as a diversionary force but after four and half-hours they returned to Seething. The Group was learning to fly in England but anxiously waited to fly in combat. The chance would come soon enough.

Some personnel received a preview of the horrors of combat. On 20 December, a badly damaged B-17, flown by Lt. Charles Brown of the 379th BG, landed at Seething after receiving heavy battle damage. Despite Lt. Brown's courageous struggle the tail gunner on YE OLDE PUB was killed. The dried blood on the floor of the plane was clearly visible to all who passed nearby. It was a graphic illustration of the seriousness of combat.

The Group's baptism of fire arrived on 22 December 1943. The previous evening, Group Headquarters received field orders alerting the Group. Ground crews readied the twenty-six planes identified for the Group's first mission. Ordnance personnel loaded the 500-pound general-purpose bombs called for in the field orders. Nervous crews went to bed in an effort to sleep. Most failed, however, and spent a restless night tossing and turning.

Around 0300, CQs woke the crews. Two hundred and sixty sleepy aviators wandered to the mess hall for the standard combat breakfast. Following breakfast, the mission briefing mirrored the briefings received on numerous training missions. This time, it was for real!

The briefing outlined plans calling for the 448th to takeoff at daybreak and form at 12,000 feet in the assembly area over

When the 448th arrived at Seething, the bomb dump was not prepared to handle the requirements of a heavy bomb group. Thousands of man-hours were required to build the bomb dump as pictured. The wagon in the foreground carried the bombs from the dump to the aircraft. (Everson)

the North Sea. The Group would then rendezvous with the 20th Combat Bomb Wing (CBW) over Buncher 7. The 20th and 2nd CBWs were to follow B-17s of the 1st Bomb Division to the target area and drop over the spot marked by the B-17 pathfinders. The route of flight would take the Group over the Zuider Zee and the coast of Holland. The trip would take the fledglings only fifty miles inside Germany before they would drop their explosive loads on the marshalling yards of Osnabruk, Germany. It seemed simple enough, but reality was far from what was briefed.

As the crews proceeded to their planes, the early morning light revealed low clouds and fog. Takeoff was delayed thirty minutes while crews waited for the weather to improve. Finally, at 1035 the tower fired flares and a green light from the aldus lamp cleared crews for takeoff. At thirty-second intervals, planes lifted into the murky air. Once airborne, pilots maintained a 300-foot per minute climb at 150 mph and turned to a heading of 080. Pilots maintained this heading until reaching 5,000 feet. This route took crews over the North Sea between Great Yarmouth and Lowestoft. Upon reaching 5,000 feet, each pilot executed a one-needle width, left-hand turn to a heading of 260 and continued the climb to 7,000 feet. At this altitude a 180-degree turn pointed the aircraft on a heading of 080. Every 2,000 feet pilots performed this procedure until reaching assembly altitude.

When clear weather allowed, a more timely procedure was used. Again aircraft departed in thirty-second intervals but the lead aircraft maintained runway heading until one and a half minutes. He then executed a 180-degree turn, climbing at 300 feet per minute and maintaining 150 mph. Ten seconds after the lead turned, the second aircraft turned, followed ten seconds later by the third aircraft. Using visual references, each three-ship element was assembled before they passed abeam the airfield. They continued to the briefed assembly area and joined the rest of the formation.

On this day however, layers of clouds forced the lead plane to climb to 10,000 feet before breaking out of the thick clouds. Once clear of the pesky clouds, the lead aircraft fired yellow flares at four-minute intervals to aid the assembly. Confusion reigned as the weather hindered attempts to form the Group. Increasing the difficulty, planes from other groups were trying to do the same thing, all within a fifty square mile area. Pilots reported several near misses. Lt. A.D. Skaggs, flying HARMFUL LIL ARMFUL, took evasive action on two occasions to avoid other planes searching for their formations. Ad hoc formations materialized as the appointed time for leaving the English coast neared. Six planes formed on HARMFUL LIL ARMFUL, one from an unknown group. Crewmembers occasionally caught glimpses of other formations through the clouds, but clouds prevented the assembly of a single tight formation.

Elsewhere, eleven B-24s of the 448th tacked on individually to formations of the 446th. Poor visibility and layered cloud decks made formation flying difficult. Adding to the misery, temperatures ranged from forty below zero to sixty below zero. Oxygen masks froze, as did uncovered skin. Waist gunners suffered the most. Exposed to the blast of air from the open windows, they stood vigil searching for enemy planes. Frostbite was a common occurrence on this first mission. The cul-

On 20 December 1943, Lt. Charles Brown landed his B-17 at Seething after receiving heavy damage to his left wing and tail claiming the life of his tail gunner. This was the first exposure to the horrors of combat for most of the men at Seething. (Droiun)

The briefing room at Seething, pictured here, was similar to ones throughout England. The large map of Europe held yarn that depicted the route of flight and the large board on the right showed the formation line up for the day. Note the sign on the left, "It is better to struggle than straggle." It was a hard-learned motto for the 448th. (USAF)

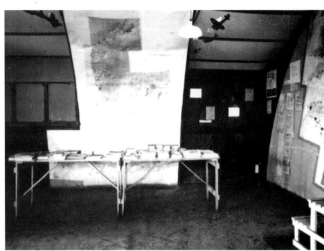

The large wall map of Europe is partially visible in this photo of the briefing room. From this map crews were first made aware of the target for each mission. (Everson)

mination of events overwhelmed nine of the original twenty-six airplanes launched. They returned to base after failing to find the formation. Two aircraft aborted before leaving the English coast even though they found a formation. Although separated, fifteen aircraft continued on to the target without the protective firepower of a complete formation.

When the formation left the Wing assembly point, the 20th CBW consisted of two widely scattered formations instead of three tidy group formations. The lead ships of the 448th joined the 93rd BG followed five minutes in trail by the 446th BG and attached planes of the 448th. The second formation failed to close the gap. As a result, two widely dispersed formations continued toward enemy territory.

As the small groups of bombers neared the Dutch coast the weather gradually improved allowing the planes to slowly climb higher. Several far-off puffs of flak were noticed but nothing of significance until the clouds scattered as they neared the target area.

As the loose gaggle of bombers neared the target area at 1405, resistance intensified. The clearing weather gave German anti-aircraft gunners and fighters better visibility. Just after the bombs fell from the belly of the B-24s, German fighters and flak hammered the new arrivals, first the flak then the fighters. Flak hit Lt. David Manning's aircraft 42-52105. As it fell out of formation, German fighters quickly pounced and shot down the stricken plane. The first 448th combat casualty crashed into the sea just west of Terschelling Island off the coast of Holland, with no survivors including Lt. Byron Lanphear who as flying as an observer. Almost immediately after Lt. Manning's plane was hit, Lt. Edward Hughey's crew in 41-28609 fell victim to a terrific flak barrage. Both the top turret gunner and the left waist gunner escaped from the stricken plane and parachuted to the relative safety of a German POW camp. All others perished.

In the melee over the target, Capt. Chester Hackett realized aircraft above his were opening their bomb bay doors and were preparing to release their bombs. Recognizing his dangerous position, he reduced power on his B-24 allowing it to slowly slip from underneath the other bombers. Not all aircraft in the formation did likewise. At bombs away he watched 500-pound bombs fall from one aircraft and impact another B-24 in the formation. The left wing immediately buckled and the aircraft flipped on its back. They saw no parachutes. (The unfortunate B-24 was not from the 448th. Since the three groups of the 20th CBW merged into two formations during assembly, it was probably from the 446th or the 93rd BG. The mission history reveals only two 448th aircraft did not return from the mission, Lt. Manning's and Lt. Hughey's. The 93rd lost five aircraft on this mission and the 446th lost two.)

Once clear of the target area, German twin engine fighters greeted the bombers by lobbing rockets into the decimated formations. Lt. Thomas Foster's plane received a burst from a rocket sending fragments into the radio operator's compartment and the number three engine. Shrapnel hit Sgt. Arthur E. Angelo in the left chest seriously wounding him. HARMFUL LIL ARMFUL received flak damage to the number four engine over the target but the engine continued to run on reduced power. The crew fought back. Sgt. Francis Sheehan, flying as a waist gunner, carefully zeroed his fifty-caliber machine gun in on a Me-110 that strayed too close. A quick burst from the

HEADQUARTERS AAF STATION 146  
Office of the Intelligence Officer  
APO 634

SOA/ald

7 February 1944.

SUBJECT: Briefing.

I. Leave screen down until briefing starts. Use a sheet in gunner's room.

II. Pilots, co-pilots, navigators, and bombardiers go to principal briefing room.

III. Radio operators and gunners go to gunners briefing room.

IV. Radio operators pick up flimies at old communications room before they go to gunners briefing.

V. Gunners are excused after briefing and radio operators are told to stay.

VI. Radio operators are then briefed by communications.

VII. After principal briefing of pilots, co-pilots, navigators, and bombardiers, navigators go to gunners briefing room, bombardiers to old communications room, and co-pilots go to ships.

VIII. Pilots, navigators and bombardiers are held up to one hour and thirty minutes before take-off.

IX. There shall be no smoking in the principal briefing room, or gunners briefing room until principal briefing is completed, and the briefing of gunners and radio operators has been completed.

X. One table for each navigator must be set up immediately after the radio operators briefing for use by the navigators.

XI. No smoking signs shall be placed in the principal briefing room and gunners briefing room as soon as possible.

Checklist used by 448th administrative personnel for the conduct of mission briefings. (National Archives)

waist gun struck the left engine of the fighter. It rolled over on its back headed toward earth trailing heavy black smoke. The 448th received its first credited kill.

On the return leg, all but three guns on HARMFUL LIL ARMFUL malfunctioned, but finally the ragged formation of planes crossed back over the coast to the safety of the North Sea. The remainder of the flight was uneventful and crews caught their breath and tried to settle frayed nerves. The first aircraft landed at Seething at 1400 but it was another hour and a half before the final plane landed. Once on the ground, Lt. Elbert Lozes counted twenty-six holes in HARMFUL LIL ARMFUL and thought to himself, "I can never make twenty-five of these missions."

Sgt. Stephen Burzenski, Lt. Manning's crew chief, watched as the B-24s appeared over the field. Everyone on the ground instinctively counted the planes as they flew overhead and prepared for landing. Sgt. Burzenski sat on the hardstand devastated when all the planes were down and his hardstand was still empty. He flew to England with the crew and they were like brothers to him. He made a vow to himself, never to become close to the crews again.

During the debriefing, crews realized the extent of damage from the first mission. Only thirteen planes were effective in dropping bombs on the target. Twelve aborted and one could not drop due to a mechanical problem. The weather played havoc with the new crews. Formation discipline suffered. The mutual fire support offered by a tight formation was lost when the formation scattered. The veteran Luftwaffe pilots recognized this fatal error and made the green aviators pay. The results were devastating. Two crews were shot down and three planes were damaged, one of which was listed as Category E. Category E denoted severe damage that was deemed uneconomical to repair. Subcategories defined it further. E1 signified damage left parts serviceable or repairable to be utilized as spare parts. E2 meant the wreck was only suitable for scrap metal.

Crews licked their wounds the next day as no mission was scheduled. However, that evening crews were alerted for a mission the next day, Christmas Eve. The great orchestra of events started again as everyone prepared for another mission. This time the target was a "No Ball" site, the code name for V-1 rocket launching sites. This particular one was located at LaBroye, France near Raye-sur-Authie. The 1035 takeoff was later than usual due to the short distance to the target. The mission proved to be a milk run compared to the first mission but black puffs of flak still littered the sky along the entire route of flight, especially along the French coast near Abbeville. At approximately 1405, the formation of twenty-five aircraft approached the target at 12,000 feet amid intense flak. P-47s nervously circled over the target providing cover while others waited at the Rally Point to drive away any Luftwaffe interference.

Capt. William Blum, 712th BS Operations Officer, flew lead of the second element. Just after the initial point, Blum's aircraft was struck from behind by another plane flown by Lt. Alan Teague. The left rudder, left wing, bomb bay, and number two engine of Capt. Blum's plane, 41-28591, received serious damage during the collision. Capt. Blum jettisoned his bombs and staggered out of formation, damaged but still in controlled flight. The collision sheared the number one propeller, bent the number two propeller, and started a fire in the bomb bay of aircraft 42-52118 flown by Lt. Teague. The crew quickly contained the fire and the plane remained airborne. Lt. Teague and crew headed west in search of the nearest British airfield. Fortunately, the short distance to the target allowed them to reach England and they quickly found an airfield at RAF Bourn. The crew bailed out leaving the pilot and copilot to land the stricken machine. After a successful landing, the salvage personnel listed the plane as Category E. Sgt. Joseph Reddit, a gunner on Lt. Teague's aircraft, suffered injuries requiring eight months of hospitalization.

Meanwhile, Capt. Blum and crew nursed their damaged plane back to Seething. The last plane landed at 1730 and crews reported to debrief to tell their findings. The official tally for

Checklist used for briefing the gunners before a combat mission. (National Archives)

the mission showed two aircraft Cat E, including Capt. Blum's, and four more damaged. The good news was no crew losses! However, the second mission was only a slight improvement over the first and it was supposed to have been a milk run.

Christmas Day offered a break from the brutal introduction to combat. Following a morning church service, the base opened its doors to the local population, especially the children, for Christmas dinner. To these children, including many who were evacuated from their bombed out homes, these men

Buncher 24 was one of the navigation aids used by the bombers for assembling and navigating. Without these radio beacons and the men that maintained them, assembling large formations in such a confined space as East Anglia would have been impossible. (Wanker)

Men enjoy Christmas away from home with their buddies. Despite the somewhat austere conditions, the men improvised with what was available. (LaPoint)

were their heroes. Just like Hollywood, these aviators spoke with a Yankee dialect and walked with a confident swagger. Airmen passed out candy and gum to the kids and the turkey and fixings seemed unreal to some kids who only knew the rationing of war. This event provided much needed therapy for both the kids and the aviators. For the kids, it allowed them to be children again and partake of the candy and fruit that had disappeared from most British households three years ago. For the airmen, the children allowed them to forget the horrors of combat they had witnessed on the first missions. The kids' excitement took their minds off being far from home during the holiday season and the devastation of war. Sgt. George DuPont took two kids to dinner. Their eyes brightened at the sight of oranges, but they politely said "no thank you" when they were offered. Sgt. DuPont quietly stuck an orange in each of their side pockets. "Just in case you change your minds," he whispered. Their smiles made him feel so good. Although saddened by the loss of his barracks mates on a mission three days earlier, Sgt. B.D. Bottoms took part in the festivities including dinner, goodies and a tour of a B-24. "I stood in the rear camera hatch and helped the little ones into the rear of the plane where they walked through and out the bomb bay."

Waldo Balzer wrote his wife after the party detailing the day's events:

"The men at this station were asked to adopt a child or two for the time they were at the base. It was indeed thrilling to see both officers and enlisted men walking around with children at their side and to see them eating together in the mess hall. Even the commanding officer had a little girl with him. At first I did not have a child but during the programme I noticed a boy who had no one to look after him so I kinda took him under my wing. The C.O. who had the little girl could not remain with her so I volunteered to foster her for the remainder of the afternoon. After the programme the children were taken to the line, that is the flight line, going through one or two of the planes. This being a bomber group the type will be self-evident. Of course the children were quite fascinated by the huge birds and at first my little girl was afraid of it, but she went through it with me, ever holding firmly onto my hand. Having completed this short tour we all accompanied the children to their various schools via trucks. It took nearly two hours to get them all back to their places but it was a lot of fun. With the boy Norman at my right and little Molly my left we rode with the other children and GIs to take them all back. Molly was such a sweet, good-

The men from crew 64 stand in front of a bomb shelter in the living site. Initially the men used these shelters for their intended purposes but they evolved into high ground from which to view the nuisance raids. (Left to right) Elbert Lozes, George Glevanik, Ben Baer, Francis Sheehan, Don Todt, and Gene Gaskins. (Gaskins)

natured girl. I was indeed pleased with her. Peculiar enough she seemed to place all her confidence in me as I took her down from the truck and carried her to her father who was at the school to meet her. Her daddy was so pleased to see her again and thanked me for having returned her safely. Before leaving Molly gave me a sweet kiss in her own girlish way."

When they returned to Seething, it was dark so Waldo Balzer headed for the mess hall concluding his first Christmas in England. The curiosity and bright eyes of the children offered the men a momentary break from their jobs and a chance to remember Christmas. Although it was not home, it was a memorable Christmas.

Not everyone got to enjoy the fun. Benjamin Everett spent Christmas Day in the hospital. Originally in the hospital for four days, he returned to duty. Still not feeling well, he requested to work indoors for a few days but due to the amount of work, he was told to go to sick call or go to work. The end result was a second trip to the hospital after trying to work in the cold British weather. This stay lasted for five days including Christmas Day. "We did get turkey for the noon meal otherwise it was just another day."

The dark morning hours of 30 December witnessed the ritual cursing and mumbling as CQs rustled sleepy

SAD SACK 41-28588. Seen here undergoing modifications (note panel added under the copilot's window for extra protection), this aircraft crash-landed in Germany on 30 December 1943 carrying Lt. Ray Gelling and crew. (Everson)

crewmembers out of their beds. The bone-chilling cold made crawling out of bed even more unpleasant. For Lt. Jack Barak and crew, this was mission number one. At the briefing, crews learned the target was the I.G. Farbenindustrie Chemical Works in Ludwigshafen, Germany. The 448th would fly tail-end-charlie for the 20th CBW and, again, formation discipline spelled trouble for the Group.

Twenty-five aircraft took off from Seething at 0823. Takeoff was delayed ten minutes after an aircraft taxied off the pe-

HARMFUL LIL ARMFUL 42-7754. This photo was taken while the aircraft was being refueled at Belem, Brazil during the Group's deployment to England. On 31 December 1943, this aircraft carrying Lt. Phillip Chase and crew was shot down by Me-109s near Cognac, France. (Everson)

rimeter track and stuck in the mud. The narrow perimeter track made taxiing difficult and a heavily laden B-24 quickly sank in the mud if a tire drifted off the concrete. The Group narrowly avoided missing the entire assembly. Twenty-one aircraft made the assembly and headed for the target with P-47s providing the escort. Clear weather prevailed but trouble arose just after crossing the enemy coast. The Group started to fall back from the larger formation, unable to maintain proper distance.

Thirty minutes after crossing into enemy territory, FW 190s attacked the now separated Group. One aircraft was damaged and turned back. The remainder of the 448th formation continued in a vain attempt to close the distance with the lead formation. Heavy flak greeted the unwelcome guests as they neared the target area. At approximately 1215, forty incendiaries fell from the bomb bay of Lt. A.D. Skaggs' plane, while ahead German fighters attacked other Groups. After bombs away, the 448th continued to have problems catching the preceding formation. German single engine and twin engine fighters quickly focused their attention on the solitary Group. Fighters lined up and in a shallow dive pressed their head-on attacks within fifty feet before peeling off. Others remained at a relatively safe distance, lobbing rockets into the formation. At 1345, four Me-109s hit the formation. One dove on the formation from 10 o'clock high with guns blazing. Lt. Thomas Foster, flying CONSOLIDATED MESS, was fatally wounded, as was the top turret gunner, Sgt. James Brant, as tracers raked the flight deck. Fire quickly engulfed the bomb bay as the copilot, Lt. Francis Rogers, pulled the plane up and to the right, away from the formation. Sgt. Francis Sheehan, waist gunner on HARMFUL LIL ARMFUL, watched as the plane exploded in mid air. Sgt. B.D. Bottoms, flying right waist in Lt. Thomas Keene's plane, saw crewmembers bailing out of the stricken plane. Only two survived: Lt. Don Hanslik, navigator, and Sgt. Chester Janeczko, waist gunner. The explosion blew both men clear of the plane. Lt. Hanslik spent three days, barefooted, evading Germans before French Underground operatives rescued him and returned him to England. Sgt. Janeczko suffered burns to his face and hands and was captured. The exploding plane crashed northwest of Crepy-en-Valois, France thirty kilometers from Soissons.

Two other aircraft, one flown by Lt. Abraham Kittredge and the other by Lt. Ray Gelling, fell prey to fighters. Lt. Gelling, flying SAD SACK, crash-landed in Germany; all the men survived as POWs except the radio operator Sgt. Harold J. Lane. Lt. Kittredge, flying 41-28599, fell out of formation after having two engines shot out, another supercharger inoperative, and the right aileron demolished. Seven of the crew bailed out while the pilots kept the aircraft flying. The copilot, Lt. Edward Fox, and the radio operator, Sgt. Kenneth De Soto, elected to stay and help the pilot. The aircraft crashed at Cul-des-Sarts, thirteen kilometers southeast of Chimay, Belgium claiming the lives of the three men that remained with the aircraft. The Group continued to fend off the determined attacks all the way to the French coast. Fortunately, the 448th tacked on to a B-17 formation for mutual fire support for the last seventy-five miles. On return, two aircraft landed at other bases due to battle damage. Aircraft 42-52116 landed at Ashford and 42-52123 put down at Gravesend.

Once on the ground after the eight hour flight, crews underwent the routine intelligence debriefing. Most men welcomed a shot of whiskey as they reeled from the loss of thirty more comrades. The sight of six empty bunks in his hut overwhelmed Sgt. Stanley Filipowicz. They belonged to his friends, the enlisted men of Lt. Foster's crew. That night Sgt. Filipowicz got drunk in an effort to erase the memory of the plane exploding after being hit by a Me-109. It had been another bruising encounter with the Luftwaffe. They received little consolation; crews were alerted for another mission the following day.

Lt. Lawson Campbell hoped for no mission as his brother, an Army Captain stationed in London, sent him two tickets to the New Year's Eve Gala at the Officers' Club at the Grosvenor House. Unfortunately, duty called and the nightmare of combat resumed. Crews sighed with relief as the target map showed the destination, two large German cruisers anchored in the harbor at La Rochelle, France. However, it was a long flight to southern France and back. Eighteen aircraft plus one spare took to the air on New Year's Eve, each loaded with 500-pound bombs, and assembled in the pre-dawn darkness.

As the formation left the British coast, the spare returned to Seething. They crossed the French coast near Caen at Grayan et L'Hopital and settled in for the long flight across the French countryside with the rising sun in their eyes. At noon, as the formation neared the target area, enemy fighters attacked the formation hitting HARMFUL LIL ARMFUL flown by Lt. Phillip Chase. Last seen lagging behind the formation with the number three engine smoking, two more Me-109s attacked HARMFUL LIL ARMFUL. The stricken B-24 entered a steep dive, immediately followed by a rapid climb. The plane stalled and spun toward the ground with the fuselage and one engine on fire. The crew bailed out before the plane crashed fifteen miles east of Cognac, France. Both waist gunners, Sgt. Thomas McNamara and Sgt. William Dunham were killed. The pilot and three others were captured but the engineer, Sgt. Arthur Meyerowitz, and the radio operator, Sgt. Joseph DeFranze, evaded and returned to Seething. (Sgt. Gene Gaskins, the

Robert Voight (left) and Earle Durley pose in front of BABY SHOES after the mission of 31 December 1943. Despite standard procedures, Lt. Durley left the formation after seeing his friend in BABY SHOES fall out of the formation due to damage. Lt. Durley escorted his friends safely to England where both landed safely. (Westgate)

COLD TURKEY 41-29248. Unknown men pose in front of the aircraft soon after their arrival in England. Lt. Max Jordan and crew were lost in this aircraft over southern France on 31 December 1943. (Everson)

aircraft's original ball turret gunner, kept a Gideon Bible in a canvas pouch near the ball turret. The Bible was found at the crash site and after the war, it caused identification problems between Sgt. Dunham and Sgt. Gaskins.)

Complete cloud cover over the harbor forced the Group to change course and hit the secondary target, an airfield at Cognac. As they prepared to drop, Capt. John Grunow, flying as the command pilot in the lead aircraft of the low squadron, looked up into the open bomb bays of the high squadron. During the fighter attacks, the two squadrons became overlapped. His quick actions separating the two averted certain disaster.

Thirty-five minutes after Lt. Chase's tragic demise, Lt. Max Jordan, flying COLD TURKEY, suffered a similar fate when accurate, heavy flak bracketed the formation. Lt. Leroy Engdahl, flying as copilot on his first mission, watched as Jordan's plane, flying on his left wing, disintegrated. A direct hit in the right wing caused an explosion that ripped through the plane severing the tail and left wing. No chutes were seen.

Lt. Lawson Campbell, flying as copilot on BOMB BOOGIE, witnessed the gruesome spectacle. The explosion that ripped COLD TURKEY apart propelled pieces of airplane in all directions. The flying debris damaged BOMB BOOGIE; flak created the remainder of the damage. The pilot, Lt. Robert Martin, and Lt. Campbell struggled with their damaged aircraft and successfully returned to southern England. With no hydraulics, no flaps, and no brakes they crash-landed at Predannack. One hundred and fifty flak holes scattered over the aluminum skin of their plane mocked those who predicted a "milk run." Salvage crews condemned it as Category E. Despite the horror of the mission, Lt. Martin and Lt. Campbell kept to their plan. They thoughtfully carried their class A uniforms with them on the flight. Once on the ground, they changed into their clean uniforms and caught a train to London. They arrived at 1900 hours and spent the rest of the night welcoming the New Year. It was a much-needed respite from the war.

The plane flown by Lt. Earle Durley and Capt. Grunow was also damaged and could not maintain formation. Flak disabled the number one supercharger forcing them to fly at a lower altitude. After clearing the target area, the pilot slid the plane away from the others and headed for the sea. Seeing his friend in trouble, Lt. Robert Voight in BABY SHOES followed Lt. Durley and acted as an escort for the damaged airplane and crew. Upon reaching England, both planes were low on fuel and typical, misty English weather prevailed. Searching through the fog, Lt. Durley found a dirt strip at Larkhill Camp and landed. Lt. Voight landed BABY SHOES at Brize-Norton, a British glider-training base in southern England, and refueled before returning to Seething. The crew of BABY SHOES spent New Year's Day dressed as English officers as guests of the RAF.

Seventy-five mile per hour headwinds made fuel conservation a priority for the returning crews. Furthermore, low ceilings and poor visibility in England hampered crews attempting to land. CRAZY MARY, damaged by flak over the target, limped back to southern England. The crew landed the crippled plane at Yeovilton. Salvage crews listed the plane as Category E several days later. Lt. Ed Chapman, landed in southern England at Tarrant Rushton to refuel. After landing, the copilot, Lt. Engdahl, surveyed the damage. He found twenty-five holes larger than his fist but they were able to return to Seething the next day. Lt. Elmer Hammer and crew, flying LADY FROM BRISTOL on their maiden mission, suffered damage to their plane. Exploding shrapnel wounded a crewmember. On return, fuel starvation forced them to land short of Seething in a field at South Store near Oxford. RAF personnel carried the shaken crew to nearby RAF Mount Farm where they spent a week before returning to Seething without their plane. Eventually their plane was repaired and returned to Seething.

The almost eleven hour mission took its toll. Operations clerks at Seething posted two more crews as missing in action. Two damaged planes were beyond repair; five others were repairable. Only two planes made it back to Seething; the rest were forced to land short of Seething to refuel. Many of the planes did not return to Seething until the next day, New Year's Day.

As the New Year started, the twenty-five missions required to complete a tour seemed impossible for the crews. Statistically, odds did not favor the crews. During the first month of combat, Luftwaffe fighter and anti-aircraft units handled the Group roughly. Of the sixty aircraft originally available to the Group for combat operations, seven were lost over enemy territory. Several others would never fly again due to combat damage. It was not an enviable record for a new combat unit.

four

# THE ENEMY ... WEATHER:
# January 1944

Bombing procedures varied greatly depending on the weather. The summer months with longer days tended to produce more hospitable weather. The winter months were another story. On rare occasions in the winter, high pressure produced CAVU (ceiling and visibility unlimited) conditions, but normally low clouds prevailed over northern Europe.

When weather permitted, crews utilized the classified Norden bombsight. In the hands of a trained bombardier, this high precision instrument provided unprecedented accuracy in bombing. Developed by Carl Norden, the football, as it was commonly called, allowed the bombardiers to find the exact point to release the bombs in order to hit a specific target four miles below. However, to do so required numerous inputs from the bombardier.

"Since an aircraft pitches, rolls, and yaws or turns, a gyroscope was needed to hold the bombsight optics in a stable position, regardless of aircraft movement. This gyro was located on the left side of the bombsight. A large silver knob just above it was used to cage or lock the gyro until the bombardier started on the bomb run. Once uncaged, the gyro stabilized the optics against pitch and roll of the aircraft. It then established a vertical reference to the ground, which was necessary to align the course and measure the dropping angle. There were two "bubbles" much like those used on a carpenter's level. One was placed parallel to the fore and aft portion of the aircraft and the other laterally. Two silver knobs on the left of the bombsight head were used to center these bubbles. Once the gyro was leveled during the bomb run, it would tend to remain level. Leveling of the bubbles was critical. If an aircraft was flying at 10,000 feet and the top of the gyro was tilted one degree to the rear, the bomb would hit about 214 feet short of the target. This was "bubble error." While the Norden bombsight stabilized the optics for pitch and roll of the aircraft, the yaw or lateral movement of the aircraft had to be compensated for also. This was accomplished by the stabilizer, the box-shaped equipment on which the bombsight fit and to which it was connected. When the bombsight was properly fitted and the nemesis of the bombardier-the dovetail locking pin-was inserted to complete the connections, a directional gyro stabilized the entire sighthead and gave the bombardier control of the aircraft through the autopilot. With the bombsight connected to the stabilizer and accurate data set into the bombsight, the bombardier was ready to synchronize on the target. Two sets of large silver knobs on the lower right of the bombsight were used. These "course knobs" were used to make drift and course corrections. When these corrections were made precisely, the vertical crosshair would remain stationary. The other set of knobs on the lower right were the rate knobs, which synchronized the lateral crosshair on the target, controlling groundspeed or rate of closure. Groundspeed varies with the heading of the aircraft, since wind causes the differences between true airspeed and groundspeed. Therefore, the aircraft had to be on correct heading before the rate of closure could be accurately synchronized. When this occurred, the lateral crosshair remained stationary.

All of this was accomplished by the bombardier in the few minutes prior to bomb release while flak exploded nearby and fighters pressed their head-on attacks. Despite these deadly distractions, the bombardier had to concentrate and precisely manipulate his controls. A slight error at high altitude would send the bombs hurtling somewhere beside the target. Regardless of how well the mission progressed, the ultimate outcome of the depended on how well the bombardier performed his job. When the target crossed the stationary crosshairs of the Norden bombsight, the bombardier toggled the bomb release switch and called out the long awaited words, "bombs away." (*Air Force Magazine*, Sep. 1981)"

Obscured conditions severely limited the bombers' ability to use the Norden bombsight. To counter this, several methods were employed. The most common method employed by the 8th Air Force was known as H2S. Referred to as "Stinky," this self-contained system was designed by the British and used an on-board radar to send pulses that reflected off the ground. The return signals painted a "picture" on a scope in the aircraft. Built up areas and mountains reflected the best while water provided the least reflection. With training, navigators learned to identify targets and navigate using this primitive radar.

The U.S., impressed by early tests, received a license to build and improve a similar system. This system, known as H2X or "Mickey," operated much the same as the H2S system but with some improvements. Both systems were fitted to B-17s and B-24s. On B-24s, ball turrets were replaced with retractable cylindrical domes, called dustbins that housed the scanner. These were retracted while not in use or on the ground and were extended inflight. The extra drag reduced the top airspeed by ten knots. After hard-earned lessons, the best results were obtained using the two systems, H2S and H2X, in conjunction with each other. The different wavelengths usually meant when one failed the other still worked. As a result, when conditions permitted and aircraft were available, the lead aircraft carried H2X and the deputy lead carried H2S or vice-versa.

Another system for beating the clouds involved using a system known as GEE-H or G-H. This system relied on radio beams sent from permanent stations on the ground to the aircraft. Using time and distance calculations the navigator could tell if he was flying a prescribed course and how far he was from the station. At the prescribed time and distance, bombs were released. This system proved more accurate on smaller targets than H2X but was limited by distance from the ground stations. A fourth system combined G-H with H2X providing

Almost all the buildings at Seething were prefabricated huts including the Aero Club. There was always a bomb shelter close by, notice the one to the left of the club. (LaPoint)

the advantages of both systems. Known as Microwave-H or Micro-H for short, it was not used until late in the war and primarily by B-17s. (For further information refer to *The Mighty Eighth War Manual* by Roger Freeman.)

Due to the scarcity of aircraft with these new systems, they were pooled into one Group, the 482nd BG and only one squadron, the 814 BS, flew B-24s. This group provided lead aircraft to the 2nd Bomb Division until it was removed from operational status on 26 March 1944. The 482nd BG shifted to a teaching role and trained all 8th Air Force personnel in the use

Ed Drouin and Roy Wisch of the 1596th Ordnance Company stand in front of SWEET SIOUX. Men like these were the unsung heroes of the Group. They toiled day and night to gas, load, prepare, and repair the aircraft in preparation for the missions including the numerous ones that never went. (Drouin)

of this new technology. With the loss of the 482nd BG and still short of these specialized aircraft, the 2nd Bomb Division centralized all PFF aircraft into the 389th BG on 18 March 1944. Four days later they started providing lead aircraft for the entire Division, a job that lasted until the following winter, when enough lead aircraft and qualified crews were available for each Group.

These aircraft, known as the Pathfinder Force or more commonly PFF, provided navigation in case of overcast weather utilizing H2X, H2S, G-H, or a combination of systems. In the winter of 1943-44, only one PFF aircraft was assigned to fly with each Wing due to the high demand. The plane and crew arrived at the host Group before the morning briefing. After the briefing, the PFF aircraft with the host Group command pilot on board took off and assembled with the host Group. The mission proceeded normally until reaching the target. Then the command pilot decided on what type of bombing to use based upon the weather.

If visual conditions existed over the target, the lead aircraft signaled the deputy lead by aldus lamp and the deputy lead assumed the lead position. Visual procedures required bombing by individual squadrons (also known as sections). The lead squadron continued on course while the high right squadron maneuvered to fall in trail of the lead squadron. The low left squadron also maneuvered to follow the high right squadron. The squadrons crossed the target separately and dropped on each squadron lead's cue. After clearing the target, the formation reassembled at the rally point.

If weather conditions obscured the target, different flares and code words indicated instrument bombing. As the formation passed the initial point, the PFF fired yellow flares and at the release point red flares were fired. Also, smoke markers

One of the most common pastimes during January 1944 was trying to stay warm. The metal huts provided little protection from the elements and the small stove heated only the area immediately surrounding it. Here Everard Wandell (left), Don Todt (middle) and Elbert Lozes try to coax more warmth from their stove. (Skaggs)

were dropped to identify the release point. The Group maintained its integrity, each squadron in its respective position, and dropped on cue from the lead PFF aircraft. The following aircraft dropped over the designated point. If there was not a PFF aircraft leading the Group, they dropped on smoke markers released from the previous group's PFF aircraft. Instrument procedures produced a much larger bombing pattern; however, it allowed the bombers to bomb with precision, although degraded, when the target was not visible. As PFF aircraft became available to each squadron later in the war, instrument procedures became identical to visual procedures. Squadrons crossed the target individually regardless of the weather and dropped on cue.

As more radar equipped aircraft became available, they trickled down to the Groups. First these aircraft were assigned

Smoke markers arc toward earth. These markers, dropped by the lead aircraft, identified the release point for the trailing aircraft. (Bailey)

The Tumble Down Dick pub in nearby Woodton was one of the favorite watering holes for the GIs from Seething. (LaPoint)

to Wings and eventually they were found in each Group. By the end of the war, each squadron was led by radar equipped aircraft in the lead and deputy lead position. However, in January 1944 only B-17s were equipped as PFF aircraft. As a result, when weather dictated, the B-24s followed a B-17 Wing and dropped their bombs on smoke markers released by the preceding Wing's B-17 PFF aircraft.

Bad weather forced the Group to wait until 4 January to commence operations in the New Year. Thirteen aircraft took off from Seething leading the 20th Combat Bomb Wing (CBW) to the submarine pens at Kiel, Germany. Two aircraft returned early due to mechanical problems. Layered clouds and poor visibility created a hostile environment for formation flying. Once again these conditions complicated the assembly and scattered the formations reducing the effectiveness of the Group.

Leading the 20th Combat Bomb Wing (CBW), the 448th followed a formation of B-17s to the target. On the run to the target, a cloud deck concealed the area; however, it also hid the bombers from flak batteries on the ground. A few radar-guided guns feebly attempted to discourage the bombers with heavy, but inaccurate flak. Taking cues from the smoke markers dropped by the B-17 Pathfinder, eleven Liberators of the

The Mermaid pub in Hedenham was another favorite spot. The pub's sign of a partially clothed mermaid led to the pub's unofficial christening as the "Swingin' Tits." (LaPoint)

SEQUOIA GAL lies in a twisted heap with its fuselage broken after crashing on takeoff on 5 January 1944. (Everson)

Below: Armorers load bombs on ICE COLD KATIE in preparation for a mission. The cart the men are standing around was used to transport the bombs from the bomb dump to the aircraft hardstands for loading. (LaPoint)

448th loosed their deadly load of general purpose and incendiary bombs. Results were unobserved due to the cloud cover.

Cold weather again took its toll during the seven-hour flight. This time the Group Commander, Col. Thompson, received frostbite injuries while flying as command pilot on Lt. A.D. Skaggs' aircraft. Capt. Robert Lambertson's plane received minor damage from flak but the Group suffered no losses. On landing, crews were promptly alerted for a mission the following day. Unfortunately, the results would be much different.

As 5 January 1944 dawned, fourteen aircraft lifted off from Seething for a return trip to Kiel. SEQUOIA GAL returned to base at 0900 to replace a faulty oxygen mask for the bombardier. The pilot overshot on landing and could not stop the B-24 on the ice-covered runway. They slid off the end of the runway and passed through two hedgerows before stopping. Damage to SEQUOIA GAL was too great and she was salvaged. Two more planes aborted the mission during assembly for various mechanical difficulties. Another plane left the formation and returned to Seething before the Group reached enemy territory. The B-24s arrived over the target area just before noon and dropped their mixed load of general purpose and incendiaries. Good visibility and lack of clouds made acquiring the target easy but also made it easy for German fighters to attack the planes. Immediately after bombs away, German Me-109s and FW 190s attacked the formation in successive waves for thirty-six agonizing minutes.

In the melee the 448th lost four planes and crews. While still in the target area, the B-24 flown by Lt. James Curtis, aircraft 42-7712, exploded in mid air after being hit by flak. Lt. Curtis, Lt. Donald Clift, and Lt. Emmett Moore all survived; the others perished in the crash. At the same time, fire from a German JU-88 hit 42-7722 flown by Lt. Walter Yuengert. The engine cowling flew off the number two engine and fire consumed the engine, but the plane slipped away from the formation under control. The crew reached the North Sea in a vain attempt to reach England. They crashed into the icy waters with no survivors. Lt. Graham Guyton and crew on 41-29230 succumbed to a head on attack by a FW 190. Only the copilot, Lt. Thomas Allen, and the navigator, Lt. Richard Wheelock, survived. MAID OF TIN, flown by Lt. William Ferguson and Major Kenneth Squyres, 715th Bomb Squadron commander, also fell prey to the vicious and continuous fighter attacks. Again, there were no survivors. When the decimated formation arrived over the field at Seething, the welcoming party gathered around the control tower was shocked to count only six B-24s.

After landing, stunned crews rode trucks to the de-briefing. The Red Cross served coffee, sandwiches, and cake and

Two men from the 58th Station Complement Squadron in the green house above the control tower. From this vantagepoint they controlled all of the air traffic into and out of Seething. (Everson)

Cpl. William Schwinn of the 58th Station Complement Squadron uses the light gun to signal an aircraft from the second floor of the control tower. (Everson)

on rough missions like this one, medics provided crews with a shot of scotch or whiskey. The chaplain was in his makeshift chapel for crews needing comfort. Intelligence personnel pored over the crews gathering information: position flown, bombs dropped, target attacked, method of bombing, flak encountered with location and color, fighters encountered with location and color, information on missing crews, and aircraft malfunctions. After rough missions, debriefing was usually pretty somber; today was no different. Crews individually returned their parachutes and flying equipment to their lockers. Outside, as the men hitched rides to their quarters, the sun was starting to set. The mess hall was a popular first stop for hungry crew. Some men stopped briefly by the various clubs. If they were alerted for a mission the following day, it usually meant an early night. Fortunately, the 448th did not launch planes the next day.

The next opportunity for combat arrived on the morning of 7 January. The formation again battled poor weather in assembly. Unable to reach their assigned altitude due to weather, the Group turned back at the Dutch coast when the other Groups outdistanced the formation. There would be no repeat of the first mission when scattered formations continued to the target without the full protective firepower of the combat box.

Poor weather restricted all Eighth Air Force operations during a three-day period starting on 8 January. It was not until 11 January that the Eighth put up missions against Hitler's Reich. On this day, the 448th launched twenty-four aircraft destined for Brunswick, Germany. Lt. John McCune and crew, flying their first mission, delayed their takeoff to correct a radio malfunction. Finally after fixing the problem, they departed late. After climbing to the assigned altitude, they broke into clear weather only to find an empty sky. They continued the climb to 22,000 feet searching for the Group. Back in the clouds, they decided to descend again when they received the recall message. Deteriorating weather in England necessitated a recall of all 2nd Air Division bombers. Eleven 448th planes, including THE PROUD WANDERLOST flown by Lt. Bill Bonner, dropped their bombs on a target of opportunity at Meppen, Germany. Other planes hit a target of opportunity at Zundberg, Germany on the Dutch border.

Again, a lack of formation discipline resulted in a fragmented formation and German fighters attacked the stragglers. A formation of eight bombers, separated from the main formation, received the most attention. Twelve Me-109s attacked approximately ten minutes after bombs away and continued the attacks for fifteen minutes. At 1155, THIRTY DAY FURLOUGH, flown by Lt. James Urban, crashed near Exloo, in the Dutch province of Drenthe, as a result of damage received during the attacks. All ten crewmembers died in the crash. Five minutes later, fighters shot down PRODIGAL SON near Dwingelo in northern Holland. The pilot, Lt. Donald Schuman and four crewmen survived, but the other five died. The lead aircraft of the low left squadron, flown by Lt. A.D. Skaggs and Capt. Jack Edwards, took a 20-millimeter shell in the left wing, rupturing a hydraulic system. Fortunately, the crew nursed the airplane back to Seething despite a crippled hydraulic system that required the landing gear and flaps to be manually cranked down. Despite the damage received on the mission, the crews flew an instrument letdown and planes pulled into their hardstands six hours after takeoff. Exhausted crews returned the airplanes to the crew chiefs who immediately started preparing for the next mission. It was a vicious cycle. Intelligence later learned the target of opportunity was actually a sugar beet factory.

When low clouds and poor visibility prevailed over Seething, crews were required to fly instrument letdown procedures. This was just as harrowing and nerve-wracking as the preceding instrument departure. Once again planes filled the skies of East Anglia trying to return to their own bases. Seething, identified by the large, white letters SE prominently displayed on the ground in front of the control tower, lay two minutes from Buncher seven at Hardwick on a 055 heading. Pilots flew to Buncher seven, then turned to proper a heading and descended to a specified altitude. After two minutes, they either broke out of the weather and saw the field or executed a go around. Approximately 1,000 feet from the runway, directly in line with the runway heading, ground personnel fired flares to identify the field and aid pilots in their alignment. Later in the war, a more advanced Instrument Landing System was installed which incorporated a more sophisticated and accurate means for descent. Harrowing enough as an isolated event, the instrument letdown was the conclusion of a combat mission filled with its own dangers.

Bad weather forced cancellations of all missions until 14 January. Twenty aircraft launched to bomb a V-weapon site at St. Pierre-d'Jonguret, France. The formation departed the English coast at Brighton at 1449. Crossing the initial point, the Group took evasive action due to the flak but they still bombed the target from 12,000 feet. All aircraft landed safely at Seething without battle damage. Strike photos revealed elements of the Group badly missed the target.

Crew chiefs and aircraft mechanics worked endless hours preparing the aircraft for missions. Engine changes, tire changes, and battle damage repairs were all accomplished with no protection from the elements. The bone-chilling, wet, English weather numbed fingers. Bare skin stuck to the cold metal of the aircraft. Yet, the required number of planes were always

Two men watch as B-24s taxi past and takeoff for another combat mission (Everson)

ready for takeoff. Once all the aircraft were airborne, these tired men sought hot food and a quiet place to catch up on over-due sleep. An hour or so before the scheduled return, they meandered back to the tower and hardstands. Eyes nervously searched the horizon for any tell-tale sign of returning aircraft. Men occupied their idle time with various activities. When the planes roared overhead, men scattered to their stations. For these men, their work started when the planes landed.

Other personnel congregated near the control tower. Firemen in asbestos suits stood by their vehicles ready to respond if needed. Ambulances or "Meat Wagons" staffed with medics lined up beside the tower ready to meet any aircraft needing medical assistance. Planes requiring medical assistance received priority in the traffic pattern by firing red flares. Medics watched the planes carefully as they returned, hoping not to see flares. Unfortunately, business was booming lately.

Since commencing combat operations less than one month prior, the Group experienced a bloody introduction to combat. Thirteen crews and aircraft were missing in action. Seven other aircraft were damaged beyond repair and were salvaged. The future looked bleak.

As a result, the men sought diversions from this bleakness. Letter writing to loved ones back home provided some connection to their past lives. For many men, a day did not pass without writing a letter to a wife or girlfriend. Sleep was another popular pursuit as the long, nerve-wracking days that started in the middle of the night deprived the men of normal sleep. For many of the ground crews who worked tireless hours preparing the planes, sleep was a luxury. Passes were the most popular means of taking a break from the war. Men visited London if time permitted or nearby Norwich, Great Yarmouth and Bungay for shorter excursions. Sgt. Wally Balzer wrote his fiancée describing one of his trips to London:

"To my disappointment the place was filled up (the American Red Cross Club). I was directed to the American Red Cross Headquarters near Piccadilly Circus where I would be able to secure lodging and where I suspected my friend Ray would be hanging out. I was told to report back at 10 that evening, so I left to get something to eat

B-24s from the 448th outbound to targets on the European mainland. The weather seriously hampered the bombers' accuracy forcing them to rely on primitive airborne radar to identify the target area. (Bailey)

after having posted a message in case my friend Ray would come in my absence. I dined at a pretty nice joint and enjoyed a social time with an English businessman who sat across from me at this small table. He was in London and I found him to be very interesting. After dinner he tried to find me a room but no luck, so after thanking him I rushed back to the American Red Cross Headquarters so that I would not be late for the 10 o'clock assignment. After looking around a bit I decided to go to the snack bar in the basement of the building because there was nothing doing at the time. There in the line to purchase something to eat stood my friend Ray, so we had a bit to eat together while discussing arrangements for the following day. Well to make a long story short, the American Red Cross Clubs were full up and about 400-500 GIs, soldiers, sailors, so on were homeless. But, we were taken care of. Trucks hauled us to an air raid shelter in one of the outlying districts. The air raid shelter was constructed during the Battle of Britain but had never been used during an air raid. It was only 130 feet beneath the earth's surface and thirty to forty feet below the 'tube' or subway to the Yanks. A circular staircase, which wound around a small sized elevator, furnished the means of descent down in the shelter. It was well lighted and ventilated. The immensity made it seem that we were in some great Army hospital and the Public address playing the American popular tunes greeted us as we neared the entrance to the catacomb. It almost seemed as though we were on some other planet, Mars perhaps. Off in one of the wings was a snack bar, which

Men enjoy coffee and snacks as they huddle around a stove for warmth following a combat mission. (LaPoint)

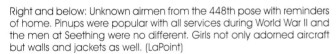

Right and below: Unknown airmen from the 448th pose with reminders of home. Pinups were popular with all services during World War II and the men at Seething were no different. Girls not only adorned aircraft but walls and jackets as well. (LaPoint)

was serving sandwiches, cakes, sweets, tea and coffee at a minimal fee. The beds were pretty fair and two blankets served as a covering. I slept comfortably all night although I woke up a few times. At 7 am the news was broadcast on the PA system which served as the eye-opener cause we all had to clear out of bed at 7:30 am. There was no charge for this lodging and it certainly was a good solution to the housing problem besides offering a chance to sleep in a British air raid shelter like moles in their palaces."

Later in the war, Capt. Richard Kimball, a navigator on a combat crew, was made an honorary member of the non-flying personnel's theater club. "They went to the theater in Norwich once a week, as I recall, usually on Tuesday. The theater manager reserved two rows in the second balcony for this club and the girls in the second balcony bar saved the beer until they arrived on 'their' night. From the theater they often went across the street to a pub and would help the owner close the place. Sometimes he would take the men into a back room and keep serving them after closing time. We liked to shout out, 'Time gentlemen! Act of Parliament you know!'"

Closer to home were the local pubs. Some of the popular ones were "The Swan" at Loddon, "The Mermaid" at Hedenham and "Tumble Down Dick's" at Woodton. "The Mermaid", commonly referred to as the "Swingin' Tits" in reference to the partially nude mermaid on the swinging road sign, sat on the road between Norwich and Bungay just over a mile from the base. Sgt. Floyd Marroon of the 58th Complement Squadron found the pub with the help of his commander. "I'm not sure, but it was one of the first few nights at the base that we were escorted by our great C.O. down through a farmer's field and along a dirt road to a pub that will always have a fond place in my heart. It was called the Mermaid Inn. We had many a good time there with Mac at the piano and the 'mild and bitters' flowing freely." Despite these diversions, the realities of war awaited the men when they returned.

The next week proved to be an effort in futility for the Group. Mission after mission was scrubbed due to poor weather. The sun rarely broke through the cold, ominous clouds. The 15 January mission to an airfield at Gotha, Germany was scrubbed before any planes got airborne. The mission to Frankfurt, Germany on 17 January, and the 18 January and 19 January missions to Raye-sur-Authie, France all had similar fates. Finally on 21 January 1944, weather cleared enough to launch a raid against Raye-sur-Authie, France. Twenty-seven aircraft took to the skies and only one returned early. Misfortune struck the Group five minutes before reaching the target. A smoke marker in the lead aircraft, 42-52145 flown by Lt. Earle Durley, acci-

Elbert Lozes (left) watches as his crewmates suit up in preparation for a mission. Part of the noseart from their aircraft THE STURGEON can be seen in the background. (Gaskins)

dentally ignited in the bomb bay sending acrid smoke pouring throughout the plane. Thinking the plane was on fire, four crewmembers, Sgt. Joe Ford, Sgt. William Hackney, Sgt. William Walker, and Sgt. John Stemmerman, bailed out over France and were taken prisoner. Clouds covered the target preventing most of the formation from bombing; however, five aircraft in the last squadron visually acquired the target through breaks in the clouds. They dropped their bombs. On return, Lt. Durley landed his aircraft with the ball turret extended. When the tips of the twin fifties contacted the runway, a shower of sparks and smoke filled the air. Although unable to taxi, no one was injured.

Poor weather continued to hamper bombing efforts. The next day, Headquarters scrubbed the mission before any planes got airborne. Not until 24 January did the weather clear enough to permit an attempt at flying operations. Ground crews prepared twenty-seven Liberators for combat only to have all planes recalled before the mission to Frankfurt, Germany actually got underway. English weather scrubbed two more attempts at missions to Frankfurt on 26 and 28 January. No planes left Seething in either effort. Ordnance personnel worked over-

Snow covers huts at one of the living sites at Seething. (Everson)

time. After every scrubbed mission, they down-loaded each aircraft they had spent the previous night loading.

Finally on 29 January, after trying for a week to launch a mission, the bombers headed for the railway marshalling yards at Frankfurt. After a 0430 briefing, twenty-seven planes lifted off, the first at 0730. Five planes returned early for various reasons. One, SAD SACK with Sgt. Stanley Zabrowski on board, returned after they failed to keep up with the formation. Rather than fly alone, they wisely aborted while the Group continued into Holland. Lt. Gail Sheldon led the Group with Capt. Jack Edwards, Group Operations Officer riding as command pilot. The flight was routine until reaching the target area where German flak batteries threw up an ugly wall of flak.

Lt. John McCune, flying in MAID OF ORLEANS, watched the preceding Group endure a beating. Somehow, all twenty-two planes of the 448th squeaked through without taking any damage, including Lt. Russel Reindal and crew on their first mission. On smoke markers from the PFF aircraft, the Group dropped their cargo through a complete undercast. After leaving the target area, German fighters milled around the area but did not attack. Lt. McCune, also on his first mission, was all too aware of the proximity burst that rocked the plane on the bomb run. It was just the beginning of many more harrowing experiences. Several black puffs of flak were noticed near the coast on return, but nothing threatening. Upon landing at Seething, inspection of the MAID OF ORLEANS revealed several flak holes in the tail. Although several close calls rattled nerves, the Group suffered no losses.

The following day, the Group participated in its eleventh combat mission. The target of the day was the Muhlenbau Industries in Brunswick, Germany. Twenty-four planes left Seething on the morning of 30 January 1944 in two sections. The first joined a formation from the 93rd BG and the second flew as the high right section in a 446th BG formation. Five re-

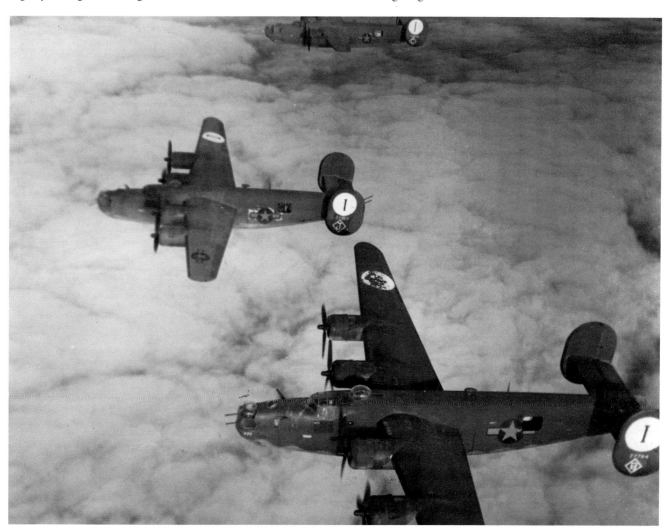

PROBLEM CHILD leads CARRY ME BACK (right wing) and BAG O' BOLTS (left wing) on a combat mission to Germany. The undercast clouds complicated the missions especially during the winter months. (LaPoint)

A special treat for crews was a pass to London where they could enjoy the relative comforts of the big city, see the sights, and forget about the war for a few hours. George Glevanik (left to right), A.D. Skaggs, Ben Baer, Francis Sheehan, and Gene Gaskins pose in wartime London. (Gaskins)

It was not uncommon for brothers to serve in nearby units. Francis Sheehan, left, a gunner with the 448th poses with his brother, a gunner in the 379th BG, in front of a B-17. (Gaskins)

turned early including the lead aircraft of the first section. Nineteen airplanes dropped their bombs through 10/10 cloud cover on smoke marker cues dropped by PFF aircraft. Again, fate smiled on the Group as no losses were encountered. The final day of the month reflected the frustration encountered the entire month. A return trip to Frankfurt was scrubbed due to weather.

Weather frustrated the Group's efforts in January. The 448th flew only four missions the first two weeks of the month and suffered six aircraft and crews lost. Continuing low clouds and poor visibility allowed only three missions to launch the remainder of the month. On a bright note, the last three missions were conducted without losses. To date, this was the longest string of missions without combat losses.

five

## ... AND MORE WEATHER:
## February 1944

February started without relief from the bitter elements of the English winter. Planners again selected Frankfurt, Germany, as the target for the mission scheduled for the first day in February. Continuing the string of weather cancellations, Headquarters scrubbed the mission before any B-24s from the 448th took to the air. Cold, tired crews trudged back to their huts in search of warmth and sleep. They could at least catch up on some sleep. For the ground crews, their work had just started. All of the bombs loaded during the night now had to be unloaded and trucked back to the bomb dump. Guns and ammunition had to be taken off the planes. Mechanics, armorers, ordnance and the rest of the ground crews did not get the day off.

On 3 February, twenty-eight airplanes loaded with 100-pound incendiaries took off bound for Emden, Germany. The Luftwaffe could not stop all attacks, but the weather could and did. High cumulus clouds over the Zuider Zee forced crews to abandon the mission and all planes returned with their cargo. Four and one half hours later, crews climbed out of their planes still no closer to finishing their missions than when they left.

Snow fell during the night but the weather improved enough on 4 February for twenty-nine planes to takeoff. For the crews, the day started at 0330 in preparation for a 0800 takeoff. Nine aircraft aborted for various mechanical problems and another erroneously followed an aborting aircraft but nineteen B-24s of the Group pressed the attack using 100-pound general purpose and incendiary bombs. Sgt. Stanley Zabrowski, aboard HELL'S BELLE, dreaded flying with the M-47 incendiary bombs. They only needed one piece of flak to set them off and turn the plane into an inferno. The formation crossed into enemy territory at 1130 and encountered light flak, but the real "fun" started thirty minutes later. Frankfurt escaped the wrath of the bombers thanks to an overcast cloud deck, but Russelheim, Germany sufficed as a target of opportunity.

Sgt. B.D. Bottoms, flying as waist gunner on his fourth mission, witnessed the intense flak over the Ruhr Valley, commonly referred to as 'Happy Valley.' Before reaching the target his aircraft, NO NAME JIVE, shuddered as flak struck the number three engine and right wing. The pilot successfully feathered the engine and managed to stay with the formation until over the target. Things continued to worsen as the left bomb bay door failed to open. Upon releasing the bombs, twenty-six of the fragile incendiaries fell onto the closed door. Despite the forty-seven degree below zero temperature, the bombardier, Lt. Edwin G. Moran, threw thirteen of the dangerous bombs out the right door with one hand while hanging on to the bomb rack with the other. Unable to remain in the bomb bay with the doors open, Sgt. Bottoms closed the doors, shielding him and Lt. Moran from the bone-chilling wind. Lt. Moran passed the remaining deadly cargo to Sgt. Bottoms who in turn handed them one by one to the other waist gunner, Sgt. Charlie Blanton, who dropped them through the camera hatch. Luckily, the only injury was frostbite. Flak damaged other aircraft, including COMMANCHE, but everyone returned without further incident. The damp, numbing British winter welcomed the crews home.

The following day, the Group completed its thirteenth mission since arriving in the ETO. An airfield on the outskirts of Tours, France, at Parcay-Meslay, was the target of choice. Fifteen planes left Seething and attacked with 500-pound bombs. As the formation made a wide sweeping turn onto the bomb run, a FW-190 pumped twenty-millimeter cannon shells into the left wing of FEATHER MERCHANT. Lt. Leroy Engdahl flying as copilot on his fifth mission, could clearly see a hole in the left aileron and wing left by the attacker. Fearing damaged landing gear, they elected to land at Seething with full flaps and with extra speed. Fortune smiled on the crew as the landing gear withstood the stress of the landing.

Strange looking objects appeared with increasing frequency on reconnaissance photos taken of the Pas de Calais area of France. These ski-jump shaped objects baffled photo-interpreters initially. Intelligence sources eventually determined they were launching sites for Hitler's infamous V-weapons, in particular the V-1. These sites, code-named "No Ball" targets, received increased attention starting in late winter. Fearing another "Blitz", political pressure dictated these sites be destroyed before the Germans completed the launch facilities.

On 6 February, twenty-six B-24s from Seething took off bound for a "No Ball" site near Siracourt, France. This target would be revisited many times in the coming months. Unfortunately, like many of the sorties in February, most planes returned without bombing. Lt. John McCune, flying MAID OF ORLEANS, endured the thick flak only to find clouds obscured the target. "It is the most disappointing feeling in the world to fight one's way clear to the target through flak and fighters, and then be unable to release the bombs!" The crew of SWEET SIOUX dropped their ten 500-pound bombs somewhere over France. The excellent escort provided by RAF Spitfires kept the German fighters at bay, but several of the planes picked up flak holes, although none were serious. Upon return, crews learned a crew had accidentally released a bomb over England; it failed to explode and no one was hurt.

A briefed milk-run to Calais was scrubbed just before take-off due to weather on 8 February and not until the following day, 9 February, did the weather allow planes to take to the skies. Even then, after two hours of flying, twenty-six crews

William Blum (left) and Jack O'Brien on their way to the control tower during the winter of 1943-44. This photo provides an excellent view of the back of the control tower as well as the "greenhouse" on top of the tower that offered a clear view of the entire airfield. (Everson)

While the control tower was the operations focal point, Group Headquarters at Seething was the nerve center that kept the base running. Notice the sandbag emplacements on top of the buildings. (Everson)

answered the recall during assembly and returned to base. Ordnance personnel unloaded the 500-pound bombs and returned them to the bomb dump.

The following day, 10 February, weather begrudgingly permitted completion of a mission although not without tragedy. Sgt. Dale Bottoms sat in his hut pitying the boys who were flying. After just returning from a forty-eight hour pass, he found Seething covered in snow. Rain, sleet, and snow discouraged everyone from going outside. Still, after several hours of delay due to high winds and blowing snow, crews received orders to takeoff. Snow flurries forced intermittent cessation of the takeoffs due to inadequate visibility. Despite the horrible conditions all the aircraft departed and climbed toward assembly area. Poor weather, characterized by ice and clouds, offered hostile conditions for assembly. It proved to be deadly.

Two aircraft returned to Seething early due to mechanical problems but the remainder of the 448th struggled to assemble, as did the other Groups. After takeoff, Sgt. Edward Schroeder, waist gunner on THE BOOMERANG, crawled to the cockpit

Men try to stay warm as they relax in their hut. Notice the black out shades covering the windows. The lack of sunlight made these buildings colder and drearier than they already were. (Lozes)

to turn off the auxiliary power unit used during ground operations. The winds tossed the aircraft around so his feet hardly touched the catwalk as he made his way back to the waist. Once there, he performed a life-saving act; he hooked on his chest pack. Climbing through the weather, everyone peered through the soup searching for other aircraft.

Continuing in these conditions for an hour, Lt. Robert Ayrest, piloting THE BOOMERANG, searched for the Group. Spotting the Group at two o'clock high, Lt. Ayrest initiated a turn to join the Group when a flash of another aircraft forced him to violently push forward on the controls to avoid a collision. Pulling the aircraft back to level flight after the abrupt maneuver surpassed the structural limits of the aircraft and the tail broke off between the waist and tail gunner. The airplane pitched up into a stall, then the wingtips folded, and THE BOOMERANG entered a spin. The centrifugal forces of the spin pinned Sgt. Edward Schroeder to the floor of the aircraft with another person pinned on top of him.

As suddenly as it started, the powerful forces subsided. The waist section of the airplane disintegrated freeing Sgt. Schroeder from its death-like grip. Slipping backwards, he fell out of the aircraft where the tail section and the tail gunner used to be. Falling free of the aircraft, he groped for his chest pack. Finding the ripcord between his legs (he had connected only one buckle of his chest pack), he pulled it. Following the opening shock, he looked above his head to see the parachute. In the clouds, he watched as falling debris showered down around him. Slowly, the clouds below gave way to an inky black surface. Water! Unable to control his chute due to the high winds, he watched and prayed as the winds pushed him over land.

Doing his best to avoid several trees and a building, Sgt. Schroeder cleared the obstacles but hit a pond embankment with a jarring impact. Numb with pain from the landing, he unsuccessfully tried to collapse his chute. The unforgiving wind caught it and pulled him over the embankment and across a field before the parachute entangled itself in a hedgerow. Covered from head to toe with mud, he lay there trying to catch his breath. An eighteen-year-old English girl witnessed his land-

Right and below: The crew of FEATHER MERCHANT after crash landing on 5 February 1944. The hole in the wing is from 30mm cannon fire from a FW 190. Although heavily damaged, the left wing was replaced, the landing gear repaired and FEATHER MERCHANT returned to combat. (Engdahl) (Sheehan)

ing. To Sgt. Edward Schroeder, she was an angel. She stood silent, waiting for him to speak and confirm his nationality. After he spoke, she readily held out her arm and helped him to a gate where her mother was waiting with a car. After getting in the car, they started driving toward the billowing smoke of the crash located at Oaken Hill Farm, Badingham. After telling her he did not wish to go to the crash site, she took him to their home. The Chinnerys lived on a nearby farm and they told him how they had watched two chutes descending from the falling wreckage. After some time, an ambulance carrying Sgt. Leonard Snell, one of the waist gunners, picked up Schroeder and carried them to Seething. He later learned to his surprise that Lt. Ronald McAllister had also survived with only a broken ankle. It was later learned the ball turret gunner broke free of the airplane but his damaged parachute only partially opened. Seven crewmembers lost their lives.

During this same time, HELL'S BELLE piloted by Lt. Edward Markewicz, experienced severe icing conditions. At approximately 18,000 feet the crew felt the first hint of trouble. A sudden shaking indicated a stall. Lt. Richard Nardi instinctively clipped his parachute on and freed himself from his oxygen mask and intercom connection. "The navigator seemed to be frozen so I thought I would open the emergency exit and pull him out. I dove under the navigation table and at the same time the plane must have flipped over for I found myself halfway up the tunnel. Somehow I managed to get back to the emergency handle and pulled it but the doors would not open. My first thought was to get to the bomb bay which I started out to do but as I got about half way up the tunnel a terrific blast of red flame hit me in the face. The plane again must have flipped over for I wound up on my back and tangled in a mass of wires. I managed to put the fire out on my face and clear myself of the wires. I ended up at the emergency exit and at that time noticed the doors were open so I crawled out but a wire caught my chute so I was hanging out headfirst and I could not free myself anyway. At this point I was about to give up but I gave one final try and managed to pull myself back in and cleared my chute from the wire and fell out of the flame."

Sgt. James Whyte, a gunner on the doomed plane experienced a similar escape. "I called the navigator for our altitude and we were at 18,000 feet. We were still in the clouds and couldn't see any break in them. Immediately after this the ship started vibrating violently, like an earthquake. This was throwing us all about the ship so we put our parachutes on expecting to use them. After this vibrating we seemed to fall off on the right wing, straighten out, and go into a power dive with the engines screaming. When we fell off on the wing I was thrown against the right side of the plane and believe that is when I

FEATHER MERCHANT 42-73477. After being repaired following the 5 February 1944 mission, it resumed flying and was eventually transferred from the Group. (Everson)

dislocated my shoulder. It seemed as though the pilots were trying to pull the ship out of the dive but it didn't seem to help. At this time the right wing broke off probably due to the speed and the pilots' efforts. Immediately following was an orange sheet of flame which blew the door to the bomb bays open. This flame struck me in the face and that is all I remember until I was falling through the air. I believe I was blown from the ship because I didn't jump. I regained consciousness and pulled the ripcord. While coming down on my parachute I saw pieces of the plane falling around me. I again lost consciousness before hitting the ground."

The excessive forces encountered ripped the airplane apart and at least two bombs detonated causing the blast and flames. Sgt. Stanley Zabrowski watched in horror as the aircraft in which he flew six days previously plummeted to earth in thou-

Standing at attention before receiving their first air medal. (left to right) A.D. Skaggs, Don Todt, Elbert Lozes, George Glevanik, Stanley Filipowicz, Francis Sheehan, and Gene Gaskins. (Lozes)

sands of pieces. The wreckage was widely scattered over a large area near Badingham. Only Lt. Nardi and Sgt. Whyte escaped from the stricken plane.

Despite these tragedies, the mission continued but clouds made assembly impossible. Three separate formations resulted and they all set course for the target. A total of nineteen aircraft crossed the target, an airfield at Gilze-Rijen, Holland, but only seven were able to attack due to the cloud cover. Crews encountered very little flak, but contrails increased the difficulty of formation flying. For Sgt. George Glevanik, engineer on Lt. A.D. Skaggs' crew, this almost five-hour mission was the coldest yet. Eight of the returning bombers landed at other bases because of the inclement weather at Seething. The next day the headlines of "Stars and Stripes" read, 'Fortresses Fight Biggest Battle Over Germany.' No mention was made of the B-24s.

Following the tragedy, twenty-six aircraft of the 448th left the runway at Seething on 11 February leading the 20th CBW to a "No-Ball" target at St. Pol, France. Problems with the PFF ship leading the 448th forced them to pass lead of the Wing to the 93rd BG. One aircraft aborted and another failed to drop due to a broken bomb bay latch spring, but the remainder dropped 500-pound bombs on the V-1 launch facility. Sgt. Bottoms, recovered from frostbite received on an earlier mission, flew his fifth or air medal mission.

During the early years of the war, aerial opposition was usually intense and the losses were correspondingly high. As a result, it was assumed crews demonstrated exemplary courage, coolness and skill sometime during these five missions. To boost morale, orders were issued to decorate crews with an air medal after every five missions. Thus the fifth mission became known as the air medal mission. By the summer of 1944, losses diminished and the requirements for an air medal changed. Only those crews cited for specific accomplishments were awarded the medal.

Allied pressure on the "No Ball" targets continued. Although short in duration these "No Ball" missions were full of action. As the frequency of these missions increased so did the number of German Luftwaffe flak units assigned to protect these targets. On 13 February, Sgt. Bottoms logged his sixth sortie. CQs hurried to find crews for a short notice mission. Twenty-seven crews rushed to make an 1150 takeoff. It was delayed fifteen minutes while ordnance personnel loaded the last few aircraft. Three B-24s from Seething returned and only twenty-four aircraft revisited St. Pol, France. Crossing the target at 1500, only twelve aircraft dropped their bombs. The remainder of the bombers failed to positively identify the target and intentionally did not bomb. Some elected to jettison their bombs

Col. Thompson pins an air medal on Gene Gaskins following his fifth mission. The air medal was awarded initially after every five missions to boost morale among the bomber units. (Gaskins)

while others returned to Seething with full bomb bays. Two aircraft suffered flak damage and landed at other bases. Flak ruptured the hydraulic system on SAD SACK and the pilot, Lt. Robert Carroll, elected to land at the emergency field at Woodbridge. Lt. Robert Martin also elected to land at another airfield, this one at Headcorn, after flak damaged his aircraft, WABASH CANNONBALL.

On 15 February, the 448th Bomb Group assumed operational control of Seething air base from the British. The now fully operational heavy bomber base consisted of four bombardment squadrons, 862nd Chemical Company Air Operations, 123rd Quartermaster Company, 58th Station Complement Squadron, 262nd Medical Dispensary, 459th Sub Depot, 2102nd Engineer Aviation Fire Fighting Platoon, 1193rd Military Police Company, and the 1596th Ordnance Supply and Maintenance Company.

During the middle of February, Seething and the 448th struggled as the bitter English winter continued to make life and work difficult. Crews received unpopular news when Headquarters increased tour lengths from twenty-five to thirty missions. On the bright side, the first 448th crews enjoyed the flak home at the Stanbridge Earl Rest Home near Eastleigh, Hampshire. These flak homes, several elegant English estates converted for use by the military, provided aircrews with an opportunity to leave the war and enjoy a normal life for a week.

An unidentified Sergeant receives a customary kiss on cheek after receiving an air medal. (USAF)

NO NAME JIVE 41-29230. One of the original 448th aircraft, it completed its combat missions and returned to the U.S. after the war. (LaPoint)

While the flak homes offered a break for a few lucky crews, the war continued for everyone else. On the 20th of February, crews groaned as the briefing officer unveiled the target: the Messerschmitt plant at Gotha, Germany. A fortnight had passed since the 448th had attacked targets in Germany. Unknown to the crews, this was the first mission marking the beginning of what historians would call Big Week. Weather forecasters predicted good weather over Germany as a high-pressure system settled over central Europe. This provided the perfect opportunity for planners to launch an all-out assault on the German Luftwaffe. To the crews of the 448th, it only meant dangerous missions into the heart of Germany.

During the cold, wet English night, crew chiefs prepared thirty-six aircraft for the effort. Despite their best efforts, three planes returned early due to mechanical problems. Lt. A.D. Skaggs, flying aircraft 42-73512 as deputy lead, watched the English coast slip underneath the nose of the aircraft and disappear. The enemy coast loomed ahead. Sgt. Stanley Zabrowski, flying in THE PROUD WANDERLOST, found the flak to be less imposing than expected. They bombed the target with 500-pound bombs despite the overcast weather. German opposition was negligible. Lt. Lewis Sarkovich was overjoyed by the lack of opposition encountered on his fifth mission. The eight-hour mission concluded with all aircraft returning safely despite bad weather at Seething. Two of the bombers landed at other bases and waited for the weather to improve before returning. The next day, "Stars and Stripes" trumpeted the success of the mission. "Allied Air Forces based in Britain struck the greatest blow of the war at German aircraft production in the twelve hours ending yesterday afternoon." While aircrews flew over German territory, salvage crews completed the grisly task of clearing the wreckage of HELL'S BELLE. No one wrote about this ugly facet of war.

Clearing weather allowed a return trip to Germany the next day, 21 February. Planners targeted the airdromes at Vorden, Hesepe, and Diepholtz. The 448th assembled on the yellow and black, checkered assembly ship, YOU CAWN'T MISS IT, and then fell in behind the 93rd BG which was leading the Wing. Clouds obscured the original targets forcing the Group to search for a target of opportunity. They found Achmer airfield on the outskirts of Hesepe. On the first pass they were unable to bomb but on the second pass over the target they dropped their cargo. German fighters rose to challenge the bombers. On THE LADY FROM BRISTOL Sgt. Stanley Zabrowski managed fifty-caliber hits on a FW-190 but only damaged it. Other FW-190s responded by shooting down two

On 15 February 1944, British troops lowered the Union Jack and officially transferred the base to USAAF control. (LaPoint)

B-24s and damaging several others, all on the second pass. The first aircraft lost fell to flak at 1445. CARRY ME BACK, piloted by Lt. Stanley C. Cooper, crashed at Westerkappeln, ten kilometers northwest of Osnabruk, Germany. The copilot, Lt. Wesley V. Helvey, died as a result of parachute failure and Sgt. Joseph F. Nickerson's body was recovered from the aircraft crash. All other crewmembers survived and served the remainder of the war as POWs.

Five minutes after CARRY ME BACK succumbed to flak, THE PROUD WANDERLOST, flown by Lt. Clair W. Cline, lost its battle with three FW-190s. Exploding twenty-millimeter shells killed the ball turret gunner, Sgt. Robert Yarnell, and wounded the radio operator, Sgt. Guy Padgett. Sgt. Ira Loyd died after falling from the aircraft without a parachute. The remainder of the crew survived. THE PROUD WANDERLOST impacted the ground in the woods outside Horstel, Germany. At approximately the same time, another 448th crew struggled to maintain control of their airplane while simultaneously fighting off relentless fighter attacks.

An unnamed B-24 piloted by Lt. Harvey Broxton received the full brunt of the Luftwaffe's defensive power as the Group arrived over the initial point. With the bomb bay doors open, the planes flying a steady course for the bomb run presented an easy target for the flak gunners. They zeroed in on the high right element of the 448th. A well-aimed flak burst disabled the number three engine of Lt. Broxton's B-24. The crew partially feathered the propeller but with the increased drag from the slowly spinning propeller, they slowly dropped behind the formation. Reaching the target alone and on three engines, the bombardier, Lt. Claire Sharp, dropped where he thought the main group had released their bombs. Leaving the target, five FW-190s spotted the straggler and dove on the sitting duck from five o'clock high. Without the defensive support from the other bombers and with no fighter protection, Lt. Broxton maneuvered the aircraft violently and headed for the relative safety of a cloud deck at 10,000 feet. Thinking the end was near, one of the waist gunners, Sgt. Henry Kubinski, jumped from the aircraft. Sadly, he did not survive; a Dutch family recovered his body from a river near Ommen, Holland.

The clouds offered a brief respite from the fighters, but they soon disappeared and the fighters returned. To make matters worse, the bomb bay doors only partially closed after the bomb run, slowing the plane. The Focke-Wulfs closed in on their damaged prey. Machine gun and cannon fire quickly destroyed the number four engine and riddled the B-24 from nose to tail. The disabled engine ignored repeated attempts to feather and the propeller continued to windmill slowing the flying wreck further.

British and American soldiers congratulate each other in front Group Headquarters following the change of command. (USAF)

Lt. Broxton and the copilot Lt. Dwight Covell struggled against mounting odds to keep the crate flying. The rapid descent caused the windscreen to fog reducing visibility to zero. When Lt. Covell opened the side window for visibility, machine gun fire entered the open window and exploded in the dash destroying much of the flight instruments. Relying on their training from primary flight school, they flew by needle, ball, and airspeed.

During these savage attacks, the crew suffered a beating. The top turret gunner, Sgt. Donald Birdsall, narrowly escaped injury when machine gun fire smashed into his top turret rupturing hydraulic lines and covering him in the red fluid. Others in the plane were not as lucky. The tail gunner, Sgt. Irving Elba, fought off the fighters until machine gun fire ripped into his turret disabling his guns and seriously wounding him. Sgt. Robert Hudson firing from his waist gun position noticed the injured tail gunner and stopped firing long enough to extricate his wounded comrade from the tail turret and administer first aid. After receiving treatment for his wounds, Sgt. Elba returned to his guns only to find them out of commission. He quickly returned to the waist where things were very busy.

With two engines disabled and the airplane sinking to the ground, Lt. Broxton ordered everything not attached to the airplane thrown overboard. Guns, ammunition, and parachutes went overboard to the unsuspecting Dutch citizens below. A GI Thermos used as a relief tube went out the window. Meanwhile the vicious attacks continued as the B-24 sank lower and lower. Cannon fire ignited hydraulic fluid that puddled in the bomb bay. The flash fire exploded out the open door with a chilling noise. Sgt. Hudson, with the help of Sgt. Elba and the

YOU CAWN'T MISS IT or CHECKERBOARD 41-23809. This veteran B-24, originally HELLSADROPPIN of the 44th BG, completed its combat tour and was recruited by the 448th as the formation ship. After aiding in numerous assemblies, it finally was salvaged in January 1945. (Everson)

bombardier, Lt. Sharp, put out the fire. Without warning, a bursting shell dug into the back of Sgt. Hudson knocking him to the floor seriously wounded.

The German fighters, seeing the stricken plane on fire, withdrew momentarily to admire their handiwork and watch the end of another B-24. It was not to be; the thoroughly riddled machine kept flying. With only two engines on the left side working and one on the right side still windmilling, both pilots strained to displace enough rudder to keep the airplane skimming in level flight 100 feet above the countryside. Villages, factories, church steeples and trees all streaked by at that low altitude. The five FW-190s continued to pelt the plane with fire until they exhausted their ammunition. With nothing else to do, the fighters pulled alongside the stricken plane in amazement. How could such a damaged plane continue to fly? Lt. Broxton looked across at his counterpart in the cockpit of the German fighter. The German pilot smiled and waved then the fighters peeled off to return to base, out of fuel and ammunition. Now all alone, the ominous North Sea loomed ahead, but home was only 120 miles away.

Passing over the sea wall, they surprised a flak battery that fired hurriedly at the Liberator but to no avail. Quickly out of range and over the North Sea, thoughts turned toward the two good engines. Red-lined to maintain flying airspeed, the engines hummed along while both pilots' cramped legs strained at the rudder pedals trying to keep the plane in level flight. In the back, the less injured administered to the more seriously injured, especially Sgt. Hudson whose injuries appeared life threatening. Loose items that escaped the first pass quickly reached the cold, black water of the North Sea. The completely stripped B-24 gradually climbed to 500 feet by the time they reached the English coast. British gunners were rumored to shoot at everything below that altitude.

The navigator, Lt. Robert Fauerbach, flawlessly brought the stricken plane to the coast conveniently aligned to runway two-five at Seething. Lacking hydraulic power, one crewmember hand-cranked the gear to the down position. With no radios or means of communication, Lt. Broxton entered the traffic pattern just as the last squadron set up for their landings. While the engineer worked the throttles, both pilots struggled with the controls to keep the plane aligned with the runway. Legs quivered with fatigue as the engineer added power on short final to clear the runway threshold. With this sudden application of asymmetrical thrust, the airplane veered to the right as if it were standing on its wing. Using sheer will power, the pilots coaxed a final rudder input from their exhausted legs to right the plane as the wheels touched down.

Due to the awkward landing and lack of brakes, the B-24 swerved into the muddy in-field and stopped. Rescue personnel quickly extinguished a fire that continued to burn in the number four engine and attended to the wounded. Ambulances rushed the wounded, including Sgt. Hudson, to the hospital. The remainder of the exhausted crew piled into trucks and told their story during the ritual debrief. The battered plane contained over 400 holes, an unexploded shell in one fuel tank, and miraculously, a severed right rudder cable. A visiting war

Above: The 448th chose a black and yellow checkered paint scheme for their first assembly ship. The result was named after a popular British phrase used in giving directions, YOU CAWN'T MISS IT. However, most of the men simply referred to it as CHECKERBOARD. (LaPoint)

Right: YOU CAWN'T MISS IT slips out of the formation after assembly is complete and returns to Seething. (LaPoint)

correspondent quickly named the unadorned Liberator BAG O' BOLTS. It was a short-lived name as the mangled Liberator never flew again.

The outstanding flying skills of Lt. Broxton and Lt. Covell allowed them to fly the plane home with only half of the rudder working! Furthermore, they did it with two engines out on the same side. As a result of their valor and heroism, Lt. Broxton, Sgt. Elba and Sgt. Hudson received Distinguished Flying Crosses. Unfortunately, Sgt. Hudson succumbed to his wounds and died 10 January 1945.

21 February proved to be a tough day for the 448th. Of the thirty-seven aircraft dispatched, thirty-two attacked their target. German fighters and flak downed two planes and crews and another was listed as Category E. Two other aircraft landed at other bases due to damage. Tomorrow was another day and the war continued.

Planners targeted Gotha, Germany, as the 448th's next target and its twentieth mission but the dubious European winter played havoc with the desires of the tacticians. Thirty-six aircraft left Seething on 22 February with clusters of incendiaries intended for Germany. Sgt. Stanley Zabrowski flew for the third consecutive time, this time on RAB-DUCKIT. Poor weather over the target foiled attempts at dropping and a target of opportunity was sought. Thinking the formation was still over Germany, the lead aircraft released its deadly incendiaries followed on cue by thirty-one other B-24s. Unfortunately, faulty navigation placed the formation over the Dutch border town of Enschede, not Germany.

A short respite from combat followed the unfortunate bombing of Enschede. The Group did not participate in operations on 23 February but once again took to the skies on 24 February. The Messerschmitt factory in Gotha, Germany, saved by poor weather two days prior, 'welcomed' the Liberators with thick flak and menacing fighter attacks. Lt. Lewis Sarkovich, flying his sixth mission, watched as three Me-109s circled above his formation. Two of the menacing fighters stayed in the sun as one streaked a mile in front of the bombers, then turned around. The German fighter pressed his head-on attack to the point the smoke from his twenty-millimeter cannon fire was clearly visible. Sgt. Alfred Maino, the ball turret gunner, unleashed a flurry of lead from his twin fifties and received a probable kill. A second fighter dove out of the sun mimicking his leader. Firing his cannons, he rapidly closed on the formation then rolled onto his back and dove away without inflicting any damage.

Crossing the initial point with the bomb bay doors open, the bombardier in the lead aircraft passed out from oxygen starvation after a leak developed in his mask. He accidentally salvoed his bombs and seven other bombardiers followed his cue. Lt. James Misuraca, bombardier on MAID OF ORLEANS, toggled his bomb release after seeing bombs fall from the lead aircraft. He quickly realized his mistake as the bombs fell well short of the target. Amazingly, a FW-190 rolled in behind the B-24s to attack just as the bombs fell. The arcing bombs from MAID OF ORLEANS intercepted the flight path of the attacker resulting in a sudden explosion of debris and smoke. Using probably the most unconventional method, Lt. Misuraca received credit for one FW-190 destroyed.

Despite the stiff opposition, the remaining nineteen Liberators unleashed their fiery load of fragmentation cluster bombs, although not without cost. The returning crews watched running dogfights between their "Little Friends" and the Luftwaffe. Fighters were not the only menace. Luftwaffe flak crews near Brussels, Belgium, zeroed in on the formations with multi-colored flak bursts. ABIE'S IRISH ROSE, flown by Lt. Ronald D. Warnock, succumbed to this silent, sometimes spectacular killer, after a direct hit in the nose. Fellow aviators watched seven chutes blossom in the hostile sky before the plane crashed at 1445 on the outskirts of Brussels. Half of the crew perished including Flight Officer Morgan Goodpasture, Lt. Sam Bullard, Lt. Frank French, and Sgt. John Feyti.

Lt. A.D. Skaggs endured several close flak burst from the same batteries. One burst exploded immediately in front of the aircraft sending deadly shards of shrapnel into the nose of the plane. Lt. Elbert Lozes, at his bombardier station, took the brunt of the impact. Miraculously, the deadly flying metal missed him, but shattered an ammunition box showering him with plexiglass and wood. His flying suit protected most of his body and only his uncovered face suffered from the effects of flying debris. Lt. Don Todt, navigator, carefully administered first aid to Lt. Lozes' face and stopped the profuse bleeding. Thankfully, his eyeglasses saved his eyes from permanent injury.

Six hours after lifting off British soil, the more fortunate crews landed at Seething quite battered. Many aircraft received considerable damage from the flak and fighters, but crew chiefs and engineers could fix those wounds. No remedy existed for lost comrades.

The next day, 25 February, the Group launched a maximum effort despite the damage received the day prior. The formation ship, YOU CAWN"T MISS IT, took to the skies first at 0915. With visibility limited to one mile, thirty aircraft rumbled down the runway at forty-five second intervals taking to the overcast sky. Pilots relied on now familiar but still unnerving instrument procedures to climb through the overcast. By 0940, all 448th aircraft were safely airborne. Once on top of the

448th Liberators head toward Germany while figure-eight contrails from the escorting fighters are visible above. Typical winter weather blankets the ground below the bombers. (LaPoint)

clouds, mayhem ruled. All over East Anglia, aircraft from other Groups searched for their particular formation ship. The pilots of YOU CAWN'T MISS IT searched in vain for the 93rd BG formation ship, BALL OF FIRE, which plans called for the 448th to follow. After forming behind the wrong aircraft, the pilots of YOU CAWN'T MISS IT recognized their mistake and quickly found BALL OF FIRE. The error delayed assembly but the Group completed assembly at 9,500 feet and joined the 20th CBW over Buncher seven on time. Mechanical problems forced an unprecedented fourteen aircraft to return to Seething. The remaining seventeen aircraft proceeded to Splasher six then on to Colchester and Dover. Crews watched as the English coastline at Dungeness slipped by and the enemy coast appeared on the horizon. Once over water, crews test fired guns and prepared for the worst.

Nasty black puffs of flak explode off the left wing of a 448th Liberator. This was a dreaded sight for the crews. Each tiny explosion sent thousands of metallic shards flying into the nearby skin of aircraft or men. It was the deadliest defense employed by the Luftwaffe. (Bailey)

... And More Weather: February 1944

Bombs fall toward earth on 22 February 1944. Unfortunately, the Dutch town of Enschede was mistaken for a German town and was accidentally bombed as a target of opportunity after the original target was obscured by clouds. (Ray)

Controllers prepare to fire flares indicating aircraft are cleared to take-off for the 25 February 1944 mission to Furth, Germany. These men communicated to the tower by telephone and passed information to the bombers by prearranged signals. This maintained radio silence preventing an early warning to the Germans about the upcoming mission. (Maxwell)

Intelligence officers and planners planned the route well as only meager flak was encountered. Enemy fighters were present in only small numbers and attacks were not pressed home due to excellent fighter protection. Events continued to unfold in favor of the attackers; the target was visual and conditions were CAVU. At 1405, crews loosed their loads of 500-pound bombs over the aircraft plant at Furth, Germany from 20,000 feet. Bombs exploded throughout the hangars of the airfield. After reassembling following the bomb run, things continued to go as planned until just prior to the French coast when the good luck ran out.

Approaching Amiens, France, a wing of B-17s forced the B-24s of the 448th to take evasive action to avoid a collision. At 1613, with the formation fifty miles southeast of Amiens, Lt. John Williams flying in the second element of the high squadron recognized a 448th aircraft in trouble. LADY FROM BRISTOL, flown by Lt. Lawrence Edman of the 714th BS, slowly started lagging behind the formation and losing altitude. The number two engine appeared to feather but other witnesses reported seeing all engines operating at the time of disappearance. The original copilot, Lt. Lawson Campbell, did not fly with his crew on this mission. The flu saved his life. Running a high fever, he was hospitalized and Lt. Edward Meents, a replacement, flew as copilot. The crew was never heard or seen again. (The mystery of LADY FROM BRISTOL puzzled officials. Air-Sea Rescue received no Mayday calls. Later, the body of Sgt. Jack Ghormley washed onshore in France. Sgt. Harold Delay's mother and brother received a letter smuggled through the French Underground stating they parachuted to safety and hoped to be home soon. It is possible they were betrayed by someone in the Resistance and possibly executed. Also, some speculation centered on operations of KG 200, a German unit that flew captured Allied aircraft. An unidentified B-24 was reported in the area on this day and could possibly have been involved.)

Immediately after crossing the enemy coast, crews started descending and once over the British coast at Beachy Head, pilots pointed their aircraft direct to Seething. Due to the length of the mission, fuel supplies ran perilously low. Navigators feverishly figured the time remaining and the crew of 42-100342 elected to land at Biggin Hill because of gas shortage. Aircraft 41-29235 crash-landed at Chipping Ongar because of battle damage; fortunately no one received injuries. Over Seething, weather required an instrument letdown but by 1750 all planes were safely on the ground.

The Group's second full month in the ETO concluded with a "milk run" to a "No Ball" target at Escalles-sur-Buchy, France. Twenty-five Liberators dropped their 500-pound bombs on the V-1 launching site with no resistance from flak or fighters. Even though the month ended on a light note, February proved to be a tough month for combat operations. Six crews were killed or missing in action in twelve missions. After twenty-three missions in the ETO, crew losses totaled nineteen crews killed or missing in action. In addition to the nineteen aircraft lost, salvage crews deemed five more aircraft total losses. Combat was taking its toll. Of the sixty-four aircraft that left the U.S., only thirty-six remained.

six

# COMBAT LEGS:
# March 1944

Winter weather in northern Europe continued to undermine and wreak havoc with the Allies' attempt to disrupt the German war effort. While Allied fighters, led by the new North American P-51 Mustang, started to gain the upper hand from their German counterparts, the strategic bombers of the U.S. Eighth Air Force continued to battle the inclement weather. When the weather cleared, massive formations hit vital targets with increasing accuracy. However, during poor weather the shortage of PFF aircraft hampered their efforts and effectively neutralized the bombers.

The 448th's first mission of March highlighted the plight of many bomb groups. Twenty-nine aircraft launched from Seething on 2 March to bomb Frankfurt with incendiaries. Intense flak over the city pounded the bombers. Clouds obscured the target except for occasional breaks but required the use of a PFF aircraft to bomb. Unfortunately, the bomb racks on the lead aircraft malfunctioned and the rest of the formation was unable to identify the target and release their bombs. They all returned to Seething with their bombs except the deputy lead aircraft, 42-100356. Flak damaged the aircraft requiring the pilot, Lt. John Bringardner, to jettison the bombs in the Channel.

Twenty-seven B-24s took to the skies of East Anglia destined for Berlin on 3 March. Although clear weather prevailed over England, the formation encountered deteriorating conditions as they approached the enemy coast. The formation flew S-turns in an attempt to climb above the worsening weather, but strong winds, thick clouds and contrails rendered formation flying nearly impossible and too hazardous to continue. The Group answered the recall message but not before a near tragedy occurred.

Flying tail-end charlie in nice weather was a challenge, but in poor weather with thick contrails reducing visibility to near zero, it was almost impossible. Lt. William Brown, pilot, and Lt. Kenneth Barnett, copilot, struggled to keep CAROL-N-CHICK in formation despite the deteriorating conditions. Prop wash from the preceding aircraft buffeted the plane and forced it into an uncontrollable nose-up position. The severe turbulence forced it over onto its right wing then back onto its belly. The pilots tried in vain to regain control as the invisible force pushed the plane onto its left wing then back onto its right wing in a manner similar to a leaf falling. As an apparent coup de grace, CAROL-N-CHICK flipped onto its back and entered a spin at 20,000 feet. Fighting the odds, the pilots regained control of the possessed aircraft and recovered from the spin at 10,000 feet. All alone, the crew maintained their composure and dropped their bombs on the runways at Heligoland, a German island in the North Sea. Two P-51s joined the Liberator and escorted it back to Seething without further incident. Meanwhile, the formation executed the recall and returned to Seething without dropping its bombs. Crews reported CAROL-N-CHICK missing in action. Ironically, CAROL-N-CHICK was one of only four B-24s in the entire Division to drop its bombs.

Crews wearily left the huts in the dark morning hours of 4 March to find a raging blizzard. Sgt. Stanley Filipowicz could not see across the airfield because of the heavy snow. Yet the mission to Berlin continued as crews started engines and taxied for takeoff. Just before takeoff, the mission was scrubbed. Crews breathed sigh of relief. Combat was one thing; flying in atrocious British weather was sometimes worse. It had already claimed two crews. Although no crews flew a mission on 4 March, many were on the move. Four officers and 202 men of the 715th BS transferred to the 855th BS as part of the new 491st BG that was preparing to enter combat. These men provided much needed combat experience to the new Group.

After a day of rest, the Group resumed the offensive on 5 March by launching twenty-nine aircraft against a Luftwaffe airfield at Bergerac in southern France. Lt. Lawson Campbell cringed at his bad luck. His crew was supposed to leave on a three-day pass. Instead, they were riding bicycles through the darkness on their way to breakfast and another mission. They were flying as a replacement for another crew and were promised their pass when they returned from the mission. Clear weather made assembly easy compared to the previous days. As the formation crossed the English coast south of London at Beachy Head, friendly fighters filled the sky creating a false sense of security. The bombers first flew east to the Zuider Zee before turning south in an effort to lure the Luftwaffe up for a fight. The long trip across France continued without a glitch until the white-capped Pyrennes Mountains came into view. The beautiful mountains reminded Lt. John McCune of the Atlas Mountains in Africa. As the formation commenced the bomb run, clouds from the Bay of Biscay crept overland and obscured the target. With no PFF aircraft in the formation and too little gas to fly a second bomb run, the formation turned north.

As the formation approached Bordeaux, the command pilot ordered the formation to prepare to bomb a nearby airfield at Bergerac as the secondary target. As the formation prepared to bomb, B-24s of the 446th BG attacked the same airfield and forced the 448th to cut short the bomb run. Fourteen aircraft of the second section dropped but the first section did not because of the short bomb run and fear of hitting the town. Lt. Lawson Campbell watched the B-24s and also watched the flak. His nightmare continued as heavy flak struck his airplane, 42-100414, in two places. The pilot, Lt. Robert Martin, and the copilot, Lt. Campbell, struggled with the controls of their damaged aircraft. One flak burst damaged the number three engine and a second shell left a gaping hole in the flight deck before exploding 100 feet above the plane. Lt. John McCune, flying MAID OF ORLEANS, watched them slowly fall out of formation and lag farther behind the Group.

For the crew of the damaged plane, the horror was just beginning. Once clear of the protective umbrella of the formation, German fighters swarmed over the wounded Liberator. They succeeded in destroying much of the tail assembly before Lt. Martin found the relative safety of clouds at 12,000 feet. A U-turn in the clouds successfully lost the pursuers but left them heading south and away from England. Breaking out of the clouds at 7,000 feet, they vainly searched for a place to crash-land. The radio operator, Sgt. Richard Thalhamer, bailed

Many of the tasks required to make Seething an operational base fell on the 58th Station Complement Squadron. They, along with all units at Seething, responded admirably completing these jobs while simultaneously conducting combat operations. Here the 58th Station Complement Squadron poses for a photo during March 1944. (Bollschweiler)

out but the remainder of the crew stayed with the aircraft. They found a large wheat field near the town of Chef-Boutonne, northeast of Cognac, France. After a successful wheels-up landing in the green wheat, all of the crew were whisked away and hidden in nearby woods by the local villagers. The wheat field unexpectedly offered great camouflage for the downed aircraft. The height of the grain concealed the aircraft for three days before the Germans discovered it. (After hiding in a Chateau for 3-4 days, the men were separated into small groups and hidden in the area for about a month. They then walked from Bergerac to Toulouse where the Resistance once again hid them this time in an apartment directly across from a German Headquarters. Lt. Campbell made it to within a half of a mile of the Spanish border in the Pyrennes Mountains when he was captured by a German patrol.)

Overhead, the 448th, flying as the last Group in the entire Division, searched for another target. Another airfield, this one near Cognac, France, was selected as a target of opportunity.

Men build a wall at one of the living sites. First priority of construction projects were those that affected operations. With time, men were able to complete projects that improved their quality of life. (USAF)

The heading of the formation did not properly align the bombers with the target so the decision was made to circle and re-attack. The 448th started their turn while the remainder of the Division continued toward England. Flak immediately filled the sky around the bombers. Despite the flak, the bombers dropped their bombs but failed to hit the airfield.

Now all alone in the sky, the bombers from the 448th set course for home. Then, the fighters struck. Two squadrons of tan and blue FW-190s with twelve planes each, hit the 448th formation at 1320. Gunners aboard the bombers fought back. Sgt. Francis Sheehan, aboard 42-100342, destroyed one fighter but the fighters continued their attacks. Sgt. George Glevanik manning the top turret of the deputy lead aircraft watched as the fighters raked the formation with machine gun fire. After enduring the attacks for almost ten minutes, Lt. Cherry Pitts saw Lt. William Ross' ship, 42-100430, suddenly climb. The stricken plane, with the number two engine smoking, entered a steep turn and crossed over the top of Lt. Pitts' aircraft. It continued to roll until it was on its back. Four chutes appeared just as the plane dove into the clouds. It suddenly reappeared above the clouds and then dove back into the undercast with three Me-109s in hot pursuit. They were last seen sixteen miles north of Niort, France. Despite the sudden end, all of the crew bailed out although several were injured. Sgt. Arthur Mied, the engineer, suffered the most serious wounds. He suffered shrapnel wounds to the chest and his foot was shot off. Upon landing, he was taken to a hospital in Niort where he died two days later. (He was given a full military burial by the Germans. French patriots secretly put flowers on his grave even after German authorities threatened serious retaliation if they were caught.) Lt. George Wenthe and Sgt. Charles Susino were also wounded and captured by the Germans. They survived as did all the other members of the crew who evaded with the help of the French Underground.

Aboard aircraft 42-109793, Sgt. Stanley Zabrowski saw the two airplanes of his Group fall. One he noticed lagged behind the formation with fighters following it like a pack of rabid dogs. When one of the German fighters passed within range of his guns, he got revenge. "I just started shooting until I saw the aircraft falling apart." The ball turret gunner confirmed the menacing fighter exploded thus crediting Sgt. Zabrowski with one enemy aircraft destroyed.

After bombs from Lt. John McCune's MAID OF ORLEANS impacted harmlessly in an open field, the Luftwaffe attacked. Successive waves of FW-190s swarmed from head-on in a line abreast formation of three. Lt. McCune described the thirty-minute attack as "the worst thirty minutes I have ever lived through." Cannon fire pierced their plane in numerous

A battle-damaged B-17 sits on the tarmac at Seething on 4 March 1944. Despite the coming of spring, cold weather and snow persisted late into the season. (LaPoint)

places. The number three engine caught fire and started trailing a dense white smoke. To Lt. McCune, it was a nightmare:

"The poor 'MAID' began to leap and quiver like a thing alive, shuddering as the slugs began to tear into her vitals and explosive forces of the twenty-millimeter and thirty-seven millimeter cannon shells rocked her in a hellish lullaby. In the cockpit it sounded as if the plane was disintegrating behind me ... all I could think of was the sheet of armor plate behind my back ... the bandits were firing tracers and a constant stream of them flew over the canopy like some obscene 4th of July fireworks display ... mixed with the tracers' trails were the dirty yellow puffs from exploding cannon shells ... my thoughts turned to my Maker, and I prayed to God for deliverance ... I thought none of us would ever see home again ... "

The tail gunner, Sgt. Willard Cobb, had an unwanted front seat. Six FW-190s attacked from the rear of the bombers showering them with lead. Sgt. Cobb hit two of the attackers sending them out of control but they also hit back. A twenty-millimeter shell entered the tail turret hitting Sgt. Cobb's leg seriously injuring him. He ripped his communication cord and electric suit cord from their connections and tied them around his leg as a tourniquet to stem the profuse bleeding. Three more FW-190s rolled out of the sun pressing their attack. Sgt. Cobb fought back knocking a third fighter from the sky but once again they hit him, although not as seriously as in the previous attack. There was no time to rest as two more FW-190s pressed their attack. Unsure of his remaining ammunition, Sgt. Cobb valiantly fought off the attackers sending a fourth fighter diving toward earth.

Finally, the relentless attacks subsided but over an hour still remained before they reached the English Channel. Crews surveyed the damage to their aircraft. MAID OF ORLEANS had suffered serious damage. The hydraulic system was destroyed, control cables were frayed, numerous holes perforated the skin of the aircraft, the wing and left flap were heavily damaged and the number three engine burned up. Despite the damage, Lt. McCune and crew returned to Seething. Once overhead, the landing gear and flaps were lowered using hand cranks and red flares were fired to receive priority for landing. Without brakes, Lt. McCune deftly landed THE MAID and held the nose in the air as long as possible. The aircraft came to a stop just off the end of the runway on the perimeter strip. Medics quickly loaded Sgt. Cobb into an ambulance and rushed him away for treatment. His heroic actions did not go unnoticed. Although recommended for a Silver Star, the award was downgraded and he received the Distinguished Flying Cross several weeks later. In February of the following year, General Charles de Gaulle awarded Sgt. Cobb the Croix de Guerre avec Etoile de Bronze on behalf of the French government for his heroic actions.

Bombs fall from a 448th Liberator on the mission to Bergerac, France on 5 March 1944. (Everson)

The MAID OF ORLEANS required a new tail section, new bomb bay doors, new flaps, two new propellers, new control cables and numerous patches for the holes in the skin. PROBLEM CHILD, flown by Lt. John Daley, also struggled to return to England after receiving extensive damage from the intense flak. The pilots successfully landed their heavily damaged plane at Knettishall. Two others planes landed at different bases due to fuel shortages and flak damage. The long nine-hour haul across France resulted in two more crews listed as MIA while 448th gunners claimed ten enemy aircraft destroyed, five probable and one damaged.

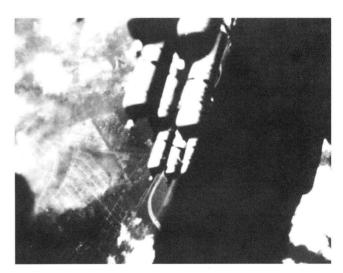

Incendiaries fall over Berlin just after bombs away on 6 March 1944. (Everson)

The following morning, 6 March, crews listened intently as the target was briefed: Berlin! In reality, the target was twenty miles south of Berlin at Genshalen, but it was the first large-scale attack on the Berlin area. Twenty-one aircraft left Seething and twenty successfully bombed the target. HELLO NATURAL, flown by Lt. Charles York, was last seen over the target at 1347. Flak damage forced the crew to head north for the safety of neutral Sweden instead of risking the arduous trip back to England.

Two days later, 8 March, the Group revisited the Reich capital. Twenty-three Liberators from Seething, led by a 389th BG PFF ship, took to the morning sky in the vanguard of the 20th CBW. Only seventeen aircraft departed the British coast bound for the ball-bearing plant at Erkner in the southeastern suburbs of Berlin. Early spring weather provided clear skies, perfect conditions for navigating and flying formation, but it also allowed fighters and flak batteries unobstructed views of the bombers. The route across the continent brought the Group

Minutes later smoke rises from the target area. (Everson)

A damaged HELLO NATURAL sits in the grass after landing in Sweden on 6 March 1944. Crews landing in neutral countries spent the remainder of the war as internees; however, some escaped and returned to England. (LaPoint)

to a point southwest of the city followed by a left turn to align the formation for the forty-second bomb run. The sky erupted with thick flak as the 448th crossed over the industrial area at 1440.

A flak burst exploded in the number one engine of Lt. John Bringardner's aircraft, 42-100342, just after "bombs away." Lt. Bringardner and Capt. John Grunow, on board as a command pilot, kept the airplane flying, but the high altitude coupled with a destroyed engine made maintaining formation impossible. Slowly, the damaged aircraft fell out of formation. Once alone and without the protection of the other bombers, Me-109s attacked the crippled bomber like vultures. The pilots attempted to reach the Netherlands. At least there, they could try to make contact with the Underground and evade the Germans. Cannon fire from the fighters shattered these hopes. Lt. Bringardner rang the abandon ship bell and all parachuted to safety. They were quickly captured near Hannover, Germany, and spent the remainder of the war as POWs.

Several other aircraft suffered damage from the thick flak over the target. TWIN TAILS, piloted by Lt. John Daley, endured flak damage to the number two engine and punctured fuel tanks. Oil from the damaged engine sprayed onto the windscreen where it quickly froze severely limiting visibility. Lt. George Allen, copilot, opened his side window and wiped away some of the grime. Fearing a mid-air collision, they dropped out of formation and descended. Evading the menacing fighters they reached Holland without incident. Fuel continued to leak from the punctured fuel cells making their fuel situation critical. As they crossed the coast heading over the North Sea, the radio operator, Sgt. James Nugent, sent SOS calls on the Air-Sea Rescue (ASR) frequency. Twenty-five miles off the coast of Great Yarmouth, England, Lt. Daley ditched the aircraft with empty fuel tanks. Rough seas broke the aircraft in half and it quickly sank. Air-Sea Rescue launches from Great Yarmouth arrived on the scene fifteen minutes after the ditching. Lt. Daley was thrown through the windshield and drowned. The navigator, Lt. Aaron Bramhall, was rescued from the wingtip but died before reaching the shore. Sgt. Nugent and the top turret gunner, Sgt. James Hood, were also rescued. Aircraft from RAF 279 Squadron continued searching for the missing crewmembers the next day to no avail.

Despite the savage flak, bombs from the seventeen aircraft of the 448th bracketed the Erkner ball bearing works. Results were outstanding. Unfortunately, two more crews were missing making a total of four crews in the last two missions. Stateside, forty telegrams arrived at the homes of grieving families announcing the tragic news. It was a steep price to pay.

Lt. Col. Hubert Judy, Air Executive Officer, used this P-47 to monitor formation assembly and aid in the difficult procedure. Tight formations kept the Luftwaffe at bay while loose formations and stragglers seemed to attract the wrath of the potent Luftwaffe. (Bailey)

Seventy-two engines sputtered to life shattering the early morning quiet of 9 March. The B-24s left Seething to bomb Nienburg, Germany, near Hannover. Just prior to the target the 20th CBW executed a 360-degree turn to fall in behind the 14th CBW after the Wing lead aircraft suffered PFF failure. Since clouds prevented visual bombing several more turns were accomplished as they tried to fix the problem. With fuel running low and clouds obscuring the entire area, crews were ordered to jettison their bombs. Again the murderous flak from Luftwaffe flak batteries took their toll. Lt. Everett Musselman, flying BABY SHOES, took several hits from the exploding shells. They valiantly turned for home but the damage was too severe. The entire crew bailed out and BABY SHOES crashed at 1630, six kilometers north of Arendonk, Belgium, near the Dutch border. Six of the men were captured but the other four, including the Group First Sergeant James Brown, evaded.

Morale improved dramatically during the month. Numerous dances were held, two for the officers and six for the enlisted men. The only complaint seemed to be the lack of representation from the WACs. As the Group diary noted, "A WAC necessarily comes from somewhere back home – and a touch of home, although vicarious, is much to be desired." Several visiting troupes entertained the men as well during March but possibly the largest boost came from the construction of a baseball field in a nearby pasture. Funds raised from the Officers' Club were used to lease the field and construct a backstop. The field quickly became a gathering place for the men.

Several weekly functions provided entertainment for the men. Lt. Col. Hubert Judy, Group Executive Officer, led current event discussions and showed strike photos from the week's missions. Also, every Sunday night, the Aero Club director, Miss Harriet Thompson, provided classical music for those men interested in a more soothing environment. Contact with the local British population continued to strengthen American-British ties in a more formal manner than the dances. Capt. Robert Lewis and Lt. Newton McLoughlin visited the Jenny Lind Children's Hospital in Norwich carrying gifts from the base to the unfortunate children.

Closer to home, the warming weather with less mud improved life for the men. Construction crews poured over 111,000 feet of concrete improving roads and building new facilities. Slowly, Seething dried out from the wet winter and the men were able to spend more time outside, especially at the ball diamond. Despite these improvements, the war was the only reason they were there and with the improving weather came an increase in the operations tempo.

The sign welcomes visitors to the living area of the 714th Bomb Squadron. Notice the bomb shelter as well as the bikes leaning against the huts. (Bailey)

On 10 March, bad weather forced a recall of a mission to Essen. The Group did not take to the skies again until 13 March. The target, a "No-Ball" site near St. Pol, France, promised a short mission. Twenty-six B-24s left Seething but due to poor weather in the assembly area, only eighteen completed the process. Six others flew with other Groups and two aborted for mechanical reasons. Escorted by a host of P-47s the bombers flew to the French coast near Calais, but positive identification of the target eluded the Group and the planes returned with

An unknown aviator poses in front of TONDELAYO. (Everson)

their bombs. The small, tactical nature of the V-1 launching sites proved to be difficult for the large formation to locate. Fearing unnecessary collateral damage to the French population, positive identification of the target was required to drop bombs. As a result, the bombers returned to Seething with full bomb loads.

Three days later, 16 March, the B-24s from Seething headed to southern Germany to attack the Dornier assembly plant at Friedrichshafen on the shore of Lake Constance. Eight aircraft aborted during assembly, but twenty-three made the long trip. Clouds prevented bombing by visual methods so bombardiers released their bombs on PFF smoke markers released by the Wing leader. Meager, inaccurate flak offered little resistance to the Group over the target but it did damage FEATHER MERCHANT. Lt. Thomas Apple landed his damaged plane at Wattisham without further incident. All the other attacking aircraft returned to Seething.

The mission the following day was recalled, but on 18 March the Group revisited Friedrichshafen with different results. After a later than normal takeoff, the 448th led the entire 2nd Bomb Division to the target on the Swiss border. As the twenty-two aircraft of the Group crossed the target at 1507, heavy flak erupted hitting 42-100284 flown by Lt. Robert Carroll. Capt. Jack Edwards, 715th BS commander, was also on board as command pilot. With the plane heavily damaged and leaking fuel, the crew decided to turn for Switzerland. They left the formation fifteen miles southwest of Strasbourg, France and safely navigated to the neutral country. Once in Swiss airspace they found a grass strip at Dubendorf, Switzerland, near Zurich. They counted over 100 holes in the fuselage of the aircraft, but the most serious damage was to one of the gunners, Sgt. Robert Miltner. Swiss medics rushed him to a hospital but he died from wounds inflicted by the flak. Despite three operations, the navigator, Lt. Castleton Smith, died from wounds received in the incident on 20 July 1944.

Flak inflicted other damage as well. Aircraft 42-52116, flown by Lt. A.D. Skaggs, suffered battle damage to the right wing and fuselage. Once the formation turned for home, fighters pressed home attacks on the Group but did little damage although Lt. Skaggs' aircraft began leaking fuel. The engineer, Sgt. George Glevanik, started transferring fuel and preserved enough to complete the eight-hour mission. Flak punched twenty holes in the thin aluminum skin of Lt. Russel Reindal's WABASH CANNONBALL. Despite the damage they, too, completed the long trip. The following day's Stars and Stripes reported thirteen aircraft had landed in Switzerland while three others crashed. A total of forty-three bombers from the 8th Air Force were reported lost.

After a day off, twenty-five aircraft from Seething headed for Frankfurt, Germany. Lt. Robert Voight with Maj. Robert Campbell, 712th BS commander riding as command pilot, led the Group in a PFF aircraft, 42-109808. Leaving the English coast clouds separated the Group from the rest of the formation. They continued on course flying below the clouds until a hole developed allowing them to climb above the bothersome cloud deck. Lt. Voight and Maj. Campbell tried to catch up and continue as briefed but the setback was too great. As they crossed into France near Dieppe, Maj. Campbell decided to abort the mission and the Group executed a left turn in order to cross the coast at the briefed exit point.

The returning formation crossed the coast near Abbeville, France, at 1246 when flak suddenly erupted. Sgt. James R. Bricker, engineer on FASCINTAING LADY, watched the demise of the lead aircraft. "I was flying in the deputy lead position, rear and left of the lead ship 42-109808 piloted by Lt. Voight. As engineer of 42-72981 I was in the top turret when I

Left: Several 448th men with musical talents formed the 448th Band. Here they perform for an unknown event. (Bollschweiler) Right: Men and women enjoy one of the numerous dances held at Seething. Dances were held regularly for both officers and enlisted men and were a welcome break from the rigors of preparing and executing the almost daily combat missions. (LaPoint)

North American P-51 Mustangs fly low over Seething in the spring of 1944. These new fighters provided long range escort for the bombers and swept the Luftwaffe from the skies neutralizing it as a threat for the bombers. The 448th formation ship is in the background. (LaPoint)

saw a flash which came out the top of the lead ship. I believe that this was the explosion of oxygen and hydraulic fluid. The skin above the bomb bay melted, and the aircraft settled in a glide to the left. After the ship had settled approximately 1,500 to 2,000 feet the bombs dropped free, the tail section twisted off, and the wings split at almost dead center. I saw three men come out of the plane, one from the right waist and two from the nose. No chutes were seen to open. The flak at this point was accurately aimed and heavy."

The force of the explosion blew Lt. Voight and Lt. William Edwards, the copilot who was riding in the top turret since Maj. Campbell was in the right seat, clear of the aircraft. Both deployed their parachutes but were captured after landing. Maj. Campbell survived the explosion but died several days later from wounds received in the explosion. One other crewmember, the group bombardier Capt. Frank Phillips, escaped the fiery explosion but died from wounds received after landing in a tree. The crew's original bombardier, Lt. Robert Ash, learned of the tragedy at Seething. He survived because he was attending PFF school when the crew flew their thirteenth and last mission.

Flight Officer Stu Barr, the copilot on THE COMMANCHE, saw four huge flashes on the ground near Dieppe. "I had a premonition they were shooting at us and put on my steel helmet ... Then I waited for the bombs to go off. There were two explosions in front of us. The next shell hit us and went right through the plane. It did not explode but cut all the fuel lines. Thick blue fuel was running all over the inside of the plane which was out of control. We had no way to communicate. As copilot I made the decision to salvo the bombs and get

Lt. Robert Voight, Capt. William Blum, and Maj. Robert Campbell stand in front of BTO. (Everson)

Lt. Robert Voight's crew pose for a crew photo after receiving the Air Medal. Unfortunately, most of the crew were killed on 20 March 1944 when their aircraft exploded after being hit by anti-aircraft fire. The pilot, Lt. Voight, and the copilot Lt. William Edwards survived the explosion. The crew's bombardier, Lt. Robert Ash, was at PFF school on the fateful day and was not flying. (Westgate)

rid of all the excess weight to try to get the plane home. We threw out ammunition, guns, everything, but it was useless. The crew bailed out, with me last, going through the ship to make sure they had all jumped. Then the pilot and I jumped and the plane flew on." The plane eventually exploded and fell to earth near Pommereval, France. Everyone survived the nightmare. Seven, including the pilot, Lt. Myers Wahnee, were captured near St. Agathe, twenty kilometers southeast of Dieppe. But two, the copilot, Flight Officer Stuart Barr and a gunner, Sgt. Richard Elliot, evaded and returned to England. The last man captured, Lt. Royal Goldenberg, almost escaped; he was betrayed by a girl riding the train from Dieppe to Paris.

At the same time, Lt. Paul Harrison jettisoned the bombs from LITTLE SHEPARD after the flak severely damaged his plane. However, despite the damage they returned safely. In the Seething control tower, clerks posted two more crews on the status board as Missing in Action.

On 22 March, the Group set out for another trip to "Big B." Twenty-three crews completed the long, tortuous trip to the Reich capital without incident. Even though the weather was improving, the dreadful cold at altitude hampered operations and claimed numerous frostbite casualties. Several men spent days in the hospital recovering from painful frostbite injuries. Heavy, accurate flak exploded in the sky over the German city but caused no damage. The result of this mission was a much-needed victory compared to the results during the last two weeks. The Group had dropped bombs on the heart of the Third Reich and returned with no losses!

The next day, 23 March, the Group prepared for its thirty-fifth mission in the European Theater of Operations, this one

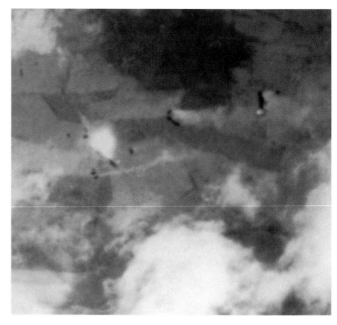

This fateful picture shows aircraft 42-109808 exploding over Abbeville, France after being hit by anti-aircraft fire on 20 March 1944. Amazingly, two men survived this frightful incident but claimed the lives of the others. (USAF)

to Munster, Germany. Twenty-two aircraft departed Seething but Lt. A.D. Skaggs, flying VADIE RAYE, quickly aborted due to an engine fire. He safely landed the heavily laden aircraft at Seething with a full tank of gas and five 2,000-pound bombs. The remainder of the Group continued to assemble and the Group left the English coast with twenty airplanes in the formation.

The formation crossed into enemy territory at Egmonde, Holland, and headed for the target. Although heavy flak filled

THE CAROL-N-CHICK was the only D model B-24 to fly combat with the 448th. It flew numerous missions until it was finally salvaged as war weary in February 1945. (LaPoint)

Lt. Myers Wahnee, a native American pilot with the 448th, admires the noseart on his B-24, THE COMMANCHE. Lt. Wahnee and crew were shot down on 20 March 1944 over France. Fortunately, everyone survived; two evaded and returned to England and the other eight including Lt. Wahnee were captured. (Everson)

Smoke erupts from targets as bombs fall around Nancy, France during a mission on 24 March 1944 (Bailey)

This view from atop the control tower is looking toward technical area in the general direction of Group Headquarters. (Everson)

the sky over the Dummer Lake area, the excellent escort of P-38s, P-47s, and P-51s kept the Luftwaffe fighters at bay. Clear weather permitted visual bombing of the target despite a heavy concentration of flak over the city. Leaving the target, MAID OF ORLEANS, flown by Lt. Robert Pinkus, started lagging behind the formation without obvious damage. A fire in the bomb bay eventually forced the crew to bail out. Lt. Pinkus died when his parachute malfunctioned. Sgt. Julius Rebeles never got to know his new barracks mates; it was only their second mission. The Group completed the mission with outstanding bombing results.

The airfield at Nancy/Essey, France was the target on 24 March. Twenty-three aircraft completed assembly but REPULSER, flown by Lt. Thomas Keene, took off late because of a mechanical problem and failed to catch up with the formation. Heavy flak struck the formation as they crossed the French coast five miles south of Dieppe. Three of the B-24s suffered damage, including the Group lead WABASH CANNONBALL, and returned early. Two men on TONDELAYO suffered injuries from the exploding shells. Lt. Gail Sheldon flying VADIE RAYE assumed the lead when WABASH CANNONBALL aborted. The mission continued.

Patchy clouds and haze over the target caused indecision about whether to bomb the primary or secondary target. Finally, red flares were fired from the Wing leader directing the formation to bomb the primary target. As they were turning on the bomb run, clouds thickened below the bombers obscuring the ground, then suddenly scattered, allowing a quick peek at the target. In the confusion, the lead 448th squadron did not bomb but nine aircraft in the low left section did. The lead squadron attempted to bomb the secondary target but smoke caused by the bombing from another Group covered the airfield at St. Dizier, France, and only two aircraft were able to bomb. Flak hit SKEETER II damaging a tire and the landing gear. Despite the damage they landed safely.

Heroism was not reserved for the crews fighting over Europe. Cpl. William Baukus risked his own life to save an aircraft at Seething. While retrieving the flare load from an aircraft, a flare cartridge was mistakenly left in a flare gun. When the gun accidentally fired in the cockpit, the flare ignited, setting fire to the flare box. Cpl. Baukus grabbed the box with his bare hands and removed it from the aircraft. Although slightly burned, his quick thinking and courageous action saved the aircraft. The citation for his Bronze star succinctly described the event. "William J. Baukus, 31324649, Sergeant (then Cor-

A British Lancaster sits on the perimeter track at Seething on 26 March 1944 after returning from a night mission. (LaPoint)

poral), Army Air Forces, United States Army. For heroism displayed at a bomber station in England on 24 March 1944. With disregard for his own safety, Sergeant Baukus entered a burning aircraft and extinguished the blaze. Entered military service from Connecticut." The award was bestowed on 5 June 1944.

A briefed mission to Munich on 25 March was scrubbed but on 26 March the Group returned to France hitting a "No Ball" site at Moyenneville. Just prior to "bombs away," the lead aircraft's bombsight malfunctioned and its bombs were jettisoned. The entire lead squadron followed on cue and released their bombs early. The second squadron dropped on the target area as briefed. All twenty-five aircraft suffered battle damage and flak injuries were numerous. Eight aircraft returned with feathered engines caused by the ferocious flak and two landed at other bases before returning to Seething.

Two days later, the airfield at Lille, France, was the prize but a recall cut short the mission. Lt. A.D. Skaggs, flying VADIE RAYE with Col. Thompson as Command Pilot, led the 2nd Bomb Division in the vanguard of the entire 8th Air Force. Just prior to reaching the Dutch coast, Sgt. Stanley Filipowicz, radio operator on Lt. Skaggs' aircraft, received the recall message and relayed it to Col. Thompson. They turned for home.

The next day, 29 March, the Group again headed for France. The "No-Ball" site near Watten, France, was the target. The lead squadron dropped on cue from the lead G-H ship but the G-H aircraft leading the second squadron suffered a malfunction. Some aircraft in the second squadron dropped on the smoke cues from the lead ship but in the confusion five did not bomb. SQUAT-N-DROPPIT suffered a punctured fuel cell caused by flak but all twenty-one B-24s of the 448th returned safely to Seething.

During the month of March, the 448th recaptured its fighting form after operations the preceding three months had decimated the Group. Many of the original cadre were either POWs or dead. Their legacy and sacrifices laid the foundations for many crews joining the Group. As new faces swelled the ranks of the Group, a growing fighting spirit passed from old to new. The Group had faced the Luftwaffe's challenges so far and in return dealt their own blow to the German Reich. The Group found its combat legs!

By March, morale improved in a large part due to the services at Seething and the improving weather. Red Cross plays such as this one, "Show a Leg", entertained the men during the month. (USAF)

The cast of the play "Show A Leg" pose after a performance on 28 March 1944 at Seething. Plays and shows like this were quite popular. (LaPoint)

seven

# LUFTWAFFE OVER SEETHING:
# April 1944

After finding their feet in March, the 448th received a bloody nose on 1 April. Starting at 0630, bombers rolled down the runway at Seething bound for Ludwigshafen, Germany. All planes were airborne by 0700. The planes from the 448th rendezvoused with the two PFF ships from Hethel over Buncher seven as the 448th was leading the 20th CBW. The 448th, with Col. Thompson in the lead PFF aircraft 41-28763, departed the English coast leading the Wing and the entire Second Bomb Division. A dense haze forced the formation to descend from 22,000 feet to 14,500 feet where they crossed the enemy coast. At 0909, Col. Thompson received an optional recall signal. Based on his judgment and experience, he elected to continue with the mission. The 446th flying high right became separated from the formation due to unfavorable weather conditions and turned back. The 448th continued. The haze that forced the 446th to return made navigation extremely difficult. Further complicating matters, PFF equipment in the lead and deputy lead aircraft failed. After much troubleshooting, the lead aircraft revived their PFF equipment but the deputy lead's PFF remained inoperative.

Stronger than forecast tail winds forced the formation to zigzag in an effort to stay on schedule which they did with great difficulty. Approaching the target, red-yellow flares floated down from the lead aircraft indicating bombing would be by PFF instead of visual. At the initial point, bomb bay doors opened and the formations maneuvered into position for the bomb run. Once established on the bomb run, lead notified deputy lead his PFF equipment was again not functioning properly. Twenty minutes after the initial point, the lead aircraft directed the deputy lead to assume lead of the formation. After discovering deputy lead's equipment was not working, Col. Thompson, the command pilot in the lead aircraft, decided to maintain the lead. During the confusion, the 14th and 2nd CBWs diverged from the 20th CBW course and flew further to the east. Seeing this error and without operable PFF equipment, Col. Thompson ordered a 180-degree turn in order to rejoin the other two Wings. As luck would have it, these two Wings also experienced PFF failure. They closed their bomb bay doors after failing to visually acquire the target and headed for home.

The 448th formation eventually rejoined the two Wings and the lead aircraft again called the deputy lead requesting him to take the lead. The two aircraft quickly changed positions and Capt. Heber Thompson, the command pilot in the deputy lead aircraft, directed the formation to set a course for home. He told the navigator to find a target of opportunity on the route home. He quickly picked out a town and the radio operator sent the visual code word over the radio. Twenty-one aircraft of the 448th released their bombs at 1104 over the town of Pforzheim, Germany. The excessive maneuvering consumed more fuel than planned and the stronger than expected tailwinds now became headwinds. As they turned for home, twenty-one crews faced a daunting trip through enemy territory with perilously low fuel tanks.

Low fuel was not the only obstacle facing the crews. Clouds and haze that had plagued them on the ingress further deteriorated and forced the returning bombers to fly at 14,000 feet into the teeth of the headwinds. Crews faced tough decisions. Lt. Harrison Mellor, flying 42-110087, advised Col. Thompson at 1315 that only forty gallons of fuel remained in his tanks. After one engine stopped, they turned south away from the heavily defended Pas de Calais region. Fifty miles north of Paris, Lt. Mellor gave the bail out order. Everyone safely exited the doomed plane but the parachute of Sgt. William Warren failed to open. Lt. Mellor faced a similar occurrence after pulling his ripcord. The parachute did not open and only after multiple attempts did it work. His chute opened a mere four hundred feet above the ground. Four of the crew successfully evaded capture and returned to Seething.

Lt. Alan Teague, flying the lead PFF aircraft, 41-28763, quickly calculated his plane lacked the fuel to return to England. After Col. Thompson relinquished the lead to his deputy, Lt. Teague fell out of formation and slowly drifted aft of the formation. Three P-47s provided close escort. Reaching France, where the possibility of making contact with the French Resistance existed, he ordered the crew, including Col. Thompson, 448th Group Commander, to bail out. Thinking the plane was empty, Lt. Teague prepared to crash-land the plane and then destroy the aircraft on the ground. Col. Thompson suddenly reappeared on the flight deck and returned to his seat for awhile as the plane descended. After a brief time, he announced he was leaving and once again climbed out of his seat and went to the back of the plane. It was the last time anyone saw him alive. Unfortunately, he died after his parachute failed to open due to insufficient altitude.

Lt. Teague successfully crash-landed the airplane in an open field and scampered away from the wreck before a German patrol arrived to inspect the downed plane. After a cursory glance for survivors they left, obviously searching for the crew. Eager to destroy the plane, Lt. Teague returned but too soon. Someone in the German patrol spotted him and promptly captured him. They also captured an intact set of pathfinder equipment that was installed on the aircraft although the German authorities failed to recognize the significance of the equipment. It was left on the aircraft under the care of two guards. Two days after the crash, three of the crew, Lt. Jesse Hamby, Sgt. John Dutka, and Sgt. Simon Cohen, who were in care of the French Underground, returned to the crash site and overpowered the guards. They destroyed the equipment denying the Germans access to the highly classified system.

At approximately the same time, the pilot of CRUD WAGON, Lt. Charles Knorr, contemplated his options. The copilot, Lt. Herb Bunde, explained to the crew over interphone that they were leaving the formation in an effort to conserve gas. Moments later it was apparent they could not make it to England and everyone prepared to bail out. The navigator, Lt. Stanley Baranofsky, was the first out. The bombardier, Lt. Charles McBride, watched him fall feet first toward earth. No one else followed him for over five minutes. Trying to build up the courage to jump, Lt. McBride stood over the open bomb bays looking down. Suddenly a loud, staccato noise sounded above the whine of the engines. "Standing on the catwalk in the bomb bay, I could see only downward, but it was not diffi-

BLACK WIDOW 42-100356. This aircraft ditched in the North Sea after running out of gas while returning to Seething 1 April 1944. (Everson)

cult to visualize the source of these sounds. I knew unmistakably they had come from an enemy aircraft attacking us, and months later this was confirmed to me. The attacks had scored hits on the hydraulic lines which accounted for the half closed bomb bay door. This experience thoroughly frightened me and provided all the impetus for jumping. Without further thought I closed my eyes and went out the right side, head first." With the exception of Lt. Baranofsky, they landed in the vicinity of St. Pol, France. For Lt. McBride, it was the start of a five month ordeal that happily ended with his return to Seething via the French Underground.

Returning crews last saw Lt. Kenneth Weaver's aircraft, EASTERN QUEEN at 1330 while still over France. Nine minutes later he and his crew bailed out rather than risk a dangerous ditching in the Channel. Nine men of the crew safely landed in France but the tenth, Sgt. Harvey Dickey, died after his parachute failed to open. The rest of the crew were captured immediately.

Flak damaged an engine of BLACK WIDOW, piloted by Lt. Jack Black, over Roubaix, France, but Lt. Black and crew elected to try to make England. Over Pas de Calais, another

Airmen carry one of their wounded comrades from a B-24 to a nearby ambulance. (USAF)

engine failed and they started descending. A third engine quit shortly afterwards. Small arms fire from troops along the coast peppered the plane as they headed out to sea. At 1357, unable to fly any further, Lt. Black ditched BLACK WIDOW fifteen miles off the coast of Dunkirk. An after action report described the ordeal:

> "The Liberator was traveling at eighty to ninety miles per hour when it touched water, and Black said it was like hitting a brick wall. Neither he nor his copilot Lt. Joseph Pomfret, were wearing shoulder harnesses since none of the B-24s in the 448th Group were then equipped with them. Instead, they had only the normal seat belts to rely upon in anchoring them in position, and these were not enough for the unique landing they were about to execute. The force of the sudden stop rammed Lt. Black's head into the instrument panel causing several deep scalp gashes. His legs were suddenly wrenched at the knees so badly he was unable to walk without a cane for a long period. The copilot, Lt. Joseph Pomfret, was actually thrown out of his seat through the cockpit to the outside. Miraculously, the pilot did not loose consciousness on the plane's impact, and though momentarily stunned and suffering from head lacerations, clambered out of the cockpit and down into the cold, choppy sea, along with the rest of his men. There was actually no other choice to make in braving the water's chilling numbness rather than clinging to the bomber and risking being carried under for what they believed would be a matter of five minutes or less."

With the aircraft floating, the crew tried to deploy the inflatable life rafts but were unsuccessful. With the rafts being their only chance of survival, the radio operator, Sgt. Eugene Dworaczyk, swam inside the half-filled aircraft and worked underwater until the life rafts were deployed. Amazingly, the aircraft stayed afloat long enough for him to complete his lifesaving work. For his selfless actions, Sgt. Dworaczyk received the Soldier's Medal.

While Sgt. Dworaczyk worked to free the dinghies, the rest of the crew floundered in the water trying to help the wounded. The navigator, Lt. Peter Wermert, held the injured copilot who could not swim. Finally, weakened by the cold waters of the North Sea, he was forced to release him or go under himself. Lt. Pomfret cried frantically for help, but sank before anyone could reach him. He never resurfaced.

Once in the life rafts, it soon became apparent Sgt. Charles Nissen suffered injuries and was suffering from shock. Unfortunately, without the necessary medical attention, his friends

THE IMPATIENT VIRGIN 41-29231. One of the original aircraft assigned to the 448th at Herrington, Kansas, it was transferred to the 44th BG shortly after its arrival in Europe. Another 448th aircraft by this name landed in Bulltofta, Sweden on 9 April 1944. (Everson)

The Air Executive officer, Lt. Col. Hubert Judy, stands in front of FASCINATING LADY. Lt. Col. Judy was awarded the Silver Star for his actions leading the 10 April 1944 mission. FASCINATING LADY ditched in the English Channel on 20 April 1944. (Everson)

A stricken B-24 falls out of formation while men bail out of the doomed 448th plane. Notice the man jumping from the aircraft. (Bailey)

A formation of 448th B-24s, identified by the large I on the tail, pass a growing thunderstorm on their way to a target in occupied Europe. The aircraft in the foreground is PEGGY JO, one of the aircraft lost during the night of 22 April 1944. (LaPoint)

could not help. He died of his injuries. Emotional crewmates buried him at sea amid the recitation of the Lord's Prayer and Hail Mary. The ball turret gunner suffered injuries but his broken jaw and collarbone were not life threatening. Just off the coast of France, the men started paddling toward England. The helpless crew drifted for forty-four hours until Bert May, the skipper of the English fishing boat *The Three Brothers*, spotted the eight surviving airmen. Miraculously, favorable winds blew the dinghies across the Channel to within five miles of the English coast when they were rescued.

As the B-24s of the 448th neared the Channel, Group integrity went by the wayside as fuel-starved aircraft scattered searching for a place to land. Lt. Russel Reindal elected to leave the formation prior to the French coast to conserve fuel after his number four tank ran dry. After evading several flak burst at the coast, they threw out all loose items to lighten the ship. It worked as they reached England and landed at a British fighter base. Other crews also landed at the first available field; only five aircraft returned directly to Seething. They landed over one hour past their scheduled return time. Gradually, all the aircraft returned but only after refueling at various bases.

The decision to search for the target following the PFF failure proved costly. On long penetration raids like Ludwigshafen, fuel consumption was critical. The tragic indecision cost the 448th four aircraft and crews. Furthermore, the Group lost their commander and two others, the Group Navigator, Capt. Robert Thornton, and Capt. Minor Morgan, who had also flown on the lead PFF ship.

The Group stood down the next day as it waited for word on the fate of their commander. On 3 April, the Group experienced its first change of command since its inception. Col. Thompson built the Group from scratch and led it through its bloody introduction to combat. The early days claimed many of the original crews and, ironically, just as the Group found its feet, Col. Thompson died not at the hands of the Luftwaffe but as a result of poor planning. Col. Gerry Mason, a West Point graduate and experienced single and twin-engine pilot, assumed command in his absence. Prior to taking command of the 448th, Col. Mason had served as the Wing Executive Officer of the 96th Combat Bomb Wing and the deputy to Brigadier General Peck. Before arriving in England, he saw action as a fighter pilot in the China-Burma-India Theater. Now he was given the tough job of motivating a Group ravaged by combat.

After the tragedy of 1 April, the 448th did not return to combat until 6 April. Two aircraft joined a G-H aircraft over Buncher seven destined for a "No Ball" site near Watten, France. Clouds pestered the small formation the entire route but good fighter cover kept the Luftwaffe at bay. Inaccurate flak burst around the formation as they neared the target. They dropped their loads of 2,000-pound bombs and returned without incident.

Two days later, the Group returned to the heart of Germany. Twenty-eight aircraft left Seething starting at 1015 bound for Brunswick, Germany. After some difficulty completing the assembly, the 448th formed to the high right position in the 93rd BG formation. Good weather along the entire route allowed for visual bombing but it also offered Luftwaffe flak crews near Dummer Lake a chance at the Liberators. Fortunately, their aim was bad and the puffs of flak fell harmlessly back to earth. Once over the target, the flak crews proved more proficient. Smoke screens placed around the target were very effective and bombs were released over a factory north of the assigned target; smoke from the smoke screens and the preceding bomb strikes obscured the original target. Enemy fighters attacked a B-24 formation five minutes in front of the 448th and shot down two planes. Fortune, however, smiled on the 448th as all aircraft landed safely at Seething by 1630.

The next day, the Group would not be as fortunate. Easter morning dawned in East Anglia as thirty-one crews listened to the briefing for another deep penetration raid into Germany. The target was an aircraft park outside of Tutow, Germany. Once again, poor weather hampered the assembly. The 448th attempted to form in the low left position of the 20th CBW but clouds required the Group to fly slightly high and in trail of the 446th BG. Another Wing flew head-on through the formation forcing the 448th aircraft to scatter to avoid a collision. Miraculously, everyone survived and the Group reassembled. As the formation left Buncher seven, two elements of the 446th climbed through an overcast cloud deck while the low section of the 446th attempted to remain below the clouds. Not eager to lose sight of the lead Group, the 448th followed the low section and remained below the clouds. The other two elements of the 446th remained hidden by the clouds. Finding a hole in the cloud deck, the formation climbed to altitude and returned to the briefed course. The unplanned maneuvering due to the clouds and reassembly after the near collision delayed the Group and they departed the English coast forty minutes late.

Shortly after this mad scramble, the Group abandoned the mission, but not all the aircraft received the recall message. Seven aircraft continued to the target with other groups. The nineteen crews that answered the recall returned to Seething and found marginal weather and low ceilings. Deteriorating visibility made landing conditions hazardous at best. Two aircraft, 42-11069 flown by Lt. Joseph Liebich, and 42-110079 flown by Lt. Earle Durley, suffered damage on landing and were salvaged.

Lt. Vincent Liedka, pilot of PICADILLY PETE, failed to get the recall. On the way to the target over the Baltic Sea, FW-190s attacked and severely damaged PICADILLY PETE. With leaking fuel and a damaged wing, Lt. Liedka and crew decided to turn toward Sweden. Without sufficient fuel to return, they landed at Bulltofta near Malmo, Sweden and spent the remainder of the war as internees. THE IMPATIENT VIRGIN flown by Lt. Kent Moseley, also failed to receive the recall and con-

A.D. Skaggs smiles from the cockpit of his B-24 VADIE RAYE as he prepares for the mission to Hamm, Germany on 22 April 1944. His flying skills were tested on their return to Seething that evening. (Sheehan)

Lt. Elbert Lozes, Sgt. Francis Sheehan, Sgt. Ray Lee (left to right) and Capt. Jack Edwards (kneeling) pose in front of VADIE RAYE on the morning of 22 April 1944. Little did they know what excitement the upcoming mission held for the crew. (Sheehan)

ICE COLD KATIE 41-28595. Another original Group aircraft, this B-24 was involved in a pileup at the end of the runway during the night of 22 April 1944 and was salvaged. (LaPoint)

tinued. Flying with an unknown Group, he was forced to land at Bulltofta, Sweden after the Liberator's engines were damaged and fuel tanks were punctured.

Lt. Robert MacKenzie flying CAROL-N-CHICK was another pilot that did not hear the recall. Unable to find the formation during assembly, he joined the 93rd BG. An uneventful trip to the target ended just after dropping their fragmentation bombs on parked aircraft at the Tutow airfield. Passing near Kiel, JU-88s and Me-410s stood off from the formation and lobbed rockets into the bombers. Immediately following the rockets, Me-109s swooped down with head-on attacks. Firing from the nose turret, one of the bombardiers set a fighter on fire forcing the pilot to bail out. Withstanding this attack, the formation faced another type of trouble as they neared the English coast. A heavy weather front settled over East Anglia closing all of the airfields in the area. As the returning bombers reached the Channel, they scattered in all directions searching for clear airfields. One of the 448th planes headed toward London and landed at Bovingdon. After spending the night, the weather cleared the following morning allowing them to return.

On rare occasions, crews flew a "milk run." Generally, little or no opposition characterized these missions. The 448th participated in very few of these missions during the winter and spring. For Sgt. Julius Rebeles, the 10 April mission to Bourges, France, fit the "milk run" mold, but even a "milk run" required attention. Carelessness or inattention was deadly. Twenty-six aircraft left Seething at 0630 with Lt. Col. Hubert Judy, Deputy Group Commander, flying as command pilot with Capt. A.D. Skaggs' crew in the lead aircraft. The 448th led the

TONDELAYO 41-29240. An original B-24 received by the Group before going overseas, it was one of four aircraft involved in a pileup at the end of the runway during the night of 22 April 1944. Crew 33 poses in front of the plane before it was transferred. (LaPoint)

20th CBW but followed another wing that led the Division. Soon after takeoff, Sgt. Ray Lee, waist gunner on Capt. Skaggs' crew, noticed fuel siphoning from the left inboard tank and notified the engineer, Sgt. George Glevanik. After visually confirming the problem, Sgt. Glevanik unsuccessfully tried to stop the leak. Informing Capt. Skaggs of the problem, Sgt. Glevanik stated if the problem continued they would have to return to Seething. In no uncertain terms, Lt. Col. Judy said the mission would continue and they would worry about the fuel problem after 'bombs away.' The formation was well over France before Sgt. Glevanik stopped the siphoning.

Clouds at the cruising altitude and an undercast deck from the Wing initial point until several miles prior to the target made

Three of the four aircraft involved in the pileup at the end of the runway during the night of 22 April 1944 are examined the day after the incident. THE RUTH E.K., ICE COLD KATIE and TONDELAYO are visible while the fourth aircraft, SKY QUEEN, is on the far side of the pileup and is not visible. (LaPoint)

SKY QUEEN rest in the foreground while the other three aircraft lay twisted together in the background. SKY QUEEN was the first aircraft that ran off the runway but avoided most of the damage suffered by the next three that departed the runway. (LaPoint)

navigation and target acquisition difficult. Only occasional breaks in the clouds allowed the navigator, Lt. Don Todt, to update their position and cross check the route of flight. Prior to the target, Lt. Todt noticed the lead formation drifting right of course. Lt. Todt notified Lt. Col. Judy of the error while Capt. Skaggs maintained course. After an intense three-way conversation between Lt. Col. Judy, Capt. Skaggs, and Lt. Todt about their position, Lt. Col. Judy decided to break radio silence and notify the lead of their error. After a brief discussion, the 448th assumed the lead and the other formations fell in behind the 448th. Lt. Todt navigated perfectly to the initial point. Just prior to the target, the clouds broke allowing the bombardier, Lt. Elbert Lozes, to positively identify the target and drop the bombs. The outstanding work caused considerable damage to aircraft and the airfield at Bourges.

Leaving the target Lt. Todt calculated the duration of the flight and Sgt. Glevanik figured fuel requirements. Sgt. Glevanik asked Capt. Skaggs to lean the mixture as much as possible without overheating the cylinder heads in order to conserve fuel. After crossing the French coast, they made a beeline for Seething. On final approach, two engines sputtered and quit. The other two engines quit as they taxied into the hardstand. What a 'milk run!' For his exemplary leadership, Lt. Col. Judy was awarded the Silver Star. Lt. Todt got a pat on the back. No one held a grudge but it did provide good material for teasing the newly promoted Lt. Col. Judy.

Twenty-six aircraft participated in the mission bound for Bernburg, Germany just south of Magdeburg, on 11 April. As the formation crossed the Zuider Zee into Holland, the lead aircraft flown by Lt. Earle Durley dropped out of formation and returned to England due to an oxygen regulator problem. BUSTED FLUSH, flown by Lt. Paul Harrison with Capt. Lester Miller on board as command pilot, assumed the lead. "Little Friends" provided excellent escort to the target and although enemy fighters operated in the distance, none molested the bombers. The ritual flak dotted the sky near the formation as they crossed over Dummer Lake. Although at times very colorful, crews feared the flak. Each dirty puff of smoke indicated thousands of shards of shrapnel flying through the air. One of the sooty puffs hit Lt. Harrison's aircraft rendering all instruments for the number three and four engines inoperative. Anticipating dropping out of formation, the lead aircraft radioed the low section lead, Lt. Harvey Broxton, to stand by for Group lead. After further examination, the damage was not as severe as first thought and Lt. Harrison maintained his position. Although the flak scattered the formation, only one aircraft suffered damage.

Beautiful weather over the target allowed the navigator and bombardier perfect conditions on the bomb run. The aircraft released their fiery incendiary bombs over a factory and executed a wide right turn to avoid further flak. The return to England concluded with an instrument letdown into Seething. For Sgt. Stanley Zabrowski, only eleven more missions re-

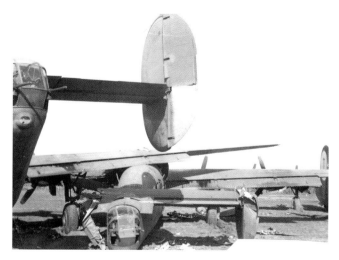

THE RUTH E.K., ICE COLD KATIE, and TONDELAYO sit in a pile the morning after the attack. Daylight revealed the true extent of the damage inflicted during the chaos the night prior. (LaPoint)

mained before going home. That night a lone raider flew over the airfield and dropped several bombs near the bomb dump. Fortunately he missed.

The next morning, the 12th, the Group returned to the same area around Magdeburg. This time it was Oschersleben. Starting at 1030, twenty-three aircraft lifted into the sky to begin the now routine assembly procedures. Although routine, it was never easy. Passing 13,000 feet, vapor from the engines condensed into unwanted contrails decreasing visibility for trailing aircraft. As they crossed the enemy coast contrails and clouds reduced visibility to the point where visual contact with the Wing Lead became impossible. In the vicinity of Brussels, the 20th CBW leader announced his intention to abort. Poor visibility and a depleted formation made continuing the mission into Germany a deadly proposition. Some aircraft dropped their bombs on a target of opportunity and all 448th planes returned to Seething. Not all groups were as lucky. Sgt. Stanley Zabrowksi witnessed two Liberators from another group fall to enemy aircraft. On return to Seething, Capt. A.D. Skaggs learned the tragic news of his brother's loss. A B-24 pilot with the 466th BG, he was shot down on the previous day's mission, ironically, to Oschersleben.

For the sixth straight day, ground personnel serviced and prepared aircraft for a mission. They toiled to prepare eighteen aircraft for a 1030 takeoff on the 13th. The tasks were endless: engine changes, loading the bombs, refueling, cleaning the guns, patching flak holes ... Nonetheless, thirteen aircraft completed assembly and headed for the target at Oberpfaffenhofen, Germany. No flak or fighters disturbed the Group enroute to the target but the ever-present flak appeared as they neared the target. The lead group flew 'S-ing' maneuvers on the bomb

Men work to clear the debris the next morning. ICE COLD KATIE received the majority of the damaged after being sandwiched between two other B-24s and was salvaged. THE RUTH E.K. and TONDELAYO were repaired and returned to combat. (LaPoint)

run in an attempt to locate the target. The erratic maneuvers on the bomb run caused many aircraft to be out of position. At the last minute, a turn into the formation by the low group put aircraft beneath the 448th. Recognizing the danger, only one 448th plane dropped its bombs on the primary target. Lt. Ed Chapman, flying COMMANDO, jettisoned his bombs leaving the target in order to conserve fuel. The rest of the formation departed the target with their bombs. With dwindling fuel and poor weather on the horizon, Capt. William Blum, the command pilot, advised the formation to drop on the first available military target without deviating from the Wing formation. A small factory area south of Heilbronn, Germany fell in the crosshairs of the Group bombardier and the bombs from eleven Liberators arced toward earth. COMMANDO landed at Manston to refuel before returning.

TONDELAYO sits among the wreckage at the end of runway 25 at Seething. Despite the heavy damage it was repaired and later transferred to another Group. (LaPoint)

VADIE RAYE 42-73497. A victim of the German intruders during the night of 22 April 1944, friendly fire damaged it before a marauding Me-410 inflicted further damage. The crew landed at Seething but the aircraft burned uncontrollably illuminating the entire field for the attackers. (LaPoint)

The 448th earned a much deserved break following six consecutive missions. Since the disaster on 1 April, the Group conducted five successful penetrations into Germany without loss. Anticipation swelled as rumors of the upcoming invasion of Europe grew. No one knew when or where it would occur, but the increased tempo of operations promised action soon.

The next 448th mission added fuel to the rumor mill. For the first time, the 448th flew two missions in one day. The first mission on 18 April launched at 1100. Twenty-six crews were briefed to attack a training plane factory at Rathenow, Germany, thirty miles west of Berlin. Maj. Chester Hackett flew as command pilot with Lt. Gail Sheldon in 42-73512. The formation assembled using instrument procedures and departed the English coast on time. The trip was uneventful until reaching the target area. Once again, flak dirtied the sky with black puffs. The formation successfully dodged the flak and each squadron moved into a trail position for visual bombing.

Complicating matters, a dark, foreboding weather front moved into the target area, directly in their line of flight. Despite the weather, Maj. Hackett led the Group across the target and struck the aircraft factory with excellent results. Just as the formation cleared the target, they initiated a turn to avoid the weather. While the lead aircraft and others on the inside of the turn missed the clouds, the less fortunate ones on the outside of the turn flew into the dark clouds. For a few terrifying moments, several of the aircraft were lost from sight. They quickly returned to clear skies and the formation reassembled and headed for home.

During the bomb run, Sgt. Julius Rebeles watched heavier flak pound groups attacking targets closer to Berlin while the skies over Rathenow were less hostile. Lt. John McCune witnessed an extraordinary event as they left the target. A damaged B-17 fell out of formation and immediately started calling for fighter cover. Fighters attempted to find the damaged plane but were unsuccessful and the B-17 seemed doomed to the fate of a straggler. Seeing the helpless B-17, a B-24 from a nearby formation pulled out of its protected position and escorted the stricken plane home.

Flak struck the formation again as they flew near Hamburg. Shore batteries and ships in the harbor opened fire on the planes. It was accurate and intense. Flak struck Lt. Gail Sheldon's left wing puncturing a fuel tank. Fuel siphoning out of the hole produced a vapor stream visible to other planes in the formation. As a result of the damage, they feathered an engine and slipped out of the formation passing the lead to Flight Officer Karl Schlund flying SLEEPLESS NIGHTS. Flak also knocked over thirty holes in the right wing of Lt. Russel Reindal's aircraft, BIM BAM BOLA. Upon landing, damage

The burned remains of VADIE RAYE sit between the runways the day after the deadly night attack by German intruders. (LaPoint)

was discovered to be so great the wing had to be replaced. The long trip over the North Sea produced no further excitement and everyone successfully landed at Seething.

As the first formation proceeded toward Germany, ground crews prepared four aircraft for the second mission. At noon, four aircraft took off from Seething and rendezvoused with two G-H aircraft from Hardwick. This formation, two three-ship elements, attacked a "No Ball" site at Watten, France through 10/10 cloud cover. A blessing in disguise, this cloud cover hampered the flak crews' aim and all planes returned safely. When the day ended, all the aircraft safely landed at Seething despite two missions, one of which flew within sight of the Reich capital. The Luftwaffe failed to provide any opposition during either mission.

The following day a larger formation visited the same "No Ball" site. After three planes aborted, the remaining twenty-five Liberators met the G-H lead ship over Buncher seven. The formation, five six-ship elements all in trail, departed the English coast at 1627, later than usual. The target was clearly visible as crews maintained a tight formation on the bomb run. However, three minutes prior to the target, accurate flak erupted around the bombers. Lt. Roy Davis' plane, MAID OF ORLEANS II, immediately caught fire after being hit. The formation dropped the bombs at 1710 and turned back over the English Channel. Dropping out of formation, Lt. Davis attempted to extinguish the fire by diving but eyewitnesses watched as the plane exploded. Three parachutes blossomed in the sky as the flaming wreckage fell to the water. Crews called Air-Sea Rescue and when boats reached the area they rescued Sgt. Ernest Robinson from the water. Lt. Albert Charette's body was also recovered but no one else was located. Except for Sgt. Robinson, this was the crew's first mission!

Sgt. Stanley Zabrowski slept late on 20 April welcoming the thought of no mission. However later, he sat in the briefing room surprised as they were briefed for a dusk takeoff to bomb another "No Ball" site, this one located at Bonnieres, France. With the sun low on the horizon, twelve aircraft took off from Seething on the Group's fiftieth mission. The 448th led the Wing with the 446th in the low left position and the 93rd in the high right. The formation reached the initial point uneventfully but as the Groups prepared for the bomb run, clouds reduced visibility. The first two elements dropped over the target but the third failed to visually acquire the target and did not drop. The fourth element of 448th aircraft did not bomb after the lead aircraft suffered flak damage on the bomb run. Dropping out of formation, FASCINATING LADY, flown by Lt. William Martin, headed toward the Channel. A flak burst exploded in the nose killing the navigator, Lt. William Cuthbert, and the nose turret gunner, Sgt. Anthony Kuzminski. When a fire broke out in the bomb bay, the rest of the crew bailed out over France and were taken prisoner.

The Luftwaffe, recognizing the significance of the "No Ball" sites, reinforced these sites with more anti-aircraft batteries. As the bombers returned to these sites, more and more flak units responded. The once benign targets were becoming increasingly dangerous. The Germans also were trying to make Seething a dangerous place. Early the next morning, about 0500, German aircraft bombed and strafed the airfield. Although the attack rattled nerves and interrupted sleep, no one was hurt. Sgt. Stanley Filipowicz wrote of the attack, "He came pretty close – business at the air raid shelter outside is picking up these days."

A deep penetration to Brux, Czechoslovakia, was recalled the following day, 21 April but not without loss. Lt. Jack O'Brien struggled with LITTLE SHEPPARD through the bad weather. "We were climbing on instruments and had accumulated clear ice to the extent where de-icing equipment was rendered useless. On approaching an altitude of 13,000 feet the aircraft stalled out and fell into a tailspin. I rang the bail out bell for the men to jump. The navigator let the bombardier out of the nose turret, then pulled the emergency door release and jumped. We were over the English Channel. The B-24 was brought back into normal flight at 1,400 feet still over the Channel and immediately we began to search the water for the navigator. The copilot had seen the chute hit the water but by the time we reached the spot, the chute had disappeared. We searched the area for the next four hours but were unable to find him."

Lt. Arthur Steele, the bombardier in the nose turret, heard the single ring from the bail out alarm. "I started unplugging things and I felt the door give way. Ausie had released the turret door. By the time I had popped the release on my flak vest and lifted myself out of the turret, I could see Ausie exiting through the nose wheel well. I did not hear the second bell ring for bail out, in fact it was not given. I buckled on my chest pack parachute and had my feet pressuring at the slipstream ready for a push-off into space. I had my escape kit in my left rear pants pocket. It caught on a canvas tie down post. The

Sgt. George Glevanik examines what remains of his engineer's position on VADIE RAYE. The oxygen bottles that saved his life lay scattered on the ground behind him. (Sheehan)

Sgt. William Jackson stands in his burned out tail turret of VADIE RAYE. Although he miraculously escaped this encounter, he was killed by a single piece of shrapnel during a mission on 9 May 1944. (Sheehan)

canvas wall, which separated the nose wheel well from the rest of the forward compartment, was removed for combat missions. I was getting set for another try and I saw the bomb load being salvoed, so I knew I was not alone in the aircraft. I pulled myself away from the bail out position and crawled through the tunnel to the flight deck where things were very tense, but no one other than Ausie had left the ship." Although Air-Sea Rescue launches arrived on the scene they, too, failed to find Lt. Seymour "Ausie" Ausfresser. All twenty-six aircraft returned safely. The reality of the loss struck Lt. Steele when he returned to their Quonset hut. There on Ausie's cot lay an unopened letter from his bride of six months.

CQs first awoke crews for a mission to Hamm, Germany at 0200 on 22 April. Bad weather forced leaders to postpone the mission numerous times. The "off-then-on-then-off-then-back-on" mentality played havoc with the flight crews. Adrenaline rushed as crews prepared to takeoff then subsided when the mission was canceled. Just as suddenly as it was scrubbed, crews were told to prepare to go after all. The adrenaline raced again only for the mission to be postponed again. Finally, at 1615 the first aircraft took off. Missions had departed Seething this late in the afternoon before but never with a target so far away. Usually a late takeoff meant a target in France, not in Germany.

Unknown to the B-24s, the B-17s from the 1st and 3rd Bomb Divisions launched that morning and attacked targets in occupied France. Originally scheduled for the morning, the Second Bomb Division mission was postponed until the afternoon. With all coordination between attacks lost, the Luftwaffe focused their attention on the B-24s.

Two squadrons of the 448th executed assembly without incident and departed Buncher five leading the 20th CBW. The route to the target was unopposed and at the Wing initial point the formation prepared for a visual bomb run. As was the custom, a well-aimed flak barrage greeted the Liberators on the bomb run but all planes dropped their loads on the marshalling yards with excellent results. German fighters passed overhead attacking the 93rd BG and 446th BG but spared the 448th. As the formation turned for home, they headed south to avoid the flak concentrations of the Ruhr. Soon afterwards, a large formation of fighters, mostly Me-109s and FW-190s, hit the entire 20th CBW. Sgt. Julius Rebeles on NO NAME JIVE watched as the escort of Lockheed P-38 Lightnings drove away the attackers. All of the 448th planes escaped damage but one 93rd BG B-24 fell to the fighters. This attack was only an omen!

Once clear of the Ruhr and established on a westerly heading with the sun in their eyes, the planes encountered a stronger than anticipated head wind. The planes' ground speed dwindled and when the bombers crossed the French coast between Ostend and Calais, the sun was setting. Escorting fighters dropped down and strafed areas along the coast. The first hint of trouble came when crews saw German twin-engine fighters taking off from bases in Belgium. Capt. Blum, command pilot flying on VADIE RAYE, broke radio silence and requested fighter support and notified the British coastal gunners. As a precaution, gunners were ordered to stay at their positions until landing.

At Seething, ground crews, unaware of the brewing storm, started the migration to the airfield to 'sweat' the returning crews home. Sgt. George DuPont sat in a bomb crate shack with several other guys waiting for the aircraft to return. The setting sun illuminated the sky with a beautiful pink hue. Guys took a moment out of their nervous work to enjoy the sight. As the fading sky gave way to darkness, someone made a foreboding statement. "Boy, wouldn't this be the time for the Luftwaffe to show up." Another person chimed in, "They wouldn't dare!" Little did they know how insightful they were.

Jim Turner was one of the many local children who spent their evenings around the local GI hangout at Mundham Garden House trying to bum candy, gum, or ration cards off the GIs. Saturday night, 22 April was no different. As it turned dark, they all headed for their homes. Situated near the approach end of the runway, Jim Turner's home was perfect for airplane watching. Tall poles in a field across the road held approach lights for the returning bombers. They were illuminated and that meant only one thing. The bombers were coming home. Suddenly, the lights flickered then went out. Machine gun fire reverberated overhead as two aircraft approached the field. Flashes on the horizon glowed then quickly disappeared. Tracer bullets arced from one aircraft to another. One flew ahead for several seconds then caught fire and crashed. From his home Jim Turner wondered what was happening. (This was Me-410, 9K+HP, serial number 420458 flown by Oberleutnant Klaus Kruger and his observer-gunner Feldwebel Micael Reichert. It was shot down by a 458th BG aircraft and crashed near Ashby St. Mary five miles north of Seething.)

In the darkening skies, the crews wondered the same thing. Standard identification procedures required crews to cross the coast at 1,500 feet then head back toward the sea from Great Yarmouth. After firing the colors of the day, crews would then perform a 180-degree turn and proceed to their base. All of the airplane lights were on to aid identification as was the IFF (Identification Friend or Foe) equipment. As the formation crossed Great Yarmouth at 2200, British coastal gunners opened fire on the B-24s. VADIE RAYE shuddered as friendly flak riddled the number two engine setting it on fire. A worst fate befell their wingman. They watched in horror as aircraft 42-52608, piloted by Lt. Cherry Pitts, first erupted in flames, then exploded and fell into the sea. Air-Sea Rescue scoured the area and found a large oil slick one mile off the coast of Hopton-on-the-Sea but no survivors. One of the men, Sgt. Ernest Robinson, was rescued from the sea after ditching just four days before. This time he was not as fortunate.

Lt. Eugene Pulcipher and crew flying REPULSER also suffered damage from the friendly fire and crashed near Kessingland at 2220. All on board perished when the B-24 crashed into the marshland. Simultaneously, PEGGY JO, flown by Lt. Melvin Alspaugh, shook from the impact of cannon fire from a Me-410 that destroyed the number three engine and set the plane on fire. The tail gunner, Sgt. Ray Chartiers, fired his guns in retaliation as the fire spread. With one wounded waist gunner and a raging fire in the waist and bomb bay, Lt. Alspaugh told everyone, "This is it boys. Hit the silk." By the time the crew left the aircraft it had descended to 2,500 feet. As Lt. Alspaugh released the controls to bail out himself, the plane tried to enter a spin. After righting the ship after several attempts, he engaged the autopilot and set a course toward the North Sea. He quickly dove from the flight deck and out a partially open bomb bay. The stricken plane crashed on a rail line at Worlingham, Suffolk but everyone survived. The only injuries were to the waist gunner who was hurt in the fighter attack and the engineer who suffered a broken ankle on landing.

The crash site of PEGGY JO near Worlingham, Suffolk reveals little of the 448th B-24 that was shot down during the night of 22 April 1944. The pilot, Lt. Melvin Alspaugh, and crew were able to safely bail out before the crash. (Rebeles)

Eager to get away from the British gunners and with an engine on fire, Lt. A.D. Skaggs headed VADIE RAYE for Seething, twelve miles away. Sgt. George Glevanik left the top turret to inspect the damage. He suggested to Lt. Skaggs to close the cowl flaps, add power, and dive while activating the fire extinguisher. This worked momentarily but when they leveled off at 1,000 feet the fire re-ignited. They quickly feathered the engine. As Sgt. Glevanik climbed back to his position, thirty-caliber bullets from a German fighter ripped into the airplane. The tracers missed Sgt. Glevanik but hit Sgt. Francis Sheehan in the left leg and caused considerable damage to VADIE RAYE. Already struggling with an engine fire, the crew now fought a flash fire in the bomb bay. 'Friendly' anti-aircraft fire and enemy cannon fire had knocked off a bomb bay door, damaged the fuel transfer pump and severed two main fuel lines. The high-pressure lines acted as a blowtorch spraying fuel and flames around the bomb bay.

As the engines sputtered from fuel starvation, the plane slowly descended. Seeing the smoke and fire and with no intercom, the gunners in back prepared to abandon ship. The ball turret gunner, Sgt. Gene Gaskins, and the tail gunner, Sgt. Bill Jackson, helped the wounded Sgt. Sheehan out of the waist window. Sgt. Gaskins immediately followed as the aircraft descended through 800 feet. Sgt. Bill Jackson, the tail gunner, recognized their low altitude and deployed his chute in the airplane then threw it into the slipstream as he jumped out. Sgt. Ray Lee, the other waist gunner, climbed into the window to follow his friends but his parachute harness became tangled inside the aircraft and he could not free himself.

After reaching the ground, Sgt. Gaskins tended to Sgt. Sheehan's wounds and tried to stop the bleeding with his scarf. Spotting a light in the distance, he left to find help. After searching, he met two young ladies from the Women's Land Army. They returned to Sgt. Sheehan and carried him to a nearby farmhouse. An ambulance later carried them to Seething. Sgt. Jackson landed near a Military Police station at the 446th BG base near Bungay. After convincing them of his nationality they also returned him to Seething.

Meanwhile, Sgt. George Glevanik contained the fire in the bomb bay and re-established some fuel flow to the engines on the left side by physically holding the fuel lines in place. This action allowed Lt. Skaggs to add power on the engines and level off. Hearing the engines come to life and feeling the aircraft level off, Sgt. Ray Lee pulled himself back into the aircraft. All of this occurred in a matter of minutes as they lined up for their final approach.

Confusion reigned as German fighters continued to attack the B-24s as they tried to land. With one engine still on fire, Sgt. Glevanik continued to hold the fuel lines in place while he straddled the open bomb bay. Fearing another engine failure Lt. Skaggs kept additional airspeed as he prepared for landing. After touching down Lt. Skaggs swung VADIE RAYE onto the muddy infield in an effort to clear the runway for other aircraft. The aircraft continued across the grass, struck a ditch and came to rest on the other side of the crossing runway. The jar of the impact with the ditch knocked Sgt. Glevanik off his perch. The electric cord on his heated flying suit caught on the bomb rack and he was dragged along underneath the aircraft. When the aircraft stopped, Sgt. Glevanik was trapped in the furrow dug by the nose gear with the aircraft pinning him face down to the ground. Seeing only darkness and not able to move, he feared his back was broken and the end was near. Those remaining inside the aircraft crawled out any available hole. Afraid the burning plane might explode at any second, they wasted no time exiting the plane. Lt. Don Todt helped the radio operator, Sgt. Stanley Filipowicz, out the top escape hatch then followed. Sgt. Filipowicz jumped from the wing and landed in a ditch severely spraining his ankle. Lt. Todt climbed off the wingtip then the two men ran toward the safety of the control tower.

The engine fire quickly spread throughout the aircraft. For Sgt. Glevanik, the fire was ironically his savior. As the fire reached oxygen bottles in the rear of the plane, they exploded

Sgt. Walter Johnson and another airman examine a .50 caliber machine gun mounted in the waist window of a B-24. Sgt. Johnson was credited with shooting down a Me-410 during the night of 22 April 1944. (USAF)

rocking the plane up on its nose and freeing him from his tomb. Seeing daylight, he started digging and freed himself before fire consumed the rest of the plane. As he ran from the burning plane, an unknown person tackled him just as a German fighter swooped in low strafing the airfield with aid from the blazing inferno.

Overhead, chaos continued. Controllers in the tower turned off the airfield lights and told all aircraft to leave the area. However, lights were not needed as the fire from VADIE RAY provided a beacon for both B-24s and the enemy. Seeing this, Lt. Frank Gibson circled the field in NO NAME JIVE weighing the options. Low on fuel, he elected to land despite no run-

RUM RUNNER 41-28583. One of the original 448th aircraft, it sits on a snow covered hardstand at Seething during the winter of 1943-44. It was one of two 448th aircraft that landed in Switzerland on 24 April 1944. (LaPoint)

way lights. Using only their landing lights and the eerie glow cast by the burning plane, they landed and taxied off the runway. From there, they watched as the show continued in the air over the field. Tracers filled the sky and aircraft fell to earth. Unable to distinguish friend from foe, Sgt. Rebeles could only wonder. They gathered their gear from the plane and headed across the field to the briefing room. As they stumbled along in their flying boots, the burning Liberator cast the only light on the entire airfield.

Lt. Tom Apple, flying ICE COLD KATIE, witnessed Lt. Cherry Pitts crew's fiery demise just off the coast and now watched the confusion unfold at Seething. He decided to circle the field until the situation cleared. They finally decided to attempt a landing and lined up on the long runway, two or three planes behind VADIE RAYE. After they watched VADIE RAYE land on fire, Lt. Apple decided to 'go-around' and try later. Lt. Cater Lee and others in ICE COLD KATIE moved to the catwalk and prepared to bail out, just in case. The smoke and flames from VADIE RAYE forced Capt. Bob Lambertson, flying SKY QUEEN, to 'go-around' as well. The smoke from the fire and the bright glare from the flames effectively closed the main runway.

Adding to the bedlam, low fuel states on some aircraft required them to utilize the shorter, crossing runway, runway 30, or divert. Landing on the shorter runway was not an impossible feat but since it was unlit, the difficulty drastically increased. Capt. Lambertson landed SKY QUEEN uneventfully on runway 30, but unable to see, he ran off the end of the runway and the wheels of the big bomber mired in the mud. The crew quickly evacuated the aircraft.

Lt. J.M. Williams landed long also and his aircraft THE RUTH E.K. stopped in the mud off the end of the runway as well. Lt. Apple and crew on ICE COLD KATIE were unaware of the two aircraft as they lined up to land on the short runway. After landing longer than normal, they prematurely breathed a sigh of relief. Suddenly the dark hulks of the two parked aircraft appeared in their windscreens. Lt. Tom Apple quickly warned everyone over the intercom of the impending collision, an action that surely prevented injuries. He guided ICE COLD KATIE between the parked B-24s but the wing tips of ICE COLD KATIE each clipped the parked aircraft shearing off both wings at the outboard engines. Lt. Cater Lee quickly vacated his aircraft and landed in ankle deep mud but did not stop running until he was well clear of the scene. Sgt. Jim Pegher jumped out and looked back toward the burning aircraft that guided them in. To his astonishment lights from another aircraft lit up the area as it barreled down the runway. He yelled as loud as possible warning his crewmates of another collision and then ran for his life.

Just like Lt. Apple on ICE COLD KATIE, Lt. Jack Barak, flying TONDELAYO, surveyed the situation and decided to wait for things to calm down. After circling the field several times, the fire from VADIE RAYE started to subside and he set up for the landing. Touching down on the short runway, he too was unaware of the aircraft pileup at the end of the runway until it was too late. The brakes squealed but the B-24 was going too fast and slammed into the aircraft. The right wing of TONDELAYO sheared off the top of the rudders of ICE COLD KATIE before coming to rest with its right wing resting on the fuselage of ICE COLD KATIE just aft of the wings. The weight of the aircraft broke the fuselage of ICE COLD KATIE at the waist gunners' position. The tail turret of ICE COLD KATIE rested grotesquely on the ground.

In the mad scramble to clear the scene, Sgt. Pegher noticed another man passing him as they both raced from the area on foot. In the shadows of the night, he noticed the stranger trailing a deployed parachute. Laughing to himself, he thought "Man, that guy could run faster if he did not have that parachute holding him back." Later he discovered this stranger was his best buddy and crewmate, Sgt. Furman Powell. The M.P.s promptly arrived on the scene announcing to the men "Take cover, there is an air raid on." Sgt. Powell quickly responded, "Hell, it is safe down here, you should be up there." Eventually they all found their way to the de-briefing where medics carried them to the hospital to check for injuries. It took several shots of rum to soothe the frayed nerves.

After crossing the enemy coast, Sgt. Walter Johnson left his waist gun position to release hung bombs in the bomb bay. With the bomb bay lights on, Sgt. Johnson unsuccessfully tried to free several bombs which jammed the bomb bay doors open. After spending a considerable amount of time in the bomb bay, he heard the unmistakable chatter of gunfire followed by explosions. Looking through the open doors at his feet, he saw what was supposed to be friendly land. He hurried back to his position. The other waist gunner, Sgt. Al Kohl, pointed to numerous holes in the rear fuselage. Fortunately, the tail gunner had left his turret in preparation for landing and no one was injured.

Unable to land in the melee, Lt. Marion Peek circled in the darkness waiting for a chance to land and ordered all the gunners to remain at their guns. This proved fatal for one unsuspecting Me-410. Deciding to take a chance, Lt. Peek prepared aircraft 42-52435 for landing. Sgt. Walter Johnson stood at his waist gun searching the night sky when he noticed a fast-moving object low to the ground and drawing considerable ground fire. Making a wide, climbing turn, the enemy plane approached within 250 yards of the B-24 as they lined up on

Crew 46 pose beside a B-24. Front row (left to right) Albert DiLorenzo, navigator; William Rogers, pilot; Raymond Cohee, bombardier; Wentzel, copilot. Back row (left to right) unknown, radio operator; George Robicheau, ball turret; Johnny Jones, tail gunner; Grady Howell, engineer; Bordie Haynes, waist gunner; Ralph Meigs, waist gunner. (DiLorenzo)

final. Sgt. Johnson fired his single fifty-caliber gun and watched the tracers impact the fuselage and cockpit. The top turret gunner and engineer, Sgt. Bob Kerrick, clambered back into his turret just in time to send a burst into the fighter from his twin fifties. The enemy fighter veered to the right and disappeared. (This Me-410 was probably flown by the commander of KG-51 "Edelweiss," Maj. Dietrich Puttfarken, and his radio operator/gunner, Oberfeldwebel Willi Lux. They never returned to base and were presumably lost in the North Sea.)

Still in danger, the hung bombs dangled precariously from the bomb bay rack. The armed bombs needed only a slight jolt to ignite them. The result would have eclipsed the still burning VADIE RAYE. Lt. Peek deftly landed the B-24 keeping all of the bombs in place and quickly taxied clear of the runway. Everyone scrambled out of the plane and raced for the nearest foxhole or bunker. After a short time wandering around in the darkness, they found the interrogation room and accounted for everyone. Lt. Jack Parker flying HELLO NATURAL II landed in a rage. He asked to be refueled and rearmed so he could wipe out some British anti-aircraft guns.

For the ground personnel peacefully waiting for the bombers to return, mayhem quickly took hold. The bright flashes on the horizon initially created confusion. Was it flares or gunfire? German fighters strafing the airfield quickly put all doubt aside. As night closed in so did the bedlam. The noise and gunfire increased the confusion. There was air-to-air, air-to-ground, and ground-to-air fire. The noise was unimaginable. Sgt. George DuPont initially found himself jammed head and shoulders into a brick trench but quickly decided to leave his haven for the apparent safety of the wide open area in the middle of the airfield. He returned to the hardstand area after the continuing gunfire made his open-air theory seem ill advised. As the battle dissipated, he wandered back toward his barracks reasoning there was nothing he could do until daylight. Passing an officer's hut, the door opened and a shadow asked what day it was. Answering the unknown shadow, he realized it was his birthday.

At Group Headquarters, Col. Mason ordered a truck and driver to pick up the ladies working on base and take them to safety. A short while later, he found a truck full of ladies parked in front of the Headquarters building. Interrupting their busy conversation, he demanded to know where the driver was. "Hiding under the truck" was the quick reply!

Sgt. Stanley Zabrowski raced out of his Quonset hut at the sound of air raid sirens and gunfire. Eager to see what was happening, he climbed an embankment instead of the safer air raid shelter. What he saw made him thankful he was not flying. The landing lights and airfield lights silhouetted the B-24s perfectly for the enemy fighters. They made perfect targets. They were sitting ducks!

After a seemingly endless onslaught, the German night fighters left. The night shrouded the destruction but the following morning the Group came face to face with the devastating damage. After all was tallied, the 448th had lost three planes to a combination of enemy fighters and friendly anti-aircraft fire. Another crash-landed from damage received from the combination and completely burned. Four other aircraft lay in a twisted pile at the end of the runway. Salvage crews were required for two of the aircraft but the other two were eventually repaired. Although eight aircraft lay damaged or destroyed, crews were miraculously lucky. Clerks posted two crews missing in action and five others had narrowly escaped death. VADIE RAYE burned until 0200 serving as a ghastly reminder of the night's action. The next day, the Second Bomb Division did not launch any aircraft as Groups throughout East Anglia struggled to recover from the blow dealt by the Luftwaffe. All of this was caused by only four Me-410s. What would have happened if there had been more?

On 24 April, after only a one day recovery, the Group resumed offensive operations with a mission to a repair facility at Gablingen, Germany. The 20th CBW launched an equivalent of two Groups. Each Group provided two squadrons with the 448th's two squadrons being split between the 93rd BG and the 446th BG. The aircraft took off during the normal morning hours beginning at 0715. Assembly was good and the route to the target was uneventful. Friendly fighter escort again was excellent and no enemy fighters were seen but the dreaded flak peppered the sky as the formation neared the target. Sgt. Ray Waters, from the waist window of SKEETER II, could see B-17s bombing Fredrichshafen far to the south. Dark spots

CHUBBY CHAMP 42-7655. An unknown airman poses in front of his plane. Transferred to the 448th from another Group, this aircraft was one of six that failed to return from the mission on 29 April 1944. (Engdahl)

in the sky indicated heavy flak to the south as well as over the target.

Lt. John McCune, flying THE FLYING SAC, suffered engine problems before reaching the target. An engine change just before the mission produced higher than normal cylinder head temperatures that increased during the flight. It worsened until the propeller started to overspeed. Lt. McCune managed to control it for awhile until finally the engine disintegrated and they quickly feathered it. The heavily laden bomber slowly fell out of formation. Lt. McCune attempted to tack on to another B-24 group but was not able to do so. He also tried unsuccessfully to tag along with a B-17 formation. The high power setting consumed excessive fuel that ruled out a return trip to England. Without protection or other options, they jettisoned their bombs and turned toward Switzerland. Crossing Lake Constance, three Moraine fighters of the Swiss Air Force met the bomber. Lt. McCune rocked his wings and lowered his landing gear in an act of submission. They followed the fighters and landed at Dubendorf but not before destroying their bombsight and other classified equipment. The war was over for them.

Very accurate and intense flak surrounded the bombers over the target but quickly subsided once away from the target. The remainder of the flight was uninterrupted by enemy action but unfortunately it was not without incident. Once the aircraft were over the English Channel, gunners began clearing guns to prevent inadvertent firing upon landing. Sgt. Stanley Zabrowski watched in shock as a turret from another aircraft became jammed and began firing wildly. The tracers ripped into a B-24 in the formation forcing it to ditch in the channel. (It was not a 448th aircraft but it happened to an aircraft in the same formation from an unknown Group. The pilot and five enlisted men were rescued. The rest of the crew perished.) As numerous crews could attest, combat was an unforgiving business; tragedy did not discriminate.

The following day, the 448th would again smart at losses inflicted by accident and the enemy. The 448th readied twenty-four aircraft for a morning mission to Mannheim, Germany. Takeoff and assembly progressed with no problems. Clear weather allowed the Group to position itself in the low left position from the 93rd BG which was leading the Wing. As the formation crossed the enemy coast between Dieppe and Le Treport, France, broken clouds gradually formed a 5/10 deck below the bombers. Flak, at times intense and accurate, badgered the Liberators enroute to the target. As the bombers flew the steady bomb run, clouds completely obscured the target. Unable to bomb visually, bombs were held and they picked up a heading to the secondary target. It too was shrouded in clouds and the bombers held their bombs. As the formation turned for home, two aircraft from the 448th were missing.

COMMANDO, flown by Lt. Henry Schroeder, suffered heavy flak damage to the number two engine. Shortly after feathering the damaged engine, another engine started losing power and they fell from the protective umbrella of the formation. Unable to stop descending, they turned for Switzerland. After crossing the border, Swiss fighters rendezvoused with them and escorted them to Payerne airfield. After a steep approach, they overshot and ran off the end of the runway and collided with the boundary fence. Lt. Schroeder and the copilot, Lt. Lewis Sarkovich maneuvered the aircraft between two trees but the wings were ripped from the fuselage. The plane quickly caught fire but not before the crew escaped unharmed.

The second missing aircraft, RUM RUNNER, also turned for Switzerland. Lt. Robert Lehmann landed at Dubendorf with a full bomb load. The Swiss used this undamaged aircraft on numerous occasions as part of the Swiss Air Force. The remainder of the 448th aircraft returned safely to England with their full bomb loads. No bombs were dropped, but two more crews were missing.

Once again the great orchestra of events commenced in preparation for a mission the following day, 26 April. Another early takeoff was planned for twenty-four aircraft. Precisely at 0600 aircraft rolled down the runway led by the formation ship, CHECKERBOARD. During assembly, THE STURGEON, flown by Lt. Marion Peek, experienced a tragic mishap. Vibrations shook the nose turret loose and as the heavy turret shifted it crushed the nose gunner Sgt. Anthony Wasalik. The pilot immediately departed the formation in the vicinity of Bungay and landed at Seething. Sadly, Sgt. Wasalik was dead by the time they arrived. The remainder of the formation completed assembly and the 448th led the Wing to the target at Paderborn. Without the aid of PFF or G-H aircraft, the bombers were forced to bomb visually. Once again clouds completely covered the target area and the bombardiers held their bombs. Again, the Group returned to Seething with all their bombs. For two consecutive days the Group failed to get bombs on the target due to weather but the price of the two missions was high. Two crews MIA and one person killed. For Sgt. Stanley Zabrowski it was mission number twenty-three, only seven to go!

On 27 April, the Group flew two missions. The morning mission visited a "No Ball" site at Wizernes, France. Twenty aircraft hit the target with a combination of 500 and 1,000-pound general purpose bombs despite intense flak over the target. These short missions to the coast of France, once considered "milk runs", were becoming increasingly dangerous due to the heavy flak. Fortunately, all attacking aircraft returned to Seething.

The afternoon mission started in disarray. Plans called for the 448th to lead the Wing and Division to the marshalling yards at Blainville-sur-l'Eau, France, near Nancy. The Group took off starting at 1531 with only minimum information concerning the mission. Field orders did not arrive at Seething until forty minutes after takeoff. By then, all the planes were airborne executing their assembly procedures. Further complicating matters, after the first couple of aircraft took off, the wind shifted, forcing a change to the active runway, and caused the remaining aircraft to takeoff late. The rest of the Wing encountered difficulty also. Only nine aircraft from the 446th BG made it to assembly. The Wing completed assembly with a reduced force and left the English coast at Beachy Head on time. After crossing the enemy coast, the formation ran into heavy flak. The Group turned on to the bomb run without problem but haze in the area made it difficult for the bombardier to precisely locate the target. As a result, the first section's bombs were short but the second section's plastered the marshalling yards. After "bombs away," the Group made a right turn to help the rally and avoid flak. B-17s bombed targets in the vicinity and as they left their targets, they conflicted with the 448th's route of flight. Considerable maneuvering was required to avoid a collision. Sgt. Julius Rebeles on THE SAD SACK watched P-51s fight off attacking Me-109s. While the attackers did no damage, P-51s shot down two of the enemy aircraft. Flak however, claimed the life of Sgt. Joe Corziatti aboard 42-109793 flown by Harvey Broxton.

No aircraft participated in missions the next day, but Seething welcomed much needed replacement crews. Intense operations over the last week severely strained the crews and aircraft. It was a rough week and unfortunately the hard luck continued.

On the 29th, CQs woke cursing crewmembers in the wee morning hours. "Briefing at 0430" was the wake up call. Crews enjoyed fresh eggs for breakfast instead of the customary powdered variety. It was like fattening a pig for slaughter. When the curtain went up in the briefing rooms, crews groaned with anguish as the map showed the target: Berlin. The target was not on the outskirts this time, but the Daimler-Benz Works downtown. Crews manned their stations at 0650 and the first crew, Lt. A.D. Skaggs flying as group lead with Maj. Chester Hackett as command pilot, lifted off at 0720.

The flight to the target although long was uneventful with good fighter escort. As the formation crossed the initial point near Hannover, the terrifying and always present flak blossomed in the sky. Deadly black puffs dotted the horizon and as the planes neared downtown Berlin the flak intensified. Crewmembers cloaked in flak vest and helmets cringed as the exploding metal showered the planes. Sgt. Stanley Zabrowksi wondered if his time had come. He had been briefed to make this trip three times before, but they all were scrubbed. Now he was over Berlin; it was the most intense flak he had experienced. The flak barrage lasted for what seemed like hours and a 120-mile per hour headwind did not help. Group integrity was maintained as the target was covered with clouds and bombing was done by PFF. Immediately after "bombs away," a steep turn toward the rally point by the lead aircraft caused the first squadron of the 448th to fall behind and out of position. It would prove to be a cruel twist of fate. Meanwhile flak continued to hound the formation until they reached the Dummer Lake area.

The good fighter escort on the way to the target now completely evaporated. As the flak subsided, enemy fighters joined the fray with quick, devastating results. FW-190s hit the out-of-position bombers in waves as they neared Dummer Lake. SAD SACK fell first at 1335. Twenty-millimeter shells exploded in the cockpit setting the front end of the plane on fire. The plane entered an out of control dive and crashed west of Nienburg, Germany at Pennigshel. Six of the crew died in the

SWEET SIOUX 42-7683. An unknown man poses with another of the Group's original aircraft sometime before it was lost on 29 April 1944. The crew bailed out of the stricken plane before it crashed into a swamp in Switzerland. (Everson)

SAD SACK 41-28588. Another of the original Group aircraft rests in the mud after sliding off an icy runway. It was the first of six aircraft that fell to the enemy on 29 April 1944. Six of the crew perished after cannon fire from FW-190s set it on fire. (LaPoint)

fiery crash including the pilot Lt. William Ponge. Miraculously, three men survived, the copilot Lt. Edward Snowbarger, radio operator Sgt. Haari Halvorson, and the right waist gunner Sgt. Henry Maynard.

Lt. Max Turpin in MISS HAPP fell to fighters at 1340. A flak burst in the nose of MISS HAPP killed the nose gunner, Sgt. George Daneau, but the rest of the crew bailed out before the plane crashed at Pennigsehl. The tail gunner, Sgt. Donald Elder broke his leg after landing on a house but everyone else escaped unscathed. However, they were captured and served the rest of war as POWs.

Flak damaged SWEET SIOUX, flown by Lt. William Rogers, and now the incessant FW-190 attacks knocked out three engines and wounded Sgt. Ralph Meigs. At 1348 the aircraft fell out of formation and rapidly lost altitude. Lt. Albert DiLorenzo was sitting at the navigator's table when a twenty-millimeter shell exploded in the nose knocking him unconscious and trapping the nose gunner, Lt. Raymond Cohee, in his turret. As the airplane lost altitude, Lt. Cohee banged on the plexiglass in an effort to awake Lt. DiLorenzo who was bloodied and unconscious under the navigator's table. He finally regained consciousness and freed Lt. Cohee from his nose turret just as the bailout alarm rang. Both hastily donned their parachutes and jumped out of the stricken plane. Once on the ground, angry civilians quickly captured and detained them. Thankfully, Lt. DiLorenzo was taken to a hospital in serious condition but recovered and spent the rest of the war as a POW. SWEET SIOUX crashed in a swamp near Goldenstedt and although everyone bailed out, Lt. Rogers was killed during the process. German authorities from Vetcha, Germany, rounded up the scattered survivors.

Fighters also heavily damaged 42-52435, piloted by Lt. John Cathey. They left the formation near Osnabruk, Germany, on the return trip. They bailed out at 1356 and the plane crashed four kilometers west of Elbergen, Germany. Sgt. Jack Arluck perished in the crash but the remainder of crew successfully bailed out. However, German guards shot and killed the bombardier, Flight Officer Carl Carlson, when he refused to pack his parachute. Everyone else was captured.

Two minutes later Lt. James Clark and crew aboard CHUBBY CHAMP fell victim to the fighters. They left the formation approximately fifteen minutes after bombing after flak and fighters pounded the rear of the plane incessantly. Five men, the copilot Lt. Lawrence Anderson, navigator Lt. Ralph Casey, bombardier Lt. Alfred Tellman, engineer Sgt. Manuel Caballero, and the radio operator, Sgt. Robert Harrison, bailed out of the plane. As the plane descended through 4,000 feet it suddenly entered a spin and crashed at Wildeshausen, Germany, thirty-three kilometers southwest of Bremen. The five men who bailed out were quickly taken prisoner and the other five men perished in the crash including Lt. Clark.

Lt. Orland Howard, flying BIG BAD WOLF, departed the formation over Berlin after they were damaged on the bomb run. Flak destroyed three engines, damaged the nose turret and right stabilizer, and wounded Sgt. Harold Nininger. They flew the damaged ship to the coast of Denmark where they bailed out over the Danish island of Bornholm in the Baltic sea. Sgt. Harry Ambrosini died after his parachute malfunctioned but the remainder of the crew survived. Sgt. Nininger and the copilot, F/O Thomas Verran were captured but the rest evaded to Sweden with the aid of the Danish Resistance. (Sgt. Ambrosini was buried in the Cemetery of Pedersker on Bornholm and is still there today.)

The remainder of the formation continued to fight their way home. Sgt. Stanley Zabrowski manned the tail turret on THE RUTH E.K. as wave after wave of fighters streaked through the formation. Cannon fire raked his plane hitting the nose, top turret, number three engine and wounded three of the crew. Two P-47s finally fought off the enemy fighters but the damage was already done. Another aircraft, 41-28784, managed to fly within thirty miles of the British coast before falling. The plane and crew from the 389th BG had a special connection to the 448th. The crew, led by Lt. Alfred Locke, was one of the original 448th crews. After flying several missions with the Group, they were transferred to the 389th for PFF training. After leading the 448th on this mission they ditched in the cold North Sea. The plane quickly sank. The impact of the ditching broke the back of the copilot, Lt. Errol Self and fractured the skull of tail gunner, Sgt. Dale Van Blair. Although seriously injured, they survived although five others did not. Royal Marine Air-Sea Rescue launch #498 picked up the seven survivors and the bodies of the less fortunate.

After the planes landed at Seething, aircrews reeled at the loss of seven more crews. Six planes from the 448th failed to return plus the lead aircraft from the 389th BG aircraft ditched with casualties. April ended with the Luftwaffe rekindling the fear it once instilled in the bomber crews. Since the first of April the 448th posted eighteen crews missing in action plus two PFF crews that carried 448th personnel. It was the largest monthly loss yet. The total crews lost since leaving the States reached forty-nine. The Group could not endure continued losses at this pace. Who could survive the attrition, the bombers or the Luftwaffe? The increased tempo of operations supported the rumors of the upcoming invasion but no one knew when. A decisive moment in the air war over Europe approached.

eight

# BUILDUP: May 1944

As the air war continued into May, control of the air remained undecided. Although Allied fighters swarmed over western Europe, the Luftwaffe still provided a potent adversary. Air superiority, a key tenet in the invasion plans, was required over the invasion beaches. Luftwaffe activity over England similar to the previous two months would wreak havoc with the invasion forces. For the 448th, the date and time remained unknown.

Deep penetrations into Germany continued but an increasing number of missions in May visited Luftwaffe airfields and marshalling yards in France. Also, the month of May witnessed a large increase in aircraft at Seething. The month started with forty-two aircraft assigned to the 448th. By the end of the month seventy-two B-24s were scattered around the airfield. All the hardstands were full. Adding to the congestion, a section of the north perimeter track filled with water making a 100-foot section impassable. As a result, runway 01/19, which needed repairs to patch holes, was closed and utilized strictly for parking and as a taxiway.

Ominously, the first mission of the month was recalled. Twenty-two aircraft returned to Seething without hitting their assigned "No-Ball" target in the Pas de Calais region of France. Recovering from the morning's setback, the Group sent twelve aircraft against the marshalling yards north of Brussels in the afternoon. They waded through the flak and dropped their bombs despite several other groups converging on the target at the same time. Excellent fighter support during the entire mission ensured the safe recovery of all the B-24s. The following day "Stars and Stripes" heralded the mission as the "seventeenth straight day of the pre-invasion aerial offensive." In an effort to overshadow the heavy losses in April, they reported 1,300 German aircraft were destroyed the previous month and 24,000 pounds of bombs fell on twenty-eight separate German aircraft factories stopping production at twenty-one of them. The paper also quoted Under-Secretary of War Robert Patterson who "revealed that U.S. production of 100-octane aviation gasoline has in the past three years increased from 40,000 to 400,000 gallons a day." These statistics meant nothing to the crews that faced the dreaded flak and deadly fighters mission after mission.

Again on 4 May, all aircraft answered the recall message and returned to base. Not until 6 May did the Group manage to fly a complete mission. The increasingly popular "No Ball" sites near Siracourt, France, were the target. Plans called for three nine-ship sections each led by a G-H aircraft to attack the target in trail. However, the third lead ship experienced engine problems after takeoff and returned to Seething. The crews improvised and assembled on the two remaining G-H Liberators and continued to the target. Bad weather continued to hamper operations as clouds completely obscured the target area. Bombs were released on smoke markers dropped by the G-H aircraft. Friendly fighters roamed the skies allowing the bombers to fly unmolested by the Luftwaffe. Even though British coastal gunners opened fire on the returning Liberators, all 448th aircraft returned without a scratch. Only five more missions remained before Sgt. Stanley Zabrowski could go home.

The following day, the Group flew on the first deep raid in a week. Lingering weather over Europe required crews to drop on smoke markers dropped from the lead PFF aircraft. The original target was an airfield at Munster but, due to the cloud deck, the center of the city took the majority of the impact. Sgt. George Glevanik watched the heavy flak from his top turret. These dirty black puffs brought more fear into the hearts of the crews than the Luftwaffe. "Which random flak burst was destined for your aircraft?" It seemed to be the question always on the mind of the crews. In this instance, the flak exploded below the formation thanks to chaff, code-named 'window.' Crews of selected bombers dropped containers filled with chaff or thin metallic strips of foil. Upon hitting the windstream, the containers disintegrated scattering the chaff. It provided thousands of radar returns effectively hiding the actual bombers in the clutter of false radar returns. Thanks to this new tactic, thirty-two aircraft from Seething completed the almost six-hour mission without Luftwaffe interference.

On 8 May, the 448th returned to the Reich homeland. Planners briefed two targets during the morning briefing, Brunswick and Munster. Shortly after takeoff DASIY MAE lost an engine. At 0720 Lt. Jack O'Brien landed DASIY MAE at Seething with only three engine operating. The brakes failed and the aircraft went off the end of the runway into the mud collapsing the nose gear. Fortunately, no one was injured.

Meanwhile, thirty-one aircraft climbed into the British sky and set course for Germany. Approaching the target area flares from the lead aircraft burned brightly as they fell to earth marking the initial point. The lead aircraft also sent the code word for PFF across the VHF radio. It echoed in the headsets of every radio operator as the B-24s started down the bomb run. Red flares spouted from the PFF ship over the target and twenty-eight planes from the 448th hit Brunswick with incendiaries. The accurate flak again prodded at the bombers filling the sky with black puffs. Sgt. Stanley Zabrowski, flying his third mission in three days watched helplessly as three bombers from a nearby Group burst into flames. It was gut-wrenching to watch as the burning planes fell apart without any parachutes visible. Enemy fighters appeared but the good fighter escorts kept the Luftwaffe away and all planes returned to Seething safely.

The 448th flew again on 9 May. Clearing weather allowed the bombers to continue applying pressure to the Luftwaffe. Twenty-eight aircraft left Seething leading the 20th CBW to the marshalling yards at Liege, Belgium. The route kept the formation over water for the majority of the flight. The Group crossed into enemy territory but the sky stayed ominously quiet. Even on the bomb run, the usually heavy flak strangely stood silent. Only occasional puffs littered the otherwise beautiful sky. Lt. Elbert Lozes, flying in the nose of BARFLY as bombardier, took control of the ship as they approached the initial point. Just as he opened the bomb bay doors, the peaceful sensation of flying dissolved into a sudden eruption of flak. A tremendous, calculated flak barrage bracketed the entire formation. Crews cringed inside their inadequate flak vests and helmets as the flak peppered the sides of the aircraft with a metallic clang.

Seething airfield seen from above. It was a welcome sight for crews returning from a combat mission. (LaPoint)

The control tower and T2 hangar are prominent in this aerial photo. The nearby hardstands are full of parked Liberators. (Everson)

At 0926, the deputy lead aircraft BUSTED FLUSH (also known as MARGARET L) flown by Capt. Robert Lambertson, shook as flak penetrated the wing between the number three and four engines. The punctured fuel tanks in the wing quickly caught fire. Recognizing the hopeless situation, Capt. Lambertson rang the bail out alarm and everyone parachuted from the stricken plane including Maj. Ronald Kramer, command pilot for the Group. As Lt. Ralph Brown floated to earth, he watched as the burning aircraft continued to fly for almost four minutes before it exploded in mid air. The wreckage rained down on the town of Riksingen near Tongeren, twenty-two kilometers northeast of Liege. Tragically, the parachute of the right waist gunner, Sgt. Andrew Long, collapsed during the descent. German troops captured the remainder of the crew alive.

For the men that serviced, maintained and prepared the aircraft there were not many breaks in their day. One of the highlights however was the arrival of a Red Cross mobile. (USAF)

Men enjoy coffee and doughnuts from a Red Cross mobile and take advantage of the rare break before returning to the business of war. (USAF)

Lt. Melvin Alspaugh, flying NO NOTHNG, skillfully avoided a similar fate. The horrendous flak severely damaged two engines. Making a left turn direct to England, he departed the formation and quickly feathered the two engines. The crew flew the crippled B-24 to Seething where they successfully completed a challenging two-engine landing.

The crew of BARFLY was not quite as lucky. Four flak rounds burst almost simultaneously in close proximity to the right vertical stabilizer, violently shaking the aircraft. The pilot, Capt. A.D. Skaggs, breathed a momentary sigh of relief as the aircraft continued flying. However, in the back of the plane, Sgt. Gene Gaskins quickly realized things were not right. From his right waist gunner position, he saw the tail gunner, Sgt. Bill Jackson, fall backwards out of his turret. Realizing his friend was hurt, Sgt. Gaskins moved him to the waist area to attend to him and notified the crew over the intercom that Sgt. Jackson was hurt. He fumbled in the cold but connected Sgt. Jackson's oxygen mask and turned it to pure oxygen. He started searching for wounds but the thick clothes concealed any signs of injury. Unable to find any wounds, he unsuccessfully searched for a pulse. The bombardier, Lt. Elbert Lozes, crawled to the back to help Sgt. Gaskins. He too could find no signs of a pulse. Once clear of the Belgian coast, Capt. Skaggs relinquished the squadron lead to the deputy aircraft and left the formation. With the throttles on the firewall, he headed straight for Seething.

Receiving permission from the pilot to leave his turret, Sgt. George Glevanik crawled through the cramped bomb bay to help his friend. Being the oldest crewmember, he had taken Sgt. Jackson, the youngest crewmember, under his wing. As a result, he had grown fond of his crewmate and felt a special responsibility for the "kid." Now he was hurt and he intended to take care of him. When he got to the waist, there was nothing to do. Despite the presence of a flak vest, a single piece of flak had punctured his heart. In a vain attempt, Capt. Skaggs ordered red flares fired as they flew over Seething alerting the ambulances. Doc Joseph Kaiser, the flight surgeon, climbed aboard the aircraft almost before they stopped rolling at the end of the runway. He, too, could do nothing.

Even though all twenty-eight aircraft dropped their bombs on the target, the Group paid a pretty heavy price. Operations clerks posted another crew missing in action; letters had to be written to next of kin explaining the death of their loved ones. Another crew narrowly averted disaster. Although the Luftwaffe failed to show again, the flak crews let their presence be known.

The following day, twenty-five crews headed for Paderborn, Germany only to be recalled. Sgt. George Glevanik struggled with the death of his friend the day prior. The rest of the crew also struggled with the intense emotion associated with the loss of a comrade. The last several weeks stressed the crew to the breaking point, starting with the close call on 22 April and ending with the death of a crewmember. The Group flight surgeon, Maj. Joseph Kaiser, noticed the strain on the crew and grounded them. They were given a week of rest and recuperation at a flak home. Although the officers and enlisted men went to separate facilities, they both enjoyed the luxury of civilian life with no wartime distractions for a week.

"When the bosses decided that a crew was sufficiently combat weary (flak happy), they allowed the crew a few days rest and recreation at one of the rest homes. We didn't think we were really that bad off, but finally agreed to take off a few days from combat to join the other 'flak happy' troops. We were really pleased to find our rest home was a renovated mansion on a large estate. There were two to four men assigned to comfortable bedrooms. There was a nice dining room and lots of recreation facilities.

The American Red Cross and their British counterparts managed and staffed the rest home. We were not allowed to wear uniforms. In fact, we were issued an old pair of blue jeans, a shirt, and a pair of tennis shoes and that is all we had to wear while we were there. Recreation facilities included tennis, golf, horseback riding, punting on the Thames, British croquet, and many kinds of games (checkers, dominoes, cards, etc.). We were particularly delighted with one special treat – breakfast in bed most any time we wanted it from 0600 to 0900. This was served by one of the waitresses, who would knock on the door at the time we requested the night before. In establishing this time, however, it did take us a while to get accustomed to the British use of the English language when the young lady at the desk would ask, "What time would you like to be knocked up in the morning?"

DAISY MAE sits in the mud off the end of the runway with a collapsed nose gear on 8 May 1944. After suffering an engine failure, the aircraft returned to Seething only to have its brakes fail after landing. The nose wheel was repaired and DAISY MAE flew many more combat sorties with the Group. (Sheehan)

For everyone else, the war continued. On 11 May, the Group flew its sixty-third mission. For Sgt. Ray Waters, it was his seventh mission in eight days (one was a practice mission). The 448th was split between the other two groups in the Wing. Nine aircraft flew low left in the 446th BG formation and the remaining aircraft flew in the 93rd BG formation. All twenty-four aircraft attacked the marshalling yard in the French town of Mulhouse with little resistance. The flak was very light but the weather was poor. Haze severely limited visibility and intermittent clouds forced the bombers to change the inbound course on the bomb run. Still, they dropped their bombs on the target. Leaving the target, Me-109s half-heartedly attacked but all 448th planes returned to Seething without damage.

For Lt. Leroy Engdahl, this was his thirtieth and final mission. When he set foot on terra firma on return history was made; he was the first 448th crewmember to complete a tour. Elated personnel treated him to a rousing welcome. A special dinner in his honor was held to commemorate the event. His accomplishment provided hope to other aircrews; someone had finally made it through the seemingly impossible thirty missions. He remained at Seething until 26 July flying weather ships and test flying aircraft. His presence undoubtedly offered hope to those still flying missions that they too could make it.

The next day, 12 May, the Group returned to Germany. The joy of seeing a comrade complete his tour quickly slipped into the sub-conscious as the Luftwaffe returned to the skies.

MARGARET L 42-100287. Also known as BUSTED FLUSH, it was lost on 9 May 1944 over Liege, Belgium to flak. The crew led by Capt. Robert Lambertson and the commander of the 712th BS, Maj. Ronald Kramer, bailed out. (Everson)

Nineteen Liberators left Seething for the oil refineries at Bohlen, Germany. Once again, the Group was split between the 93rd and the 446th. Only fifteen aircraft completed the entire trip, as four returned early for various reasons. Sgt. Julius Rebeles scanned the horizon from his precarious perch in the ball turret searching for the Luftwaffe. Blazing fires and boiling smoke from the target area assured the crews the bombs had hit the target. Immediately after leaving the target, Sgt. Rebeles watched in shock as B-24s started falling to a black wall of flak. Three planes succumbed to the thick flak. One simply disappeared in a puff of smoke and another lost a wing, gyrating to the ground like a falling leaf. Fortunately all 448th crews avoided the flak, only to face enemy fighters on the other side.

From the belly of his plane, Sgt. Rebeles observed FW-190s forming below the formation as they prepared for the attack. Escorting P-51s quickly dispersed them by diving into the formation scattering the enemy fighters. While the Mustangs dealt with the single-engine fighters, about fifteen twin-engine fighters, including Me-410s and Ju-88s, hit the formation. Sgt. Rebeles dispatched a few rounds at passing fighters but to no avail. Another B-24 fell but again the 448th was spared. Somehow, all fifteen B-24s returned to England despite a coordinated effort by the Luftwaffe fighters and flak batteries.

On 13 May, Sgt. Julius Rebeles again headed for the skies over Germany. Mechanical problems forced his crew to abort but not before they loosed their 500-pound bombs on the German airfield on the island of Heligoland. The remainder of the formation led the 20th CBW to Tutow, Germany, which they successfully bombed without loss. It was a long, tiresome mission for Sgt. Ray Waters.

"The CQ came around this morning about five o'clock and woke us up before briefing. We ate breakfast, which

Several large estates in the countryside were enlisted in the war effort as rest homes, commonly referred to as "Flak homes." These homes provided much needed breaks from the war for crews that were flak happy. (Gaskins)

was better than usual. We had fried eggs, sausage, and cereal. Briefing was at six-thirty. Everyone was in on time, waiting to see where we were going. The screen was lifted and there on the opposite side of the map from our home field was the target, Tutow, Germany. The roll was called, then we had a short prayer by the chaplain, afterwards the briefing continued. The red ribbon seemed to have no end. Starting at Norwich and leaving the English coast at Great Yarmouth it continued in a northeastern direction over the North Sea. We entered the German coast south of Denmark heading for Berlin. Then we turned east on the initial point, making the bomb run. The target had a 6/10 cloud cover, but we dropped our bombs, which was ten 500-pound G.P.s. Ship 013, SWEET SIOUX TWO, had plexiglass waist windows so I could not stick my head out to watch the bombs hit. Later we made a left turn and I could see smoke rising from a large hangar which we had hit. Other groups behind us bombed the same target, which was an airfield and factory that manufactured about 200

Several 448th men enjoy their time away from the war at a Flak home. They enjoyed clean sheets, hot meals, and a choice of games and activities with no mention of the war. (Gaskins)

A life preserver from Moulsford Manor Rest Home hangs on a tree at a nearby lake. (Gaskins)

Above: Friends help Lt. Leroy Engdahl out of his flying clothes after his return from his last mission on 11 May 1944. This significant event marked the first time an airman from the 448th completed his combat tour. (Engdahl)

Right: Lt. Leroy Engdahl poses in front of his aircraft, LITTLE JOE, wearing striped pajamas and a sign commemorating the event. The ribbon pictured is the Distinguished Flying Cross given to crews finishing their combat tour. Hence the sign reads "I finished my DFC mission." (Engdahl)

Focke-Wulfs a month. On the route to the target little flak was encountered, except for a barrage they threw up, about thirty minutes inland. I was wondering what they were shooting at, when all of a sudden about fifty-five or sixty FW-190s came zooming out of it. Was I scared? YES! This is it, I thought, but they did not attack us. Why remains unknown. What few P-47s we had for escort attacked them and there was some fierce dog fights. I saw one go down. No flak was encountered coming out. The route out led us over the Baltic Sea. We could see Sweden north of us. We crossed many Danish islands then Denmark. We traveled something over 1,400 miles and were gone eight hours and thirty minutes from takeoff to landing. Had a headache before we got back and was very tired." (Clouds obscured the target. Just as the PFF turned toward the secondary target the number three aircraft accidentally released and all but two aircraft in the formation dropped. The PFF aircraft and the other two returned with their bombs.)

Not until 19 May did the 448th venture out again, this time to Brunswick, Germany. Maj. Glassel Stringfellow flew as command pilot on the lead aircraft. The mechanical releases on their aircraft malfunctioned on the first pass and since clouds obscured the target, all the bombers withheld their bombs. The formation executed a left turn and prepared for another bomb run but other formations starting their runs prevented the 448th from bombing on the second pass. As the formation turned around for a third pass the lead aircraft of low section was damaged by flak causing the entire section to fall out of position. Finally, on the third pass, the lead section released their bombs on cue from the PFF ship and the low section dropped with the 14th CBW.

Heavy flak and fighters greeted the Group on each pass. Sgt. Stanley Zabrowksi, flying on DUAL SACK, tried desperately to line up one of the enemy fighters in his gun sight. Each time he got close, the shadow of a friendly B-24 entered the sight as well. Frustrated, he held his fire. Escorting red-nosed P-51s of the 4th FG successfully fended off two waves of attacking fighters. During the thirty-four minutes over the target and in range of the flak batteries, numerous aircraft suffered battle damage. SKEETER II, flown Lt. Robert Silver, succumbed to the fighter attacks leaving the target area. After the third wave of fighters hit the plane with cannon fire, the entire crew bailed out of the stricken plane at 1316 and the burning aircraft crashed near the German town of Helmstedt. All but three of the crew escaped but were later captured. Sgt. Charlie Blanton, Sgt. John Foss and Sgt. Robert Spruill were killed. Ironically, Sgt. Blanton replaced Sgt. Edward Owen, one of

Col. Gerry Mason, seated left of Lt. Engdahl, hosted a dinner in honor of the happy warrior following the mission of 11 May 1944. This successful completion of a tour gave hope to all airmen who flew combat missions; it was possible to finish and go home!"(Engdahl)

Part of the air defense at Seething included several anti-aircraft guns. The presence of German intruders over East Anglia required constant vigilance by the crews that manned these guns. Here two GIs man one of the guns at Seething. (USAF)

the crew's gunners, for this particular mission. On return, ground crews reported ten aircraft with battle damage. The Luftwaffe fought like a caged animal.

Following an all night and all day train ride, Sgt. Fred Kerniss arrived at Seething around 2000 as part of a replacement crew fresh from training in Ireland. They were quickly assigned to the 715th BS and trucks dropped them off at their new home, Nissen huts in the 715th BS area. The next morning, they underwent physicals and spent the afternoon visiting nearby Norwich. Meanwhile, a mission to Reims, France, was recalled due to heavy cloud cover.

On 22 May, B-24s of the 448th hit a "No Ball" target at Siracourt, France. Cloud cover forced the bombers to drop by PFF. There were no fighters and only four lonely puffs of flak, perfect for Sgt. Stanley Zabrowski's thirtieth and last mission. Lt. Cater Lee and Lt. Bruce Winter also joined the "Happy Warrior" fraternity on this mission. They completed the impossible and were headed home. What a feeling!

The 448th returned to France the following day. Operating in a more tactical mode, the bombers dropped general-purpose bombs on a JU-88 base at the Bricy-Orleans airfield. A total of 167 B-24s from the Second and Third Bomb Divisions hit the airfield. Elements of the 448th plastered the bomb and ammunition dump north of the field. Only light flak greeted the bombers as the Luftwaffe again was clearly absent from the skies.

Tensions ran high as invasion jitters increased. One of the precautions taken was posting of guards at certain aircraft in case German fighters strafed the field at night. Sgt. Fred Kerniss pulled this long, monotonous duty the night of 23 May. One person manned the turrets of the aircraft while another stood guard outside the plane. No one attacked.

On the 24th of May, the 448th continued their tactical operations hitting the airfield at Orly, France. A force of 151 B-24s attacked the Orly airfield while 168 B-24s simultaneously hit the nearby airfield at Melun. Again, crews only saw flak and no fighters. However, the 448th ran into problems over the target. The high squadron did not release their bombs since the low squadron was directly below them at the release point. Spacing problems between the Groups jammed the 448th placing the sections out of position. Still, the skies over France were quickly becoming Allied territory and the Luftwaffe airfields took a beating. Air superiority, a prerequisite for the invasion, was becoming a reality. The only question was when would the invasion take place and where.

The 448th flew to France the next day although in a slightly more strategic role. The marshalling yards at Mulhouse, France, near the German-Swiss border fell under the cross hairs of Norden bombsights. Sgt. Julius Rebeles survived the moder-

A gunner aims his gun at a P-38 Lightning that is visiting Seething. This gun site, located near the control tower, provided defense against German aircraft that continued to conduct nuisance raids against the base. (Bollschweiler)

ate flak and again breathed a sigh of relief as the Luftwaffe failed to respond. Enroute to the target, the 448th watched from a ringside seat as B-26 Marauders plastered targets in the Calais sector. While the main force of bombers flew toward Mulhouse, a smaller force of B-24s, including six from the 448th, took part in an experiment related to the pending invasion. Although the skies were clear, they attacked gun emplacements at Fecamp, France, using H2X equipment as part of the trial. Both missions returned safely without loss or damage.

On 27 May, the 8th Air Force finally returned to targets in Germany. The 448th attacked marshalling yards at Konz on the outskirts of Trier. As the command pilot, Maj. Glassel Stringfellow, described in his report after the mission, the Group experienced problems over the target. "After turning at CP #4 (control point #4) another Wing cut across the path of the 20th CBW at the same altitude and interfered seriously. This may have accounted for the Wing lead turning short of the IP and consequently passing ten miles south of the target. When the target was observed to the right of the course the lead of the 448th turned away and passed over the target. The first section meanwhile continued on and after being notified that the target had been missed, made a 360-degree turn and made a second run on the target. This maneuver, occurring as it did while the second section was on the bomb run, resulted in the 448th losing track of the lead section of the 20th CBW." After bombing, the Seething Liberators returned with the 14th CBW. Despite the snafu, all the bombers returned to Seething after the seven-hour flight with no damage.

The next day the 448th flew as part of a large force assigned to attack oil facilities at Merseburg, Germany. Lt. Russel Reindal, flying DUAL SACK, narrowly averted disaster on takeoff when they sluggishly lifted off the ground and the aircraft refused to climb. Unable to abort, they continued the takeoff crossing the end of the runway only a few feet off the ground. They clipped the top of a tree with the landing gear but somehow avoided damage to the aircraft. After what seemed like an eternity, the aircraft started to climb and they continued on the mission. Two aircraft turned back to England inside enemy territory. One, DAISY MAE, piloted by Lt. Jack O'Brien, hit a target of opportunity at Vetcha, Germany while the other returned with bombs intact.

When the bombers turned onto the bomb run, thick smoke and haze obscured the target forcing them to hold their bombs. Moderate flak rose to meet the B-24s as they crossed the target and when they turned south away from the area. The lead bombardier located another oil refinery twenty-five miles south of

Men carefully unload a wounded comrade from a 448th aircraft while crewmates watch. For men that were wounded over Germany, it was a long trip back to Seething. (USAF)

Col. Gerry Mason awards the Distinguished Flying Cross to Maj. Glassel Stringfellow, the commander of the 714th BS. On several occasions Maj. Stringfellow successfully led the Group as command pilot on missions over the continent. (USAF)

SKEETER II 42-52638. An unknown airman poses beside SKEETER II before it was lost on a mission on 19 May 1944. (LaPoint)

Lt. Robert Silver and crew pose in front of a B-24 probably during training in the States. The crew was lost on 19 May 1944 after fighters attacked their aircraft SKEETER II leaving the target. (Bottoms)

the primary target and the formation commenced another bomb run. No opposition was encountered as they hit the oil refinery at Zeitz with excellent results. Oily smoke, visible from as far away as Frankfurt, billowed from the heavily damaged area. Again the Liberators returned to Seething without loss.

They returned to Germany again on 29 May. Attacks against the air depot at Tutow, Germany, continued the unrelenting pressure on the Luftwaffe. Twenty-eight aircraft were split between two sections, one flying with the 93rd BG and the other with the 446th. The eight-hour flight succeeded in its mission, destroying the airfield. On 30 May, the Luftwaffe again fell under the cross hairs as the 448th targeted the airdrome at Rotenburg, Germany. Capt. A.D. Skaggs, in THE STURGEON, led the Group on its seventy-fifth mission. No flak or fighters were encountered and the twenty-four bombers landed at Seething unscathed.

Flying their fifth mission in five days, the 448th returned to a more tactical role leading the entire 2nd Bomb Division to the rail yards near Metz at Woippy, France. As the formation approached the French coast, towering cumulus blocked the route. Unable to climb above the clouds, the entire Division abandoned the mission. They performed a wide sweeping turn that carried them thirty miles into enemy airspace. Although small amounts of flak were encountered, no other hostile actions were experienced and everyone received credit for the mission. Lt. Frank Gibson and crew on BIM BAM BOLA dropped on a target of opportunity at St. Omer, France. Everyone returned to Seething unmolested.

Although the month started with a flurry of tactical missions indicating the proximity of D-Day, it ended with the 448th resuming strategic operations aimed at the oil industry and Luftwaffe. Although Luftwaffe fighter activity noticeably decreased during the month, they still posed a serious threat claiming one of the Group's two losses during the month. Flak knocked the other B-24 from the sky and the flak gunners seemed to improve with time. The gunners threw up a curtain of steel on almost every mission. Whether it was Berlin or a 'No Ball" site in France, flak always greeted the bombers.

An overhead view of the hangar area on the 715th BS area shows B-24s on their hardstands. The B-24 in the center, lower left is the assembly ship CHECKERBOARD. (Johnson)

Five hundred pound high explosive bombs fall from the belly of a 448th Liberator. Notice the smoke rising from the target area after a previous Group's bomb run. (LaPoint)

Twenty B-24s from the 448th required major repairs due to flak during the month of May.

May was the best month to date. Only two crews were missing in action and despite the best German efforts, the American war machine finally reached its stride. The number of B-24s at Seething swelled to seventy-two by the end of the month. The number of crews grew as well despite crews completing their tours or being lost in action. The fortunate ones were headed home. New crews, eager and well trained, stepped in to fill the void. They underwent two days of training following their arrival. To hone their skills, they had access to three link trainers, a British bomb trainer, two American bomb trainers, four turret trainers, and other aids. The huge influx of men and equipment strengthened the notion that the invasion was near. Still, everyone wondered when the big day would arrive.

Right: This aerial view shows one of the numerous living sites at Seething. In the foreground between the huts is a bomb shelter for the inhabitants. Another living site is visible in the background. (Everson)

nine

# INVASION: June 1944

For what would be the turning point of the war, June started off like many of the previous months. Cloud cover reduced aerial operations over Europe to only meager weather reconnaissance sorties. The 448th did not fly on 1 June but took to the skies on the following day. Targets in the Pas de Calais region of France received the attention of the morning mission. Thirty B-24s of the 448th hit heavy gun emplacements near Beauvior, France, on 2 June. No Group aircraft suffered damage on the mission. This was the first of twenty-five tactical missions flown by the Group during June.

On 3 June, the Group returned to the Pas de Calais region this time hitting targets along the coast at Stella-Plage. Crews left Seething at 0800 and landed by 1300. Due to the close proximity of the target to the coast, they spent less then ten minutes over enemy territory. Twin-engine P-38s provided close escort just in case the Luftwaffe decided to play. They did not. General Eisenhower's promise of air superiority for the invasion was becoming a reality. Although heavy flak struck the following group, the 448th was spared. Aboard RED SOX, Sgt. Ray Waters departed the hostile skies of Europe for his thirtieth and final time.

Again on 4 June, the 448th returned to the same region of France. This time twenty-one B-24s dropped their loads on coastal installations at Sangatte near Calais. Once again the target was near the coast and the bombers were gone for only four hours. The next day was a repeat as crews stumbled out of bed at 0230. Sgt. Fred Kerniss was awakened for his first mission. After a breakfast of two eggs he attended the briefing and prepared for his mission. Twenty-two crews took off at 0640 and hit the same area near Sangatte. This time three gun emplacements on the coast, five miles west of Calais, were the targets. Everyone was safely on the ground by 1100. Sgt.

Kerniss enjoyed sandwiches, coffee and cake served by the Red Cross ladies and savored the completion of his first mission. After four consecutive missions to the same area of France, crews felt certain the invasion would occur at Pas de Calais – so did the German generals.

That evening orders quickly circulated throughout the base restricting everyone to base. Those personnel on passes were quickly ordered to return. Shows were cancelled and everyone was told to stay near their barracks. Sgt. George DuPont and other men received instructions to line up aircraft on both taxiways leading to the main runway. Orders called for aircraft to be filled with 2,000 gallons of fuel indicating a short mission (long missions typically called for a full tank of 2,800 gallons). Aircraft were required to be ready by 2200 for an unspecified takeoff time.

Standing outside the engineering office, Sgt. George Dupont asked Capt. Clifford Gaither, the engineering officer, if this was the day. The short reply was "YES":

> "We saw an endless parade of Lancasters towing large Horsa gliders heading south along the coast. This was a momentous day we had all prayed for and now it was here. I don't recall any bravado or cheering. Mostly I guess we were each praying and hoping against hope for success because it seemed all civilization hung in the balance. We all became busy and did not have much time to dwell on the matter."

CQs woke the first crews up at 2330. "Briefing at 0030!" Although most people suspected what was about to transpire, Col. Mason confirmed the rumors when he stated the 448th planned to hit targets in France in support of D-Day the next morning. Crews listened to the details of the invasion. H-hour was briefed as 0630. Crews received instructions not to drop bombs after 0628. Their target, gun emplacements at Pointe et Raz near Vierville, overlooked Omaha Beach where the U.S. 1st and 29th Infantry Divisions were scheduled to land. Bombs dropped after H-hour would wreak havoc with troops storming the unfriendly beaches. Thirty-six B-24s left the English coast flying carefully prescribed routes to avoid unfortunate friendly fire losses. Sgt. Fred Kerniss flew as a ball turret gunner on this mission:

"The invasion barges were to be about seven hundred yards away after we bombed the installations on the coast. I never saw so many barges in the channel in all my life. There were thousands and thousands of planes in the air. It was a beautiful site. When we reached our IP I turned my turret forward and watched all the bombs fall. Wow, what a sight. Invasion coast was Point Saint Pierce off Caen. At one time I looked down and saw all the German guns firing. You can see the flashes. I thought they were firing at us but they were firing on the barges. Flak nearly got us, but – ! We hear Germany every night and they said before when we start D-DAY they will be waiting for us- HA-I guess they still don't know that we invaded them because they weren't waiting when we went over."

General Eisenhower kept his promise of air superiority over the beaches. Only friendly aircraft flew over the beaches of Normandy on 6 June. There was only one exception when two German FW-190s strafed Sword Beach. Bad weather hampered the efforts of the first mission. Six of the bombers failed to find the formation and returned to Seething. Six others did not release their bombs when clouds obscured the target. Twenty-four however dropped their bombs on the beach installations just before the first wave of troops hit the shore.

As the first mission headed for France, ground crews at Seething continued to prepare for the second mission. The second target was a road junction at Coutances, France. Capt. A.D. Skaggs in THE STURGEON, with Col. Mason flying as command pilot, led twelve aircraft from the 448th on the second mission. Twelve aircraft from the 93rd and twelve more from the 446th flew as the high right and low left squadrons respectively. Clear skies over the channel revealed thousands of ships in the water. The radio operator on THE STURGEON, Sgt. Stanley Filipowicz, witnessed this great event. "This is some-

An excellent view of a North American P-51 Mustang as it performs a wing over away from a bomber formation. The combination of this great aircraft and the concentrated bombing tactics against German aircraft facilities completely overwhelmed the Luftwaffe. (LaPoint)

thing I would never have liked to miss. We had no opposition – let's hope the boys going in on the boats have as much luck. We get the glory and they do all the dirty work-all the credit to them. Hope the little we do will be of some help to those boys on the beaches." Once over land clouds thickened and obscured the ground. Poor weather completely obscured the target and all crews returned to Seething without dropping their ordnance.

The Group's third mission of the day was again in support of the invading troops. This time the target was Caen, France, in the British sector. Poor weather in the assembly area forced the bombers to assemble enroute to the target. The eleven planes from the 448th flew in trail of the 93rd and 446th. A complete

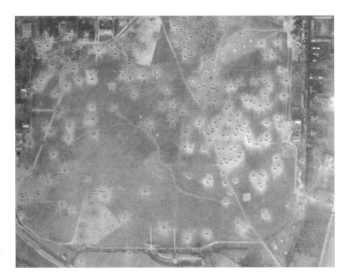

A 448th strike photo shows a bomb-cratered airfield following a mission. The extensive attacks on airfields and aircraft during the spring completely neutralized the Luftwaffe for the invasion of France. (Ray)

*Invasion: June 1944*

B-24s fill the perimeter track in front of the tower in preparation for a mission. The large SE letters on the ground in front of the tower identify the field as Seething. (Everson)

undercast covered the target and when the G-H equipment failed in the lead aircraft, the bombers were forced to hold their bombs. They tried three times but on all three passes the weather hid the target and they all returned to Seething with their cargo.

Sgt. Brona Bottoms flew on the fourth mission of the day aboard SQUAT N' DROPPIT, piloted by Lt. Thomas Keene. This time a railway bridge at Coutances, France, was the target. Crews were briefed at 1430 and eleven B-24s lifted off at 1700. Plans called for the 448th to fly the low left position off the 446th BG while the 93rd BG flew the high right position. After the 446th aborted their mission, the 448th assumed the lead of the formation. As the formation completed assembly, the G-H aircraft malfunctioned and the lead was transferred once more. This time the 93rd BG took the lead and the 448th fell in trail of the 93rd. Weather over the invasion beaches cleared enough to allow Sgt. Bottoms to see the activity on the ground. "…we were able to get a real good look at the greatest military operations of all times. The Channel was solid with naval craft of all sizes. The battleship Texas was seen belching flames from her sixteen-inch guns. The beaches looked like a giant bed of ants. We went in right across the beachheads in an air traffic pattern that was all one way in, turn to the right and return to England over the Jersey and Guernsey islands. The only resistance we got was from ack-ack guns on those islands. We were told at the briefing that the paratroops had taken them, but such was not the case. We received quite a bit of flak. We plastered the railway bridge we were after. We saw not one German fighter. The skies were alive with Allied planes of all sorts. Our return trip and landing was well after dark, but unlike the Hamm raid of 22 April – no enemy action." The bombers dropped visually on cue from the 93rd BG G-H aircraft. All the bombers returned safely. A fifth mission was briefed but not flown concluding the busiest day in the history of the Group.

The day following the invasion, Stars and Stripes proclaimed "Allies Driving Into France." It continued. "Between midnight and 8 am yesterday alone, 10,000 tons of steel went cascading down on German targets on the coast of Normandy. In the same period more than 31,000 Allied airmen, not including airborne troops, dominated the sky over France."

The commanding general of the 2nd Bombardment Division, General William Kepner, commended the entire Division on their efforts on D-Day. "I congratulate all commanders of the highly superior record and performance of the units of this division on D-Day. You went far beyond even the heavy commitments of you. The following facts are transmitted for your information: (a great number of) aircraft of the division were dispatched and of these (nearly all) were over the target. Of this great force, the majority of which took off at night, only one has thus far been reported lost and that by a crash landing. General Doolittle observed the bombing of the first and largest wave of this division. He has informed me that he was greatly impressed with the smoothness and orderliness of the flow over the target. (The precision with which all ships bombed was exemplary.) Information indicates that the landing beaches were terrifically beaten up with bombs. General Doolittle further informs me that ground forces were most complimentary on the support they received from the Air Force and were particularly appreciative of the absence of tragic bombing errors which

The control tower at Seething was the focal point for the base, especially during missions. Men crowded around to await the bombers return and count the planes as they appeared on the horizon. (Everson)

A British Hawker Tempest taxis on the perimeter track at Seething. This outstanding ground attack aircraft was used extensively supporting the Normandy invasion. On occasions these unexpected guests would land at Seething for fuel or repairs. (LaPoint)

it was feared might occur and which so often mar operations of this nature. This day's work reflects the greatest credit on your units. Keep it up."

The missions continued. The 448th launched aircraft to return to France on 7 June. B-24s took off at 1030 destined for Alencon, France. Clouds from 500 feet to 14,000 feet hampered the assembly. Clouds continued over the target area. The formation was recalled but some of the bombers hit a target of opportunity at Conches, France in the rear of the German lines.

Again on the 8th, 448th crews were briefed to bomb targets in France. At the 0230 briefing, crews learned the target was an airfield at Flers, France. As the aircraft lifted off, the mission was scrubbed due to extremely bad weather. Those aircraft airborne were recalled and the remaining bombers taxied back to their hardstands. After the recall, SKY QUEEN, piloted by Maj. Glassel Stringfellow experienced a near tragic incident. A fragmentation bomb exploded in the bomb bay injuring Sgt. Michael Fuller. Fortunately, the crew landed the aircraft at Woodbridge at 0735 without further damage.

Lt. John White and crew milled around the briefing room until 0800. Crews were then briefed to assemble in southern England and bomb the same target. However, before the crews left the room, the mission was scrubbed. Everyone packed up their flying gear and returned to their huts for much anticipated sleep. CQs rudely cut the naps short telling the crews a third briefing was planned for 1400. The hastily compiled briefing once again detailed a mission to Flers, France. At the last minute the target was changed to a bridge at Orleans. Lt. John White never got the word but it did not matter. One hour into the mission, the recall message sounded across the airways. Thick clouds and heavy overcast over the continent necessitated the recall.

Bad weather on 9 June prevented flight operations for all 8th Air Force units but the following day the 448th returned to France in support of the ground forces. CQs roused the participating crews at midnight, then the sleepy crews stumbled to the briefing at 0145. Once in their seats, they learned the target was a Me-410 airfield at Evreux-Fauville, France. The formation assembled and proceeded to the target without incident. Light but accurate flak streaked up from the airfield as the formation approached. At 0914, just after bombs away, a tremendous explosion rocked the formation. Black smoke billowed from below. Although the concussion damaged some aircraft, Lt. Raymond Towles' aircraft, 44-40107, suffered the most. Flak ignited a fire in the rear bomb bay. They continued flying for a short time before the plane broke in half. Five of the crew bailed out before the flailing aircraft disintegrated. Five men were killed. The ball turret gunner, Sgt. Earl Taylor initially survived with severe burns on his face and a broken thigh. He never fully regained consciousness and died of pneumonia on 23 June. It was the Group's first loss in a month.

On 11 June, the 448th returned to Seething after a heavy overcast and menacing cloud cover prevented bombing the target. However on the 12th, the weather permitted missions against rail and road bridges in France. The 448th's target was a bridge at Redon, France near St. Nazaire. The day started ominously for Lt. William Bailey and crew as they aborted their first aircraft for a mechanical problem and returned for a spare. They quickly prepared SQUAT-N-DROPPIT for takeoff but failed to catch up with the rest of the Group. They tacked onto the 446th formation hoping to complete their ninth mission.

The tactical nature of bombing bridges required expert navigation and bomb aiming skills. The 448th formation arrived over the target but required three passes through intense flak before releasing their bombs. No one suffered any serious damage. However, leaving the target faulty navigation led the Group over a nearby city. Thick flak pounded the formation riddling almost all the planes with holes but none seriously. Sgt. Fred Kerniss, a gunner from Brooklyn, New York, watched in amazement as no B-24s fell to the murderous fire. Unlike the 446th, he saw no enemy fighters.

Similar to the 448th, the 446th required numerous passes over the target, a rail bridge at nearby Pocaro. The numerous passes and accurate flak increased the difficulty of formation flying and highlighted the bombers for the gunners on the ground. Just after bombs away on the fourth pass the formation passed over the town of Rennes. Thunderous flak erupted hitting the tag-along, SQUAT-N-DROPPIT. The crew struggled with the wounded craft but slowly fell out of formation. The Luftwaffe, seeing the straggler, quickly arrived on the scene. Two Me-109s ravaged the damaged plane. The pilot Lt. William Bailey continued to fly as the copilot, Lt. George Cooksey, and Sgt. Kenneth Zierdt fought a fire in the bomb bay. They assessed the confusing situation: oxygen gone, intercom gone, number three engine destroyed, right vertical stabilizer missing, and fire in the bomb bay. At 3,500 feet, the nose of the aircraft pitched down and Lt. Bailey sounded the bail out alarm. Everyone bailed out successfully but the radio operator, Sgt. Kenneth Zierdt, died after his parachute collapsed during the descent. The bombardier, Lt. Benjamin Isgrig, narrowly escaped the guns of a strafing Me-109 thanks in part to escorting P-51s from the 352nd Fighter Group that drove off the attacking fighters. SQUAT-N-DROPPIT crashed in a fiery explosion at 1050, two kilometers north of Bonnemain, France.

As the survivors floated to earth, their attention turned to landing. Lt. Ben Isgrig landed in an apple tree. A passing Frenchman, sympathetic to the Allies, gave him his clothes and hid him. It took over two months for him to finally make it back to England. Lt. Cooksey checked his billowing parachute for damage, then looked at the quiet and peaceful countryside. He landed in a small wheat field with a thud. Scrambling to his feet, he was surprised by a Frenchman standing on a road watching his every move. Fortunately, the Frenchman hid him and started the treacherous, yet successful journey to American lines. Six of the survivors evaded and returned to England. The other two, Sgt. Leslie Fischer and Sgt. Vladimir Kovalchick, were captured.

Back at Seething, crews prepared for an afternoon mission to Conches, France. Twelve aircraft, led by Capt. Lester Miller, flew the high right section as part of a 446th BG formation. Unlike the morning mission, this almost six-hour mission concluded safely as everyone returned to Seething.

Carl Ahrendt, a mechanic with the 448th, sits on his bike near the technical supply area. (Ahrendt)

Despite the fast pace of operations, the men retained their sense of humor. The Group diary recalls one such incident. "With fresh milk and beef so scarce, an altruistic group of combat personnel, glowingly returning to base from a nearby pub one beautiful June evening, decided to safeguard the future of an all-white cow which passed before them on the road. After a hurried consultation, they agreed to help the cow blend itself with the surrounding countryside as much as possible and thereby be less of a target for enemy aircraft or paratroops. They proceeded to paint the helpless cow a pea-colored green that made the men proud of their camouflage work. The next day men rubbed their eyes when they saw the green cow and thought different section numbers when they took a second look and still could see a green cow."

Even though the green cow did not improve American-British ties, two marriages during the middle of June added a new facet to the relationship. Capt. Edward Israel the base technical inspector married a young lady from Norwich and one week later Sgt. Jesse Kain, having just completed his combat tour, married a local woman. They formalized the otherwise social, good-natured relationship between America and Britain.

Bad weather crept into East Anglia during the morning hours of 13 June as crews settled in the familiar briefing room at 0300. Details of the mission were discussed but at the last minute the original target, Leipzig, Germany, was changed to an airfield on the outskirts of Paris. Eight aircraft lifted off before the mission was recalled due to the unrelenting weather. Lt. John White was airborne only twenty minutes before returning to Seething. The original gas load for Leipzig coupled with a full bomb load made the landing challenging. Not often did crews land the B-24 weighing 64,000 pounds.

Crews received an early morning wakeup on the 14th. The 0130 briefing was the first for Lt. Carl Eggert and crew. The first bombers left Seething in the pre-dawn hours carrying fifty-two 100-pound bombs to the target, an airfield, test laboratories and munitions dump at Bricy, on the outskirts of Orleans, France. The night takeoff and assembly was harrowing as B-24s from around England muddled around the dark skies searching for their respective formations. Finally, the Liberators approached the target at 17,000 feet and waded through the ever-present, deadly flak. Thankfully, none found its mark but the bombers did. Smoke rose to a height of 15,000 feet testifying to the damage caused by the attack. Lt. John White and crew on 42-95186 returned safely to Seething, as did the other B-24s.

The following day, 15 June, was a mirror image of the preceding day, another early morning departure and harrowing, night assembly to attack a tactical target in France. Two railroad bridges over the Loire River at Tours garnered the attention of the 448th. Clouds covered the target so the Group hit their secondary target, bridges at Cinq-Mars-la-Piles and le Port Boulet just down river from Tours, with good results. Flak again unsuccessfully jabbed at the attackers. The almost seven-hour flight was void of German fighters.

A scheduled afternoon mission was scrubbed as was the next morning's mission. On the afternoon of 16 June, the 448th returned to the Calais region of France to hit a "No Ball" site located at Renescure, five miles east of St. Omer, France. A complete cloud deck obscured everything so bombardiers utilized PFF for bombing.

The 448th again did their part in keeping the Luftwaffe at bay. On 17 June, they hit the airfield at LeMans, France. Eleven Liberators took to the sky but mechanical problems forced three of the bombers to return before the formation left the English coast. Two others failed to find the formation during assembly and also returned early. Six B-24s completed the mission and landed at Seething six hours later without casualties.

A Luftwaffe Control Center at Fassberg, Germany offered a less friendly target for the Group the following day, 18 June. A one-hour delay before takeoff provided the bombers the luxury of a daylight assembly. The briefed route over the Danish peninsula skirted the known flak areas but unfavorable weather over the target forced the bombers to switch targets. Turning away from Fassberg, the PFF aircraft led the formation directly over the heavily defended town of Hamburg. Flak thick enough to walk on surrounded the bombers as they passed over the city. At 0954, just six minutes from the coast, flak

SKY QUEEN 42-110026. This aircraft experienced a near tragic freak accident. Following a mission recall on 8 June 1944, a bomb exploded on the aircraft wounding one man. Miraculously, the crew landed the damaged aircraft without further damage. (LaPoint)

crippled the aircraft flown by Lt. Leland Beckman. The first burst ruptured the number four oil tank resulting in the engine becoming uncontrollable. Shortly afterwards, flak damaged flight control cables in the bomb bay followed by another hit that disabled servo controls to the flight controls. In a matter of seconds, tail number 42-52119 succumbed to the dreaded flak. All ten crewmembers, on their fifth mission, bailed out.

Sgt. George Copeland exited the plane through the camera hatch located near the tail. After hitting his back on the plane as he jumped, he pulled the ripcord almost immediately fearing he might pass out from pain. The right waist gunner, Sgt. Dewy Conn, suffered a flak wound in the leg before jumping and Sgt. Michael J. Eannone, tail gunner, jumped from the stricken plane with his silk parachute in his hand. Somehow it deployed in the airplane prior to jumping. Despite these near

SQUAT N' DROPPIT 41-28710. This aircraft was lost on a mission to France on 12 June 1944. Flak damaged the aircraft then fighters finished the job forcing Lt. William Bailey and crew to bail out over France. (Everson)

disasters everyone survived the exit. As Sgt. Copeland floated earthward, he heard gunshots. Assuming they were aimed at him, he plotted his escape. Spotting a forest near his landing area, he planned to hide in the woods. However, after narrowly missing power lines, he landed only to find "a pistol in my face in very nervous hands." He spent the remainder of the war as a POW. Less fortunate was the right waist gunner, Sgt. Dan Wais. Angry civilians killed him after he landed.

While the downed crew tried to evade in Germany, the remainder of the Group bombed an airfield near Hamburg as a target of opportunity. After a seven-hour flight planes landed at Seething where awaiting medics in the meat wagons tended to the wounded aboard 42-51079.

More fortunate crews avoided the gruesome flak of Germany. While the larger formation invaded German airspace, a smaller force of ten Liberators took off from Seething at 1030. They led a formation augmented by planes from the 446th BG and attacked a "No Ball" target at Watten, France. They dropped their large 2,000 bombs on cue from the lead G-H aircraft. Two bombers from Seething failed to drop after the squadron lead did not drop. Everyone returned to Seething without a scratch three hours later.

On 19 June, crews grumbled as CQs woke them at 0015. The briefing, originally scheduled for 0115, was rescheduled for 0515. Poor weather forced a delay in the mission and it was briefed again at 1030 and then again after lunch. Mercifully, at 1245 the planners scrubbed the mission due to towering clouds. Weather reports indicated clouds reached heights of 30,000 feet. Thankful crews returned to Nissen huts around the airfield for much needed sleep. Their sleep was short-lived. Twelve bombers led by a PFF aircraft persevered and took off at 1535 bound for a rocket plant at Haute Cote, France. Heavy accurate flak hounded the bombers flying low left on the 93rd BG from the coast to the target and back but failed to cause damage. Although a PFF ship led the formation, clouds scattered over the target allowing for visual bombing. Everyone returned safely.

They received very little reprieve as field orders arrived on base initiating preparations for another mission on 20 June. Ground crews busily prepared thirty-three Liberators for the upcoming mission. Crews were awakened at 0100 and the first aircraft lifted off at 0430 destined for the Hydrierwerke Politz AG, a synthetic oil plant at Politz, Poland. Col. Mason flew in the lead aircraft, 41-28779 a 389th BG H2X-equipped plane; Maj. Chester Hackett, 715th BS commander, flew deputy lead in another H2X equipped aircraft also from the 389th BG. The force left the British coast at 0715 for one of the longest raids of the war.

Unidentified airmen from the 448th pore over target photos and maps in preparation for a mission on 13 June 1944. Intelligence crews maintained a vast array of photos, maps, and information for the crews to use in preparation for a mission. (USAF)

For Maj. Hackett, "the mission was uneventful enroute to target with the usual flak and German fighters. We had escort fighters for the entire trip to Stettin and over the target. The weather was clearing as we proceeded across Germany and by the time we changed course just north of Stettin the sky was absolutely clear. We could look over to the south and watch each twelve-ship formation make their bomb runs on Politz. We were flying at 20,000 ft and the anti-aircraft fire over the target was heavy. Before turning south to the IP I called Col. Gerry Mason and asked him to consider changing our altitude to avoid some of the flak. This was my 24th mission and Col. Mason recently joined our Group to replace our previous Group Commander. He called me back having decided that we would not change altitude. So we made our run at the same altitude as the formations ahead of us." It proved costly.

CAVU skies provided no concealment and well-aimed flak ravaged the 448th as they crossed the target at 1000 in the morning. Lt. Leroy Conner flying as lead of the low left element witnessed the brutal display of firepower. Aircraft throughout the formation shuddered as the murderous black puffs sent shrapnel flying through the thin metal skin of the planes.

Capt. Thomas J. Keene, flying aircraft 42-110044, felt a burst just under the bomb bay. They kept the aircraft flying straight and level until bomb release. Once the bomb load exited the aircraft, the plane nosed-over and entered a steep dive. Unknown to the crew, flak damaged the pilot's oxygen system. As a result, Capt. Keene passed out from lack of oxygen and fell forward on the controls. The top turret gunner pulled Capt. Keene off the controls as the copilot recovered the aircraft at 18,000 feet. The radio operator attempted to call for fighter escort but the radio was not working. Also, the flare gun was damaged and could not be used to summon help. Fortunately, Maj. Lester Miller, flying another aircraft noticed the stricken plane and radioed for fighter escort. Two P-51 Mustangs, one named "Galveston Gal", appeared off the wing allowing Capt. Keene and crew to breathe a momentary sigh of relief. The battle for survival, however, had just begun.

Now over the Baltic Sea with Sweden in sight, the crew debated their options. After surveying their damage more closely, they determined the engines were unharmed and a fuel leak in the bomb bay was repaired. They all elected to try for England. Keeping their fighter escort, they dumped all loose items overboard and headed for home.

Just after bomb release, the aircraft flown by Maj. Hackett seemed to stop in mid air. Multiple flak hits in the nose, bomb bay, fuel tanks, and waist reduced the sleek B-24 to a barely flyable wreck. With reduced power on all four engines, Maj. Hackett ordered everyone to bail out. Ten of the thirteen-man PFF crew jumped. However, one of the waist gunners called on the interphone and explained one man was badly hurt. He instructed the gunner to attach a static line to the parachute and help the wounded man out the waist window.

Unbuckling, Maj. Hackett headed toward the bomb bay when he saw the bombardier, Capt. Garland East of the 389th BG, walking from the rear of the airplane without a parachute. An exploding oxygen bottle had shredded his parachute. He searched for another parachute in the rear of the airplane but found nothing. With no other options, Maj. Hackett climbed back in the pilot seat and Capt. East sat in the copilot's. They proceeded to shut down all power to extinguish the fire and set up a shallow glide. Two Me-109s attacked the gliding B-24 but there was nothing left to damage. The fighters disappeared as suddenly as they arrived; one unexplainably crashed.

At about 2,000 feet, Maj. Hackett located a grain field and prepared for landing. After touching down, the plane hit a ditch and sheared off the nose gear. The tail of the airplane lifted skyward as the nose buried in the ground. After teetering momentarily, the aircraft fell back to the ground without flipping over on its back. Maj. Hackett and Capt. East quickly destroyed the bombsight and H2X equipment. As they climbed out of the wreckage two P-51 Mustangs flew low overhead in a salute to the downed crew. It was now obvious why one of the attacking Me-109s mysteriously crashed while they were preparing to crash land. Nevertheless, the German populace swarmed over the area and quickly captured the two aviators. (The damaged Liberator, 41-28779, was repaired and flown by the Luftwaffe as part of KG 200, a clandestine unit flying captured Allied aircraft. It was destroyed in a landing accident on 13 April 1945.)

British-American relations were good, but some men and women carried it to a more formal level by getting married. (USAF)

Crew 55 pose by the waist of a B-24. Back row (left to right) Harold Smith, James Bettcher, Thomas Keene, Edwin Moran. Front row (left to right) B.D. Bottoms, Grover Bingham, Charlie Blanton, George Sansburn, Frederick Krepser, William Demetropolous (Bottoms)

Three other 448th Liberators suffered overwhelming damage. Aircraft 42-51079 flown by Lt. Aldrich A. Drahos and crew of the 712th BS, suffered flak damage forcing them to land in Sweden. SWEET SIOUX II, piloted by Lt. Robert Nimmo, also sought haven in Sweden after suffering flak damage over the target. They landed at Bulltofta, Sweden as did a third crew, led by Lt. Raymond Wermeyer, flying FLAK JACK.

Meanwhile, Capt. Thomas Keene and crew struggled with their damaged aircraft. They started a gradual letdown as they neared England with their trusted "Little Friends" providing a vigilant guard. Hand cranks were used to lower the landing gear and flaps. The crew utilized an unorthodox method of stopping the aircraft. Once the main gear touched down, everyone quickly moved to the back of the aircraft forcing the tailskid to drag the ground. The trusted airplane brought everyone home despite over 200 flak holes. Flak shattered all the plexiglass in both the tail and nose turrets. The eight-hour flight cost the Group dearly. Despite the success story of Capt. Keene and crew, three crews and a lead crew were posted as MIA.

While the main force attacked deep into Germany, a smaller force hit the familiar "No Ball" site near Siracourt, France. Ten B-24s from Seething led the 20th CBW to the target. Unlike their brethren on the morning mission, the flak did not cause damage. The fortunes of war were not fair. Although everyone returned unscathed, poor weather made their return interesting. Low ceilings as low as 400 hundred feet challenged pilots and navigators. They were up to the task.

As Allied ground forces struggled in the hedgerows of Normandy, the Allied air forces resumed strategic attacks

HEAVEN CAN WAIT 44-40875. An unknown airman poses with nose art and his pistol. This B-24 served with the 448th until it was salvaged on 17 July 1944. (LaPoint)

against the Third Reich. On 21 June, the 448th participated in a mission to the Reich capital. B-17s of the Third Bomb Division also participated but instead of returning to England they continued eastward landing in Soviet-held territory. For the 448th, it was a long trip to Marienfelde, Germany, a suburb of Berlin, and an equally long trip back. Three 448th crews did not make it back.

The mission started ominously as the crews took off in the rain at six in the morning. Lt. Leroy Conner experienced a wildly oscillating propeller and an inoperative supercharger forcing them to abort. Another aircraft aborted before reaching the target but otherwise the trip was uneventful. Over the target at 1015, the menacing flak stabbed at the planes with inten-

As the war progressed, the ball turret was removed from the B-24s to reduce the weight of the aircraft. As a result, the bombers could carry larger payloads, fly higher and faster. In this photo personnel use a crane to carry one of the turrets. Notice the bombers are parked on the unused runway 01/19. (LaPoint)

sity and accuracy. Enemy fighters patrolled in the distance but did not attack.

Lt. Ralph Welsh, flying BIM BAM BOLA, on his fifteenth mission watched the murderous flak. "They opened up on us with big guns - bigger bursts that we'd ever seen before, probably 105s and 155s, and maybe 240s. We could really hear those bursts explode. It was practically deafening and we almost had to fly instruments through the smoke. Then we got hit in the number one engine, and the oil leaked and down went the pressure. Before I feathered it we got a close burst on my side and it blew out both of the four-inch thick glass sections, and through the steel window division, and hit me in the back. Glass was scattered all over the cockpit, and my copilot - who was flying off a left wing - told me later he thought I was dead. I could feel my left shoulder hurting but couldn't see, so I asked the radio operator if my clothes were ripped. He said no, so I felt better." Lt. Welsh returned safely, thanks to bulletproof glass. Although the air depots were adding the thick glass to planes when they were brought in for repairs or modifications, not all were retrofitted. Fortunately, BIM BAM BOLA had received the new windows; others were not as lucky.

Flak damage forced DUAL SACK, piloted by Lt. Roland Fox, to seek refuge in Sweden. On landing at Bulltofta, they ran off the end of the runway and hit an embankment crushing the nose of the aircraft. They joined twelve other bombers from the Eighth Air Force that were forced to land in Sweden on this mission. They made the seventh crew from the 448th to land in Sweden, four in the last two days.

Lt. Cleve Howell and crew aboard 42-95186 did not reach the safety of a neutral country and went down over the target. HAPPY HANGOVER carrying Lt. Jack Mercer and crew succumbed to the flak at 1017 after the flight controls were destroyed. The plane crashed in the outskirts of Berlin at Schulzendorf. Four men in the crew were killed in the encounter and a fifth, the copilot Lt. John Masters, died on 3 July in an Air Force hospital in Berlin due to wounds received in the attack. The remaining five men survived as POWs. Over eight hours after takeoff, B-24s entered the pattern at Seething and prepared for landing. Red flares arced skyward from MY BABY indicating wounded on board. Ambulances rushed to reach them as they taxied clear of the runway. Despite the decreased presence of the Luftwaffe, flak crews continued to make their presence known, especially over important targets such as Berlin.

While crews battled for their lives over Berlin, a smaller force of seventy 2nd Bomb Division B-24s, including five from the 448th, revisited the "No Ball" target at Siracourt, France. Unlike the mission to Berlin, this force encountered no opposition. What a way to complete the Group's one-hundredth mission in the European Theater. There was no time to stop and celebrate; field orders arrived setting the great sequence of events in motion for the next day's mission.

Crews left Seething in the afternoon of 22 June leading the 20th CBW to an airfield at Guyancourt, France in the outskirts of Paris. The bomb run was cut too short and the crews were unable to release their bombs on the first pass so the formation executed a 360-degree turn and re-attacked the target. Thick flak greeted the crews but failed to inflict any damage. As they cleared the target area, Lt. John White, flying DOWN N' GO, breathed easier. Usually a second bomb run proved disastrous as the gunners on the ground perfected their aim.

A 448th strike photo reveals heavy damage to the docks in Hamburg, Germany following the mission of 18 June 1944. Notice the bomb craters around the waterways. (Ray)

The briefing room awaits the mission return on 19 June 1944. Men from the intelligence sections debriefed all the crews about the circumstances of their mission. (Everson)

This time the bombers escaped unharmed, they thought. Five minutes after bombs away, an unknown flak battery targeted the Group. Approximately twelve bursts rocked DOWN N' GO. Fearing damage, Lt. White asked for a damage report:

"In the waist and tail they reported holes torn all over the thing, Paladino said the engines were hit. Bush said the tail looked like a sieve and that a piece had hit him in the foot. Part of the interphone was shot out and we had what Vic said amounted to about fifty holes in the bomb bay. He told me gas was leaking in there so I had him open the doors and when I looked around I just about fainted. Gas was just pouring from the wing tanks into the bomb bay and waist. About that time our control cables broke and I had to set up the A-5 to fly the plane. The servo units in the tail had been hit so the A-5 was not working very well. Dick knew we were in trouble so he gave me a heading to the beachhead and my original intentions were to land it there. However, fire broke out in the number one engine and we had very poor control of the ship so I decided it was time to leave it. It was just a question of whether we should bail out over enemy territory or wait and take a chance on making the beachhead. There was not any question in my mind she was going to blow. Looms gave me a position so I called some P-47s and they came over and gave us excellent fighter cover all through the experience. I called all the boys out of their turrets and told them to stand by to bail out. Bob was flying and was working his head off to keep the plane on an even keel. Everybody was anxious to leave and I was amazed at how calm everybody was. Our training stood us in good stead and there seemed to be no excitement."

With the bomb bay doors open, Sgt. Victor Cieslewicz walked out on the catwalk without oxygen or gloves in an effort to plug the leaking fuel tanks with a piece of cloth. The cold at 21,000 feet froze his hands but his valiant efforts failed to stem the fuel leak. As they started losing altitude, flak again erupted around the plane. (It was later determined the flak was from British guns.) Thinking the fire was from German units, they delayed bailing out in an effort to reach friendly lines. Once sure they were over friendly forces, Lt. White rang the bail out alarm. After everyone else left the ship, he pulled the A-5 autopilot release just before jumping himself. Shortly afterward, the aircraft exploded hitting the ground near an airfield being hastily prepared by 9th Air Force engineers. Surprisingly, no one on the ground suffered injuries. All the crew survived their parachute landings and were greeted by members of the U.S. 29th Infantry Division.

DUAL SACK 42-95089. On 21 June 1944 this aircraft, flown by Lt. Roland Fox and crew, was one of thirteen 8th Air Force bombers to land in Sweden that day. (LaPoint)

Col. Mason congratulates Capt. James Sullivan, lead pilot of the 713th BS, after the successful completion of the Groups 100th mission. Capt. Sullivan was one of the Group's original pilots. (USAF)

The following day, a medical officer carried the crew to the site of the crash. Nothing from the aircraft was identifiable. They saw first hand the destruction wrought by the invasion. They toured the beaches and witnessed freshly dug graves, demolished equipment both German and American, bunkers and pillboxes, and the now infamous hedgerows. It amazed Lt. White that people survived such a brutal battle but instilled a new confidence in the U.S. Army. They flew to Seething aboard a C-47 arriving back at 'home' on 24 June just in time for an upcoming party. However, they received a better deal, a week at a rest home in southern England.

The base's daily bulletin, *Liberator*, announced on 23 June, "Hardwick's Hill-Billy band fresh from three nights at the Rainbow Corner, London, will entertain at the Aero Club, Saturday night, 24 June 1944. This is a preliminary celebration of the completion of the Group's 100th mission." The advertisement continued with the main event. "A big celebration, including dancing, GIRLS, good eats, decorations, etc, commemorating the completion of the Group's 100th mission will be held at the Aero Club, Thursday evening, 29 June 1944. You may secure your tickets from Miss Thompson." Although Guyancourt was the one hundred and first mission in Europe, returning crews prepared for the festivities commemorating one hundred missions. The road from Boise, Idaho to Guyancourt was difficult and costly. Sixty-one aircrews were missing or killed and seventy-five aircraft were missing or destroyed. What would the next hundred missions hold?

A CQ woke Sgt. Fred Kerniss at 0100 on 24 June for a mission to an airfield at Melun, France. They departed Seething around 0400 but quickly lost superchargers on two engines and aborted. The rest of the formation continued and joined the 93rd BG and 446th BG. Due to the close proximity of the airfield at Melun to Paris, visual conditions were necessary to bomb. Unfortunately, clouds obscured the area. The bombers flew twenty-eight miles south searching for another target before they returned to Seething with their full load of bombs.

Just after lunch the same day, crews were briefed for another tactical mission to a "No Ball" site near Haute Cote, France. The fourteen aircraft did not takeoff until 1630 with twelve aircraft flying in trail of the 93rd BG. Two 448th planes flew in the 93rd BG formation. Trouble started just after assembly when the formation started to climb. The 448th got ahead of the 93rd BG and despite zigzagging in an effort to fall into position, the 93rd remained too far behind the Group. Leaving the British coast, the 448th tacked on to the 14th CBW formation with the 93rd BG in trail. Over the target, three rounds of flak burst in the sky around the bombers. One aircraft, TANGERINE, flown by Lt. Elwyn Palmerton, in the 448th formation suffered heavy damage. They immediately dropped out of formation and turned for the coast. However, the battle damage was debilitating and the B-24 entered a spin and crashed near the coast one kilometer northeast of Beussent, France. Amazingly, all the crew successfully bailed out but were captured.

The formation continued to the target but minus one crew. Things continued to go wrong as they prepared to drop their ordnance. The lead aircraft's bombsight malfunctioned after

Unidentified men from the 448th enjoy a small party and a rare break from the action. The stand down following the 100th mission and subsequent party allowed many men the first real break since combat operations started. (LaPoint)

condensation inside the sight fogged up the optics. Eleven aircraft returned to Seething with their bombs after the lead aircraft did not bomb. Only the two aircraft in the 93rd BG formation bombed at the cost of one crew missing.

The following afternoon crews returned to France to bomb three separate targets. Capt. A.D. Skaggs, flying BARFLY with Maj. Glassell Stringfellow as command pilot, led the 20th CBW with thirty-four bombers from the 448th. Marginal conditions existed over the target but after the initial point the three squadrons uncovered in trail in preparation for visual bombing. The first squadron bombed a highway bridge three miles south of Orly airfield at 1931 while the second squadron hit Orly air-

Men and women enjoy one of the parties celebrating the Group's 100th mission. (Bollschweiler)

field at the same time. The third squadron hit the airfield at Bretigny two minutes later. Unfortunately, all three squadrons' bomb results were poor due to the marginal weather conditions that hampered the bombardiers' aim. Heavy flak pummeled each group as they crossed Chartres, France. Sgt. Fred Kerniss watched several aircraft from other groups fall victim to the deadly artillery. On return, aircraft 44-40099 crash landed at the emergency airfield at Manston, England, after the pilot, Lt. Wayne Garceau, suffered flak wounds to his left arm. All other aircraft returned to Seething after the six-hour mission.

Heavy rains on 26 June grounded the 448th and a planned mission was cancelled. However, on the next day, the 448th resumed tactical operations in France. Again bad weather forced a change of plans. Three crews failed to find the formation and returned to Seething. Another crew joined a different group but engine problems eventually forced them to abort the entire mission and return to Seething. Thick contrails swirling in the cold air created nightmarish conditions for formation flying and for the crews, especially the waist gunners standing in open windows. Adding to the discomfort, Sgt. John Shia's flying suit failed. The heavy fleece-lined leather provided very little warmth against the thirty below zero temperatures. Skirting the large flak emplacements near Brussels, the formation set course for a "No Ball" supply site hidden in a railroad tunnel at Cilun, France just north of Paris. Clouds obscured the site when the bombers arrived.

Unable to bomb the primary target, the formation sought a target of opportunity. Despite the thick contrails generated by the preceding aircraft, they located the airfield at Creil, France. As the 448th crossed over the target at 1055, a tremendous flak barrage pounded the formation. Flak struck RED SOX between the number one and two engines as well as in the nose. The crew, led by Lt. Peter McVean, on its twelfth mission, turned around and headed for the sea. Unable to reach the relative safety of the ocean, the crew bailed out of their plane fifteen miles southwest of Soissons, France, near Vez. One crewmember, Sgt. Robert Slack, bailed out but his chute caught on the tail of the aircraft killing him. Another, Lt. Lawrence Carney, also bailed out but was killed by ground fire during the descent. Four others were captured, but the remainder of the crew found refuge with the French Underground and successfully evaded. They eventually returned to England.

Flak also found its mark on 41-29004, piloted by Lt. Harold Turpin. It crashed at Genthin killing everyone on board. Lt. Seymour Jarol, flying 42-50369, also fell to flak and crashed at Villers. Once again, there were no survivors.

Lt. William Dogger's crew, flying COME ALONG BOYS, endured six hits, one passing within inches of the waist gunner's

Maj. Glassel Stringfellow, commander of the 714th BS, poses in front of the operations area. Maj. Stringfellow led the squadron from its inception until October 1944. (Bottoms)

foot. Their number one engine smoked from flak damage for the remainder of the flight and finally seized just prior to landing. The five and a half-hour flight cost the 448th three crews and numerous damaged aircraft.

Crews received an unwanted surprise as they left the briefing room in the early morning hours of 28 June. A single German Ju-88, using the cover of darkness, attacked the airfield at Seething dropping three bombs. They exploded on the perimeter track just as crews left the briefing room. It caused minor damage to a plane scheduled to fly the upcoming mission and rattled the nerves of the men leaving the briefing. Shortly after the menacing attack, pilots lifted their Liberators into air just before 0500 and climbed to assembly altitude over the East Anglia countryside. The target, the marshalling yards at Sarrbrucken, lay just across the French border but the Luftwaffe did not wait for the bombers to invade the Fatherland before putting up resistance. Fifteen minutes after crossing the enemy coast, Me-109s attacked the Group. Escorting P-38 Lightnings quickly drove off the attackers before they caused any damage. The Liberators crossed the target at 0842 amid the customary flak and dropped their bombs. On return, Lt. Leroy Conner lost an engine as they crossed the Channel. Despite a low fuel state and the loss of an engine they safely landed at Seething.

On 29 June, the Second and Third Bomb Division sent 591 bombers against strategic targets in Germany. The 448th participated in a large force assigned to attack Bernberg, Germany. Trouble started early in the flight when the lead aircraft, LITTLE JOE flown by Capt. A.D. Skaggs, aborted before reaching the enemy coast due to an oxygen leak. Continuing on, the new lead aircraft in the lead squadron suffered equipment failure and did not drop. Seven other aircraft followed their leader and did not drop; however, seventeen other aircraft, recognizing the primary target, dropped their bombs on the airfield at Bernburg, Germany, amid heavy flak. Flak struck the lead aircraft of the second squadron, 42-50300, in the bomb bay. The pilot, Lt. William Warke, quickly jettisoned the bombs twenty seconds from the target. Despite these actions, all ten men perished.

Eight aircraft that did not drop their bombs on the primary target bombed an airfield at Strinum, Germany. Just northeast of Bernburg, it provided a perfect target of opportunity for the bombers. Departing the target area the bombers committed a tragic mistake; they flew within range of the heavy flak guns ringing Brunswick. Accurate flak exploded around the bombers. Aircraft 42-95326, flown by Lt. Glenn Jones, suffered damage. They kept the plane flying until they reached Holland where they crash-landed near Alkmaar. Everyone survived the bumpy crash landing but they were promptly captured.

Bad luck continued its curse on the mission as they descended over the Zuider Zee. Light but extremely accurate flak knocked holes in several aircraft as they descended below 16,000 feet. Sgt. James Donovan on HELLO NATURAL II suffered flak wounds to both legs. Two less crews landed at Seething than took off.

On return, crews celebrated for several reasons. Everyone enjoyed the celebration at the Aero club in honor of the Group's 100th mission. Planners added to the festive occasion by scrubbing the mission that was scheduled the next day. Payday, 30 June, brought a close to the busiest month in the ETO. Morale

BATTLIN BABY 42-99971. This B-24 J model sits on a hardstand at Seething before being transferred to another Group. (LaPoint)

remained high, thanks to recreational pursuits. The station's baseball team with an undefeated record of 6-0 led the Eastern League through the first half of the season. Pvt. Vincent Padilla from the 1193rd Military Police Company placed second in the 135-pound division of the 8th Air Force boxing finals. Food quality improved but the most popular improvement was the construction of a bar in the Post Exchange that sold beer. Although it was British brew, the only complaint was the shortage of it.

The thirty-one credited missions in June were not without cost. The increased tempo of missions severely damaged the perimeter track. A company of men from the 844th Engineer Aviation Battalion led by Lt. Obed Fitzgerald arrived and started emergency repairs in order to keep the field in operation. Despite the busiest month of the war to date, his men completed their temporary repairs and allowed the bombers to continue their work.

The most important damage was reflected by a chalk board in the control tower that listed fifteen crews as missing in action in June. Despite this high cost, more aircraft and crews called Seething home than ever before, 106 crews and an equal number of aircraft. Finally, the economic might of the U.S. reached the front lines. It continued until the end.

ten

# ENGLISH SUMMER: July 1944

The first day of July witnessed the 448th returning to a familiar target, a "No Ball" target, this time at Fiefs, France. Crews received their briefings at 1400 and did not takeoff until 1700. Forty aircraft took off and assembled at 24,000 feet. Over the target, bad weather forced a recall of the formation and all aircraft returned to Seething with their bombs. However, the following day weather permitted the 448th to return to Fiefs with different results.

This time crews conducted a more normal pre-flight routine with a 0700 briefing. Twenty-four crews left Seething at 0940 but Lt. Edward Malone quickly aborted due to a gas leak in the bomb bay. Trouble also started for Lt. Billie Blanton and crew during assembly. They tried unsuccessfully to feather the number four engine that developed a runaway propeller. The wild gyrations threatened to destroy the aircraft but somehow Lt. Blanton maintained control of 42-109793. The wild propeller tore from its moorings and destroyed the adjacent engine. After calling the control tower and advising them of their situation, Lt. Blanton ordered everyone to evacuate the aircraft. Everyone successfully bailed out but two men suffered injuries. The copilot, Lt. James Schierbrock, suffered a broken back upon landing and died from these injuries five days later. The bombardier, Lt. Fred Bernard suffered a severe laceration and was hospitalized but recovered. The aircraft crashed three miles north of Boreham between Braintree and Chelmsford.

Heavy rains forced cancellation of all missions scheduled the next day and it was not until 6 July that the 448th participated in a mission. Once again, planners assigned the 448th a tactical target in France. To further isolate the Normandy bridgehead, all bridges over rivers in western France were earmarked for destruction. The 448th was assigned the railroad bridge at Saumur. Liberators took to the skies late in the afternoon on 6 July. The formation accomplished its now standard assembly at 10,000 feet before departing the English coast. Fifteen minutes from the target, clouds formed below the bombers and when they reached the target the clouds completely hid the ground. The formation departed the area in search of a target of opportunity. Twenty aircraft dropped their 2,000-pound bombs on the nearby bridges at Gien and another dropped its bombs with the 446th BG at Sully-sur-Loire. However, the bombs overshot the intended target leaving it undamaged. Sixteen aircraft from the group that were unable to drop on the bridges attacked an airfield at Chartres. Returning unscathed, the bombers landed at Seething in the dark.

On 7 July no 20th CBW aircraft participated in a mission but the following day the Wing including the 448th returned to France. The increasing threat from German V-weapons required increased efforts to destroy their launch and support facilities. The supply emplacement at Rilly-la-Montagne on the outskirts of Reims was just such a target. CQs awoke Lt. William Snavely's crew at midnight in order for them to attend the 0100 briefing. Sgt. Marvin Hicks ate fresh eggs at breakfast then headed to the briefing room. The briefing called for crews to take off at 0500 and assemble at 18,500 feet before crossing the Dutch coast. Everyone executed the mission according to plan until they crossed into Holland. Ten miles into Holland thick clouds made formation flying impossible and forced the lead aircraft to issue a recall message. Lt. Leroy Conner jettisoned his bombs in the North Sea on return to Seething while others returned with full bomb bays.

Beautiful early morning skies gave way to horrible weather on 9 July and it persisted through the following day. No aircraft launched from Seething either day but on 11 July, the B-24s of the 2nd Bomb Division resumed strategic operations against Germany. Crews were awakened around 0400 and were briefed on the target, Munich. Shortly after 0800 Liberators rolled down the runway bound for the southern German town where visual bombing conditions were expected. After forming at 15,000 feet the formation encountered a high wall of clouds. Although able to penetrate the wall, the clouds scattered the formation leaving the Group in disarray. Eight bombers, including ROSIE RIVETS flown by Lt. Leroy Conner, banded together and bombed a target of opportunity at Seringen, Germany. Despite the heavy flak over Germany and the French coast and short on fuel Lt. Conner landed safely at Manston, England.

The larger force continued as briefed until thirty-eight minutes from the target. Three generators failed on OL' BUDDY forcing the crew to abort. Leaving the formation, the crew failed to fire red flares indicating an abort. As a result two other aircraft followed. Once they realized their mistake, the rest of the formation was lost in the clouds. They returned to England.

For the remaining twenty-seven bombers, a combined force of P-51s, P-47s, and P-38s swept the skies clear of enemy fighters. The fighters could not suppress the heavy flak that once again surrounded the bombers. Flak damaged the lead PFF aircraft, 41-28776 from the 44th BG. Onboard as command pilot was Maj. James Conrad, the 715th BS Commander. Over Ghent, Belgium, they reported they were low on gas and were taking a shorter route back. Eyewitnesses watched as the damaged aircraft left the formation at 1520. They reached the safety of Switzerland. INCENDIARY BLONDE, piloted by Lt. Marcus Horton, also suffered flak damage but unlike the previous crew they reached England. Their crash landing near Gravesend claimed the lives of two crewmen, both waist gunners, Sgt. Harold Fowler and Sgt. Kenneth Prieb. Three other men suffered various injuries.

The officer's club at Seething was a place for entertainment and relaxation. (Everson)

"Happy Warriors," airmen who have completed their tour of combat missions, buzz Seething following the completion of their last mission. By July 1944, more and more men were completing their missions and this unauthorized flyby quickly became common leading to a Group directive that prohibited these maneuvers and threatened the offender with a court martial. (Bailey)

On 12 July, the 8th Air Force returned to Munich in another attempt at a visual attack. Poor weather again hampered the attack, as did deadly flak barrages. As the bombers crossed the target at 1407, the clouds scattered enough for crews to clearly see the Isar River snaking through the city. Flak found its mark on several aircraft. FAT STUFF II, flown by Lt. George Wilson, shuddered from the impacting shrapnel. Unable to return to Seething, they sought haven in Switzerland. They landed in a swampy field at Altenrhein near Lake Costance. BIM BAM BOLA, piloted by Lt. Billie Blanton, suffered similar consequences after flak destroyed their number three engine. They turned toward Switzerland. Over Lake Constance, Lt. Blanton ordered the crew to leave the aircraft. Six of the crew, Lt. George Klein, Sgt. Adrian Denbroeder, Sgt. Paul Sherlock, Sgt. Robert Larson, Sgt. Bernard Stelzer, and Sgt. Salvatore Sparacio, landed near Bregenz, Austria, and were captured. The pilot as well as Lt. Edwin Hewitt and Sgt. Armor McKain landed in Switzerland and were interned. The plane continued until crashing at Fideris-Kublis at 1530.

This aerial view provides an excellent view of the operations area around the control tower at Seething. Group headquarters is in the upper right. (LaPoint)

Over the target, flak destroyed the number four engine of SLEEPLESS KNIGHTS and set the plane on fire. The pilot, Lt. Michael Kuchwara, rang the bail out alarm. His quick action saved most of the crew because immediately after everyone bailed out the plane exploded in midair and crashed at Feldmoching in the northern suburbs of Munich, Germany. Unfortunately, Sgt. Calvin Crumbley's parachute caught fire and he perished in the descent. Lt. Aurel Popa and Sgt. Richard Suhay suffered broken legs and were taken to an Air Force hospital near Munich. The rest of the crew were also captured.

Leaving the target and turning toward the rally point, Sgt. Marvin Hicks heard a loud bang as a ten-inch piece of flak ripped into the left wing. Fortunately, Lt. William Snavely kept the damaged aircraft in the formation but problems remained. Fuel in the tanks was insufficient for the return trip. Then just over an hour after bombs away, the number four supercharger failed. Unable to keep up with the formation due to the degraded performance at high altitude, they slipped behind the formation and descended from 23,000 feet to 16,000 feet. Lt. Snavely asked if the crew wished to bail out or try to land the stricken plane. Everyone chose the latter option. All loose equipment went overboard as they attempted to lighten the aircraft and improve its performance. Low on fuel, they headed for the Normandy beachhead. They landed safely on a steel-matted RAF fighter base on the outskirts of Caen, France. Talking with the commander, Lt. Snavely asked for 500 gallons of fuel for the return trip to Seething. The commander laughed stating he could fly his Spitfires for a week on that much gas. However, he reluctantly gave them the fuel. Flight Officer William Morris helped form a human chain to refuel the aircraft using five-gallon Jerry cans. Arriving over Seething at 2130, well beyond the expected return time, tower refused to let them land. Finally, after some discussion and identification procedures, they were allowed to land.

The crew of 42-50475 with Sgt. William Gamble was not as lucky. The horrendous flak over Munich disabled two engines and seriously wounded Sgt. William Maynard in the right leg. Enduring the return trip on two engines, they navigated successfully back to England only to have their two good engines quit. Lt. Richard Moody landed the aircraft in a wheat field near Hardwick. The crash-landing threw the navigator, Lt. Jack Glicksman, from the aircraft but miraculously he survived. Medics carried Sgt. Maynard to a nearby hospital but even after thirteen pints of blood, they could not save his mangled leg.

Flak ravaged the bombers during the nine and one-half hour flight. Three crews never made it back to Seething and were listed as MIA and another aircraft was destroyed after

Phil Ray and Tom Allen take advantage of warm days and sunshine by writing letters outside their hut. (Ray)

crash-landing. Three other aircraft landed at other airfields due to battle damage. The nature of the air war was changing. Attacking Luftwaffe fighters no longer posed the primary threat to crews. The black, oily puffs of flak were the main hazard. At least with the fighters, the gunners shot back. Although air superiority was reality, there was no defense against the flak.

At 0130 on the morning of 13 July, crews walked through the darkness toward the mess halls. An hour later they sat quietly as the briefer discussed the specifics of the target. Today, the marshalling yard in the German border town of Saarbrucken was the target. B-24s lifted into the early morning sky bound once more for Germany. A solid overcast increased the difficulty of navigation but failed to shield the target. Navigation was perfect and the formation commenced the bomb run arriving over the target at 0907. Once again, a heavy flak barrage greeted the bombers. Flak destroyed aircraft 42-94989's number three engine and another engine lost power. Lt. Dale Grubb turned his B-24 toward Switzerland. A fire developed in the number four engine and continued to burn despite the crew's efforts to extinguish it. Shortly after crossing the Rhine River into Switzerland, Lt. Grubb ordered everyone out of the aircraft. All nine men bailed out before the fire engulfed the plane. It exploded over Batterkinden, Switzerland, showering the town with debris.

On return to Seething, Capt. A.D. Skaggs performed the ritual 'buzzing' of the tower to commemorate the completion of the thirtieth mission. Most of his crew were going home!

Despite the war and the hectic operations around the airfield, the adjacent fields were still cultivated and farmed. A parked B-24 is visible over the tractor while a man and woman work. (Everson)

MARY MICHELE 42-99993. This Seething Liberator proudly displays thirty-seven bombs representing combat missions. Although she never fell to the enemy, combat took its toll and the aircraft was eventually salvaged. (Everson)

Coincidentally, Col. Mason released a memorandum addressing flying violations. "Due to recent violations of the flying regulations at this base, there will be no-repeat-no 'buzzing' or low flying of any kind by personnel of this command. Any pilot violating this order will be subject to trial by court martial."

Weather on 14 and 15 July again stymied operations but on 16 July the Group returned to Saarbrucken to attack the rail marshalling yards and depot. Weather conditions similar to previous trips persisted as clouds completely obscured the ground during the entire flight. Light flak peppered the sky as they crossed the French coast but the worst lay ahead. Just as before, the German gunners flung a curtain of shrapnel at the bombers. Nonetheless, bombs fell from the big bombers at 0939. The aircraft flown by Capt. Alfred Fox experienced a bombsight malfunction and they did not drop. Four other aircraft held their bombs as well after their element leader did not drop. The formation quickly executed a turn away from the target without loss; however, several aircraft suffered damage. Unfortunately, returning to Seething proved more deadly. Lt. George Booth, flying HELLO NATURAL II, turned off the active runway onto the crossing runway, 01, where he stopped on the left side of the runway to unload his wounded ball turret gunner, Sgt. Stephen Lawnicki. Lt. Melvin Alspaugh returned to Seething from a practice mission just as the planes from the combat mission landed. He landed his aircraft, 42-51221, and stopped adjacent to HELLO NATURAL II on the crossing runway. BETSY JAY, also returning from a practice mission landed next and without warning slammed into the stopped aircraft killing Sgt. Edward McGinnis, the nose gunner of 42-51221. Lt. Joseph Borsch, a crewmember board BETSY JAY, was wounded in the collision.

The next day, early morning fog scrubbed a mission to France but crews were quickly gathered at 1100 for another briefing. This mission was also canceled. Finally, at 1630 crews listened to the details for a mission to a "No Ball" site near St. Sylvestre, France. Two six-ship formations tried to bomb the target amid accurate flak. Lt. Leroy Conner's aircraft fell out of formation after suffering an engine problem prior to the target. Plagued by the malfunctioning engine, they followed the formation and dropped their bombs as planned. Five aircraft failed to drop after the lead aircraft's bombsight malfunctioned and forced them to hold their bombs. The heavy flak wounded Sgt. John Kushner on SONIA and another aircraft, 44-40875, suffered flak damage to the hydraulic systems. Upon landing at Seething, the brakes of the damaged aircraft caught fire. The fire quickly spread as the crew hastily evacuated their positions. Men from the 459th Sub-Depot salvaged the useable remains of the burned-out aircraft the following day.

The 448th flew in direct support of ground troops around Caen, France on 18 July. Bombers from Seething hit the French town of Grentheville in support of Gen. Bernard Montgomery's troops as they tried to break out of the Normandy beachhead. The B-24s left Seething at 0530 then assembled before turning toward the rail yard at Grentheville. Smoke from exploding bombs and simultaneous artillery barrages obscured the area making target acquisition difficult for the bombardiers. Despite the difficulties, explosions filled the target area. The Group endured sporadic flak and although no aircraft were lost, flak wounded Lt. Alfred Cannon on BARFLY.

When the men returned from the mission, they found a boxing ring set up. That evening, men from around the base auditioned their pugilistic skills. The purpose was to arrange a boxing card for a future date. Men packed the building to watch prospective boxers show their skills.

*English Summer: July 1944*

THE MENACE 41-29232. One of the original Group aircraft flown to Britain from the States by the 448th. It flew numerous missions with the Group before it was salvaged. (LaPoint)

Greasy puffs of dreaded flak explode around a 448th formation as they cross the coast. Although enemy fighters garnered most of the attention, flak proved to be the most deadly of the Luftwaffe's defenses. (LaPoint)

Early on the morning of 19 July, a British Lancaster returning from a mission crashed near the airfield at Brooke. Military Police responded to the accident and posted guards around the crash until the proper authorities arrived. Fortunately, the entire crew bailed out safely and were taken to the station hospital for observation.

Capt. A.D. Skaggs flew the formation ship, STRIPED APE, as the 448th assembled for a mission to Germany on 19 July. During assembly, Lt. Leroy Conner discovered a fuel leak on their aircraft forcing them to land at Seething. Quickly changing to a spare aircraft, they took off again but the delay proved too long and they failed to catch up with the formation. They returned to Seething. Unable to hit the briefed target at Eisenach, the formation did find Koblenz and dropped their bombs on the city as a target of opportunity. Fifteen aircraft failed to drop their bombs after the lead aircraft failed to release. All dispatched aircraft landed safely after the six-hour flight.

The next day, the Group again searched for the engine factory located at Eisenach, Germany. Following another early morning takeoff, the crews completed assembly and started for enemy territory. One aborted for mechanical problems but the others continued to the target. Mistaking the town of Schmalkalden for Eisenach, eighty B-24s from the 2nd Bomb division dropped their bombs. Once again Eisenach was spared at the expense of another city.

On 21 July, the Group flew its sixth consecutive mission. The object for the 448th was a jet propulsion factory in the Bavarian town of Munich. The day started with the early morning briefing and a dawn takeoff. Low clouds hampered the takeoffs and assembly but the crews completed the assembly and joined the larger Wing formation enroute to the target. Rising cloud decks forced the formation to climb higher than anticipated. The thin air at 25,000 feet strained the engines as the pilots horsed the bombers over the clouds. As a result, formation integrity disintegrated as each pilot struggled just to keep up with the formation. When the formation crossed the target at 1036, the largest flak barrage Sgt. Fred Kerniss ever saw buffeted the formation. Multiple flak batteries ringed the city of Munich and they all unleashed their fury on the B-24s. Black, oily puffs from the exploding anti-aircraft shells literally darkened the entire sky. Despite their best efforts, the bombs from the ragged formation overshot the intended target.

As the formation executed a turn away from the target, numerous aircraft surveyed their damage. Lt. James Beaver and crew, flying LITTLE SHEPPARD, endured several hits from the flying shrapnel. Unable to continue on the return trip, they turned their damaged Liberator toward the nearby haven of Switzerland. Lt. Melvin Alspaugh and the crew of DEAD END KIDS experienced a similar incident with different results. Prior to the target, flak severely damaged both engines on the left side making their return to England impossible. Skillful piloting by Lt. Alspaugh allowed them to remain in the formation until after the bombs were dropped. Then, they slipped from the formation and turned south toward Switzerland. After quickly conferring over the intercom, the crew voted unanimously to try to make the 400-mile trip to Italy instead of sure internment in Switzerland.

Once clear of the Alps and German fighters, they encountered another problem. "The markings on our ship were from another theater of operations and our radio wouldn't work," explained Lt. Alspaugh. "We lowered our landing gear to indicate a friendly ship in case the gunners were thinking of taking pot shots at us." Despite their apprehension, they received a warm welcome and spent thirteen days in Italy while DEAD END KIDS underwent repairs before returning to England. For his outstanding airmanship Lt. Alspaugh received commendations from the commander of the U.S. Strategic Air Force in Europe, General Spaatz, the 8th Air Force commander, Lt. Gen. James Doolittle as well as Maj. Gen. Kepner the 2nd Bomb Division commander.

Damaged mounted when the bombers returned to Seething. Not only did two aircraft fail to return but numerous others suffered damage. During the bomb run, flak knocked out the pilot's oxygen on Sgt. Marvin Hicks' aircraft. The pilot, Lt. William Snavely, flew the return trip without oxygen but landed safely at Seething. Lt. Leroy Conner and crew also returned unhurt despite the numerous flak holes in their aircraft. On landing, the gear of PICCADILLY PAT collapsed. In the days that followed, salvage crews combed through the broken aircraft in an effort to recycle as much as possible.

Following the pummeling from the flak guns, bad weather provided a much-needed respite on 22 July. The dreary weather grounded all 8th Air Force bomber operations. The following

The crew of FAT STUFF II led by Lt. George Wilson are escorted by Swiss guards after landing near Altenrhein on 12 July 1944. (Everson)

day, however, the aerial onslaught resumed. Capt. Aubrey Cates lifted TROUBLE N' MIND into the air for his final mission on the afternoon of 23 July. Nearby, engine problems cut short Sgt. Robert Kessler's trip on EAGER ONE. Immediately after takeoff, the number three engine spun out of control forcing the pilot to shut down the engine and feather the propeller. Another problem forced them to shut down another engine making the heavily laden bomber difficult to handle. Against

FAT STUFF II 42-7591. This aircraft was interned after landing in Switzerland. During a happier time the original crew (not the crew interned) pose in Dakar enroute to England in November 1943. Standing (left to right): Joseph Bushek, Walter Garland, Jay Dempsey, Thomas Abott, Charles Hutton, Jerome Haas, Brown. Kneeling: Harold Podolsky, Arthur Steele, Raymond Boll, Jack O'Brien, Seymour Ausfressor, and William Blum. (Boll)

*English Summer: July 1944*

A picture of a taxi accident on 16 July 1944 shows BETSY JAY sitting between the tail of HELLO NATURAL II on the right and 42-51221. The unfortunate accident killed on man and injured another. (LaPoint)

The tail of HELLO NATURAL II lies on the perimeter track after being sheared off in a collision with BETSY JAY. A third aircraft involved in the accident is barely visible over BETSY JAY. (LaPoint)

overwhelming odds, the crew circled and safely landed on only two engines. The remainder of the force took off without incident. Despite it being summer, the temperatures during the trip plummeted chilling the gunners standing in the open waist windows. It was a long cold trip to the target. As they neared the target Sgt. Marvin Hicks watched from inside COME ALONG BOYS as light flak dotted the sky. Although still dangerous, it failed to compare to the deadly barrage he had experienced two days earlier. Using G-H equipment to identify the target, the 448th hit the airfield at Laon/Athies, France. All aircraft and crews returned safely but Lt. William Dogger experienced problems after arriving at Seething. He called tower stating they were having difficulty locking the landing gear of THE MENACE into place. After a short time, they appeared to correct the problem and prepared to land. When the wheels of the plane touched the runway, the right main landing gear collapsed outward. Although injuries were avoided, the plane was salvaged.

In France, General. Omar Bradley prepared for one of the decisive operations of the war. His plan, code-named "Operation Cobra" called for a massive carpet bombing of German forces located around St. Lo, France. Following this aerial onslaught, a concentrated attack along a very narrow front by the U.S. First Army would rupture the German lines and hopefully break the stalemate in Normandy. Aircraft from the 9th and the 8th Air Forces provided the muscle for the aerial onslaught. On 24 July, Liberators from Seething took to the sky at 0900 as part of the great armada bound for Montreuil, France on the outskirts of St. Lo. Poor weather earlier in the morning delayed the launch, but clearing skies over England finally allowed the planes to take off. Due to the proximity of Allied troops, clear weather was required to drop bombs. As usual, the weather failed to cooperate. A complete undercast forced the bombers to return to Seething with their loads. Even though fighter escort was excellent and flak was scarce, the trip was not uneventful for Sgt. Robert Kessler flying in OUR BABY.

HELLO NATURAL II 42-52606. This Liberator was salvaged on 16 July 1944 after its tail was sheared off in a taxi accident. (LaPoint)

He suffered burns on his left shoulder and back from a loose wire in his heated flying suit.

Due to the uncooperative weather, Gen. Bradley postponed "Operation Cobra" for twenty-four hours. Early on the morning of 25 July, CQs woke crews for a return trip to Montreuil, abeam the St. Lo-Perriers-Lessay road. After feathering an engine due to a malfunction, one crew departed the formation and salvoed their bombs into the North Sea. The rest of the Group crossed the French coast at 0753. Crews quickly realized the Germans had moved in flak batteries during the night. Heavy flak littered the sky as the Group searched for the target. Adding to the melee on the ground, American artillery units fired red smoke markers to mark the front lines. Despite this precaution some units dropped short of the target into American troops. On the run to the target another Group forced the lead 448th aircraft to salvo its bombs in order to avoid a collision. The premature release caused the bombs to fall into friendly lines. Other Groups also dropped short including a B-26 unit from the 9th Air Force. This unfortunate incident claimed the life of Lt. Gen. Lesley McNair, a senior U.S. Army officer who was there to observe the attack. Flak pounded the bombers. Although the 448th survived the gauntlet without loss, other units were not as fortunate. Sgt. Fred Kerniss witnessed a B-24 in another group explode in mid air. Four other aircraft from various groups fell to the guns. Despite the unfortunate case of friendly fire, the carpet bombing was a tremendous success as Gen. Bradley's troops punctured the German lines and set in motion the great chase across France.

Crews were awakened just after midnight on 27 July for a mission to a target south of Munich. Bad weather scrubbed the mission thirty minutes before takeoff and sleepy crews headed to bed. The following day mirrored the previous one. This time, fifteen minutes before engine start, the cancellation order was

PICADILLY PAT 42-50470. This aircraft was salvaged on 21 July 1944 after its landing gear collapsed on landing following a mission. (LaPoint)

Above: DEAD END KIDS 42-94992. This aircraft suffered severe damage on a combat mission but the crew flew it to Italy for repairs. It returned to Seething and survived the war. The nose art is barely visible. (Johnson)

Right: LITTLE SHEPARD 41-28711. Flak damage over Munich, Germany forced the crew to land this Seething-based B-24 in Switzerland on 21 July 1944. (LaPoint)

received. Once more crews returned to their huts, no closer to finishing their tour than when they woke up.

On 29 July, the 123rd mission started like all the others, an early morning wake up followed by breakfast then a briefing. The designated target was an oil refinery in Bremen. This time they took off. An uneventful trip to the target quickly became exciting as the formation approached the target. Once again the ugly puffs of black smoke signaled the presence of dreaded flak. From 22,800 feet the bombers released their deadly cargo with the aid of PFF but the Germans returned fire with a vengeance. Aircraft 42-100435 flown by Lt. Leroy Conner shuddered as flak ripped through the thin aluminum in the waist of the bomber. Amazingly, no one was injured! A second, then third burst quickly followed the first. The hits severed the aileron cables and jammed the rudders sending the B-24 plummeting to earth. After falling 4,500 feet, the crew regained control of the battered ship by using the autopilot and differential power settings. It required twenty degrees of left bank to keep the aircraft in level flight. Unable to contact fighters for support, they headed for home. Fortunately, they encountered no opposition enroute and made it safely to England. Once over Seething, they discussed their options with the tower and their probability of a safe landing. Six of the crewmembers bailed out near the base; only the nose gunner suffered injuries. The remaining four airmen stayed with the aircraft. They headed for the longer emergency runway at Woodbridge with intentions of crash-landing. Due to the expert skills of Lt. Conner and the copilot, they successfully landed the damaged aircraft with only minor injuries. They counted over 200 holes in the skin of their B-24.

Other aircraft also suffered damage. LITTLE JOE, piloted by Lt. John White, absorbed three bursts in the nose. Flying debris showered the inside of the plane but failed to injure any-

Three unidentified men stand on the control tower at Seething scanning the horizon for aircraft returning from a mission. Meanwhile, an ambulance, commonly referred to as a "meat wagon" sits poised by the perimeter track. (LaPoint)

one. Exploding shells also damaged an engine and knocked holes in the waist and the right rudder. On SONIA, a waist gunner Sgt. Chester Anderson, also suffered flak wounds. Sgt. Robert Kessler counted thirty holes in OUR BABY after landing. Despite the heavy flak, the 448th recorded excellent results using PFF equipment. The early experience with PFF started paying dividends and it would blossom in the upcoming winter months. Upon landing at Seething, crews were re-

A 448th Liberator is visible below another B-24 as they cross the English Channel. Notice the numerous wakes from boats in the water. (LaPoint)

General-purpose bombs fall toward their target. (LaPoint)

warded with a day off the following day and free beer that night.

After a day off, the Group resumed the strategic offensive with a strike at the chemical works located in Ludwigshafen, Germany. Heavy flak again pounded the bombers but the 448th escaped damage. They bombed by PFF with good results and all bombers landed at Seething at 1600 without damage.

July proved costly for the Group. Twelve aircraft and crews were lost during the month, six from the 712th BS. Twenty-eight aircraft also suffered damage. Despite the heavy losses in the air, the contributions made by the heavy bombers paid off on the ground. The breakout of General Omar Bradley's forces at St. Lo sent the Allies racing across France. July ended with the German Western Front in complete disarray and the Luftwaffe totally absent from the skies over the battlefield. Flak, however, continued its deadly assault reminding the airmen the war was not over. On a bright note, the Group's baseball team continued their unbeaten streak with four more victories.

THE MENACE lies broken after its landing gear collapsed on return from a combat mission on 23 July 1944. Damage was too great and the aircraft was salvaged. (Everson)

eleven

# HOME BY CHRISTMAS?:
# August 1944

As the English summer moved into August, the 448th continued their fast-paced operations in the air war. Maintaining the pressure on German forces in France, the Group targeted a fuel dump in the heart of Paris on 1 August. The 448th flew in two separate formations. The main formation took off just before noon followed by another squadron that flew with the 93rd BG. Poor weather over the French capital forced the bombers to change to their secondary target, an airfield at Melun. Once more weather prevented bombing but did not suppress the intense flak. After they were unable to bomb the secondary target, the Group scattered looking for targets of opportunity. Twelve 448th aircraft hit the airfield at Villaroche, France while nine other 448th crews bombed the airfield at Orleans/Bricy, France. Yet another ten aircraft from the Group attacked a highway bridge near Auberneaux, France. Only light inaccurate flak opposed the Group over these targets of opportunity and a tight escort of P-51s and P-38s discouraged any Luftwaffe aerial activity.

Sgt. Robert Kessler viewed the carnage wrought by the savage ground war and previous bombings. The French towns of Caen and Bayeux lay in ruins. The month-long fighting devastated the entire area. As the bombers crossed the channel, a large umbrella of barrage balloons appeared on the horizon. Despite the Allied air superiority, the Luftwaffe still visited London on occasion so these metallic balloons remained in place guarding the British capital. All 448th crews landed at Seething around 1800 without loss.

The next day, the Group briefed for a return trip to the Paris area. The target, a German Army supply dump, lay near the town of Pacy-sur-Armancon southeast of Paris. Crews stumbled out of bed at 0500 only to learn the early morning takeoff was delayed due to bad weather. Finally, after lunch the weather cleared enough for the bombers to take to the sky. Enroute to the target, the lead navigator became disoriented and the Group failed to locate the target. After numerous futile turns searching for the target, the Group sought a target of opportunity. As the formation passed over the town of Chaumont, one aircraft accidentally released its bombs. Eight other aircraft seeing bombs fall, dropped their loads on cue thinking they were over the target. Sgt. Robert Kessler watched as the bombs from his aircraft fell harmlessly into a wooded area near Chaumont, sparing a nearby hospital.

The formation continued its winding trip through France. Lt. Leroy Conner dropped his bombs a little later with the rest of the formation. Twelve aircraft hit the bombed out airfield at St. Dizier and four others hit the airfield at Creton. Thirteen other crews including Lt. William Snavely's, returned their bombs to Seething. The lead aircraft, FEUDIN REBEL, was forced to land at another base for fuel before returning to Seething.

The confusion proved costly. German flak crews zeroed in on the lost formation severely damaging aircraft 42-50566, flown by Lt. Joseph Madden. The bombardier, Lt. Jack Werts, heard the audible splats as the exploding shells ripped into his aircraft. German gunners poked over 100 holes in his aircraft including the hydraulic system. Hydraulic fluid soaked the interior of the plane. Arriving at Seething, they attempted to lower the landing gear but without hydraulics they had to manually lower the gear. This failed when the cable from the hand crank broke. Much discussion ensued between the crew, Col. Mason, Maj. John Laws, the engineering officer, and the tower weighing the options. They eventually agreed that a landing was out of the question and Col. Mason ordered the crew to bail out. Lt. Madden climbed the aircraft to 5,000 feet and turned the plane toward the sea. On the pilot's command, everyone hit the silk. Sgt. Fred Kerniss hesitated as he looked down from his stricken plane but a quick shove from behind sent him falling to earth. After pulling his ripcord, he landed safely near the airfield at Hardwick as did the remainder of his crew. The damaged aircraft continued flying, re-crossing the North Sea until it eventually crashed at Hagestein, Netherlands.

For the third consecutive day, the 448th flew to a target in France. Once again the target was fuel supplies, this time at Douai, France. Thirty-four B-24s left Seething in the Group's main force on 3 August. Three of the Group's planes joined the 93rd BG formation to complete another formation. All bombers dropped their twelve 500-pound high explosive bombs despite fires belching dense black smoke over the target. Only a few meager, inaccurate bursts of flak threatened the formation and no enemy fighters darkened the sky. Allied fighters continued to sweep the skies clean of any menacing planes and everyone returned to Seething without incident.

On 4 August the Group returned to Germany as part of a maximum effort aimed at aircraft factories throughout Germany. The 448th's target, an aircraft factory at Rostock, produced the Heinkel He-111. Lt. Donald Briola and crew left Seething aboard 42-50648 at 1030 leading the low left squadron. For the navigator, Lt. Walter Kurk, this was his twenty-third trip over German-occupied Europe. Lt. Leroy Conner flew LITTLE JOE loaded with a belly full of incendiary bombs designed to burn the target. Unlike the controlled fall of the larger bombs, these smaller incendiaries tumbled out of the bombers and fell 21,000 feet to the target area. Smoke and flames erupted as they struck their targets. Sgt. Donald Zeldin watched as other bombs fell short of their mark. Red-nosed P-51s of the 4th Fighter Group provided escort to northeast Germany and once again enemy fighters failed to materialize in the beautiful, clear skies. However, flak did appear. Lt. Briola's crew on aircraft 42-50648 suffered damage from the German guns that prevented a return trip to England They turned north toward neutral Sweden and safely landed at Sovde, Sweden where Swedish authorities impounded the plane and interned the crew.

Leaving the target area, navigators throughout the formation figured the fuel requirements to return to England. The long trip to eastern Germany taxed the limits of the B-24 and now strong winds blew in their faces. Amazingly, all but Lt. Briola's crew returned safely to Seething.

That day Stars and Stripes reported that U.S. troops occupied Rennes, France and the Russians were on the outskirts of Warsaw. The rapid advance of Allied armies created a sense of euphoria and hopes for a quick end to the war. Thoughts of Christmas at home crept into everyone's mind. This unexpected

An unidentified man stands in front of a UC-64 utility aircraft used by the 448th for various purposes. (LaPoint)

fortune of war fueled a false sense of hope that spread rapidly throughout the base.

Early the next morning, 5 August, thoughts returned to the job at hand as crews listened to briefers who outlined the mission and provided details of the target, a Ju-88 parts plant at Fallersleben, Germany east of Hanover. Weather, route to the target, and expected enemy opposition were discussed and final words from the Group Commander, Col. Mason, concluded the familiar sequence. Crews collected their combat gear and headed for their aircraft in preparation for a 0830 takeoff. Just after takeoff, Flight Officer Hosea Matthaes encountered mechanical problems with 42-94809 and aborted the mission. Returning to Seething, they overshot the runway and crashed. Fortunately the bombs in the bomb bay remained intact and did not explode. No one was seriously injured but the force of the crash landing damaged the aircraft beyond repair. Salvage crews scoured the wreckage trying to save as much as possible.

Thirty-two aircraft continued to the target, each Liberator carrying six 1,000-pound bombs. Numerous delays resulted in the Group arriving over the target one hour later than planned. A horrific flak barrage greeted them as they flew along the

When the field orders arrived at Group Headquarters they filtered down to each respective Squadron operations building where crews and aircraft were assigned. This particular building housed the 715th Operations. (Everson)

bomb run. Not only was the flak thick, but the gunners seemed to know the altitude of the bombers. They pounded the Group. As the bombs fell from 42-50443, flak struck two of its engines. Smoke billowed from the number four engine. Lt. William Snavely maneuvered the stricken aircraft out of the formation and they slowly slid aft. The tail gunner of the damaged aircraft, Sgt. Marvin Hicks, witnessed another B-24 that was not as lucky. He counted seven chutes blossoming in the sky as the Liberator from another Group fell to earth.

Nursing two damaged engines, Lt. Snavely coaxed the wounded bird to the English Channel before shutting down one of the damaged engines. The crew lightened the B-24 by throwing all loose items overboard allowing Lt. Snavely to maintain 16,000 feet until they had crossed the Channel. Safely over England, they shut down the other damaged engine and feathered the propeller. After landing, the crew climbed out of the battered aircraft at Seething, shaken but grateful to be alive.

LONESOME LOU, piloted by Lt. Donald Ginevan, faced similar problems. Damaged over the target by the deadly flak, he too set course for home. Their good luck expired and they were forced to attempt a hazardous ditching in the rough seas

A B-24 sits in front of one of the two large T2 hangars at Seething. The 459th Sub Depot performed most of the heavy maintenance on the planes here. (Everson)

of the English Channel. Air-Sea launches from Great Yarmouth answered their distress calls, but eight crew members, including the pilot, perished in the ditching. The rescuers saved two, Sgt. Robert Castell-Blanche and Sgt. Charles Jones, as well as nine other airmen from a 489th BG B-24 that ditched nearby. Another 448th crew paid the ultimate price. Elsewhere, Sgt. Ernest J. Hudgens, the tail gunner of EAGER ONE, suffered flak wounds but he was more fortunate. The crew returned the injured gunner to Seething for medical attention.

Liberators from Seething complete their assembly before departing on another combat mission to Germany. The airfield at Seething is barely visible between the clouds. (LaPoint)

The Group's 130th mission was to a rocket factory in the heavily defended city of Kiel, Germany. Another early morning briefing accompanied afterwards by the slow rumblings of multiple engines marred an otherwise peaceful morning. Pilots slowly taxied their heavily laden bombers onto the perimeter track, each plane entering the procession at well-coordinated intervals. The damp air provided a thin veil of humidity on the airfield as the rumbling bombers awaited the signal to commence their takeoffs. Nervous butterflies filled each person's stomach as they contemplated another trip to Germany. Although the events preceding a mission became routine, the mission itself was never routine. 6 August 1944 was no different. Tower fired green flares just after 0730 and the brightly painted assembly ship, YOU CAWN'T MISS IT rolled down the runway leading the procession.

After assembly YOU CAWN'T MISS IT returned to Seething and the formation headed over the North Sea avoiding the heavy flak concentrations as long as possible. Once over enemy territory, however, the flak batteries unleashed their deadly barrage. Flak was not the only problem. Smoke pots lit by the Germans belched thick smoke obscuring the target area. While several of the 448th aircraft released their bombs, most were unable to release due to the thick smoke created by the pots. The lead aircraft of the high right squadron suffered damage from exploding anti-aircraft artillery and slipped out of the formation. Lt. John Schlicher watched as the damaged aircraft fired flares. In the mayhem, the crew fired the wrong flares. Instead of firing abort flares, the crew mistakenly fired colored flares signaling the initial point. Unaware of the error, ten B-24s from the high right squadron followed their damaged leader. They unknowingly turned for home while the remainder of the formation continued on course. Realizing the mistake, Lt. Leroy Conner dropped his bombs on a target of opportunity near the coast rather than carry them home.

Two aircraft from the 448th suffered serious damage over the target. NO NOTHING, flown by Lt. Huntington Gruening, received serious damage and headed north toward Sweden. Reaching the haven of the neutral country, they landed at Sovde, Sweden, and were interned by Swedish authorities. Lt. Mauro Della Selva's aircraft, 42-51104, also experienced debilitating damage. Flak hits in several places forced the pilot to ring the bail out alarm.

For the copilot, Lt. Milt Halpern, the day proved to be a nightmare. He was not scheduled to fly but was awakened at 0030 and told he was flying as a replacement copilot. Now he was jumping out of a badly damaged aircraft over Kiel, Germany. His nightmare continued when his chute failed to open. He fell over 4,000 feet before he reached behind his back and

After bailing out of their aircraft over England, this crew's damaged plane continued flying until it crashed into a field in Holland. Standing (left to right) Madden, pilot; Lyon, copilot; Peterson, navigator; Werts, bombardier. Kneeling (left to right) Robinson, McLendon, Kramer, Bernard, Lewis, Fred Kerniss. (Kerniss)

CHECKERBOARD formally known as YOU CAWN'T MISS IT sits on a hardstand at Seething waiting for the next mission. (LaPoint)

miraculously deployed his parachute. Opening shock sent bolts of pain through his body as the loosely fastened leg straps cinched tight around his thighs and groin. Making matters worse shots fired from the ground partially deflated his chute and punctured his Mae West life vest. This was critical since it appeared he would land in the harbor. He landed in the water at 1230 and quickly started swimming toward Sweden. His aquatic trip proved short-lived as a harbor cutter fished him from the water and started his life as a POW. The remainder of his crew escaped from the B-24 but also spent the war as POWs.

Those crews that were unable to bomb the primary target turned toward the secondary target, a large oil refinery at Hemmingstedt. The bombs from FLEXIBLE FLYER missed the target after the autopilot malfunctioned on the bomb run. Regardless of the results, the crew was happy to turn for home and away from the deadly flak. Leaving the target area, the scattered formation assessed the damage. The tremendous flak dwindled the attacking 448th force dramatically. Only seven bombers from Seething dropped their bombs on the primary target. Eleven others dropped on the secondary target, an oil refinery at Hemmingtedt, but ten aircraft from the high right squadron salvoed their bombs after mistakenly following their aborting leader.

On the return trip, Sgt. Donald Zeldin aboard MY BABY, applied his limited first aid knowledge. A jagged piece of flying metal embedded itself just below the right knee of his buddy Lt. Ralph O'Neil. Fortunately it was not severe. Adding insult to injury, flak from the small German island of Heligoland, exploded perilously close to the bombers as they headed for home. Despite the unwelcome greetings, they returned to Seething without further damage.

Flak not only claimed two crews but also a waist gunner on LITTLE JO, Sgt. Earnest Easterling who was killed by a flak fragment. The navigator of LITTLE JO, Lt. John Deren, was wounded as well as a waist gunner on 42-51504, Sgt. Edward Hess. Flak damage to DAISY MAE forced the pilot to land at the emergency airfield at Woodbridge, England.

At Seething, Sgt. Julius Rebeles pondered the fate of his barracks' mates, the enlisted men of Lt. Della Selva's crew. The same hut that housed twelve men that morning now stood half-empty. Foot lockers and clothes still filled the room but the voices and laughter were ominously missing. It was a long, quiet night in their Quonset hut.

Bad weather with solid undercast prevented the Group from bombing chemical works and oil storage areas in Brussels, Belgium, the next day. All the bombers returned to Seething with their cargo. However, weather permitted bombing operations on 8 August. The entire 2nd Bomb Division attacked targets throughout France.

Two 448th squadrons supplemented other Groups; one squadron flew the high right position in the 93rd BG formation while the other flew high right for the 446th BG. Poor visibility hampered navigation and the lead navigator and bombardier were unable to positively identify the target, Villacoublay airfield. As a result of the uncertainties, they did not bomb the primary target but heavy flak barrages greeted the Liberators. The squadron flying with the 446th BG endured the most damage. Exploding flak rounds knocked several holes in OUR BABY but no one was hurt. LADY MARGARET also received flak damage and the bombardier, Lt. Robert Gardner, suffered injuries. The pilot elected to land in France to get aid for his wounded comrade and repair the damage caused from the flak. After some quick repairs, they continued back to Seething with-

MY BABY 42-95305. This Liberator flew numerous combat missions before returning to the U.S. following the war. (LaPoint)

A popular subject adorns this Seething-based Liberator. The female form was by far the most popular image on the nose of American aircraft throughout the world in World War II. (Patterson)

out their bombardier who remained in France for treatment. Bombs never fell from BARFLY; damage from the German guns forced them to return to Seething early.

Fortunately the squadron flying with the 93rd BG suffered very little damage. Unable to attack the primary target, the formation followed the Seine River until they found a target of opportunity, a railway bridge in Rouen, France. Thirteen aircraft dropped their bombs then headed for England. Despite the occasionally heavy flak, the sky was clear except for P-38s, P-47s and P-51s. Fighter escort during the entire mission was excellent; their presence always added a sense of security to the men in the bombers.

A large force of Liberators prepared to leave Seething on the morning of 9 August destined for factories in Stuttgart, Germany. Thick fog shrouded the entire base forcing the Group to scrub the mission. It was a blessing in disguise as most of the 2nd Bomb Division bombers that launched returned to their bases before reaching the target. Rapidly deteriorating weather conditions forced a recall of all B-24s.

An early morning briefing on 10 August detailed a mission designed to destroy fuel storage facilities at Pacy-sur-Armancon, France, southwest of Paris on the Loire river. Events delayed the takeoff slightly but just after 0700, thirty-six B-24s started their takeoffs. As they crossed into France, Sgt. Robert Kessler watched the battle rage below. Yellow-nosed P-51s circled above the bombers just in case the Luftwaffe decided to play games. The flight across France proved quite peaceful with very little flak. Ironically, just prior to the target, a sporadic flak burst hit 41-28648 crippling the bomber. The pilot, Lt. Parmely Ferrie, eased the damaged plane from the formation and headed for the safety of the Allied lines. Once over friendly troops near Caen, the crew abandoned the aircraft. Sgt. Charles Wolfe suffered a broken leg but everyone else landed safely. After several days, they returned to England.

Over the target, Sgt. Marvin Hicks watched the bombs fall from SONIA; not all fell as one bomb hung in the racks. All the bombs did fall from OUR BABY. Sgt. Robert Kessler watched as their bombs fell toward the target. He saw large, silent explosions as the bombs hit. Unfortunately, they overshot the support buildings and fuel tanks. It was frustrating to fly four hours to the target, miss, then fly back. On the bright side, it was one mission closer to going home.

PISTOL PACKIN MAMA. This tired-looking B-24, tail number unknown, flew with the 448th but its fate is unknown. (Everson)

A 714th BS B-24, radio call letter E, prepares to touchdown at Seething. A single letter and the shape around the letter identified each aircraft, triangle, circle, square or diamond, further identified which squadron the aircraft belonged. (LaPoint)

The next day, 11 August, the Group returned to the same area to bomb another fuel dump, this time at St. Florentin, France. Thirty-six B-24s left Seething around 1100. Moderate flak over the target failed to discourage the attackers who dropped their loads with improved accuracy. Again fighters hovered nearby providing excellent escort. Everyone returned without incident around 1900.

Airfields around Laon, France, received the attention of the 448th on 12 August. Crews stumbled through the darkness on their way to another mission briefing. The planned route of flight carried them across the Brittany peninsula, then south of Paris to the target. The outbound route crossed over Belgium and Holland before crossing the North Sea. Ten minutes after reaching the assembly area, an engine fire forced one crew to feather the engine and return to Seething. The remainder of the formation set out for France. As they neared the target area the formation split. Two formations bombed the airfield at Couvron while the other dropped their 100-pound general-purpose bombs on the airfield at Athies. Surprisingly strong flak erupted from the airfields but missed its mark. All bombers returned to Seething with only superficial damage. Crews returned the planes to the ground crews just after lunch. The devoted ground crews immediately starting fixing the problems encountered during the flight. These tireless men prepared the aircraft for yet another mission the following day.

Just as expected, field orders arrived setting in motion preparations for another mission. Ordnance personnel loaded fifty-two 100-pound general-purpose bombs on each B-24, thirty-six planes and 1,872 bombs. While the German flank reeled back in total disarray under pressure from General Patton's Third Army and General Hodges' First Army, Germans confronting the Canadians along the coast continued to put up a stubborn fight. The intended target for 13 August was choke points around Rouen, France, immediately in front of the British and Canadians. Plans called for each six-ship formation to attack three separate targets in support of the ground troops. The 448th targeted seven specific sites. Three were road junctions each packed with retreating German units. Even though the route carried the formation over enemy territory for only thirty minutes, flak damaged several planes. Sgt. Daniel Paris, waist gunner on BLUES IN THE NIGHT, suffered wounds from exploding anti-aircraft shells. For Sgt. Donald Zeldin the relatively short exposure to the flak on this trip was enough. He decided "the more times you're shot at, the unluckier you start to feel." On return the formation flew over numerous boats towing barrage balloons. The vast Allied armada on the sea and on the ground conveyed a sense of victory and confidence. Christmas at home seemed a realistic possibility.

The following day, the 448th returned to France for the fifth straight day. The airfield and associated fuel dump at

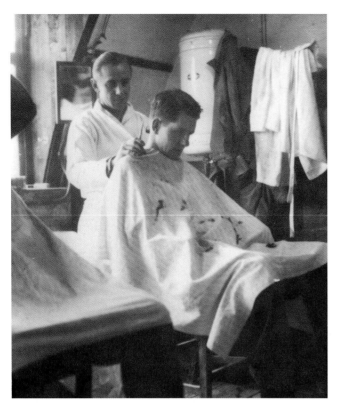

Haircuts were a staple of military life. At least one barbershop operated on base. It was located in a hut next to the gymnasium. (USAF)

The men from the base hospital staff provided excellent care for the life-threatening wounds to the incessant colds caused by the English weather. (Everson)

Longvic, France on the outskirts of Dijon was the target. To Sgt. Marvin Hicks on FEUDEN REBEL, the mission was ideal. Once again the Luftwaffe fighters failed to materialize and only a few minor flak bursts dotted the sky. Taking advantage of the light opposition, the 448th decimated the target.

Planners continued the pressure against the Luftwaffe on 15 August. Bombers from all three Bomb Divisions hit airfields in western Germany and in the Netherlands for the third time in four days. The 448th attacked the Plantlunne airfield on the outskirts of Rheine, Germany. Thirty-five Liberators departed Seething starting at 0900 but mechanical problems weakened the force by three. Beautiful weather prevailed over Germany and no flak or fighters impeded the bombers. As a result, the bombers heavily damaged the airfield. Upon return to Seething, strike photos revealed more. A large convoy of German troops, partially hidden by trees, lined a nearby road. Seizing the opportunity, a formation of medium bombers returned to the site and hit the troop concentration.

On 16 August, the Group revisited the German oil industry attacking an oil refinery at Magdeburg. Two squadrons of B-24s flew the high right position in a 93rd BG formation. Despite the summer weather, Sgt. Robert Kessler still suffered the misery of the cold at altitude. His right glove warmer stopped working and his nose ran incessantly due to the cold. His handkerchief turned into a ball of ice. The relative calm of the trip vanished as the planes neared target. Heavy, accurate flak enveloped the Group as they crossed over the target at 1105. ST. LOUIS WOMAN, flown by Lt. John Buxton, who arrived at Seething just eight days prior, exploded from flak. Flames engulfed the left wing, followed by the bomb bay, then the right wing. Nine men died in the inferno; the tail gunner Cpl. Leo Stephens survived and was captured. Sgt. Robert Kessler watched as a B-24 from another Group spun out of control with an engine on fire. Three Me-109s circled nearby but quickly disappeared when the yellow-nosed P-51s of the 20th FG sought to engage them. Following the gut-wrenching ordeal of the bomb run the planes headed for Dummer Lake and the rally point. The formation regrouped and headed for home across the Netherlands and the Zuider Zee. The ferocity of the flak, the loss of ten men and the re-emergence of German fighters served notice the war was still far from finished.

Following six straight missions, bad weather on 17 August provided a well-earned break for the Group. On the following day, the 8th Air Force reapplied pressure on the Luftwaffe by hitting airfields and fuel dumps in France. The 448th took off at 1100 with a full complement of three squadrons slated to hit a Luftwaffe fuel dump at Laneuveville, France, near Nancy. Thirty-five B-24s from Seething filled the beautiful August sky while P-38s circled overhead in anticipation of German fighters. Fighters failed to materialize as did the dreaded flak. Capitalizing on the clear skies and lack of opposition, the 448th hit the target squarely with a mix of general-purpose bombs. On the return trip, Sgt. Robert Kessler witnessed the raging ground battle as they crossed the front lines. Over the channel they saw a flight of British Halifaxes headed toward Germany as part of a rare British daylight raid. The route to and from the target took eight hours and, when Lt. Richard Moody landed at Seething, the fuel tanks were precariously low.

Bad weather followed and prevented flying operations until 24 August. The 448th resumed the offensive directed against the Luftwaffe with a mission to Brunswick, Germany. The entire 2nd Bomb Division hit various aircraft industries in the Brunswick area; the 448th bombed the airfield at Waggum. A malfunctioning propeller governor on FLEXIBLE FLYER forced Lt. Leroy Conner to abort as did five other B-24s from Seething. Good weather over the target allowed visual bombing. However, the clear skies left the bombers exposed to the German flak batteries and they offered quite a healthy appetite of lead. Although not flying in the 448th formation, Lt. Douglas Morse of the 715th BS, flying 42-95182, fell prey to the flak. Six men died but four escaped and returned to Seething.

Over the target, falling bombs narrowly missed claiming another B-24 from Seething. A damaged aircraft jettisoned its bombs directly over OUR JOY piloted by Lt. William Gilbert. Flying the left wing position of the low left squadron, he frantically maneuvered his aircraft to avoid the unexpected appearance of falling bombs. The first seven missed, but the eighth ripped a gash in the left wing puncturing a fuel tank. Unsure of the damage and leaking precious fuel, they set course for home. Once over England, they tried unsuccessfully to lower the landing gear. With one engine inoperative, they belly landed their damaged airplane at Seething without injuries with a mere twenty-five gallons of fuel remaining. A closer look at the damage revealed a five and a half by two and a half-foot hole in the left wing. The bomb completely severed one of the two main spars in the wing. It was amazing the plane continued to fly.

The next day, ground crews at Seething filled the B-24s' tanks to the top with 2,700 gallons of fuel in preparation for a mission on 25 August. Full tanks indicated a long mission and the word quickly spread around the base. Forty-six aircraft (the main effort consisted of thirty-four aircraft, twelve other 448th aircraft flew with the 446th BG) launched from Seething in a maximum effort designed to complete the destruction of the

This strike photo shows bombs exploding near the marshalling yards of Strasbourg, France on 11 August 1944. The intended target is the rail yards along the bank of the Rhine River. (Everson)

Heinkel works at Rostock. A previous attack damaged the facility but failed to destroy it. Forty-two aircraft of the 448th completed the assembly at 11,000 feet and joined the 20th CBW. Clear skies and bright sunshine provided a beautiful day for flying. Arriving over the initial point south of the city, Lt. John Rowe, flying 41-28924 in the low left squadron, saw a few widely scattered bursts. The target was clearly visible below. He watched as the bombs fell from the planes and arced toward the ground impacting the target area in fiery explosions.

Although the mission was considered a relative milk run, not everyone escaped the flak. Lt. Leroy Conner, flying EL

Although Luftwaffe fighters were scarce in August 1944, flak remained a deadly enemy. A wall of flak fills the sky around 448th planes near Ludwigshafen on 26 August 1944. (Bailey)

KORAB, in the lead squadron flew through much more intense flak. The flak knocked ten holes in his plane, none causing serious damage. Lt. Herbert Jonson flying 42-51504 also suffered flak damage and landed at the emergency field at Woodbridge with only two engines operating. For the remainder of the crews the route back across the Baltic and North Seas proved uneventful. Reconnaissance photos revealed outstanding results. The Group set a record for accuracy. The 2nd Bomb Divisions' Target Victory headlines read "448th Prize Sluggers at Rostock. Terrific damage to Heinkel works in Ace attack!" Sadly, the Group did not get a chance to savor their success.

The next day, 26 August, turned into a bloody one for the 448th. Lt. John Zima had grown weary of the incessant ground training and looked forward to his first mission. "Although I had hit the sack early I couldn't sleep at all. I kept thinking about my first mission. I had barely closed my eyes when 2:00 am hit me in the face and a hand was on my shoulder. Groggy, I stumbled into my clothes and headed for breakfast. The dawn was still far away when I gathered my flight gear and went with everyone else to the briefing room." The entire 20th CBW planned to hit the chemical industries located in Ludwigshafen, Germany. The 448th specifically targeted the I.G. Farben Chemical Works where Zyclon-B gas was manufactured for Hitler's notorious extermination camps. The immense complex sprawled for three miles along the banks of the Rhine just north of town. Intelligence sources reported 300 anti-aircraft guns in the target area capable of firing fifteen to twenty rounds per minute plus 100 German fighters in the area. It sounded ominous to Lt. John Rowe who was flying his second mission in as many days.

Thirty-four B-24s started their engines and ruptured the stillness of the quiet morning. The Group tried a new departure tactic for the first time. Two B-24s lined up on the runway at the same time. The first aircraft started its takeoff roll followed ten seconds later by the second. This new procedure expedited takeoffs thus preserving precious fuel. Five aircraft encountered mechanical problems and returned to Seething while the remaining bombers climbed to altitude and set course for Germany. The 448th flew as tail end Charlie for the entire 8th Air Force, a fact that cost them dearly. The formation approached the city just before 1030 with the target plainly visible to Lt. John Zima. "Spotting the air above our target, off to the left, was easy. It was awesome. Without exaggeration, the sky was filled with small black balls - FLAK!! ... To my uninitiated mind it was inconceivable that we were going to fly through it."

Leaving the initial point, all hell broke loose. Flak surrounded the bombers on all sides. To Sgt. Donald Zeldin flying in the lead squadron on LITTLE JOE, the flak seemed thick enough to literally walk on it. The noise reached a deafening crescendo and the flashes from the explosion caused his eyes to flinch. LADY MARGARET, flown by Lt. Francis Botkin, exploded in mid air just off their right wing; a direct hit severed the plane into two pieces and the right wing ripped from the fuselage. The wreckage crashed at Ruchheim, ten kilome-

The first bombs explode on the airfield at Plantlunne, Germany on 15 August 1944. Photo interpreters used this photo to identify a column of military vehicles on the adjacent roads which were attacked by medium bombers later in the day. (Ash)

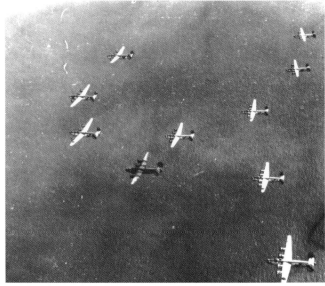

A formation of B-24s from the 448th pass over the North Sea on their way to the continent and combat. Note the single B-24 in olive drab paint. The others have a natural metal finish. (Bailey)

Crew 142 stands in front of 714th Operations at Seething before being shot down on 26 August 1944. Kneeling (left to right) Frank Bastian, pilot; Donald Disbron, copilot; David Holst, navigator. Standing (left to right) Phil DiGiacoma, waist gunner; William Wilbur, engineer; Robert Coletti, waist gunner; Frederick Theobold, nose gunner; Dewey Holst, tail gunner; William Degnan, radio operator. (Wilbur)

ters west of Ludwigshafen. Miraculously the pilot, Lt. Francis Botkin, and three others, Lt. Richard Goshorn, Sgt. James Boatright and Sgt. Harold Macauley were blown clear of the aircraft and survived as POWs but their six crewmates perished. Sgt. Zeldin instinctively ducked in his top turret as the landing gear from the exploding plane narrowly missed LITTLE JOE. The flak did not miss. It tore into LITTLE JOE wounding the navigator, Lt. Russ Vakoc, and setting fire to the number four engine. Fuel spewed from a gaping hole in the wing between the number one and two engines. Sgt. Zeldin quickly started transferring precious fuel from the damaged tanks. The remains of LADY MARGARET crashed west of Ludwigshafen, Germany, at Rucheim.

Lt. Miles Drawhorn, copilot on EAGER ONE, saw LADY MARGARET fall from the skies. His crew, led by Lt. Andrew Panicci, considered LADY MARGARET their aircraft. They flew the majority of their missions on her and were disappointed when they were not assigned to fly her on this mission. Lt. Drawhorn even complained to operations about it but was told no changes were allowed. He was promised they would fly her on future missions. There were none!

At 1030 numerous flak bursts severely damaged aircraft 42-50788 piloted by Lt. Frank Bastian. The stricken plane suddenly rolled inverted and plunged to earth. The bail out order was given and two men bailed out, the bombardier Lt. Enrico Maggenti and the right waist gunner, Sgt. Robert Colletti. The Germans quickly captured them. Struggling with the plane, Lt. Bastian regained control and returned the aircraft to level flight before anyone else jumped. He flew the heavily damaged aircraft toward France. Over Chartres, France four men, including Sgt. William Wilbur and Sgt. Dewey Holst, bailed out of the plane. Sgt. Holst's brother, the navigator Flight Officer David Holst, remained with the airplane a little longer. Over Montreuil, France, Lt. Bastian set the autopilot and ordered everyone out of the airplane. Miraculously, everyone survived and all but the first two men who bailed out returned to Seething.

Meanwhile the pounding continued. The second squadron successfully crossed the target without losing an aircraft but the final squadron was not as lucky. Lt. Edmond Postemsky's aircraft, 41-28924, took a flak round between the number three and the number four engine. The left wing folded as the plane erupted into a fireball. The burning hulk entered a mad spiral and crashed near Mannheim, Germany. The copilot, Lt. Clifford Unwin was the only survivor.

Lt. John Zima was in the last squadron. "I grew more horrified as we approached our target, and my Hail Marys were coming fast and furious. I wondered how anything, especially a four engine bomber, could get through. Our nose gunner who had the best view, said he couldn't stand it and cradled his head in his arms. As the formation ahead of us went through there was a sudden ball of fire, a puff of smoke, and chunks of airplane fell downward to earth. Then there was a lull, a suspension, followed by events unfolding faster than one can explain. It was our turn. The puffs were suddenly all around us, puffs of black smoke with centers of red fire, curling and twisting as each shell exploded. Suspended about three feet to the right, it exploded about five feet above us. It was just a blur as my eyes followed it up and watched it explode."

Lt. William Stonebraker's crew also succumbed to the horrendous flak. Two minutes after the target, exploding shells split his aircraft, 42-50443, into two pieces. Six chutes billowed from the doomed plane; four others opened without being seen. All ten crewmembers somehow escaped the death trap only to face angry civilians with iron pipes. Three of the crew suffered injuries from beatings by the civilian population. All of the crew were eventually captured.

The exploding aircraft almost claimed Lt. Zima's aircraft. "The plane was to the left, in front, and slightly above us when it was hit between the number three and four engines. As the plane rose and turned over on its back the copilot was clearly visible in his seat, looking as if nothing was happening. He was either hit or not able to grasp the calamity that had overtaken him. The plane passed over us before beginning its death dive. There was a flash of fire as the right wing parted and went over us. A piece of debris, I think it was the supercharger, came directly at us. We ducked our nose and the debris ricocheted off the nose turret, ricocheted off the navigator's dome, and then went over the cockpit. I thought that our nose gunner and navigator were dead. We kept our nose down and the command pilot jettisoned the bombs and pushed all the controls to the wall, prop pitch, supercharger, and throttles, trying desper-

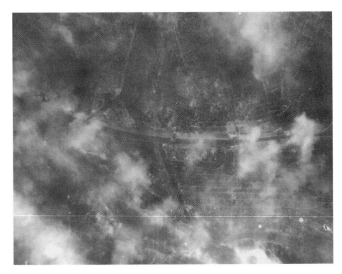

One of the aircraft lost on 26 August 1944 goes down over Ludwigshafen, Germany with a portion of its wing missing. (Everson)

ately to change altitude and get out of range of the guns below. Just then, right in front of us, a shell exploded so close that every detail was visible. Dark red churning gasses and black smoke looked like an angry living thing trying to get at us."

Smoke screens ringing the city partially hid the target but twenty-eight aircraft from the 448th dropped their bombs amid the terrifying flak. One aircraft, THE STURGEON, in the lead squadron experienced a bomb rack malfunction and was unable to drop.

For the surviving crews, they now faced a daunting return trip, many with severe damage. Numerous P-51s gathered around the battered formation protecting them from any unfriendly advances. Sgt. Donald Zeldin, the engineer on LITTLE JOE, assessed their damage. The wounded navigator appeared stable and the fuel leak ceased. The next concern was damage to the landing gear. A large chunk of rubber was missing from the right tire. They wondered if the gear would lower. Even if it did, damage might still jeopardize their ability to land. Once over England they lowered the landing gear. It locked into place without incident. Not knowing if the gear would collapse, they gently touched down and rolled to a stop. Sgt. Zeldin counted 140 holes in LITTLE JOE. What a way for his pilot, Lt. Ben Baer, to complete his thirtieth and final mission.

While Lt. Baer completed his final mission, Lt. John Zima struggled to finish his first. "When the navigator's dome was hit the navigator and bombardier were showered with the shattered glass. Our violent maneuvering threw them both to the floor, and each thought the other had been hit. When they finally found their 'mike' buttons it was a great relief to hear they were all right. Thinking of the nose turret I called to ask if the nose gunner was uninjured. 'Yes, I am all right,' he assured, 'but this turret is sure shot to hell!'" Everyone emerged rattled but miraculously unhurt.

Flak damaged the number three engine, knocked holes in the nose turret and ball turret, and punctured a hole in a fuel tank of BARFLY. Lt. John Rowe landed his aircraft at Seething with two damaged engines. OL' BUDDY landed at Woodbridge with only two engines working, but the crew was safe, unlike many others.

It was a costly mission. Numerous crewmembers required medical treatment and four crews were missing in action. The long process of notifying next of kin started. It ended with the dreaded Western Union telegram notifying loved-ones that their son, father, or husband was missing in action. Agonizing weeks, months and sometimes over a year passed before anxious families learned more details. The Group chaplain, Theodore Runyon, attempted to ease the suffering of Sgt. William Wilbur's mother, Clara, after her son was listed MIA:

My dear Mrs. Wilbur,

You know, of course, by this time, that your son, William, has been Missing in Action since August 26th. As his chaplain, I am writing this personal letter to assure you that I join my prayers with yours for his safe return. I trust that you will not give up hope that William has survived and is somewhere alive and well, until you have received definite word to the contrary.

William was one of a large force of young men over here who daily show great courage and determination in helping to overthrow an enemy who would destroy everything we hold dear and sacred. You can well be proud of him.

The next few weeks or months will be for you a time of anxiety and great uncertainty - a time when you will long for news or information that will set your mind and heart at ease. I trust your courage will not fail. You have my deepest sympathy in this time of trouble, and I hope and pray that the life of your son has been spared.

Faithfully yours,
Theodore Runyon
Chaplain

In Sgt. Wilbur's case, a second telegram brought the good news of his return to duty. More common was a second telegram notifying families of the confirmed death of their loved-one. Occasionally, it verified their status as a prisoner of war. Some families waited one long year before their missing loved-one was declared dead. It was a horrible process.

LADY MARGARET 42-95134. This aircraft exploded from a direct hit by flak on 26 August 1944 killing six of the men on board. An unknown man stands by the front of the plane. (Everson)

Fortunately for the battered crews, another deep penetration into Germany the following day, this time to Oranienburg on the outskirts of Berlin, aborted due to unfavorable weather conditions along the route. High clouds near the Dutch coast necessitated a recall. As the bombers crossed the German island of Heligoland, stray flak rounds dotted the sky. Several bombers seized the opportunity and dropped their bombs on the island.

This was the last mission of the month for the Group. Problems with the main runway and the perimeter track required the airfield to be closed for repairs. The closure was overdue as problems with the runway and perimeter track plagued operations all summer. Despite the short cessation of operations, it was a busy month. The Group flew twenty missions losing ten crews in the process. Forty-one aircraft received damage requiring repairs. Even though operations remained hectic, morale remained high.

While the progress of the war contributed immensely, several events during the month helped sustain morale. Cpl. Billy Conn and other 8th Air Force boxing champions held an exhibition at Seething on 9 August. Three days later the 8th Air Force all star baseball team, including several professional ball players, entertained a crowd at Seething narrowly defeating the home team 3-1. Despite the excellent play of the baseball team all year, they finished the second half of the season as Eastern League runner-up. In the playoffs, Hardwick defeated the team from Seething in two games to win the championship. The Group diary best explained the feelings of the team. "Seething bowed out of the show with a dejected spirit but also with the defiant cry of 'Wait till next year,' except that everyone hoped that by next year at this time he could be seated at a stadium in the States dressed as a civilian with three bottles of pop in one hand, four non-GI hotdogs (with mustard and sauerkraut) in the other, and a scorecard between his teeth roundly jeering the umpire's latest decision or calling for a home run by his favorite big league team."

A month long War Bond drive concluded with base personnel purchasing $170,470 worth of bonds. Incentives and high-pressure sales tactics aided in the drive. Incentives included a six-day pass to the highest purchasing combat crewmember, the same for ground personnel, a case of scotch to the combat crew purchasing the most, and a three-day pass to anyone who purchased a $100 bond. Winners were Lt. Grenville Baker, a 713th BS pilot with a $5100 purchase, Pvt. Ernest Pilley with a $1325 purchase and Lt. Baker's crew who combined for $6025.

Despite the once promising hopes of Christmas at home they quickly faded as resistance failed to collapse. Not only did the Germans continue to fight in the air, but the ground forces in Europe were approaching the end of their lightning campaign. Resistance stiffened and the long supply lines from the Normandy beaches plagued the Army. The following month dashed all hopes. Much fighting on the ground and in the air remained.

twelve

# VERSATILITY: September 1944

Following a busy August, the 448th commenced operations in September on a more subdued, yet versatile pace. Construction work on the perimeter track and runways suspended operations until 5 September. Over eight months of uninterrupted use took its toll. The concrete deteriorated under the constant pounding of heavily laden aircraft and changing weather conditions. Patchwork repairs enabled the airfield to remain open as engineers constantly filled the potholes and cracks. Finally, it became necessary to temporarily close the field to complete the long overdue repairs.

The first mission flown from the repaired airfield departed on 5 September. Ironically, the Liberators did not partake in destruction but flew a mission of mercy. The rapid liberation of France, an unseen event to SHAEF planners, left a large French population on the verge of starvation. To alleviate a potential disaster, B-24s were utilized in Truckin' missions to carry food and supplies. The 448th flew missions to Bricy airfield on the outskirts of Orleans, France. In a weird twist of fate, the 448th bombed this particular airfield three separate times, the last time on 14 June. Now they landed on the same airfield in a humanitarian role.

The 448th returned again the following day, 6 September. Weather prevented flying on 7 September, but the Group flew in support of 'Beans for Orleans' again on 8 September. Sgt. Donald Zeldin donated a couple of packs of cigarettes and three cakes of soap in the effort. While on the ground in France, a Frenchman surprised him by running away with his gifts and returning with a bottle of wine and a five-Franc note. The ecstatic Frenchman shook his hand, then hugged him in an emotional display of gratitude. It was a touching show of emotion in a time when horrors were daily expectations.

The Group flew its final Truckin' mission on 9 September while it simultaneously returned to its original mission, destruction of the German Reich. Sgt. Robert Kessler climbed into his bed at 0200 following a forty-eight hour pass only to be roused from his sleep two hours later by a blinding flashlight in his face. "Briefing in an hour."

Although the day was just beginning for Sgt. Kessler, many others around the base worked throughout the night to make the mission a reality. Crew chiefs prepared the aircraft, armorers loaded the bombs, planners sorted through the unending pages of field orders that spilled off the Teletype and intelligence pored over charts and reports all in preparation for the mission. The nucleus of many of these activities was the control tower. Cpl. William Schwinn was one of the men who worked at the control tower:

"The tower consisted of two sections, and my first job was Airman of the watch. I had a desk with two or three telephones on it and in the morning when a mission was to run, whoever was in charge of the mission had to bring in the paperwork to me. One of the papers would have a layout of the formation that they would be flying that day. If you dug further you would come up with the target which was always a secret. Inside the tower there was a huge flight board all marked off. You had to fill in the spaces with the pilot's name, the aircraft number, the estimated takeoff, the actual takeoff, the estimated time of arrival, and the actual time of arrival. This board was used to count the ones that were lost and identify who they were. Once in awhile they thought they had lost some, but of course some had malfunctions and landed in fields and could not get all the way back."

Men repair holes in the runway and perimeter track. The rapid construction and constant use took its toll finally forcing the airfield to close for repairs. The tower and technical area are visible in the background. (Everson)

All the necessary tasks were completed and thirty-one aircraft left Seething for the marshalling yards at Gustavsburg on the outskirts of Mainz, Germany. Bad weather over the target required bombing by PFF. Although the 93rd BG did not fly as a Group, they provided the lead as well as deputy lead aircraft for the 448th.

Poor visibility and high clouds complicated the formation flying. Summer was slowly giving way to winter over northern Europe. Although bitter cold temperatures made crews uncomfortable, they preferred the cold to the intense flak thrown up by the German guns. In an effort to fool the German radar, crews threw small foil strips, known as chaff, into the slipstream. Despite the chaff, also known by its code-name window, the intense flak continued and three seconds before bomb release, it struck with deadly results. At 1046, the lead aircraft exploded from a direct hit in the bomb bay. Adjacent crews saw the tail of the aircraft break off as an orange flame erupted out the gaping hole that once was the tail. Maj. William Blum, the Group Operations Officer, was flying as the command pilot on the lead aircraft. Despite the horrific explosion he bailed out and was later captured. The explosion also damaged the deputy lead. In the resulting confusion, only five aircraft in the low left squadron dropped their incendiary bombs. Ten minutes after it started, the flak mercifully ended.

Yellow-nosed P-51s continued their dutiful escort and again the Luftwaffe failed to pose a challenge. However, stray rounds of flak exploded as the Group passed the Ruhr on the return. The flak seemed to grow more intense with each mission. Numerous aircraft from Groups throughout the 2nd Bomb Division landed in France on 9 September due to the damage from the German guns.

The 448th returned to Germany on 11 September as part of a concentrated effort against the German oil industry. An oil refinery at Magdeburg was selected as the target for the 448th. Lt. William Snavely and his crew rose earlier than normal and after a breakfast of hot cakes, they were trucked to nearby Halesworth. They were flying as deputy lead for the 448th, a position that required a PFF aircraft. Since no operable PFF aircraft were available at Seething, they 'borrowed' a PFF aircraft from the 489th BG. Forty-six aircraft started taking off just past 0800 hours.

Normally, three squadrons comprised a Group formation. The lead squadron was flanked by one squadron slightly below and to the left and another squadron stacked slightly above and to the right. As the number of aircraft increased it became necessary to add another squadron to the formation. It usually flew high right of the high right squadron or in the bucket, inside the Vee created by the first three squadrons. On 11 September, the Group flew as four squadrons with one squadron flying high right of the high right squadron. Several aircraft aborted for various problems but forty-two aircraft departed the English coast for Germany. The route to the target carried the formation over Belgium and the front lines. A heavy pall of smoke marked the battle lines as they crossed into occupied Europe.

As the bombers passed near Koblenz at 1130, accurate flak erupted damaging several B-24s. NO LOVE NO NOTHING, flown by Lt. Richard Vogel, slipped out of formation with one engine seized. After clearing the formation, NO LOVE

Woody Spurr mans the "greenhouse" on top of the control tower. The job of air traffic controller was difficult in the tightly confined airspace of East Anglia. Radio silence, number of aircraft, and the combat environment compounded the problems. (Bollschweiler)

NO NOTHING collided with a P-51 from the 361st FG and crashed at Dalfsen, Holland. All of the B-24 crew survived except the tail gunner, Sgt. John Phillips, whose parachute failed to open. Five of the men were captured but four, including Lt. Vogel who was wounded, evaded capture. The P-51 pilot, Lt. J.H. Lougheed also died in the collision. It was only the crew's second mission. Another aircraft, GUNG HO, flown by Flight Officer Albert Lewis, also suffered flak damage. They went down over Koblenz. Five of the crew perished in the crash. LITTLE JOE experienced trouble also. Slipping out of the formation, the pilot turned LITTLE JOE for home, but not before dropping their bombs on a target of opportunity at Rotenburg, Germany.

Once they were clear of the flak at Koblenz, the Luftwaffe rose to challenge the invaders. The escorting P-51s and P-47s engaged the fighters but the German fighters still inflicted damage. Fortunately for the 448th, the fighters hit other Groups and did not attack the 448th. Crews watched as bombers from nearby Groups fell. Sgt. Robert Kessler watched the Group behind the 448th take a pounding. High clouds obscured the target but they bombed by PFF from 24,000 feet. The lead PFF ship experienced a bomb release malfunction and could not drop. Eight aircraft, including Lt. John Rowe's, in the low squadron did not drop due to heavy cloud cover. They hit an airfield at Diepholz, Germany as a target of opportunity.

The Group's Republic P-47 Thunderbolt flies close formation with a B-24. It was not uncommon for Lt. Col. Herbert Judy, the Group Air Executive Officer, to use this aircraft to monitor and supervise the assembly of the big bombers. (LaPoint)

Only the cold temperatures hampered the crews on return to Seething. Wind whistled along the entire length of the B-24s like air through a wind tunnel. Not only did air pour into the plane from the open waist windows, but it also swirled in around the poorly insulated nose turret. To allow for unhindered movement, seals around the nose turret were loose and did little to stop the rush of cold air entering the plane. As a result, crews donned numerous layers of clothes to stay warm and protect themselves from frostbite. The frigid air froze any unprotected skin. To stave off the cold, crews wore electrically heated suits under the bulky leather flying clothes. Affectionately called bunny suits because of their blue pajama-like ap-

A 448th formation flies high over Germany amidst a broken cloud layer. Notice the bombs falling from one aircraft. (LaPoint)

pearance, they offered some protection when they worked. When they failed, crews suffered. Sgt. Robert Kessler struggled with a loose connection in his heated glove. The cold temperatures numbed his hand until it was almost useless. The cold also proved too much for some of the equipment. In some cases, even the cameras used to record the bomb strikes froze. It was just September and the European winter was still in its infancy; the worst was still to come.

For the second straight day, the German oil industry was the target. Just like the day prior, Lt. William Snavely and crew loaded into a truck in the early morning hours for a ride to Halesworth. They led the Group on a mission to Hemmingstedt, Germany. Despite low clouds and rain at Seething the bombers departed. Three squadrons from the 448th dropped their bombs on the primary target by PFF. No flak or fighters opposed the bombers and everyone returned unscathed.

On 13 September, thirty-six B-24s left Seething destined for a munitions dump at Ulm, Germany. By 0930 high clouds engulfed the bombers and played havoc with the formations. Contrails added to the difficulty as they limited visibility within the formation. At 0950 as the formation passed Paris, Big Bear Blue, code-name for the 448th Bomb Group, sounded the recall and the Liberators from Seething turned for home. (The code name Big Bear Jubilee indicated the entire 20th CBW while Big Bear Blue was for the 448th, Big Bear Green for the 93rd BG and Big Bear Red for the 446th). All but one aircraft returned with their bombs. The lone aircraft bombed a target of opportunity.

The next day, 14 September, crews groaned as the briefer unveiled the map. Berlin! Lt. William Snavely's crew once again woke early and rode by truck to Halesworth. Bad weather delayed the takeoffs and at the last minute the bomb loads of some aircraft were changed. Meanwhile at Halesworth, Lt. Snavely sat in his aircraft until 1030 awaiting clearance to start engines. Finally after an interminable wait, they were cleared to start engines. Once started, they discovered three of the four generators were inoperative. As they scrambled to prepare the spare lead aircraft, tower informed them the mission was scrubbed. It was another effort in futility, a lot of work with nothing to show.

As the war continued, many of the locals befriended the men from Seething and life-long friendships developed. Families adopted 'Yanks' and made them part of their families. Sgt. Wally Balzer explained in a letter to his fiancée how he met the Mobbs family:

"About ten days ago the chaplain received a letter from an English lady in an adjoining hamlet requesting that two

WOLF PACK 42-52121. After flying its last combat mission on 6 July 1944, this Liberator was salvaged on 9 September 1944. (Everson)

soldiers be guests in her home on Sunday evening. I did not wish to go because of my duties but seemingly it was impossible to get anyone else. Most of the fellows had to be on duty and the two who had been previously scheduled were restricted for a week so I finally recruited another fellow who was able to go. So the two of us left on bicycles on the road for two miles to this nearby hamlet where after inquiries we located the home of our hosts. The name of our host was Mobbs and they lived in a tenement or something comparable to a duplex home back home. Mrs. Mobbs was so kind and hospitable and so was her husband. Both were middle-aged and the happy parents of two children, a twenty-one year old boy who was delivering milk in a nearby village and a darling seven-year old daughter named Jennifer. The boy's name was Charles. I immediately fell in love with Jennifer because she was so sweet and courteous. At 5:30 or 6:00 we had tea which consisted of various kinds of cakes and tarts with tea. It certainly was very tasty. Then after this was completed we sat in the cozy, crowded living room that also served as the dining room and talked about our homes and life in England. For a while Jennifer and I looked at picture books. One of them she had colored herself...Jennifer seemed to enjoy sitting on a lap insulated by a pillow but it was soon her bedtime. Her mother told me upon her return from having put her daughter to rest that Jennifer remarked, 'I will never forget tonight mother.' I thought it was so cute of her to say that."

This one visit started a relationship with the Mobbs family that grew over time. More and more GIs visited the friendly haven and enjoyed the hospitality of Mrs. Mildred Mobbs, the

Hedenham postmistress. Sgt. Balzer continued his fond treatment of Jennifer and helped celebrate her eighth birthday on 22 September. Despite the warm fellowship with friends, the war continued.

Poor weather hindered flying operations until 17 September when the entire 2nd Bomb Division, including the 448th BG, participated in a six-hour practice mission. The entire mission was flown at extremely low altitude while maintaining formation. It was different from the normal tactics employed but similar to what B-24s used against Ploesti. Crews did not have long to ponder why; the following day, the 448th employed this new tactic.

While the bombers were flying their practice mission, Operation Market-Garden commenced. British Field Marshal Bernard Montgomery intended to use airborne troops to seize bridges across several rivers in Holland. Simultaneously, a British armored corps attacked along a narrow corridor in an effort to link up with the airborne troops. Until they arrived, the airborne troops only supplies would arrive by air. The versatile B-24s of the 2nd Bomb Division once again inherited the role of transport. After a promising start, things faltered on the second day of the attack.

Despite the best plans, confusion reigned on the morning of 18 September. The 20th CBW received field orders for the 14th CBW and vice versa. Despite the 'snafu,' thirty-six aircraft, four squadrons, left Seething on time, each plane loaded with twenty bundles of supplies. The four squadrons each consisted of three, three-plane elements. Thirty seconds separated each element. In theory, this allowed a more accurate drop pattern as only three planes crossed the drop zone at a time instead of a nine-plane squadron crossing at once.

After assembling at 2,000 feet, the formation followed the 93rd BG and descended to 300 feet as they crossed the coast. A thick haze enveloped the planes as they descended and the 448th lost visual contact with the 93rd BG. Not only did the low altitude complicate formation flying, but also the rough air, stirred up by the preceding formations, jostled the planes. Despite flying so low, they maintained formation.

The Seething skeet team proudly displays their second place plaque from the Second Bomb Division skeet tournament. Team members are (left to right) B. D. Bottoms, W. R. Irons, Len Thornton, and W. E. Irwin. F. L. Krepser is kneeling holding the plaque. (Bottoms)

One of the deadly flak batteries that caused so much devastation among the bombers. (Wilckens)

Left: A German soldier stands among the wreckage of the 489th BG Liberator that was leading the 448th BG on the 9 September 1944 mission. (Wilckens)

As they crossed the front lines, every German with a gun turned it skyward putting up a wall of small arms fire. Numerous crewmembers suffered wounds from the hail of bullets. Adding to the fog of war, the radio locator beacon on the drop zone failed. Without visual contact with the preceding formation and no radio aids to identify the drop zone, the Group struggled to locate the assigned drop zone. As the planes neared the drop zone they dropped even lower, some as low as fifty feet. Sgt. Robert Kessler flew so low he saw a monk in a long black robe waving a scarf as they flew overhead.

Each aircraft carried an Army dropmaster who pushed the bundles out of the plane as they crossed the drop zone. Aboard EAGER ONE, a static line ripped the hoist from the ball turret. As it fell through the camera hatch it pulled Pfc. Peter Jasura of the 490th Quartermaster Company from the aircraft. No one saw him pull the ripcord on his parachute. Even if he had tried, the low altitude, 250 feet over the target, would have prevented it from deploying. Elsewhere red, white and yellow parachutes filled the sky as the bundles floated to earth. Unknown to the crews they dropped eight miles short of the intended target.

As the Group reassembled and headed home, crews on many of the planes tended to wounded friends. At least seven B-24s landed at Seething with wounded crewmembers. Lt. Leroy Conner, flying OUR BABY, landed with a wounded tail gunner. Thirty-caliber machine gun fire hit Sgt. Johnny Bretthauer in both legs just prior to the drop zone. The copilot of DAISY MAE suffered injuries as did the radio operator of PATRICK DEMPSEY. Three men on 42-51489, the pilot Lt. Carl Eggert, Sgt. Robert Lloyd and Sgt. Back, were also wounded. Damage to the nose wheel forced the crew of FLYING DRAGON to land at the emergency airfield at RAF Woodbridge. In preparation for landing, Sgt. Robert Kessler and the rest of the crew moved to the tail of the plane. When FLYING DRAGON touched down, the extra weight in the tail kept the blown nose tire from touching the ground until they had slowed. Although no aircraft were lost, personnel injuries and equipment damage were costly. The low-level route to the target was designed to maximize surprise. In retrospect, it placed every aircraft within range of even the smallest German gun.

Worsening weather grounded the Liberators on 19 and 20 September. However, the weather cleared enough on 21 September to allow the bombers to attack the marshalling yards at Koblenz. Two squadrons attacked, one with the 93rd BG and the other with the 489th BG. Cloud layers hid the ground all the way to the German border. As they neared the target, the clouds scattered slightly creating holes but they still obscured the target. All bombing was accomplished with the aid of PFF and the results were unobserved. Even though flak was en-

GUNG HO 44-10505. Flak claimed this Liberator and five of its crew on 11 September 1944 over Koblenz, Germany. (Everson)

NO LOVE NO NOTHING 42-95138. Flak heavily damaged this B-24 then it collided with an escorting P-51 and crashed in Holland on 11 September 1944. (LaPoint)

countered at three separate locations, no damage was inflicted on the bombers. On return, the Liberators found low ceilings and poor visibility over England. Crews sweated out the descent. They finally broke out of the weather over the coast and found Seething despite 1,000-foot ceilings and 140 yard visibility.

The entire 8th Air Force attacked the Henschel armor and vehicle works at Kassel, Germany on 22 September. Men loaded 500-pound general purpose and incendiary bombs on thirty-seven B-24s. Crews left Seething at 1100 and completed the standard assembly using instrument procedures. Crossing the Rhine River, the formation encountered heavy, concentrated flak which damaged FRISCO'S FRISKY. They turned for home. At the same time, the low left squadron became separated from the main body of the 448th. Unable to find the rest

This formation ship was probably the third assembly ship for the 448th following YOU CAWN'T MISS IT and STRIPED APE. It is probably aircraft 41-29489 which was believed to be named 2nd AVENUE EL while it flew combat missions. Its last combat mission was on 12 September 1944 but continued flying as an assembly ship until it was salvaged 26 May 1945. (Everson)

of the Group, they joined the 446th BG. Despite a complete undercast that concealed the target, the Group dropped their bombs with the aid of PFF. Bombardiers called 'bombs away' at 1414 and the formation turned toward the rally point. Lt. Leroy Conner saw contrails in the distance but the excellent escort of P-47s, P-51s and P-38s kept all enemy fighters at bay. On the return leg near Trier, Germany flak again dogged the planes. A gunner on BUFFALO GAL, Sgt. John Thompson, was wounded in the day's action but all the planes returned to Seething.

While the war raged over Germany, the men at Seething hosted 200 evacuee children on the afternoon of 22 September. Sgt. Wally Balzer helped with the party. "They arrived in trucks about 3:30 and after further instructions had been given the majority of them decided they wanted to see the airplanes out on the line. Most of the time I had a six-year old boy, Anthony, and an eight-year old girl, Audrey, with me. They were brother and sister and were staying with some people who had a three-year old boy. The evacuees were staying in homes and are not assembled in large buildings. They attend the public schools right along with the other children. We also watched some planes land and of course were very thrilled to see that. After returning to the Aero Club (that would be the American Red Cross Club) the children had ice cream, cookies and cakes, sandwiches and lemonade. Most of them I believe had not had ice cream for months or even years. A number of them had seconds. The ice cream is made here on the base since the freezing equipment was secured just a short while ago. That also explains why we have ice cream now once a week."

Once again, no bombers flew on 23 September or on 24 September. However, Hitler's vengeance weapons did fly. A V-1 'buzz bomb' landed near Seething on the 24th. Following constant attacks on the Luftwaffe and German oil industries, planners shifted their focus to the German transportation system. Winter weather, characterized by low clouds, made attacks on small targets such as airfields and oil refineries impossible. However, due to the technological advances associated with PFF, railroad-marshalling yards associated with larger cities were identifiable by radar. As a result, as the weather worsened, the frequency of attacks against these targets increased.

The 448th participated in a raid on the marshalling yards at Koblenz, Germany, on 25 September. Sgt. Fred Kerniss awoke at 0330 and attended the briefing two hours later. No flak or fighters opposed the bombers enroute to the target. Once again bad weather required the bombers to use PFF for bombing. Four minutes prior to the target, the bombardier in the deputy lead aircraft accidentally hit the toggle switch releasing the full load of incendiary bombs. Seeing bombs falling from the deputy lead, all the aircraft dropped their bombs as well. Only the lead aircraft did not drop early. All the bombers landed safely at Seething at 1430 after their disappointing results.

Over 300 planes from the 2nd Bomb Division took part in a raid against the marshalling yards at Hamm, Germany in the

This view of the control tower at Seething shows several B-24s on nearby hardstands. A Dodge ambulance sits on the perimeter track by the tower ready to respond to any emergency. (Bollschweiler)

heavily defended Ruhr valley. Since becoming a PFF lead crew, Lt. William Snavely's crew flew frequently. Missions usually started with a truck ride to Halesworth to pick up a PFF aircraft. 26 September was no different. Taking off just before lunch, they flew to the target without incident. As was usual, no German fighters threatened the formation. Flak, however, did! The Ruhr valley with its heavy industrial areas was the most well defended location in Europe with Berlin being the only possible exception. Heavy, accurate flak filled the sky over the target. One round hit in the radio compartment and bomb bay filling the plane with smoke. Sgt. Marvin Hicks quickly surveyed the damage. Only the IFF equipment was destroyed and no one was injured. The deputy lead was hit also. Despite the excitement, the Group dropped its bombs visually on the target, plastering the rail yards. It was the only mission in September when weather permitted bombing by visual means.

The Group returned to the skies the next day with a return trip to the Henschel tank plant at Kassel, Germany. Thirty-six aircraft departed but four returned to Seething due to mechanical problems. The remaining aircraft dropped on the target through a complete undercast cloud deck using H2X. Unlike the attack five days earlier, heavy, accurate flak exploded around the Liberators. Although no aircraft suffered serious damage, the tail gunner on DO BUNNY and a waist gunner on FRISCO'S FRISKY were wounded. Fighters kept their distance lobbing rockets at the 448th but swarmed around the nearby 445th BG. Twenty-five B-24s from the 445th quickly fell under the brutal attacks. Luckily, the 448th avoided the danger and all returned home.

Parachutes containing supplies fill the sky after being dropped by 448th B-24s during Operation Market-Garden. (LaPoint)

OUR BABY 42-50469. This aircraft was one of seven aircraft that suffered damage on 18 September 1944. It was repaired and later crashed on 28 January 1945. (LaPoint)

Possibly FLYING DRAGON. The aircraft by this name was serial number 42-95182. It was lost during a mission on 24 August 1944. (LaPoint)

BUFFALO GAL 44-10498. Although damaged during a combat mission in September and some of its crew wounded, it finished the war and returned to the U.S. (LaPoint)

In an effort to obliterate the important Henschel transportation works at Kassel, the entire 2nd Bomb Division revisited the site once again on 28 September. The 489th BG led the Wing and the 448th flew in the high right position. Flak again greeted the bombers although it was not as intense as on the previous day. Clouds hindered the bombing but breaks in the undercast improved their chances. H2X was again used as the primary means of bombing. High explosive bombs and clusters of incendiaries fell from twenty-six Liberators of the 448th. Sgt. William Gamble saw enemy fighters hitting formations behind them, but fortunately for the Liberators from Seething, they escaped unmolested.

Despite bad weather and no flying the next day, the Luftwaffe still provided excitement at Seething. Two V-1 'buzz bombs' exploded very close to Seething. Although the windows rattled, the two missiles landed far enough away from the base that no damage resulted from their blast. In another return visit, the Group hit Hamm, Germany on the last day of the month. The entire Bomb Division returned to the marshalling yards in an effort to further disrupt the German transportation system. Forecasted bad weather required PFF bombing. The twenty-eight aircraft of the 448th led the 20th CBW to the target and dropped their bombs through another complete undercast. Lt. Leroy Conner was surprised at the light flak over the target but thrilled at the lack of opposition. Enemy fighters mixed it up with P-51s but did not reach the bombers. Everyone landed at Seething without a scratch.

FRISCO FRISKY 42-51246. Another 448th veteran that survived the war and returned to the states, her scoreboard proudly displays 96 combat missions. (Patterson)

The month of September was a month of transition for the air war. The 448th proved quite versatile carrying food to hungry civilians and dropping ammunition and supplies to beleaguered airborne troops. The Group also experienced a change in tactics. Of the ten missions flown in September, only one attack utilized visual bombing methods. The other nine required PFF aircraft with H2X equipment to identify the target and drop the bombs. The weather over Europe deteriorated quickly and these methods of bombing would prove essential in the coming months. Even though it appeared the war would last beyond Christmas, the Luftwaffe mustered only a feeble defense in the air. German fighters rose to fight only once but the deadly flak continued to strike. It claimed two crews and their aircraft during the month.

## thirteen

# ANOTHER WINTER: October 1944

The new month brought thickening clouds and colder temperatures. For the ground crews this meant shorter days and long, cold shifts loading, repairing and fueling the aircraft. For the aircrews this meant already cold flights became colder. In addition to the long, cold flights, the harrowing experience of assembling in instrument conditions frayed the nerves. Mid-air collisions were not uncommon. As the formations grew in size, the potential for collision increased and the collective blood pressure of the crew doubled. With the liberation of France, the short 'milk runs' to France no longer existed; all missions hit the German Reich. The threat from Luftwaffe fighters continued to diminish. However, flak continued to claim crews. It was the most dreaded German threat as the war dragged into October.

Poor weather prevented operations on the first day of October but the Group returned to the skies of Germany on 2 October. Around Seething, the early morning ritual commenced at 0215. For Lt. William Snavely and crew, their morning ritual consisted of a short truck ride to Halesworth, home of the 489th BG, to pick up the PFF ship. After takeoff, they rendezvoused with the 448th and flew the mission as Group lead. Upon return, they landed at Halesworth and then caught a ride back to Seething.

At Seething, crews ate breakfast then meandered to the briefing room where military police checked names as the individuals entered. Lt. Leroy Conner listened as the briefing started detailing the target at Stuttgart. As crews hitched rides to the aircraft, the mission was scrubbed. The reprieve was short-lived. After a second breakfast, Lt. John Snider as well as Lt. Conner returned to the briefing room. The same crews were re-briefed to attack the marshalling yard at Hamm, Germany.

Forty aircraft from the 448th joined 264 other B-24s from the 2nd Bomb Division. The first aircraft took off starting at 1040 and proceeded to the assembly area. Four squadrons left Seething; three formed into a Group formation and one flew high right with the 489th BG formation. Crossing the North Sea, gunners test fired their guns in preparation for combat. The guns in the nose turret of BACHELOR'S DELIGHT jammed rendering them inoperative. The bombardier, Lt. John Snider, field stripped the guns in the tight confines of the turret. Despite the cramped space and cumbersome clothes he repaired the guns just as they entered enemy territory. As the formation crossed the Zuider Zee into Holland, light flak dotted the sky exploding harmlessly below the formation. They continued unhindered.

Crossing the initial point, the bomb bay doors of the Liberators slowly opened. Aboard BACHELOR'S DELIGHT, the right door jammed and only partially opened. The radio operator, Sgt. Jesse Kinsey, attempted to open the door with his foot. The bombardier, mistaking the bomb salvo button for the intercom button, sent three, 500-pound bombs crashing through the partially open door narrowly missing Sgt. Kinsey. Several B-24s, seeing bombs fall, released their bombs well before the target. The remainder of the force dropped their load of high explosive and incendiary bombs at 1348 from 23,000 feet with good results. The ever-present flak exploded around the bombers but failed to inflict serious damage. Although no German fighters resisted the attack, they were reported in the area. On the return leg, Lt. Leroy Conner was forced to leave the formation after the oxygen supply on his B-24 ran low. Fortunately, they did not encounter enemy opposition and landed safely at Seething.

Returning to England, the crew of BACHELOR'S DELIGHT discovered a bomb still hanging in the rack of the bomb bay. The pilot, Lt. Charles Platt, flew back over the North Sea where the bomb was pried loose and fell into the cold water. The excitement for the crew continued when the number two engine caught fire. Lt. Platt skillfully landed the damaged aircraft at Seething. Following a bumpy landing, the crew egressed the aircraft safely. Close examination of the aircraft revealed the fire had burned almost completely through the wing spar. BACHELOR'S DELIGHT would not fly again.

The next day's primary target was the Daimler-Benz plant at Gaggenau, Germany. Field orders directed that if weather prevented visual bombing, the crews were to bomb nearby Pforzheim, Germany by PFF. Twenty-eight aircraft left Seething but mechanical trouble cut the force by one. Bitter cold temperatures made the trip extremely uncomfortable for the men in the planes. Sgt. Joe Zonyk suffered from the extreme temperatures after his heated flying suit failed. Making matters worse, ice formed in his oxygen mask from water vapor in his breath making it difficult to breathe. As the formation passed over Amsterdam, flak struck the number one engine but they struggled to remain with the formation. Numerous unplanned turns on the bomb run by the lead aircraft scattered the formation and only the high right squadron dropped their bombs. They hit short and left of the target. Leaving the target, the deputy lead assumed command of the formation and led a second pass over the target. Unable to bomb on the second pass, the formation set course for Pforzheim. Fifteen minutes after the initial pass, the formation dropped the remainder of the bombs by H2X.

Over the target a 1,000-pound bomb hung in the bomb bay of Lt. Harold Soldan's aircraft. After numerous attempts the bomb released but the bomb bay doors froze in place and failed to close. Surprisingly, they stayed with the formation and returned to Seething. Flak damaged 42-95006 and forced them to crash land at Calais, France.

Once again bad weather scrubbed all bombing operations on 4 October. The Luftwaffe failed to stop the bombers but despite the advent of new technology that allowed bombing through the clouds, bad weather in England still grounded the bombers on numerous occasions. On 5 October the entire 2nd Bomb Division, except the 448th, participated in raids on Germany. The following day the 448th returned to action.

Thirty-eight Liberators from Seething lifted into the early morning sky destined for an oil refinery at Harburg, Germany. Several aircraft aborted with various problems but thirty waded

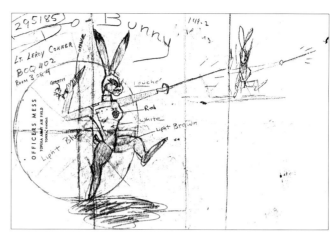

The first sketch of the noseart for DO BUNNY, 42-95185, from Lt. Leroy Conner drawn before the crew left for overseas. The caption on the stationary states, Officers Mess, Topeka Army Air Field, Topeka, Kansas. (Conner)

through the intense, accurate flak. The Group arrived over the target at noon dropping their high explosive loads from 25,000 feet. Despite the heavy barrage from the anti-aircraft guns, no one from the 448th suffered major damage. Six hours after taking to the sky, the bombers returned home.

The next day, 7 October, planners scheduled raids to revisit oil producing facilities throughout Germany. Plans called for the 448th to visually attack an oil refinery at Magdeburg. The secondary target was briefed as an explosives factory in

One of the checklist used to prepare for crew interrogations following a combat mission. (National Archives)

the twin towns of Clausthal-Zellerfeld. Another large effort, forty-four aircraft, left Seething but once again aborts reduced the size of the attacking force. After reaching the target and finding unsuitable conditions, the 448th prepared to hit the secondary target. Confusion prevailed and the 448th, led by Lt. William Snavely and crew, found themselves in front of the entire Bomb Division. Although not planned to act as Division lead, Lt. Snavely led the Group across the target and received credit for leading the entire 2nd Bomb Division. Results were excellent as the target was destroyed.

Forty aircraft dropped their bombs at 1230 with excellent results. Sgt. Robert Kessler, flying on OUR BABY in the tail end charlie position, watched twenty-two general purpose bombs strike the target. He also noticed the sky full of friendly fighters. It was a good thing. Sgt. Marvin Hicks, flying on Lt. Snavely's crew in the lead ship, saw numerous German fighters, mostly Me-109s and Me-210s, but the escorting Thunderbolts, Lightnings and Mustangs prevented the enemy from attacking the 448th formation. Although flak erupted as they crossed over Magdeburg and Liepzig, no flak was fired at the bombers on the bomb run and all aircraft returned to Seething without damage.

German vengeance weapons were the only things flying on 8 October. A V-1 landed and exploded less than a mile from the airfield. The force of the explosion knocked items off the shelves in Sgt. Robert Kessler's barracks. Sgt. Ben Johnson also felt the explosion. "One of the things that bothered us more than anything was the Buzz Bombs. We spent more time at night in the bomb shelter than in bed. I remember one night we had a warning while I was trying to write a letter. All the fellows leaving for the shelter kept saying, 'Come on Johnson!' and I continued writing. In a few minutes I heard over the speaker system, 'CRASH – TAKE SHELTER!!!' I got up and ran for the shelter, taking a short cut. Some time later I was surrounded by a bunch of fellows. I was on the ground lying flat on my back. That morning the fellows had washed their uniforms in 100-octane gasoline and hung them on a line they had strung up between their hut and another. I did not know that, and I hit it while running to the shelter. I guess I was lucky I did not break my neck. They guys would not let me forget it." Fortunately no one was injured by this buzz bomb.

On the morning of 9 October, as crews prepared the aircraft for flight, two more "putt-putts" landed nearby sending columns of smoke into the air. Neither V-1s or bad weather kept the bombers from flying but the weather did require PFF ships to lead the raid. Sgt. Marvin Hicks once again rode in a

BACHELOR'S DELIGHT 42-78481. On 2 October 1944 this B-24 suffered a severe engine fire that nearly burned through the wing spar before the crew landed. (Everson)

This strike photo of the Hamburg Ordnance depot shows smoke from the exploding bombs as well as the lazy trail from the smoke markers that marked the release point. (Ray)

Col. Mason pins a medal on an unidentified Master Sergeant. (USAF)

truck with his crew to Halesworth, the 'benefit' of being a lead crew. Takeoffs commenced at 0845. Twenty-nine aircraft battled severe weather conditions and biting cold temperatures to assemble. In a futile attempt to stave off the cold, men huddled in their heated flying suits. The cold even penetrated the heavy fleece wool. The men's breath hung in a frozen cloud in front of their faces. Oxygen masks continually froze due to condensation. Temperature gauges in the B-24 stopped at forty below zero; no one knew how cold it actually was. Only twenty-four aircraft completed the assembly in these conditions and departed for the marshalling yards at Wetzlar, Germany.

Unable to bomb the primary target at Wetzlar, the entire 2nd Bomb Division hit the secondary target at Koblenz. The large cities of Germany provided excellent radar returns and the marshalling yards in the center provided good PFF targets since they were identifiable. Koblenz was just such a target. Before reaching the target, radios crackled as the P-51 escorts described German fighters over the intended initial point. They assured the bombers no German fighters would be there when the formation arrived. As promised, the sky over the initial point was void of enemy fighters. However, the "Little Friends" could not suppress the flak and it filled the sky. Fortunately, the complete cloud cover not only hampered the bombers' ability to accurately bomb, but it also impeded the aim of the flak gunners. As a result, the heavy flak barrage was inaccurate and failed to damage any bombers.

While the main force flew over Germany, Sgt. Robert Kessler flew on an air-sea rescue mission. Four B-24s escorted by four P-51s swept over the North Sea and Holland in search of downed airmen. Although they failed to locate any survivors, they did witness a rare event. Two V-1s launched from sites in Holland and bound for unknown targets in England passed in front of the bombers.

By mid-October, winter weather completely engulfed northern Europe. Typical winters included extreme temperatures and low ceilings with snow and ice. The winter of 1944-45 would prove to be one of the worst. The days grew shorter and the skies grayer. Frost covered the ground in the mornings creating a beautiful scene but an unforgiving work place. Ground crews suffered immensely. The cold metal of the aircraft increased the difficulty of their job. Still, aircraft had to be prepared for the next mission. It would get worse before spring.

Capt. Arthur Howell transferred to Seething in July 1944 to fill the vacancy left by Lt. John Olhaber who was killed in an aircraft accident during the Group's overseas move. As the new weather officer at Seething, he was responsible for predicting the unpredictable. "As you might know, forecasting was

The 459th Sub Depot consisted of several shops that performed critical maintenance activities required to keep the bombers flying. Here men from the machine shop manufacture replacement parts for the bombers. (Hipkins)

very difficult, particularly at Seething, since it was located so close to the sea with fog sometimes drifting in from the east, affecting us but not the bases west of us. Also since it was only a little over 200 miles to the west coast of Wales, storms in the Atlantic ocean could pass through rapidly while our planes were out on a mission, with weather conditions changing markedly at our base." Twenty-four hours a day, the men from the weather office at Seething collected data including barometric pressure, temperature, dry and wet bulb, horizontal visibility, cloud cover, cloud ceiling height, wind speed and direction, and actual weather conditions. Provided to the central weather office, these readings helped build a complete weather picture. In turn, the teletype in the Seething weather office constantly chattered with weather updates.

Most importantly, the on duty weather officer made predictions which were passed to the operations officer for the upcoming mission planning. As Capt. Howell explained, "The weather officer on duty would participate in the mission briefing, by projecting on a large screen a picture of the clouds and storms enroute, and would then give a forecast of the winds and weather on this route to be taken to the bombing target, and the return to the base. His duties would start many hours before this preparing the forecast. After takeoff there was some time for rest, but upon the mission return he would participate in the interrogation of the crews to see what weather was actually encountered and then transmitted this data to headquarters." The uncooperative winter weather made their job unenviable.

Although critical by nature, the poor weather conditions made the job of the airfield controllers even more critical. These men worked from a mobile caravan at the end of the active runway. From this checkerboard-painted trailer, these men con-

The men of the 459th Sub Depot pose in front of a B-24 on the hardstand between the tower and the T2 hangar where they accomplished most of their work. These men performed a Herculean task of accomplishing major repairs and modifications. (Hipkins)

trolled all the aircraft movements on the airfield. As the days with poor weather increased, the aircrews relied heavily on these men and owed them many thanks. Cpl. Raymond Schwartz, a controller at Seething explained:

> "As an airfield controller, most of our time was spent in the black and white caravan at the touchdown point on the runway. When radio silence was maintained, air traffic was controlled by Aldis lamp signals and flares from Very pistols. We were in telephone contact with the control tower. Some of our duties were 1) To refuse aircraft if landing wheels were retracted or if there was a collision or danger 2) to refuse landing if the path of aircraft was obstructed in any manner 3) to refuse permission for aircraft to leave marshalling point or takeoff position if it would obstruct an aircraft approaching to land 4) to report any non-standard signals or unusual movements of aircraft 5) to report failure of night flying lights or equipment to the Flying Control officer and 6) to control ground movements on the perimeter if vehicles wished to cross the runway, etc.. We were on duty twenty-four hours and off twenty-four hours. When there was no traffic at night we slept in the bunkhouse in the Tower. Many nights we were alerted to bring in RAF planes with cannon flares, flare guns, and Aldis lamps. Sometime we would stand on each side of the runway and criss-crossed flares with Very pistols at the touchdown point. There were times that we worked with the Alert crew. One of their duties was to use the "Follow Me" jeep and guide visiting aircraft to the parking area and also to light kerosene flares along the runway. After the aircraft left on a mission, we had very little to do until they returned or one returned with trouble. One day we heard a fighter plane above low overcast with a stuttering engine. We contacted the Control Tower and were ordered to fire flares into the sky. Minutes later a P-51 landed and the pilot personally came out to the caravan

Line crews were the men that kept the aircraft flying. They prepared the aircraft for flight and performed maintenance tasks in all weather conditions. (Everson)

to thank us, as he was lost and out of gas when he saw our flares."

The worsening weather coupled with the slow progress of the war adversely affected morale. The slow realization that they would spend another winter in England produced tension and restlessness especially among the ground personnel.

A bright spot each day for the ground personnel slogging in the cold, wet outdoors was the visit of the Clubmobile. This converted double-decker bus, run by the American Red Cross, visited bases around East Anglia. For the men working in inhospitable conditions, this bus provided a well-deserved respite. They also offered Dick Wickham, a young English boy, a great opportunity to visit the bases around his hometown and meet his heroes, the crews:

Flying controllers from the 58th Station Complement Squadron at Seething prepare to fire a flare signaling the start of a mission. In order to maintain radio silence, flares were used to signal the start of several activities including, engine start, taxi and takeoff. (Bollschweiler)

"Its role (Clubmobile) was to visit the American Air Force bases and supply doughnuts, coffee, candy, and cigarettes to the personnel on the bases, free of charge I might add. My connection with the Clubmobile was that one of these buses was based at Harleston where I was living during the war. I had a job on this after school hours. The job was to clean out the coffee urns and make doughnuts, all ready for the next day's trip. Everything was contained in the bus for making the doughnuts and storing of them and coffee making machines. The coffee was of the instant variety and was made on the bases as required. Milk was evaporated and tinned, this being easier to keep than fresh milk. Being rationed as we were for sugar, to see all the sugar that was used to "dunk" all the doughnuts in was something else. To clean up and make doughnuts (hundreds of them), took about three hours. All the doughnuts had to be put into racks and stored in built in cupboards. The destination boards at the front and rear of the bus were made into speakers. Music was broadcast from these, played from a record player inside the bus. A small lounge was situated at the rear of the bus, this being for the two young ladies to rest in whilst travelling to and from the bases. The bus was stationed on the car park at the rear of the Swan Hotel at Harleston. The driver and the two young ladies lived in the hotel itself. The driver stayed with the bus but the two young ladies were changed to other duties after about six months. Two of us boys were employed to do the above duties and we were paid seven shillings a week. This was a lot of money in those days for schoolboys of 13-14 years of age. Not only that but we could eat as many doughnuts as we wanted."

The nearest controller uses a telephone to communicate with the control tower while a B-24 commences its takeoff roll. (Bollschweiler)

Capt. Newton McLaughlin, in charge of the Special Service, as well as the American Red Cross helped tremendously in trying to placate the men. They collected sufficient athletic equipment for the men to utilize. Football and basketball equipment was available to anyone wishing to participate. The Aero club continued to provide a hospitable getaway. Ping pong, snooker, bridge, and chess were games of choice plus the large, glowing fireplace provided much sought-after warmth. The station cinema continued showing current movies. Still, the atrocious weather remained. Adding to the men's discomfort, all overshoes were recalled for shipment to combat troops on the continent. Although everyone agreed the combat ground troops were more deserving, it did not keep their feet dry from the mud, rain, and snow.

During a break in the missions, the Group moved all PFF crews. Still assigned to the 448th, they temporarily moved to Hardwick on detached service in order to be co-located with the PFF aircraft. They would continue to lead the 448th PFF missions but no longer had to make the early morning trips from Seething to Hardwick or Halesworth.

The 448th flew again on 14 October. For the first time, a PFF aircraft led each squadron. Previously, the only PFF aircraft were Group lead and deputy lead. As a result, the formation maintained Group integrity crossing the target and everyone dropped on lead's command. The resulting bomb pattern was widely scattered and not precise. The advent of squadron PFF ships meant each squadron crossed the target and dropped on the lead's command. It provided a smaller, more concentrated and theoretically accurate bomb pattern.

CQs awoke crews at 0600 and everyone completed their customary pre-mission rituals. The long ribbon on the map stretched to Cologne, Germany, another PFF mission to a large city's marshalling yard. Such missions usually encountered heavy flak. Lt. Leroy Conner lifted his Liberator into the air just before 0900 and initiated the instrument assembly procedures. This nerve-wracking experience lasted until 19,000 feet when they finally broke into the clear. The high clouds required the assembly ship to raise the assembly altitude to allow the bombers to reach clear sky. Of the thirty-five aircraft launched, only twenty-eight departed England. Cold temperatures froze the condensation from preceding formations creating thick contrails and made formation flying incredibly difficult. Somehow, they kept the formation together and continued to Germany.

A formation of twenty 448th BG aircraft hit the primary target at Cologne at 1226 while eight others hit a target of opportunity at Euskirchen three minutes later. All bombing was accomplished by PFF and the flak was customarily heavy. JOKER'S WILD landed in Belgium with damage. The crew left the aircraft on the continent and returned to Seething unhurt.

The following day, the Group completed a return visit to Cologne. Once again cold temperatures and bad weather hampered the operations but did not stop them from dropping their bombs. Lt. John Rowe, flying an aircraft that had suffered heavy damage to the hydraulic system the previous day, aborted after the landing gear failed to retract after takeoff. The remaining thirty aircraft attacked in five waves of six aircraft each in a line abreast or "Vee" formation. They dropped their loads of incendiaries and high explosive bombs at 0953 using PFF. German gunners had learned their lessons well from the day prior and pounded the bombers with heavy, accurate fire. Sgt. Ralph Lee, waist gunner on TARFU II, suffered injuries from flak. Many others narrowly escaped injury.

Shadows grow long as men anxiously await the return of a mission. Ambulances and support vehicles are ready in case they are needed. (Everson)

This close up of SONIA shows the label present on all U.S. aircraft. The line, B-24H-20-FO, identifies it as a H model, Block 20, and FO identifies it as Ford built. The line below carries the serial number 42-95270. (LaPoint)

On OUR BABY, Sgt. Robert Kessler stood near the bomb bay throwing out chaff. The thin, metal strips were designed to confuse the Germans by clouding their radar screens with false returns. It did not work this day. As they crossed the city center, flak ripped into OUR BABY just missing Sgt. Kessler. It knocked a large hole in the left side of the bomb bay and another hole in the right side of the ball turret. It narrowly missed severing the control cables, a potentially catastrophic loss. At the same time, a bomb hung in the bomb bay doors and more flak rendered the intercom inoperative. The eight-minute barrage heavily damaged OUR BABY but the Liberator survived and brought the crew home safely.

Bad weather shut down operations for a day but the 448th returned to Cologne for the third straight time on 17 October. Again, the marshalling yards were the target. In a repeat performance, thirty-two bombers from Seething bombed the target by PFF and returned to Seething. Although crews encountered the same flak, it was not as accurate as on previous trips.

On 19 October twenty-six B-24s from Seething set out to bomb a flying bomb assembly plant at Gustavsburg, Germany. Sgt. Marvin Hicks flew again on the Group lead aircraft. Once more poor weather dogged the formation the entire trip. Thick clouds forced the Group to climb to 26,500 feet to find better flying conditions. At such high altitude, aircraft performance degraded and the heavily laden bombers wallowed in the thin air. Lt. John Rowe, flying 42-51349, aborted after a faulty radio compass directed them to the wrong buncher for assembly. The remainder of the Group battled incessant contrails created in the cold air by the engines of the preceding Groups. Crews crouched against the bitter cold trying to shield themselves from the negative forty-three degree temperatures. Although unable to hit the primary target, they successfully attacked the marshalling yard at Mainz, Germany. Even for the veteran crews, the flak barrage seemed to last forever. Crews flew through a savage gauntlet that lasted seventeen minutes. The crew of SONIA landed in France after flak heavily damaged their plane. Somehow, the remainder of the bombers completed the trip without major damage and landed at Seething.

Poor weather continued to hamper the bombing campaign but, as an omen, the first PFF aircraft were assigned to Seething on 19 October. Initially six arrived but others followed. Not until 22 October did the Liberators from Seething return to Germany. Hamm, Germany, was the target. Forty-four aircraft from Seething took part in the mission. The main formation consisting of thirty-three aircraft took off followed closely by a smaller formation of eleven aircraft. This smaller force augmented a formation from the 93rd BG. Once again bad weather hampered the force and PFF was necessary to bomb.

SONIA 42-95270. An unidentified crew pose with their SONIA. Although heavily damaged on 19 October 1944, it was repaired and flew combat missions until the end of the war when she returned to the States. (Everson)

UP IN ARMS. The tail number and fate of this 448th aircraft are unknown. (LaPoint)

WILLIES SILLIES. This 448th Liberator's tail number and fate are unknown. (Everson)

As the weather turned colder and the days shorter, indoor activities replaced the baseball games and other outdoor endeavors. Cards were a popular distraction. (LaPoint)

Bert LaPoint, Nick DeFillippo, Herb Heidrich, Carl Aquinta, and Jim Ingelsby play cards in their hut. (LaPoint)

As the main 448th formation commenced the bomb run at 1130, the lead aircraft experienced PFF failure and passed command to the deputy lead aircraft. Despite this last minute change, high explosive and incendiary bombs fell from the bomb bays at 1138 and plummeted through the cloud deck. As a morbid greeting, flak bursts dotted the sky, although none endangered the bombers. Lazy contrails above the bombers served as a reminder their "Little Friends" were there just in case enemy fighters elected to attack. Everyone returned safely to Seething without ever seeing enemy territory. Clouds covered all of northern Europe.

On 25 October, the Group targeted the Mitteland Aquaduct over the Weser river with a Focke-Wulf repair depot and airfield at Neumunster, Germany as the secondary target. Take-offs commenced around 1000. The route to the target carried the formation over the North Sea before turning south into Germany. The total time over German territory lasted only twenty-seven minutes but carried them within range of multiple flak batteries located at Kiel, Germany. As the formation passed nearby, the guns sent up a tremendous barrage but failed to inflict damage on the bombers. As the flak subsided, heavy weather forced the formation to alter the attack to the secondary target. Flak over the target was light allowing the bombers to drop by H2X. As bombs fell from one 448th Liberator the number one engine seized. Unable to remain with the formation, they drifted back. Three P-51 Mustangs quickly joined them and escorted the stragglers home. They beat the entire formation home and landed without incident. Just like the previous mission, the bombing results were unobserved due to a blanket of low clouds over the target. However, the Wing that hit Hamburg obviously hit their target. Large columns of thick, black smoke billowed through the clouds as the returning bombers passed by.

The following day, crews launched on a raid aimed at an oil refinery at Bottrop, Germany. For Sgt. William Gamble, it was a second consecutive trip to 'Happy Valley', the unaffectionate nickname for the heavily defended Ruhr Valley. Near Osnabruk, thick haze decreased visibility, making formation flying extremely hazardous. One crew lost sight of the lead aircraft and initiated a climb in order to get out of the thick soup. Already at the limits of the heavily laden B-24's performance, they jettisoned the bombs to facilitate the climb. They finally reached clear skies at 26,500 feet but only four other aircraft were visible. Unable to contact the formation and without fighter escort, they turned their Liberator for home.

Nineteen B-24s continued to the target and a guaranteed flak greeting. As expected, intense flak hounded the bombers enroute to the target and leaving the target. Customary clouds required the use of G-II for bombing. Just as the second squadron prepared to bomb, the lead aircraft suffered G-H equipment failure and failed to notify deputy lead in time. As a result, all nine aircraft in the second squadron returned to Seething with their bombs. Leaving the target area, flak severed Sgt. Gamble's oxygen supply. Thankfully, a nearby crew member found him and reconnected his oxygen. Without his help, Sgt. Gamble could not have reconnected his oxygen and would have suffered the potentially fatal consequences of the high altitude. When the crews returned to Seething that afternoon, they received official word confirming an unpopular rumor. Tour lengths increased from thirty missions to thirty-five.

During the evening of 27 October, a British Lancaster bomber crashed into the trees at Ash Tree Farm, Mundham as

*Another Winter: October 1944*

For the men working in the cold and snow on the flightline at Seething, hot coffee from the Red Cross offered a brief chance to defrost their body numb with cold. (Engdahl)

it prepared to fly a low approach to runway two-five at Seething. Crash and rescue personnel from Seething responded to no avail. The gruesome event brought a close to an otherwise uneventful day. Even on days the Group did not go to war, the war somehow always came to Seething.

On the final day of October, the 448th once again battled the flak and the weather. Thirty-three aircraft launched but only seventeen attacked the target at Hamburg, Germany. The Group battled clouds all the way to the target while trying to maintain formation. The constant flying in and out of clouds made formation flying impossible. When they arrived over the target, the formation was widely scattered and many of the bombers could not bomb due to the lack of formation integrity. Heavy flak over the target further complicated matters. Although no aircraft were damaged, the poor weather took its toll on the bombers' effectiveness.

The Group completed October without losing a crew, the first month to do so since arriving in the ETO. Flak did cause considerable damage however. Three aircraft were left on the continent after battle damage forced the crews to land short of home. Twenty-four aircraft suffered battle damage but only four endured serious enough damage to warrant sub-depot repairs. While the Luftwaffe's opposition continued to wane, a new enemy emerged during October: winter! The winter weather patterns of northern Europe created havoc with the formations and made precision bombing difficult. It was a long time until spring!

fourteen

# ALL WEATHER AIR FORCE: November 1944

As October faded into November, winter tightened its grip. Clouds filled the air challenging even the most veteran, skilled pilots. Ground crews continued to struggle with the cold trying to keep the aircraft in shape. Once pliable rubber turned stiff and as rigid as the metal skin of the plane. Refueling took longer. Loading bombs into the aircraft became more difficult. The aluminum airplanes soaked up the cold making maintenance extremely difficult. Ground crews built ingenious shacks on the flight line using bomb crates and other makeshift materials. These homemade shelters provided some relief from the inhospitable weather. Despite these adverse conditions, the men of the 459th Sub Depot and the unrecognized ground crews continued to provide capable aircraft for missions.

The first mission in November was flown on 2 November. A railroad viaduct near Bielefeld, Germany, was the target. Three squadrons of B-24s left Seething and all but one attacked. Clouds on the bomb run obscured the target initially but, when the target was finally acquired, there was not enough time to stabilize for an accurate bomb run. The 500-pound bombs all missed. The bombers returned home leaving the viaduct intact.

Lt. Leroy Conner stumbled to his feet in the darkness of his Nissen hut at 0530 on the morning of 3 November. One hour later he sat in a briefing for a mission to bomb an oil refinery at Gelsenkirchen, Germany. Bad weather prevented the mission from getting off the ground; planners scrubbed it before anyone got airborne.

The following day, 4 November, crews briefed at 0600 to attack the same oil refinery at Gelsenkirchen. Once again the Liberators attacked the most heavily defended area in Germany, the Ruhr Valley. Capt. William Snavely's PFF aircraft led a force of thirty aircraft from the 448th which was also in the vanguard of the entire 2nd Bomb Division. The weather forecast predicted complete cloud cover over the target so a PFF aircraft led each squadron.

Just as expected, heavy flak erupted as the formation flew within range of the guns in the valley. A solid cloud deck obscured the ground from the bombers as well as the bombers from the gunners, but this time the Germans used radar to direct their guns. A continuous barrage pelted the formation for almost ten consecutive minutes. Aboard OUR BABY, Sgt. Robert Kessler threw chaff out of the dispenser in an effort to confuse the German radar. Sgt. Donald Zeldin, flying on 42-50820 as a replacement, sweated as Lt. Leroy Conner took evasive action. He was nearing the completion of his tour. Now was not the time for his luck to run out.

As the bombers prepared to drop their bombs, the intense cold at 23,500 feet froze the bomb bay doors on OUR BABY. The doors were forced open using a hand crank and the bombs were dropped on schedule at 1146. Sgt. Robert Kessler watched as the bombs disappeared into the clouds. The results were unobserved. As the formation left the target, P-51s and P-47s circled overhead ready to challenge any potential airborne threat. None materialized. However, the unrelenting flak damaged the deputy lead aircraft, 42-51030, just after the initial point and forced them out of the formation. On the return trip, they landed at a 9th Air Force fighter base near Brussels, Belgium, for repairs. Unfortunately, the damage proved too severe and the plane was eventually salvaged.

On 5 November crews were briefed to hit gun emplacements around Metz, France. Heavy fortifications around the town stymied the advance of General Patton's Third Army and he sought to break the stalemate utilizing the heavy bombers in an attack similar to the one in July that preceded Operation

Cobra. The pain of friendly fire losses from this attack was not forgotten. Due to the close proximity of friendly troops around Metz, visual bombing was required. No one wanted a repeat of Americans killing Americans. If visual conditions did not exist over the target, the marshalling yard at nearby Karlsruhe, Germany, was designated the secondary target.

This mission marked the 172nd mission for the 448th and the 35th and final mission for Lt. Leroy Conner. It was almost fatal. As they ran their engines up in preparation for takeoff, one of the three 2,000-pound bombs fell from its shackles and landed on the runway. Fortunately, it was not armed and no damage resulted. They taxied back to their hardstand and changed aircraft. They took off late but caught up with the formation in time to fly their last mission.

As the bombers arrived over Metz, clouds completely obscured the ground and radio calls went out to each Group signaling a change to the secondary target. The air was more turbulent than usual and planes bounced around the formation as the pilots worked diligently to maintain formation. Sgt. Donald Zeldin, flying in the deputy lead aircraft piloted by Lt. Conner, struggled to keep from getting sick. Even after thirty-three missions, it seemed bad. For Lt. Joseph Kaiser, the flight surgeon who was flying as a waist gunner on WAZZLE DAZZLE, the rough air was too much. He spent the entire trip airsick.

As usual, the flak over the target was heavy. The bursting shells exploded in different colors, red, white, and black. A 448th plane left the formation after flak set the number three engine on fire. Sgt. William Gamble said a quick prayer for them as he watched them slide out of formation. As the bombs fell from the bomb bays, the plane shuddered slightly. Free from the heavy load, the engines no longer strained against such weight and the controls became lighter. The transition always gave Sgt. Donald Zeldin an uneasy feeling. Were they hit? Anything wrong? Fortunately, the guns were inaccurate and no one else suffered damage. Back at Seething, Lt. Leroy Conner entered the coveted world of a 'happy warrior.' He was headed home. For Sgt. Donald Zeldin, only two more missions remained.

The following day, 6 November, the Liberators left Seething for the Mittelland Canal at Minden, Germany. The Liberators crossed over the coast at Ostend, Belgium, and proceeded inland without opposition. Prior to reaching the target, as the bombers crossed north of Munster, flak struck with success. Lt. Frank Genarlsky's aircraft, 42-50820, was hit in the number two engine causing a fire. They left the formation with the damaged engine feathered and jettisoned two of the large 2000-pound bombs. Eyewitnesses reported no further sign of a fire. They were never seen again by men of the 448th. A German

Two men stop for a break in what appears to be an airplane junkyard. The aircraft fuselage in the background has obviously been stripped of all useable parts. (LaPoint)

farmer saw the plane as it attempted to land near Billerbeck, twenty miles west of Munster. The plane struck a telephone pole and crashed, catching fire almost immediately. Quickly after crashing, the plane erupted in an explosion evidently from the one bomb that was not jettisoned. All ten men perished, including Lt. Alton Kraft flying on his last mission. (Much confusion existed about their disappearance. Fourteen bodies were discovered in the area but post-war investigations accounted for all ten men from Lt. Genarlsky's crew. The other men were identified as missing from other aircraft.)

Despite the flak, thirty-two aircraft hit the target at 1045 utilizing the PFF aircraft. The flak subsided as they crossed the target but jet fighters hit. Sgt. Robert Kessler saw two twin-engine fighters attack the group in front of the 448th, but escorting P-51 Mustangs quickly drove off the attackers. Although rumors flourished about these new jets and their amazing capabilities, this was the first time the 448th encountered them. Fortunately, fifteen P-51s pursued the attackers and only the 93rd BG suffered from the attack. However, THE STURGEON

BACHELOR'S DELIGHT sits derelict in the grass at Seething after damage from an engine fire on 2 October 1944. Planes damaged beyond repair like this were cannibalized for all available spare parts. Even the distinctive twin rudders are missing from this one. (Rebeles)

suffered battle damage from the flak. The pilot, Lt. Peter Protich, flew the damaged plane to England and landed at the emergency strip at Woodbridge, England. After four hours on oxygen breathing through a rubber hose, men happily dropped their masks as they descended through 10,000 feet in preparation for landing. By 1300, the bombers were on the ground. Crews climbed into trucks and made their way to the de-briefing. There, they learned of another crew missing in action, Lt. Genarlsky's crew.

On 8 November, the 448th participated in another concerted attack on the German transportation system. The marshalling yard at Rheine, Germany, was the target. Clouds hampered the assembly of the twenty-two Liberators from Seething. Clouds and haze shrouded the bombers until they climbed above 23,000 feet, two thousand feet higher than planned assembly altitude. The flight to the target area was uneventful for Sgt. Donald Zeldin. It was not until they approached the target that the excitement started to brew. Sgt. William Gamble watched the usual flak bursts billow in the sky. Although not as thick as on previous trips, it was still just as deadly. Crossing the initial point, each squadron in the 448th formation moved in trail of the preceding squadron in preparations for PFF bombing. As the first squadron crossed the target, another crossed underneath them foiling their bomb run. Eight aircraft in the lead squadron did not release their bombs.

Leaving the target, OUR BABY, piloted by Lt. Carl Holt, threw oil from the number four engine and oil pressure dropped. After a few anxious moments, the pilot and engineer decided not to shut it down. With the excellent P-47 escort, the odds of survival increased by remaining with the formation. Aircraft that fell out of formation ran a greater chance of suffering a tragic end. When all the aircraft landed around 1340, Lt. Carl Eggert was ecstatic. This was his thirty-fifth and final mission. Although several B-24s took the bombs for a ride and returned to Seething with full bomb bays, the mission still counted in the eyes of the aircrew. One more gut-wrenching mission.

Despite previous attempts to help dislodge German troops around Metz, General Patton's Third Army remained bogged down around the fortress city. The 448th flew their 175th mission in support of renewed attacks around Metz. Crews copied down the details of the mission during the early morning briefing on 9 November. Thirty-two aircraft were planned to take-off from Seething at 0700 and attack gun emplacements at Verny, France, just south of Metz. Horrible weather conditions once again complicated the assembly. Despite the poor weather, the Group assembled and joined the rest of the 20th CBW for-

The unsung hero. The number of men required to generate a combat mission was staggering. Thousands of man-hours were required to place one Group in the air. Here one of the unidentified thousands works in one of the numerous shops repairing an engine. (USAF)

NO NAME JIVE 41-29230. Crew chief Mike Finnegan proudly stands beside his plane. One of the veterans from of the Group, it flew over fifty missions, thanks in a great part to people like her crew chief. (Burzenski)

mation. A lone weather ship flew unprotected twenty minutes in front of the main force. They relayed weather conditions and target information to the main force enabling them to prepare for the changing conditions.

Only occasional puffs of anti-aircraft fire dotted the sky as the Group crossed the French countryside. Using G-H equipment to verify their position, thirty airplanes from the 448th dropped 1,000 and 2,000-pound high explosive bombs. Sgt. Donald Zeldin, flying on Flight Officer Severyn Szudarek's aircraft, WAZZLE DAZZLE, watched a plane from another Group go down. He also saw smoke pouring from the bomb bay of the lead aircraft. A smoke bomb, intended to mark the release point, hung in the bomb bay of the lead ship. Fortunately, the smoke stopped before irreparable damage could be inflicted on the aircraft. However, the lead ship landed at an emergency field on return. The same miserable winter weather greeted the crews on return. Instrument procedures were required for the letdown.

Attacks on Germany resumed the following day. Twenty-one aircraft left Seething on 10 November destined for an airfield near Hanau, Germany, but weather forced a change of plans and they attacked the city's marshalling yard. To Sgt. William Gamble's surprise, the sky remained clear of flak during the entire trip. It seemed unbelievable. All the bombers returned safely after the seven-hour trip.

Intelligence reports indicated possible enemy attacks were planned on airfields in England. As a result, twenty-four hour guard duty was resumed on 10 November with men required to wear full gear including helmet, gas mask, web belt, canteen, first aid pack, and weapons. By the end of the month no enemy activity occurred so the requirement was eased. Men were only required to carry gas masks but around the clock guard duty continued.

Bad weather cancelled all 8th Air Force operations over the next six days. High winds, rain, fog, snow and bitter cold settled over England. Crew strength was at an all-time high making housing a problem and coal for heat scarce. The endless job of keeping the runways clear of snow gave the men another reason to complain. It was cold, backbreaking work shoveling snow off the runways.

On 14 November, Col. Charles B. Westover replaced Col. Mason as Group Commander. Col. Mason had served as the Group Commander since the first of April when Col. Thompson was killed in action. He sent a farewell memorandum to the men introducing his replacement:

DOUBLE TROUBLE. This B-24 not only carries nose art but also the names of her crew by their crew position. The fate of this plane is unknown. (LaPoint)

HOME TOWN GAL. It is obvious the artist of this nose art favored girls from Oklahoma. The fate of this Liberator is also unknown. (Patterson)

"During the past eight months it has been my happy privilege and honor to have been stationed with this group as Group Commander, and now, as in all things, the time has come when I must leave you. I can honestly tell you that I have never spent eight happier and more gratifying months than these past. It has been an honor to have been appointed your Group Commander. We have had our ups and downs. In some cases we have had our differences of opinion. But as Mark Twain said, 'It is the opinion that makes the horse race.' I hope that at some future date we may all be together again. I hate to say goodbye, but rather 'au revoir,' and until we meet again, my sincere thanks for the excellent support you have given me. I hope that you will give the same loyal support to my successor, Colonel Charles B. Westover. He is a stout lad. So long and 'Ting Hao.'"

Col. Westover, also a West Point graduate, was the son of the pre-war Air Corps Chief of Staff. Before arriving at Seething, he was the Chief of Staff at Headquarters, 2nd Bomb Division at Ketteringham Hall. Now he assumed the reins of a combat Bomb Group. Other personnel changes occurred. Lt. Col. Glassel Stringfellow, 714th BS commander, returned to the States on rest leave as did Lt. Col. Lester Miller and Maj. Leroy Smith. Although missed, they would later return to Seething.

On 16 November the 448th flew another tactical mission in support of ground forces near the German town of Aachen. The actual target was German troops in the town of Eschweiler. Low ceilings and fog hung over East Anglia as the bombers prepared for takeoff. Still, takeoffs and assembly were completed without incident and nineteen bombers of the 448th set course for Germany. The entire 8th and 9th Air Forces filled

Black diagonal bands on yellow rudders identify these bombers as 448th Liberators. They fly high above the bothersome clouds that made assembly and visual bombing so difficult. (LaPoint)

the sky over western Germany. Fragmentation bombs rained down on the German troops with excellent results. For a change, the sky was void of flak.

The real excitement for the crews started when they turned for the relative safety of England. While the planes were away, the weather worsened. When they arrived over the field at Seething, they found it closed due to the weather. The Group scattered as planes diverted to bases where the weather was slightly better, although nowhere was the weather good. The low ceilings and wet runways proved hazardous on landing. SHADY LADY skidded off the wet runway at Lisset, Yorkshire. Sgt. William Gamble and crew also diverted to a nearby British base. Three days later, he returned to Seething completing his 35th mission.

Even though the bombers eventually returned, continuing poor weather grounded the planes until 21 November when thirty-three bombers departed Seething for oil refineries in Hamburg. Just as predicted, poor weather prevailed and clouds completely obscured the target. The Group used H2X radar for bombing. Luftwaffe flak batteries threw up a tremendous barrage of metal over the target. LITTLE JOE flown by Lt. Joe Bowers fell to the heavy barrage killing eight of the crew. Sgt. Herbert Dennis escaped from the fallen Liberator but not from the Germans. Others suffered serious damage as well. FLEXIBLE FLYER received over 360 holes from the fire. Although Lt. William Hensey managed to fly the crippled airplane home, Sgt. George Allen suffered serious penetrating wounds. Two days later, he died at the Station hospital from his wounds.

Several other crewmen from various airplanes suffered serious wounds. On LITTLE JO (a similar name but different aircraft than the one lost), an unexploded anti-aircraft shell penetrated the waist hitting an ammunition box. The ammunition exploded killing Sgt. Michael Perkowski, a waist gunner, who was dispensing chaff. Sgt. John MacDonald was wounded as well. The shell continued through the aircraft and exploded after it exited the roof of the plane. The exploding ammunition severed the rudder cables and ignited a fire in the waist. The engineer, Sgt. Louis Owens, extinguished the fire and deftly spliced the cables together allowing Lt. Elliot Sidey to keep the aircraft flying. Due to the excellent teamwork, they landed their badly damaged aircraft at Seething without further injury.

Sgt. Antonio Munoz on THE RUTH E.K. was seriously injured by flak. Sgt. Clifford Blalock and Sgt. John Givens on QUEENIE were also wounded. RAB DUCKIT, flown by Lt. Charles Quirk, was heavily damaged and leaking fuel. The engineer, Sgt. John Lyles, expertly repaired a severed fuel line and despite being in danger of exploding, they landed safely at

SWEETHEART OF THE ROCKIES 44-10544. A product of the vast industrial power of the U.S., B-24s like this swelled the ranks of the bomber groups in England. This particular aircraft survived the war and returned home. (Everson)

Directly below a formation was a dangerous place. Strike cameras capture this 448th B-24 below the formation as they prepare to drop their bombs. It was not unheard of for bombs to strike a plane and one Seething crew received credit for destroying an enemy aircraft with its bombs. (LaPoint)

Seething. The 714th BS received the brunt of the enemy's guns. All eight of the squadron's participating aircraft suffered flak damage from the intense flak. The battered formation returned to Seething six and half-hours after departing. For the crews it seemed like an eternity.

Few people realized the significance of 22 November. It marked the one-year anniversary of the Group's arrival in the European Theater when the first 448th B-24 touched English soil at St. Mawgan. Now, one year later, the Group was a veteran of the terrifying battles over the German Reich. Many of the original ground personnel remained, but most of the original aircrews were either dead, prisoners of war or at home instructing new recruits. A rare few still called Seething home. Also several men had completed their tours and were back in action on their second. They were the minority. Everyone wondered what the next year held.

Terrible winter weather again grounded the Liberators, this time for four days. In the early morning hours of 25 November, men labored in the intense cold preparing aircraft identified to participate in a mission against the marshalling yard at Bingen, Germany. Twenty-one aircraft departed Seething but two returned due to various problems. The crew in the lead aircraft of the low left squadron accidentally dropped their bombs four minutes early. Six other aircraft followed their miscue and released also. The remaining aircraft from Seething crossed the target at 1220 and dropped their general-purpose bombs from 27,000 feet. Leaving the target, flak rocked the formation damaging several of the B-24s.

Sgt. Robert Kessler made it back to Seething but not without excitement. His crew left the formation early after flak damaged two engines. Continued operation of the engines was questionable. Despite this hurdle, they continued on toward Seething on reduced thrust. Somehow the two damaged engines held together and they safely landed at Seething before the other returning Liberators. Lt. Gordon Hillman and crew failed to make it back to England. Damage forced them to land FEARLESS FOSDICK in France. They returned to England the following day. Once again, German flak crews inflicted serious damage on the bombers.

The next day, the Group revisited a previous target. An attack on 2 November failed to destroy the railway viaduct and bridge at Bielefeld, Germany. So, at 0855 on 26 November, twenty aircraft left Seething in an effort to destroy the stubborn target. Once again the weather coupled with the small, tactical nature of the target proved difficult for the bombardiers and navigators. The bridge remained intact.

The mission of 27 November struck the marshalling yards at Offenburg, Germany just across the Rhine river from Strasbourg, France, in support of Army operations in the vicinity. Thirty-five aircraft departed Seething but one returned after mechanical problems forced the crew to abort. Clear skies welcomed the bombers. After multiple missions fighting the weather and the enemy, crews enjoyed the nice weather. The formation headed south toward the target with the Alps clearly visible to Sgt. Robert Kessler in MAIL CALL.

Col. Charles Westover assumed command of the 448th from Col. Gerry Mason on 14 November 1944. Col. Westover, a West Point graduate, came to Seething from the Second Bomb Division Headquarters at Ketteringham Hall. (Bollschweiler)

4-F 42-95527. Two unknown men pose with their B-24 named in jest of the fitness category reserved for those physically unfit for military service. Unlike its namesake, this B-24 served admirably, completing the war and returning to the States in June 1945. (Johnson)

SHADY LADY 42-50759. Another female form adorns the nose of this Ford-built Liberator. Unfortunately, she met her fate on 16 November 1944 when she slid off a runway in poor weather. (LaPoint)

Maj. William Searles cuts his Welcome Home cake following his return to Seething in November 1944 after a short stint in the States. (Bollschweiler)

Turning onto the bomb run, the pilot of TROUBLE-N-MIND tried to maintain his formation position. He slowed down in an effort to not overrun the leader as the formation turned in his direction. Suddenly the aircraft stalled and entered a spin. The pilots regained control and returned the aircraft to level flight a mere 1,500 feet above the ground. All alone, they turned for home. The remainder of the Group continued and dropped their bombs as scheduled. Sgt. Joe Zonyk watched the bombs fall all the way to impact where a brown pall of smoke erupted.

MAIL CALL lost an engine over the target and another engine started acting erratic. The crew initially elected to land at Lyon, France, but a closer examination of the damaged engine revealed only minor damage. They decided to continue to England. Red-nosed P-51s provided the escort home where they landed without problem. Lt. Elliot Sidey and crew, flying TARFU, landed in France after damage from the flak made their return trip impossible. The gunners returned to Seething the following day although the officers, radio operator, and engineer remained with the plane until it was repaired.

Bad weather returned the following day grounding the entire 8th Air Force. The Group flew on 29 November in a third attempt to destroy the bridge at Bielefeld, Germany. Men loaded twenty B-24s with cumbersome, 1,000-pound high explosive bombs. Clouds totally obscured the target but the formation attacked anyway. Once again fate smiled on the railway viaduct.

On the last day of the month the 448th planned to bomb steel mills on the outskirts of Neunkirchen, Germany. As the Liberators left Seething, one ran into trouble. The pilot failed to unlock the controls sending the planning careening out of

GRUJOJE. An airman leans out the window admiring the nose art. The fate of this B-24 is unknown. (LaPoint)

MISS MOUSE. This Liberator's tail number and fate are unknown. (Patterson)

*173*

Above: QUEENIE 42-50348. Despite damage and injuries to some of her crew on 21 November 1944, she survived the war and returned home. Here ground crews work on her number four engine. (Whipple)

Right: RAB DUCKIT 42-100435. This B-24 also suffered heavy flak damage on 21 November 1944 but her crew landed safely at Seething. After the war she flew a crew home to the States. (Patterson)

control and off the runway. The crash ripped the nose gear from the wheel well rendering it a complete loss. Thirty other aircraft departed and completed the assembly. Again cold weather dogged the men in the planes. Sgt. Joe Zonyk suffered another failure of his heated flying suit making it a long cold trip across France. A solid cloud deck over the target forced the crews to use G-H equipment to identify the secondary target, marshalling yards in Nuenkirchen. The falling bombs disappeared into the clouds hiding the results from the crews. Flak was light and four Me-109s dove out of the sun but failed to attack. Everyone returned to Seething for payday.

The ground war continued to move at a snail's pace due to the harsh conditions. The unprecedented weather also impacted the air war in a major way. Of the fourteen missions in November, all but two required the use of PFF aircraft to drop bombs. This new technology allowed the bombers to attack through complete cloud cover, but it adversely affected the accuracy of the bombing. Targets had to be large like marshalling yards in large city centers. Tactical targets like the bridge at Bielefeld, Germany proved too small to be successfully attacked. Despite the relative inaccuracies of this advanced technology, it allowed the Group to continue the air war despite horrible weather conditions.

Luftwaffe flak crews appeared to be unhindered by the weather. They claimed two aircraft and crews and damaged twenty-eight other planes. Nine required major repairs. Crews abandoned four aircraft on the continent after landing with battle damage. Despite the attrition the Group continued to grow in size. By the end of the month, 106 crews called Seething home.

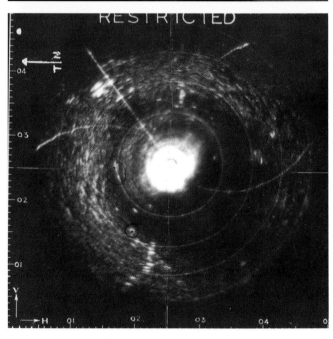

This is a photograph of a PFF radarscope image. The bright spots were identifiable landmarks that navigators used to identify the target. It required a well-trained radar navigator. (Berryhill)

fifteen

# HOLIDAY SEASON: December 1944

December started with the war grinding along on the ground, stymied by the Herculean German effort following their rout in France. Earlier hopes of a quick end to the war were shattered by the German stand on the borders of the Reich. Not only did the Wehrmacht stop the advance, but they attacked. In the air, the war also sputtered but not because of the Luftwaffe. Although flak crews continued their relentless defense of the Reich, the winter weather and associated horrendous conditions hampered aerial operations over Germany the most.

The 448th participated in their first December mission on the 4th of the month, a mission against the marshalling yards at Bebra, Germany. Forty aircraft left Seething but five returned shortly after takeoff for various reasons. Despite the great efforts of the ground crews, the cold still took its toll on many of the aircraft.

Thirty-five Liberators from the 448th reached the target. Unable to bomb the primary target, they dropped their bombs using H2X to identify the Koblenz marshalling yard. OUR HONEY, piloted by Lt. John Rowe, narrowly averted disaster over the target. The lead aircraft dropped chaff from containers. One of these large containers fell from the lead aircraft and struck the top turret of OUR HONEY. The crew thought they suffered a direct hit when the box struck with a loud bang. The turret shattered scattering plexiglass all over the flight deck. Adding to the confusion, smoke from the smoke markers dropped by the lead aircraft entered the plane. Amazingly, the top turret gunner was not injured but his oxygen hose unknowingly disconnected when he vacated his seat. Sgt. Joe Zonyk grabbed him before he passed out and fell out the open bomb bay.

Flak struck many aircraft in the formation. Lt. Marcus Horton's aircraft suffered damage to two engines. Leaving the formation, they radioed for fighter escort. Shortly afterward Sgt. Fred Kerniss counted twenty P-51s circling nearby. They landed at a forward air base near Brussels, Belgium. After spending the night in Belgium, they returned to Seething the following day. Two other aircraft, Lt. Albert Sanders in RAB DUCKIT and Lt. Charles Meining in MY BUDDIE, diverted to bases on the continent after flak damaged their planes. RAB DUCKIT lost a supercharger crossing the French coast and flak destroyed a second supercharger over the target. Flak also penetrated the plexiglass in the nose and in the copilot's windscreen. Although no one was injured, they elected to land at a mesh-covered, emergency strip at Liege, Belgium. Despite the heavy flak, all aircraft and crews eventually returned to Seething.

The Group did not fly on the 5th, but did fly on 6 December. Thirty-one aircraft attempted to bomb the marshalling yard at Lohne, Germany. Weather forced a change in the plan and the Group hit the secondary target, the Mitteland canal at Minden. Everyone returned to Seething without incident.

As America's economic might reached its full potential, new crews and aircraft poured into the bases of the 8th Air Force. At Seething, these new crews underwent indoctrination designed to familiarize the green crews with flying in the ETO. Seasoned veterans flew with new crews sharing their wealth of combat experience and tips on surviving the rigors of combat. As the night of 7 December revealed, training flights carried a very real potential for disaster. Four aircraft were taking part in night flying training missions with experienced pilots instructing new arrivals. Lt. Severn Szudarek took off at 2036 flying DRAGON LADY (also known as MAIL CALL) with Lt. Henry Mielke and four of his crew. On takeoff, they became disoriented while flying on instruments and clipped the top of two large oak trees at Green Lane near the bomb storage area of the Earsham Ordnance Depot. The aircraft crashed in a fiery explosion in a meadow behind Skinner's Farm at Bedingham. All six crewmembers were killed. Except for the instructor, Lt. Szudarek, they never flew a combat mission.

The snow and ice played havoc with the missions planned against Germany. Although training missions continued despite the weather, the Group did not fly another combat mission until 11 December. In the interim, the Group used the opportunity to consolidate its lead crews since more PFF equipped aircraft were arriving. Due to their specialized training and constant demand, all lead-qualified crews were moved from their various squadrons to the 712th BS. The 712th BS now provided the Group with its own lead crews. Even though the other three squadrons provided the aircraft and crews for a mission, a 712th BS aircraft and crew led each squadron. Crews no longer were forced to ride to a nearby base to "borrow" a PFF aircraft.

The 448th tested its new organization on 11 December when forty-six aircraft launched against the marshalling yard at Hanau, Germany. Mechanical malfunctions depleted the maximum effort by five but the remaining aircraft continued as part of the largest 8th Air Force attack to date. The usual bad weather required PFF aircraft to bomb through the complete cloud cover. The weather hindered formation flying and group integrity over the target was poor. Nevertheless, the bombs rained down on the German marshalling yard just after noon. One of the PFF ships, 42-50587, landed on the continent after being damaged by the flak. Instead of attempting to repair the aircraft, salvage crews determined it was more feasible to salvage it for parts.

Thirty-two crews flew the next day in much improved weather. After the painfully familiar early morning briefing, crews took off bound for the marshalling yards at Aschaffenburg, Germany. A few puffy clouds dotted the sky over the target. Surprisingly, only widely scattered flak bursts were noticed on the flight to the target and no flak was encountered over the target. Despite the favorable conditions, the lead and high right squadrons missed the target. The low left squadron dropped their bombs with very good results. Even though the opposition was light, one aircraft, 44-10520, from the 715th BS landed at an emergency field near Brussels, Belgium. After releasing the bombs over the target, problems on the number one engine forced the pilot, Lt. Albert Sanders, to feather it. On reduced power, they were unable to maintain airspeed and slowly fell out of formation. Clear skies gave way to increasing clouds as they headed for home. When a small break appeared in the clouds, Lt. Sanders dove the aircraft through the

RUGGED BUT RIGHT 42-94953. Shark's teeth on the nose identify this aircraft as a lead ship. Starting in December 1944, lead crews and aircraft were consolidated in the 712th BS. (Bailey)

OUR HONEY 42-50302. An unidentified airman sits on his bike while posing with his favorite baby. Despite sustaining damage from a freak accident on 4 December 1944, this Liberator completed the war and returned to the U.S. (Everson)

hole and the navigator, Lt. Joseph Nathan, provided a heading to Brussels. After spending five days in Brussels, they hitched a ride back to Seething aboard a C-47. Their aircraft was salvaged.

The following days were characterized by unprecedented bad weather creating impossible conditions for flying. Under the cover of this dreadful weather, Adolph Hitler launched an attack in the lightly defended Ardennes forest. The German

Army pierced the American lines and sent the battered Yanks reeling. German forces pounded the Americans in the sector. All the Army's air support was grounded when it was most needed. At Seething, everyone listened for more details of the attack. How could the Germans attack when it appeared they were almost beaten? Sympathetic to their Army brethren, attempts were made to launch missions but the weather spoiled each try. All available crews from the 715th BS flew a practice mission on 18 December, but the weather over the battlefield did not allow dropping bombs in close proximity of friendly troops.

Finally, on 19 December, the weather cleared enough to allow twenty Liberators from Seething to take part in a mission against tactical targets behind the German lines. The 448th was ordered to hit the German town Ehrang, a supply choke point for the German attack. Clear weather was a fantasy as thick fog enveloped Seething and much of East Anglia. Crews taxied out in preparation for takeoff with visibility as low as 150 yards. The necessity of the situation required them to continue and orders were issued to take off. The first aircraft lifted into the soup at 1030. Two aircraft failed to successfully complete the harrowing takeoff, both crashing at the end of the runway. Two squadrons, twenty aircraft, departed before the decision was made to cancel takeoff clearance for the remaining aircraft.

One of the crews making the takeoff was Sgt. George Weinberger's crew flying 42-50357. Several days prior they had bailed out over the front lines in Belgium after their aircraft suffered heavy battle damage. Now they were flying their fifth mission in support of the very troops that rescued them. Complete cloud cover over the target did not deter the German flak crews from unleashing a tremendous barrage as the bombers crossed the front lines. While several aircraft were hit, they all dropped their bombs using H2X. One plane, Sgt. Weinberger's, was heavily damaged. They headed for the front lines and quickly found themselves over Belgium. The crew crash-landed 42-50357 near Brussels, Belgium. Although no one was hurt, the aircraft was a complete loss.

The thick fog still engulfed Seething when the bombers returned. The weather in western England was clearer and the bombers diverted to bases there. Sgt. Fred Kerniss and his crew landed at Barnstaple, Devon. It was not until Christmas Eve that all the bombers returned to Seething. The valiant effort of these crews earned a well-deserved commendation from Gen. William Kepner, the 2nd Bomb Division commander. " I desire to commend all participating units on the mission of 19 December for their outstanding display of determination, courage and skill. The importance of hitting the assigned target and

DRAGON LADY 42-50457. This B-24 suffered a tragic fate during a night training flight on 7 December 1944. A new crew with their instructor pilot became disoriented and crashed just after takeoff killing all the crew. (LaPoint)

aiding our ground forces at a crucial period justifies the hazards involved, and has earned a vote of thanks from them. My heartiest congratulations."

Fog, ice and snow strangled most of northern Europe as the American troops on the ground continued to take a beating. Stories of a gallant force led by the 101st Airborne Division surrounded in the small Belgian town of Bastogne, quickly spread. A sense of urgency to help the troops was stymied by the continued bad weather. The transfer of lead crews to the 712th BS was completed during the break but this did not help the troops fighting in the Ardennes. No missions were flown until Christmas Eve. In the meantime, crews sat helpless as the miserable weather continued.

The Christmas season is known for gifts and to the troops embroiled in the intense fighting in the Ardennes, no gift was greater than clear skies on Christmas Eve. Although the hazardous fog still covered England, a high-pressure system cleared the skies over the mainland, CAVU. Throughout England, aircraft started engines signaling the start of the largest bombing mission of the war. The 448th contributed their largest force of the war, fifty-three aircraft destined for Euskirchen, Germany. The crystal-clear sky over Europe filled with aircraft as far as one could see. Sgt. Ira Welkowitz watched the bombs as they impacted squarely on the target. The unusually good visibility also allowed the flak gunners unobstructed view of the bombers. Shortly after "bombs away," PATRICK DEMPSEY suffered catastrophic damage. The intense flak started a fire in the right wing. The pilot, Lt. L.C. Barneycastle, placed the aircraft in a dive in hopes of extinguishing the fire. Unfortunately, the fire quickly reached the fuel tanks. The explosion ripped the plane apart and the falling debris crashed three miles south of St. Vith, Belgium. Five chutes descended from the falling wreckage. Only three of the men, however, survived: Sgt. John Birkhead, Sgt. Eathan Newcomb, and Sgt. Linn Garrison. They spent the remainder of the war as POWs.

In his Christmas morning letter to his fiancée, Sgt. Wally Balzer, the chaplain's assistant, described his Christmas Eve:

LIL PEACH. One of the numerous Liberators whose fate and identity is unknown. (LaPoint)

"On Sunday morning, I arose at the usual 5 o'clock, made the fires and accomplished last minute preparations. Sgt. Morgan sang a solo at the morning service, 'O Holy Night.' The attendance at the morning service was 115, a little smaller than we had anticipated but circumstances beyond our control was the chief cause of the size (that would probably be because of a briefing and a mission that was in process of taking place). After dinner I lay down on my sack for about an hour and managed to get a little sleep. Later I cycled over to the chapel to fetch some songbooks, which were to be used at the broadcast in the gym. I took the male chorus books and the folding organ to the gym and when I got there one of the Special Services officers asked me to play the organ over their public address system which broadcast to all the enlisted men's mess halls on the base. Of course I was delighted to do that although I can't play these Christmas carols too well. When I returned to the office I noticed there would not be enough time for writing a letter so I got a bit to eat and we went back to the chapel and had a good turn out there. Then the broadcast. The men gave forth with some good singing. There was a Lt. Cantz who used to be an announcer for CBS Chicago made appropriate announcements between numbers. We sang immediately after the broadcast. I cycled over to the chapel to straighten out the chairs and added a couple of more rows of chairs 'cause we were anticipating a capacity crowd and set everything in order. The men started coming in slowly and ten minutes to seven the place was filled up. Every available seat was taken and at the rear of the chapel the men were standing closely together including even the base commanding officer. Later that evening two trucks were waiting to take the chorus

PERILS OF PAULINE. This B-24's fate and tail number are unknown but the nose art was undoubtedly quite popular. (LaPoint)

Stars and Stripes fly over a snowy Group Headquarters. (LaPoint)

boys to the hospital following the service. The Catholic men were also in on this and we had a fine crowd of about twenty-four. As I remember the singing at the hospital was the best. The fellows were all with the director. Jim McConkie played one of the Christmas carols on the organ and later the patients were able to join in one verse. The chorus boys seemed to be supercharged and were very eager to sing. After all, this was Christmas Eve and back home they always used to go caroling."

Christmas Day dawned clear and cold over Europe although the pesky fog still remained over Seething. Once again the 448th conducted operations in support of the raging ground battle. The intended targets were tactical points in the rear of the German offensive sector. Snow covered the ground giving a serene look to the countryside that was ravaged with fighting. The high right squadron of the 448th attacked a communications center at Budesheim, Germany, while the lead squadron hit the jam-packed marshalling yard at Prum, Germany. The low left squadron failed to find their target at Waxweiler, Germany, and only one aircraft in the squadron bombed. Despite the clear skies over the battlefield, the returning bombers encountered freezing rain and drizzle over England, hazardous conditions for the big bombers. Despite the challenging conditions, all but one crew landed safely at Seething. Lt. Harlyn Schroeder and crew landed at the emergency field at Woodbridge.

The dreary winter weather did not subdue the Christmas spirit at Seething. The returning crews found the base filled with children. They played host to children from nearby villages for Christmas dinner. It was a memorable experience for Jim Turner, one of the local children:

"While struggling with a lesson one morning the teacher broke off to answer a knock at the front door. This was a welcome break and all heads turned towards the door. Soon a muffled conversation could be heard. Eventually the teacher stepped back inside followed by two American officers to carry on the conversation. The first thing through my mind was who was in trouble; someone has been up to something at the airfield; or was something missing from a crash site; what about all the fifty caliber bullets I had at home? Eventually the Americans left and the teacher went onto the next classroom. Shortly after she came back with the other teacher. Standing at her desk in front of the class she asked us to pay attention. She announced that the Americans at the airfield were giving a party for the school children in the area and had invited

A crew relaxes and enjoys the most welcome part of a mission, a safe return, while waiting for a ride to debrief. (LaPoint)

our school. This was met with stunned silence then followed by excited chatter as the impact of her announcement sunk in. After a pause and things settled down a bit, she carried on, and I think she said something about parents' permission. Any parent not giving permission for something as exciting as this would not be too kindly thought of. The next few days saw much discussing on the coming event. When the day of the party arrived best dress was the order of the day. My two sisters were wearing their best dresses, wool overcoats, and wool hats. I turned out with the latest fashion for school boys 1944: my best suit, a jacket and short trousers complete with three cornered tear, which mother hurriedly mended, long socks

MY GAL SYL. A crew poses with their aircraft. Unfortunately both the crew and aircraft are unidentified. (LaPoint)

Above: TEXAS PLAYBOY. The tail number and fate of this B-24 are unknown. (Everson)

Above right: PATRICK DEMSPEY 42-50799. While supporting the beleaguered troops in the Battle of the Bulge, flak knocked this aircraft out of the sky on Christmas Eve killing most of the crew. The wreckage crashed near the battered town of St. Vith, Belgium. (LaPoint)

H FOR HELEN. One of the numerous Liberators that flew during the war that is now unknown. (LaPoint)

and boots, and covered with the ubiquitous garbadine overcoat, topped off with a peaked cap. Suitably prepared we set off to the pick up point. We were taken to the airfield in a 6x6 truck fitted out with long slatted wooden seats along each side and across the front end, and covered with a canvas hood. It was noisy with everyone chatting, much to the amusement of the GI looking after us. Arriving at the airfield we were met by our American hosts for the day. My younger sister and myself were taken together with our officer guide, and my oldest sister went off with someone else. I cannot remember the exact order of events, but we were sat down to a meal of the festive season, which was really special, even though we wondered why we needed jam on the side of our plate; many years later we heard about cranberry sauce. I remember a small band playing music while we ate, and then decorated Christmas tree. We were taken to the cinema to see a cartoon film. There must have been other things we did but my memory fails me. One thing that does stand out in my mind was a B-24 flying over shooting out yellow flares and I remember asking our guide what they were for. "It's mechanical trouble", he said. Looking back on the events of the day, having a small sister five years younger than me sharing the same guide had its drawbacks. She seemed to be getting all the goodies. Most of the time her face was covered with chocolate, and her hand clutched a bag full of candy. Maybe our guide thought they would be shared, which they eventually were on a 80-20% basis. I was hoping for a much sought after pack of chewing gum, but it did not appear, perhaps I was expecting too much. Looking back, those GIs had probably given up several weeks of their candy rations to give us kids a treat, and they certainly had given us a wonderful day. This was a Christmas party that would be remembered long after others had faded into the past."

Over 300 children enjoyed cookies, candy, turkey and tours of the base. Airmen showed the children around and fielded thousands of questions. Best of all, they ate to their stomachs' were content. They enjoyed several helpings of turkey with pumpkin pie and all the trimmings. After they ate a filling meal, the base theater treated the children to a Walt Disney movie. The Red Cross club made special packages containing cookies and candies donated by the GIs from their Red Cross packages and rations. After the party, over 800 packages of cookies remained to be distributed to kids who were unable to attend. Most of the food was gone by the time the returning crews arrived. For Lt. John Rowe, it did not matter. His only regret was he did not get to enjoy the children.

The nagging weather restricted operations on the day following Christmas, allowing the Group to enjoy the holiday season for one more day. Five months earlier, hopes ran high

*Holiday Season: December 1944*

Excited kids as well as men crowd the briefing room as Santa Claus makes his appearance. Chocolate, fruit, and chewing gum, all rare commodities in wartime England, were collected from men on base and given to the kids. (Everson)

Young kids enjoy treats from Santa Claus during the Christmas party hosted by the 448th. The party provided great therapy for the men, who were spending another holiday season away from home, as well as the children who were experiencing their fifth wartime Christmas. (Everson)

for Christmas at home. Now, Christmas at home seemed far removed, a surreal life. On the mainland, Allied ground forces fought to contain the dangerous German attack. In the skies, the Air Forces fought to stem the tide on the ground. Even though the Luftwaffe remained a silent and invisible enemy, the weather was winning its own war. It kept the potent Allied air power from intervening.

The weather claimed a 448th aircraft on 27 December. Capt. Edward Malone flying 42-51505, left Seething on a practice radar bomb drop. Prior to dropping their practice bombs, they ferried two crews to Halesworth to pick up an aircraft left by the 489th BG. Then they were to continue on the practice mission culminating in dropping twenty 100-pound bombs. Unknown to the crew a thick layer of ice covered the runway at Halesworth. Upon landing, the slick ice sent the aircraft careening off the runway. Although no one was hurt, the aircraft was damaged beyond repair.

Finally, on 28 December, the weather cleared allowing the bombers to return to the skies over the Reich. Thirty-one Liberators left Seething loaded with high explosive bombs. Elsewhere in East Anglia, Groups departed for various transportation system targets in western Germany. The 448th hit the marshalling yard at Kaiserslautern, Germany. Just before the initial point the lead aircraft's H2X system failed but the deputy lead aircraft immediately assumed control. The quick reaction allowed the Group to drop on the assigned target and return to Seething as planned.

As the ground forces stopped the German attack and counter-attacked, the 448th participated in more missions designed to further disrupt the German transportation system. On 30 December, the Group hit a railroad underpass at Mechernich, Germany. Sgt. Fred Kerniss completed his twenty-seventh mission without seeing any flak. What a rare treat!

The Group closed out the year with an attack on New Year's Eve. Twenty-six aircraft attacked a little known railroad bridge at Remagen, Germany. Fortunately, no hits were recorded on the Ludendorf Bridge. Just over three months later, this bridge provided the Allies with an avenue over the Rhine when troops from the 9th Armored Division captured it intact. Thick clouds obscured the bridge as the bombers flew overhead. Despite the obvious importance of the bridge to the transportation network, crews saw very little flak on the mission.

Bad weather plagued most missions in December, especially at the height of the Battle of the Bulge. Despite horrible conditions, the 448th participated in critical attacks on the German Army and their strained transportation system. One crew and their aircraft were lost in the attacks and numerous planes were damaged, mostly in weather related accidents. The crews left three heavily battle-damaged aircraft on the continent. Although the flak was present, the incessant weather remained the main enemy. Neither Germany nor the weather showed signs of quitting.

sixteen

# ANOTHER YEAR: January 1945

The New Year started inauspiciously; no missions were flown by the Group on New Year's Day. While the bombers sat idle, the Luftwaffe undertook a major operation on New Year's Day strafing and bombing airfields on the mainland. Many Allied aircraft suffered major damage from the attacks but fortunately the airfields in England were not targeted. The New Year not only brought a change in the calendar but also a change in names. The three Bomb Divisions in the Eighth Air Force were renamed Air Divisions. The change was in name only; the command structure remained unchanged. The 448th, along with the 93rd BG and the 446th BG still comprised the 20th CBW which in turn remained a part of the 2nd Air Division, formerly the 2nd Bomb Division.

The Group departed on their first mission of the new year on 2 January. Twenty-eight aircraft from Seething attacked the marshalling yard at Neuwied, Germany, a suburb of Koblenz. German gunners failed to fire their customary flak barrage and the bombers dropped their bombs by PFF without resistance. Six groups of P-51s swarmed over the bombers providing outstanding escort. The only problems occurred when the crews returned to England. A thick fog rolled in from the sea encasing Seething in its murky grip and required pilots to execute dangerous instrument letdown procedures. All landed safely despite the thick soup.

At the same time, a 448th Liberator fought another type of battle for survival. Lt. Carl Holt arranged for a pass for his crew to spend the New Year's holiday on the coast. While ferrying COME ALONG BOYS to Wharton, the weather suddenly closed in and restricted visibility. As they climbed to gain altitude, they hit fog-obscured Burn Hill at Burn House Farm near Slaidburn, Lancanshire. The aircraft impacted the top of the hill, then skidded and bounced before impacting a stone fence and catching fire. Amazingly, the pilot, copilot and thirteen other passengers survived with various injuries. However, four men in back, Lt. Orvie Casto, Lt. James Fields, Sgt. Phillip Mazzagatti and Sgt. Edgar Lyons were killed instantly. The rest of the men, hurt and dazed, scrambled out of the wreck. Sgt. Donald Zeldin broke his back but survived thanks to a small Welshman who arrived at the wreck and helped carry the injured men down the hill to a nearby farmhouse. Without his aid, the cold temperatures would have exacerbated the plight of the injured and undoubtedly claimed more lives. Eventually, doctors arrived and carried the men to a nearby hospital.

The following day, all Second Air Division Groups participated in a mission against targets in western Germany. The 448th visited a marshalling yard once again, this time in the town of Neunkirchen. A force of twenty-nine B-24s loaded with high explosive and incendiary bombs started their takeoff roll at 0730. BUFFALO GAL, flown by Lt. William Voight, experienced an engine failure over the English Channel and returned to Seething. For the remainder of the Liberators, the flight to the target progressed uneventfully but clouds once again hid the target.

Navigators and bombardiers used H2X in combination with G-H to identify the target. Its range limited the more accurate G-H system since the required radio signals were broadcast from stations in England. Initially it only reached targets in France. However, new stations were established in liberated Europe extending the range well into Germany. Utilizing both systems, the bombers improved their all-weather capability dramatically. Although still challenging, the crippling effect of weather was reduced.

Leaving the target two aircraft from the Group sought refuge at airfields in France. DO BUNNY, piloted by Lt. George

Franklin, and aircraft 42-51666, piloted by Capt. Jack Swayze flying his second tour, both landed on the continent. Salvage crews quickly started work on 42-51666 after repairs were deemed too costly. During the debrief, Lt. John Rowe reported no flak or fighters. Later that night, however, the distinct putt-putt of V-1s overhead reminded everyone war still raged. Lt. Hershel Hausman, newly arrived at Seething, vacated his barracks rather hastily. Three of the flying bombs flew over the bomb dump and exploded just five miles away!

Poor weather grounded the entire 8th Air Force on 4 January, but offensive operations resumed the following day. The 448th launched three squadrons, each one complimenting another group. One squadron flew the high right position in a 93rd BG formation while another squadron flew high right in a 446th BG formation. The third squadron flew in a 389th BG formation.

The squadron assigned to fly with the 93rd BG experienced difficulties with the weather as they departed England. They lost sight of their formation and attempted to circle and reacquire the Group. Three aircraft stayed with the 93rd BG but the others were unable to locate their assigned Group. They tacked on to the 445th BG and continued to the target. Unable to bomb visually, the composite 445th force held their bombs and elected to bomb a target of opportunity on the withdrawal. They hit a target near Heimkirchen, Germany, with incendiary bombs.

The remainder of the Seething Liberators hit their targets despite severe cold and light flak. Temperatures plummeted in the high altitude air. Bombsights froze on the bomb run due to the extreme cold. Bone-chilled gunners manned the open waist windows straining to see a flash of an enemy fighter in the sky. Even though the Luftwaffe rarely engaged the bombers now, one could never be sure. Rumors of enemy jet planes kept the gunners on their toes. Once again the Luftwaffe stayed away and only the cold bothered the crews on their return to Seething.

Thirty-two aircraft targeted the marshalling yard at Limburg, Germany on 6 January. Mechanical troubles reduced the formation by two as they left the English coast. Lt. Bob Sampson flying THE RUTH E.K. encountered the most serious problems. Oil pressure on the number two engine dropped to zero forcing him to feather the engine. Shortly after the engine was shut down, oil pressure on the number one engine dropped as well. The second bad engine was also shut down. With two engines windmilling, the heavily laden bomber started descending as they turned for England. The crew restarted one of the engines but severe vibrations forced them to shut it down again. They quickly jettisoned their bombs into the English

Although the war continued to drag through the winter, morale at Seething remained high. Losses diminished and men completed their combat tours on a regular basis. (USAF)

Channel but THE RUTH E.K. continued descending. Seeing England through a break in the otherwise solid cloud deck, Lt. Sampson ordered everyone to bail out as he turned the plane back toward the sea. The crew bailed out of the doomed plane at 3,000 feet but when Lt. Sampson reached the bomb bay, the plane was over the water. He returned to the cockpit and circled back over land. He jumped from the plane from a mere 400 feet before it crashed at Horsey Island. The crew landed safely although Lt. Sampson spent a month in the hospital with severely sprained ankles. The second aborting aircraft, flown by Lt. Francis Piliere, also experienced oil problems. With remarkable similarity, they shut down an engine but returned safely to Seething.

Weather prevented the force from attacking the primary target so the formation set course for the secondary target, the marshalling yard at Koblenz. Just before the bomb run, prob-

COME ALONG BOYS 42-100322. This aircraft crashed into a hillside in Lancanshire on a ferry flight to Wharton. Four men in the back of the aircraft were killed. (Everson)

lems with the lead aircraft's H2X equipment necessitated a lead change. The deputy lead, Capt. William Snavely flying DON'T FENCE ME IN, assumed lead of the formation. Thirty Liberators of the 448th dropped their bombs amid scattered, inaccurate flak. Thankfully, it exploded harmlessly above the overcast. Severe temperatures froze the right bomb bay door of Lt. John Rowe's aircraft. Unable to pry it loose, they dropped only the bombs from the left rack.

Returning to Seething, crews once again flew harrowing instrument approaches into the thick English weather. Three aircraft elected to land at Rackheath where weather conditions were marginally better. One crew, led by Lt. Gordon Hillman, landed in France due to mechanical problems. Lt. John Rowe elected to land at Seething using a newly installed instrument approach system known as the SCS-51. After homing to a beacon they intercepted the runway localizer beam, which provided lateral guidance. Shortly afterward, they intercepted the glide path transmitter beam and followed it using a prescribed descent rate. They descended and followed their lateral guidance using needles in the cockpit. They safely descended below the weather and landed using this method. Although still nerve-wracking, it was an improvement over previous instrument letdown procedures.

The next morning crews returned to the skies of the German Reich. The rail yard at Achern, Germany, located just across the border from Strasbourg, France, was the target for Lt. Hershel Hausman's first mission. The route to the target was short compared to the usual, deep penetrations into Germany. Still, as results showed, it was not without peril.

Thirty aircraft continued to the target. They smothered it with incendiaries and high explosive bombs. No flak was present over the target but leaving the area heavy flak rocked the formation damaging several of the planes. DAISY MAE suffered the most serious damage. The aircraft, with Lt. Paul O'Neil as pilot, went down over France killing all nine of the crew. Lt. Francis Piliere, flying 42-51349, struggled to keep his B-24 airborne. Serious flak damage forced them to search for a nearby airfield. They found one in liberated France and landed without further incident. However, the battle damage to the plane was too severe to repair. The crew hitched a ride to Seething and salvage crews started their work on the battered plane. Despite clouds and German gunners, the bombers plastered Achern with high explosive and incendiary bombs. For their direct contribution to the success of the mission, the lead crew, led by Lt. Allen Wight, received a certificate of commendation.

The freezing conditions at Seething grounded the 448th for three days. Not only did temperatures plummet making working conditions unbearable, but heavy fog blanketed much of England producing extremely hazardous conditions. A crew of over 150 men worked during the night of 9 January to remove snow from the runway. Due to these men's hard work,

Men from Seething struggle with snow removal the hard way in order to prepare the airfield for another maximum effort mission. (Engdahl)

448th crews set out for a highway bridge at Weweler in western Germany on 10 January. Inhumane conditions greeted the crews in the skies of northern Europe. Lt. John Rowe aborted after one of his waist gunners reported his oxygen mask was frozen. They left the formation thirty miles from the English coast. Lt. Charles Platt, flying 44-48787 as the Group deputy lead, lost an engine over the North Sea and landed at forward airfield B-70 in Belgium. The crew spent several days in a little red schoolhouse near Brussels waiting for a 'ride' back to Seething. They joined several hundred others also waiting for a ride. Lt. John Snider passed the time enjoying snowball fights and treating the village children to a nearby amusement park. They returned to Seething on a C-47 just in time for the 200th mission party.

For the twenty-three attacking aircraft, the cold plagued them the entire route. Crews endured frostbite and other cold related injuries. Still, the brave men overcame these obstacles and bombs fell on the target at 1317. Over the target, hydraulic lines on MOTHER OF TEN froze, sealing the bomb bay doors shut, and forced them to return with their bombs. After a miserable six-hour flight, the crews returned to Seething to find it shrouded in a snowstorm. Nine planes landed at Seething before the storm intensified and the remaining planes diverted to other bases in Norfolk.

Earlier in the day during assembly, a B-24 from the 392nd BG lost two engines. They jettisoned their bombs in the Channel and turned for home. An inviting hole in the clouds over Seething persuaded the crew to land there. The rapid descent

Men from the 448th crowd around a table enjoying the refreshments at an unknown gathering. Events like this were important for the morale of the men especially during the cold winter months. (Bollschweiler)

Carl Ahrendt, a flight line mechanic, poses with BUGS BUNNY. The incredible work performed by these men in austere, winter conditions kept the planes flying. Without them there would have been no air war. (Ahrendt)

DASIY MAE 42-94972. Shown here with thirty-seven missions to her credit, she was lost 7 January 1945 on her eighty-second mission. All of her crew were killed. (LaPoint)

in the freezing weather created a layer of ice that covered the windscreen obscuring the pilots' view. When the wheels touched the runway, the pilots were unable to align the aircraft and the excessive side loading on the gear sheared the struts of the B-24. The gear collapsed sending the Liberator into an uncontrolled skid across the ice. One propeller, ripped from its engine, cartwheeled across the airfield. The B-24 narrowly missed the control tower before it came to a stop. Crash crews responded quickly and no one was injured.

Winter continued its brutal hold on Europe reducing flight operations to only a few weather reconnaissance sorties for the entire 8th Air Force. Finally, the severe conditions weakened enough on 13 January to allow crews to launch. Poor conditions still played havoc with the mission. Ground crews labored in the freezing weather to ready thirty-two B-24s at Seething. Only one aircraft aborted during assembly. Lt. Albert Broadfoot landed BUGS BUNNY at a forward airfield in Belgium after the number one engine failed as they crossed the North Sea. The remaining force continued toward a railroad bridge over the Rhine River at Worms, Germany. Clouds cleared over the target allowing the bombardiers a good view of the bridge. Bombs tumbled from the bellies of the bombers amid moderate flak.

Excitement started just prior to the bomb run for the crew of BUFFALO GAL. The bitter cold weather froze the bomb bay doors. Unable to open them on the bomb run, the engineer, Sgt. Earl Jordan, attempted to manually pry them open. Without the aid of an oxygen bottle, he quickly passed out due to a lack of oxygen. The radio operator, Sgt. Nick Anast, recognizing that his buddy was in trouble, donned an oxygen mask and went to help. The added burden of helping his buddy and trying to get the door open increased his oxygen consumption rapidly depleting his own supply. He too succumbed to oxygen starvation. Lt. Kay Flinders, the pilot, reconnected them to oxygen and revived them. The bombs were dropped through the closed doors as they passed over the target. Shredded pieces of the bomb bay doors dangled from the bottom of their plane as they landed at Seething. Quite an experience for their first mission!

For the rest of the crews, events quickened as they turned for home. A navigation error by the lead squadron brought the formation within range of flak batteries at Saarbrucken, Germany. The results were devastating.

Lt. Raymond Binkley left his bombardier's station on aircraft 42-50661 to inspect a hung bomb in the bomb bay. As he attempted to dislodge the stuck bomb, the flak erupted. A piece of deadly shrapnel hit him, knocking him to the floor. Fire started by the flak quickly spread throughout the ship. Both

BUGS BUNNY 42-51551. After suffering flak damage on 13 January 1945, this B-24, also known as REDDY TEDDY, landed at a forward airfield in Belgium before returning to England. (Patterson)

the nose gunner and navigator checked the unconscious bombardier for a pulse without success. The fire forced the remainder of the crew to abandon the aircraft. All were immediately captured except the pilot, Capt. Edward Wall. He evaded for four days before being captured. Ironically, Lt. Binkley was not the regular bombardier on Capt. Wall's crew. His crew received a pass and he delayed getting his. In the brief period before he got his pass, he was selected to fly as a replacement bombardier since he was available. It was a tragic case of procrastination.

Nine other aircraft received varying degrees of damage from the unexpected barrage. A fire in the bomb bay and a fuel leak forced Lt. Harlyn Schroeder to land JUNIOR at an airfield at Merville, France. Two other aircraft, OUR HONEY, piloted by Lt. Harry Mulrain, and QUEENIE, flown by Lt. Raymond Bunday, also landed at advanced airfields of the Ninth Air Force in France and Belgium due to battle damage. The crew of FLEXIBLE FLYER did not get a chance to land.

After suffering debilitating damage during the flak outburst, Lt. Harold Soldan flew FLEXIBLE FLYER back to England. After surveying the damage, they determined the aircraft was unlandable. The crew bailed out over Stowmarket, Suffolk, but the plane continued to fly until it crashed and burned near Sarratt, a small village in Hertfordshire. Four crews landed at other fields and the rest of the airplanes touched down at Seething completing the Group's 200th mission in the ETO. The war did not stop for a celebration. The Group was alerted for a mission the following day.

For their 201st mission, the Group attacked the Hermann Goering Steel Mill and Synthetic Oil Works in Hallendorf, Germany. CQs roused sleepy crews from their semi-warm huts at 0415 on 14 January. The sharp cold of the night air quickly woke the crews as they walked to the briefing. A two-squadron formation departed Seething at 0920 headed once more for the German Reich. Before reaching the target, Lt. Irwin Ruge maneuvered JOKER'S WILD out of the formation and headed for England. They landed at the emergency field at Manston due to mechanical problems.

Most of the mission was flown over the North Sea but eventually they turned south and crossed into Germany. Clear weather prevailed over the target, as did heavy flak. Sgt. Fred Kerniss sat in the plane for the thirty-fourth time as the intense flak exploded nearby. Sgt. Milton Burchett, the nose gunner on LITTLE IODINE, witnessed the heavy flak but unlike Sgt. Kerniss who was nearing completion of his tour, Sgt. Burchett and the rest of Lt. Gordon Brock's crew experienced the flak for the first time. Although the flak gunners created a thick wall of shrapnel, no bombers were seriously damaged. The high explosive bombs impacted the target thirteen minutes before 1400. On the way out, Sgt. Joe Zonyk saw P-51s strafing an airfield near Dummer Lake. They kept the German fighters occupied and none threatened the bombers. However, fearing they might follow the bombers home, all gunners remained at their guns until landing. Everyone returned home safely.

Again on 15 January, crews readied for a mission. Twenty-two aircraft left Seething but two returned. Meanwhile the formation of B-24s continued to the target. Unable to bomb the primary target, they turned toward the secondary target. Visual conditions prevailed over southern Germany, allowing bombardiers to locate a railway bridge at Kilchberg, a small town near Tubingen. Bomb blast surrounded the structure sending smoke and debris into the air.

On the return leg, SKY LARK, piloted by Lt. Hershel Hausman, ran low on fuel. About one-third the way across the channel, one engine quit due to fuel starvation. Lt. Hausman dropped out of formation and turned SKY LARK toward the continent. They called Air-Sea Rescue and prepared to ditch in the Channel. Letting down below the clouds at 6,000 feet, they started praying that enough fuel remained to allow them to reach land and bail out. Just as land appeared on the horizon, all the remaining engines quit. Lt. Hausman decided a crash-landing carried better odds of survival than bailing out over the water. Fortunately, they made landfall and crashed near the beach just south of Ostend. The aircraft was demolished but everyone survived without injury. British soldiers picked them up and after a night at a British base, they were taken to an 8th Air Force evacuation field. They returned to Seething on 20 January, just in time for the 200th mission party.

For the fourth consecutive day, men toiled to prepare the aircraft for another mission. Men struggled in the cold deli-

Lt. Forrest McCready and crew arrived at Seething on 17 January 1945 just in time for the 200th mission celebration. Back row (left to right) Harold O. Pittinger, Eddie O. McLaughlin, Forrest E. McCready, Jr., Earl S. "Pat" Patterson. Front row (left to right) Arthur J. Helganz, Claud E. Lamoy, Jr., Darwin D. Dague, Merle L. Law, Eugene T. Short, Pat H. Cochran. (Patterson)

448th BOMBARDMENT GROUP (H)

Presents

# The 200th Mission Celebration

## A MESSAGE

THE achievements of the personnel of all units on this station which have permitted in a period of slightly more than one year the accomplishment of 200 Combat Missions against the enemy cannot be summed up in one short paragraph. Volumes will be required to record in the annals of the history of the Army Air Forces the difficulties encountered, the obstacles overcome, the losses incurred, the valor and heroism displayed, and the successes attained in the execution of each of these Missions. However, the accomplishment of 200 Combat Missions is in itself a milestone of our development, the passing of which is significant of the progress we have made toward our ultimate objective ; progress made possible as a result of the teamwork and co-ordination exhibited by both air and ground personnel. As such it presents not only an occasion worthy of commemoration and honorable celebration but also one which demands from each of us a renewal of his determination to redouble his efforts in the days to come in order that we may contribute in the best tradition of the service our full share toward final victory over the enemy.

CHARLES B. WESTOVER,
Colonel, Air Corps,
Commanding Officer

SATURDAY, 20th JANUARY, 1945
and
SUNDAY, 21st JANUARY, 1945

A message from Col. Westover congratulating the Group on its milestone decorates the cover of 200th mission celebration brochure. (Everson)

cately loading the bombs into the bellies of the bombers. At 0400, CQs roused groggy crews. Two hours later, they sat on benches in the briefing room as the curtain was pulled. The target: Ruhland, Germany. Numerous flak batteries protected the synthetic oil works there. It was bad news for the crews. The weather officer explained how expected bad weather might complicate the assembly. He was right. The Liberators lifted off from Seething and quickly disappeared into the soupy weather.

The weather improved as the formation neared Germany but smoke pots surrounding the target belched a thick black smoke obscuring much of the area forcing the bombers to turn for the secondary target. As the bombers prepared for the bomb run on the marshalling yard at Dresden, flak filled the sky with increasing intensity. Ten miles from the target, flak struck the lead aircraft of the low left squadron, UNHOLY VIRGIN. The pilot, Lt. Wesley Issacson, immediately called for the bombardier to jettison the bombs. The low left squadron, seeing bombs fall from the lead aircraft, released their bombs. They landed well short of the intended target. Lt. Issacson steered the plane clear of the formation relinquishing lead to the deputy lead aircraft and started a slow descending glide toward France. Unable to maintain altitude, the crew bailed out and the plane crashed near Sebnitz, Germany, thirty-five miles southeast of Dresden. Sgt. George Suchorsky flying on his 13th mission spent the remainder of the war as a POW with the rest of Lt. Issacson's crew.

Confusion over the target due to the flak and smoke pots forced the Group to hold its bombs with the exception of the low left squadron which bombed as planned. Once clear of the flak, they set course for the secondary target, the Dresden marshalling yard. Clear weather prevailed over the secondary tar-

The two-day celebration involved numerous activities for enlisted and officers alike. The most popular attraction however were the female guests. (Everson)

get and flak was absent from the skies as the bombers dropped the rest of the bombs.

While the B-24s headed for England, several crews struggled with flak damage received over the primary target. Trim tab control cables and the hydraulic systems were completely destroyed by flak on Sgt. Fred Kerniss' aircraft. Lt. Ray Custor on ROSIE RIVETS battled with his damaged aircraft. Numerous holes in the wings and fuel tanks made a return trip to England impossible. They crash-landed in a public park in Lille, France. The crew flying 42-50809 dealt with different problems. Although the plane was only slightly damaged, several crewmembers were wounded. Complicating matters, bad weather shut down all the airfields in East Anglia. As a result, the 448th bombers scattered to land at bases in France. The weather cleared enough the following day to allow the bombers to return.

Women from the local towns and bases were invited to the 200th mission celebration and were essential ingredients for dances held around the base. (LaPoint)

Seething's own band played for several of the dances. Dick Chaney, George DuPont, James Lamb, Jack Stebbins, John Rethal, Gail Irish, Freddie James, Robert Prouty, and Ross Westphal. (DuPont)

LITTLE IODINE 44-10516. This J model Liberator finished the war and returned to the U.S. in the summer of 1945. (Everson)

RUM N' COKE. This is one of the new M models that started arriving in England in 1945. Its tail number and fate are unknown. (Everson)

Tragedy struck on 18 January for one of the returning crews. Lt. Raymond Bunday and five members of his crew from the 715th BS caught a ride on a C-47 to England after landing in France on 13 January. Unfortunately, the C-47 encountered a heavy snowstorm and crashed five miles north of Amiens, France. Everyone on board died in the crash.

Twelve consecutive days of inclement weather provided a welcome respite from combat. During this time, leaflets and advertisements appeared on billboards around the base publicizing the upcoming 200th mission party. They promised beautiful girls, dancing, food and more. Lt. Col. Frank Cruikshank acted as the 200th Mission Party chairman and despite cold weather and snow his committee provided two spectacular days starting Saturday, 20 January. The gymnasium, Officer's Club and Aero Club were converted into dance halls and three different orchestras, The Flying Yanks, The Watton Continentals, and the Seething Dance Orchestra rotated between the three venues providing the music. Trucks brought women from nearby towns and Army camps instantly improving morale. Billets were set up for the women to stay at Seething for the entire celebration. Chaplains "pulled guard duty" to ensure there was no hanky-panky. Everyone enjoyed a USO show, stage dances, shows, and food. At 2000 on 20 January at Enlisted Mess Hall number three, free beer was provided for the enlisted men and their guests. The theater ran movies continuously during the two-day event. It ended all too soon as the war still continued.

The promise of clearing weather sent twenty-six aircraft of the 448th against the rail yard and utility plants at Dortmund, Germany on Sunday, 28 January. Three of the B-24s returned early with various problems. Dortmund, located in the heavily industrialized Ruhr valley, was surrounded by the heaviest concentration of anti-aircraft artillery in Germany. The only possible exception was Berlin itself. Crews referred to the area as "Happy Valley" or "Flak Valley" because of the intense flak usually encountered. This trip was no different.

Over the target, scattered clouds allowed bombardiers an excellent view of the city. The view was just as good from the ground. Intense and accurate flak pounded the bombers as they crossed the target. The right and left squadrons released their bombs over the target but the nine aircraft in the lead squadron were unable to drop. They dropped their high explosive and

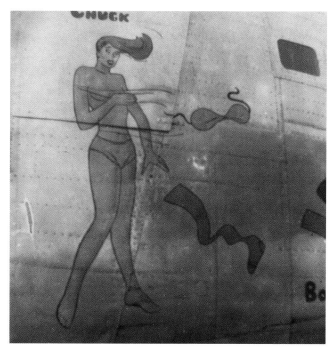

WINDY WINNIE 44-10599. On 28 January 1945, flak severely crippled this B-24 forcing the crew to crash-land in Luxembourg. Fortunately all survived and returned to Seething. (Bailey)

incendiary bombs on Lippstadt, a target of opportunity just beyond Dortmund.

The thick flak claimed several of the 448th's planes. Lt. Robert Paeschke crash-landed WINDY WINNIE in Luxembourg as a result of the barrage. The crew of LINDA MAE also landed on the continent due to battle damage. OUR BABY, with Lt. Carl Holt at the controls, suffered numerous flak hits. An unexploded shell pierced the left wing just outside of the number one engine and another round hit the right wing near the number three engine. With oil gushing out of the engine and fire igniting, they limped toward Belgium. Shrapnel shattered most of the plane's plexiglass windows and damaged the hydraulic system. After finding a Ninth Air Force fighter base near Malmedy, only thirty miles behind friendly lines, they tried to lower the gear. After two unsuccessful attempts, they finally succeeded on the third try. Flak not only damaged the hydraulic system, but it also blew all three tires which complicated the already tricky landing. Snow swirled around the area as they prepared to land. After landing they were unable to stop the battered plane before skidding off the runway and stopping dangerously close to the edge of a steep bluff. Fortunately for Sgt. Fred Kerniss and the rest of the crew no one was seriously hurt but the plane was destroyed. There was a hole in the left wing large enough for a man to crawl through. Despite the close call, Lt. John Cushman was going home: it was his thirty-fifth mission!

The returning planes fought the weather as well. Eight of the Liberators landed at Hardwick after being unable to land at Seething. A landing accident at Seething crushed the nose of aircraft 42-50676 injuring the pilot, Lt. James Blank.

While crews battled in the skies over Germany, those remaining at Seething took part in a parade commemorating the third anniversary of the Eighth Air Force. At 1045, all organizations assembled in front of the Station Headquarters to pay respect for their departed comrades. All station personnel stood at attention for two minutes of silence in honor of all those killed or missing in action. Services were held in the base chapel. After the crews returned, the station band provided musical entertainment for two hours. Following the entertainment, Lt. Leroy Middleworth opened the official ceremony at 2030. Col. Westover cut the anniversary cake shortly afterwards. Word quickly spread throughout the gathering that the Group was alerted for a mission the next day.

As salvage crews worked on the twisted hulks from the day prior, the bombers returned to the skies. Due to the large number of planes that failed to return the previous day, the 448th launched only two squadrons on 29 January. Planes from the 93rd BG filled the high right squadron. Once more, clouds

Two men (one on left is a mechanic named Tozzi) stand next to WINDY WINNIE before it crash-landed in Luxembourg. (Everson)

forced crews to takeoff in instrument conditions and fly the instrument assembly procedures until they broke into the clear. Lt. Gordon Brock narrowly averted every airman's worst nightmare. Just after clearing the clouds, another B-24 flew directly overhead. Lt. Brock pushed MY BUDDIE into a dive to avoid a collision. Despite rattled nerves they completed the assembly.

Skirting the deadly Ruhr valley to the north, the bombers hit the marshalling yard at Munster, Germany, after clouds obscured the primary target, the viaduct at Bielefeld. Weather required the use of H2X to drop the 1,000-pound high explosive bombs and the smaller M47 incendiary bombs. Heavy flak billowed over the target. The tracking radar produced an inaccurate barrage and no aircraft were damaged. All the bombers reappeared over Seething five and half-hours after departing.

Bad weather clipped the wings of the entire Eighth Air Force on 30 January. No aircraft participated in combat opera-

LINDA MAE 42-51075. Thanks in a large part to these men, (left to right) Sgt. Hammer, crew chief Wiegel, and Sgt. Chesser, this B-24 survived the war. After landing on the continent with battle damage on 28 January 1945, she was repaired and returned to combat. (DuPont)

Not all hazards were reserved for combat. The poor weather, small roads, and driving on the opposite sides of the road produced a dangerous combination. One such accident claimed the life of Sgt. George Greene on 29 January 1945. (USAF)

Snow blankets one of the living areas at Seething. On days like this the main occupation of the men was staying warm. (Everson)

tions. The next day an attempt was made to attack a tank plant in Berlin. By takeoff time, the target was changed to Brunswick, Germany. Capt. William Snavely, leading the 448th and the 20th CBW, made the necessary changes and the formation continued with its takeoff and assembly. As thirty-three Liberators from Seething pressed toward the target, bad weather descended on England. None of the designated fighter escorts were able to takeoff. Over Dummer Lake, only sixteen minutes from 'bombs away,' the Group and the entire Second Air Division received a recall message. No bombs were dropped on the last day of January.

Atrocious weather at Seething required the entire Group to divert far to the north. Some landed at a Canadian base at Skipton-on-Swale. Lt. John Rowe and crew did not return to Seething until the following day. Some of the crews were not able to return until 2 February.

Bad weather limited the 448th to only thirteen combat missions in January. However, the thirteen were costly. Three aircraft and crews were missing. Crews abandoned fourteen aircraft on the continent after suffering battle damage, eight from the 715th BS alone. Three aircraft returned to Seething which such heavy damage they were considered total losses and were salvaged for parts. The 714th BS was also hit particularly hard. They lost eight aircraft. More costly were the men wounded, missing or killed. Although the fate of the war heavily favored the Allies, it was far from over for the aircrews. They still struggled in the life and death battle in the skies over the German Reich.

seventeen

# HOW MUCH LONGER?:
# February 1945

As the days rolled into February, the German Wehrmacht retreated behind the relative safety of the West Wall following Hitler's failed attack in the Ardennes. Brutal weather conditions created an inhospitable environment for the ground forces. They slogged through the cold pressing the attack on a retreating enemy. At Seething, the bitter weather was attributed to an alarming increase in colds and bronchial infections. The medical staff spent most of the month battling these infections. In the air, the winter weather also proved unfriendly. Clouds and severe cold temperatures increased the difficulty of the missions. Yet, the 448th continued their onslaught on the German Reich. How much longer could Germany sustain the fight after taking such a beating on the ground and in the air?

Persistent bad weather grounded the bombers from Seething until 3 February when the entire Second Air Division planned to attack an oil refinery at Magdeburg, Germany. Lt. Paul Homan was awakened at 0400 to fly as copilot and observer on his first combat mission with an experienced crew. He and his crew were newly assigned to Seething and this was his introduction, and what an introduction. Intelligence briefed crews to expect heavy flak opposition as eighty-eight guns defended the refinery. Thirty-one B-24s left Seething to join the force but only twenty-nine completed assembly and left the English coast headed for Germany. One bomber, piloted by Lt. Albert Broadfoot, suffered an engine failure and dropped out of the formation prior to the target. They bombed a target of opportunity, the airfield at Vechta, Germany, then returned to Seething.

The rest of the bombers continued as briefed. As the formation approached the Initial Point scattered puffs of flak dotted the sky. Normally this indicated heavier stuff was not far off. As predicted, the intensity of the flak increased as they neared the target. Sgt. Ed Chu, in the tail of MY BUDDIE, watched as the dreaded flak intensified. "I could see the brilliant white flashes of the close flak bursts below and to the rear. At the target, flak became heavier. Fragments struck our ship – sounded like hail." Clouds and smoke screens obscured the target making the bombardier's job more difficult. Despite these hurdles, bombardiers unleashed their deadly cargo using PFF to identify the primary target. The exploding flak penetrated the aluminum skin of several of the Liberators. MY BUDDIE suffered heavy battle damage. The number three engine suffered a hit as did the nose turret and left wing. Flak punctured the aluminum skin in numerous places above the waist gunners' windows. On return, the formation picked up more flak as they crossed the Zuider Zee and Frisian Islands. Although numerous planes were damaged, everyone returned safely.

The horrible winter weather prevented all offensive operations on the 4th and 5th. Not until 6 February did marginal conditions allow for aerial operations. Lt. Paul Homan and crew, calling themselves Homan's Red Caps (Lt. Homan purchased red ski caps for everyone in Boise during training), flew SPIRIT OF COOLEY HIGH SCHOOL on their first mission.

Aircraft 42-52496 lost an engine during assembly and returned to Seething. As the Group crossed the Dutch coast, scattered bursts of flak dotted the sky but all exploded harmlessly away from the bombers. They continued on course. Heavy clouds once more concealed the primary target so the bombers returned to Magdeburg; the marshalling yard in the city center was the briefed secondary target. Heavy flak greeted the attackers again as they crossed over the city. Just after 1130, bombardiers flipped switches sending bombs arcing toward the ground. Two 448th Liberators bombed with other Groups after failing to find the formation during assembly. SLICK CHICK, piloted by Lt. Daniel Durbin, bombed with the 491st BG while the crew of PICADILLY LILLY bombed with the 93rd BG. Everyone returned to Seething without incident.

No aircraft left the ground on 7 February and although they got airborne on 8 February, they quickly returned after receiving recall instructions. They never left the English coast. On 9 February, the 448th participated in a major Eighth Air Force offensive against oil producing facilities throughout Germany. The Second Air Division returned to Magdeburg for the third time in as many missions. Three aircraft from the 448th aborted during assembly for various reasons but the remaining B-24s left the English coast at the prescribed time. Clouds over the target once again forced a change of plans. The oil refinery required visual bombing conditions so the Group reverted to the H2X secondary target, the marshalling yard. Bombs exploded throughout the target area. The same heavy flak targeted the bombers, this time with costlier results. The shrapnel penetrated the thin metal skin of SLICK CHICK injuring one man and damage inflicted by the heavy flak forced Lt. David Anderson to land BACK TO THE SACK on the continent. Meanwhile, Lt. Paul Homan watched FW-190s make a rare appearance. Sgt. Ed Chu, the tail gunner on MY BUDDIE, witnessed three of the FW-190s attacking a straggler from the trailing group. Unfortunately, the straggler did not make it.

Continuing bad weather limited the effectiveness of the bombers. Even on days when the bombers got airborne, they were usually required to bomb utilizing PFF aircraft with H2X or G-H equipment. Although capable of dropping through clouds, the precision was not comparable to visual bombing with the Norden bombsight. Although the precision was degraded, the bombers continued carrying the war to the German heartland. The entire country was now under siege not only from the air but also on the ground. The 'Bulge" was reduced and new Allied attacks sent the Germans retreating into their homeland.

The 448th flew again on 11 February when it resumed the attack on the German oil industry. The 448th's designated target was an oil storage depot at Dulmen, Germany. If by chance visual conditions existed, the Group was instructed to attack a road bridge at Weser, Germany. As usual, clouds covered the continent and as instructed, twenty-nine aircraft released their bombs at 1044 from 23,000 feet over the oil depot. Only light, inaccurate flak opposed the bombers. Just before 1300, the bombers were back on their hardstands at Seething. Only EA-

With cold winter limiting outdoor activities, the men resorted to indoor entertainment including darts to pass the idle time. (LaPoint)

GER ONE was missing. The pilot, Lt. Paul Jones, decided to land on the continent.

On Valentine's Day, the Group returned to Magdeburg. Once again they attacked the secondary target. MISS B-HAVIN experienced a propeller problem as they crossed the North Sea and the pilot decided to return rather than risk continuing. A single airplane over Germany still drew a lot of attention from the Luftwaffe despite their loss of air superiority. As the formation reached altitude, Sgt. Ed Chu, in the tail turret of WINDHAVEN, watched the unwanted contrails forming be-

Pvt. Eli Roffwarg of the 58th Station Complement Squadron pencils a sketch of a friend while some of his previous works hang on the wall. (USAF)

hind the plane. Not only did they make formation flying extremely difficult, but they pointed like a finger to the planes. Gunners on the ground as well as Luftwaffe pilots used these vapor trails to track and target the bombers. Either way, they meant trouble for the aircrews.

Flak crews learned well from the bombers' previous trips to Magdeburg. As the bombers commenced the bomb run, the gun crews opened up with a tremendous barrage. Scattered clouds provided very little cover for the bombers as the sky filled with deadly puffs of exploding shells. On the bomb run, the 446th BG course coincided with the course of the 448th. Lt. Edward Malone, flying the lead aircraft named 4-F, a reference to one of the military's medical category, saw the looming conflict and maneuvered the formation away from the potential disaster. The 448th ran the harrowing gauntlet and dropped their bombs despite the thick flak and the near mid-air collision.

BLUES IN THE NIGHT 42-50677. Despite the miserable winter weather, Sgt. Ralph Boartfield, the crew chief of BLUES IN THE NIGHT, kept her flying for seventy consecutive missions without an abort. She went on to complete the war and returned to the States. (Everson)

Leaving the target, Lt. Walter Bobak, the pilot of aircraft 42-78491, initiated a routine intercom check. Everyone checked in as expected except the navigator, Lt. Dick Wagner. Sgt. Donald Beck checked the nose turret but the doors were frozen shut. He finally broke the ice and opened the doors. There he found Lt. Wagner slumped over the guns. His oxygen mask had frozen, quietly depriving him of oxygen, and eventually causing him to lose consciousness. Sgt. Richard Mickelson brought a walk-around bottle from his position and revived Lt. Wagner. Lt. Bobak's intercom check in conjunction with Sgt. Beck and Sgt. Mickelson's timely response saved his life.

Strong winds that helped the bombers on the route to the target now battered them as they turned for home. The stiff headwinds made the leg home longer than normal. Seven hours after leaving Seething, the B-24s taxied to their hardstands where waiting crew chiefs prepared to go to work. Their work never ended; aircrews broke the planes and the crew chiefs fixed them. It was a never-ending cycle. Maintenance excelled keeping the aircraft flying. Sgt. Raymond Wood kept aircraft TROUBLE N' MIND flying for eighty-six missions without an abort and Sgt. Ralph Boartfield kept BLUES IN THE NIGHT flying for seventy consecutive missions. Also, the maintenance crew led by Sgt. Schall and Sgt. Wright completed their fifth month of abort-free missions.

On 15 February, the 448th re-targeted the oil refinery at Magdeburg, Germany. It was the fifth mission out of the last six to attack Magdeburg. The Liberators started their takeoff rolls just after 0800. Trouble quickly developed. As UMBREOGO lifted off the ground an exterior panel holding a life raft in the aircraft dislodged. Lt. John Rowe, waiting his turn to takeoff behind UMBREOGO, watched the tragic events

MY BUDDIE flies in formation on 3 February 1945 while smoke rises in the background from a freshly attacked target. (LaPoint)

unfold. The life raft on the right side of the aircraft deployed and hit the tail. The rope tethering the raft to the plane tangled itself around the horizontal stabilizer. Lt. Harlyn Schroeder, the pilot, struggled to gain altitude. The plane entered a steep climb followed by a successive dive as the pilot, copilot, and engineer pushed the controls forward. Sgt. Raymond Bailey held on as the pilots struggled with the plane. Miraculously, they gained control and tried to analyze the situation. They discussed shooting the dinghy off the tail but determined that to be impossible.

The series of climbs and dives continued as Lt. Schroeder headed for the North Sea to jettison their bombs. Sgt. Thomas Economy nearly fell out of the bomber as he kicked a hung bomb from the bomb bay just as the plane entered a steep dive. Once the bombs were released, everyone but the two pilots moved to the rear of the plane which made it more stable and easier to fly. Headed back toward the coast, Lt. Schroeder ordered everyone to bail out once over land. Sgt. Raymond Bailey was the first out. He saw another chute blossom as he floated to earth. It belonged to Sgt. Erwin Schilling. A Vicar's wife met the two airmen and helped then into her home where she made them tea. After a brief respite, she led them into the adjacent church and had them kneel at the altar. With her feeble hands on their shoulders, they prayed for the safety of their friends.

Meanwhile Lt. Schroeder and the copilot, Lt. Delwin Roorda, fought to keep the plane level. As the men left the plane, it became harder to fly, as the ballast in the rear of the plane lessened. Lt. Schroeder volunteered to stay with the plane while the last crewmembers jumped. Lt. Roorda and Sgt. Edward Vetterneck, the engineer, refused to let him. After a quick discussion, they all agreed to jump. Sgt. Vetterneck held the controls while the pilot and copilot unstrapped from their seats. Once out of their seats, they prepared to make a dash for safety. Sgt. Vetterneck went first. He was the last survivor out of the stricken plane. As his parachute opened, he saw the tail of his plane break, sending the fuselage into a spin. It crashed at Elsing near Attlebridge, only a few hundred yards from where Sgt. Vetterneck landed. Both courageous pilots died at the controls.

Meanwhile the formation continued to assemble, unaware of the life and death struggle over Attlebridge. Two squadrons from the 448th assembled but not without other problems. EL KORAB lost an engine and SWEET ROSE MARIE developed a fuel leak. Both aircraft returned to Seething. Clouds continued to plague the Group's effort to hit the primary target. Once again bombardiers used H2X to identify the secondary target. The bombs fell on the city, this time over a synthetic oil plant. Fortunately, the clouds limited the flak gunners' ability and the inaccurate flak exploded far from the bombers. When the crews returned to Seething, they learned of the tragic events that claimed the lives of Lt. Schroeder and Lt. Roorda. As crews dealt with the loss of two more friends, teletypes in the Group Headquarters rattled off the orders for the next day.

The next day, 16 February, thirty-one aircraft left Seething for the marshalling yard at Osnabruk, Germany. Low clouds threatened the mission from the start, but after a thirty-minute delay, controllers cleared the aircraft for takeoff. DEAD END KIDS lifted off and almost immediately went into the clouds. Lt. Paul Homan, at the controls, glanced at the altimeter, 400 feet! Forty-five seconds later another B-24 followed, then another. They blindly flew the prescribed assembly plan hoping the weather would clear. After an agonizing time fearing a midair collision, the planes broke into the clear. The assembly proceeded uneventfully from that point.

The formation joined the larger Division formation and left the English coast at the designated time. Clouds stretched out beneath the formation like a sea of cotton. Occasionally, breaks in the clouds provided brief glimpses of the ground but this was rare. Enroute to the target, one aircraft experienced a supercharger failure thirty miles west of Egmonde, Holland. They aborted and returned to England. The rest of the forma-

RUGGED BUT RIGHT 42-94953 takes to the air over Seething. (LaPoint)

SNUFFY'S DELIGHT 44-50976. An unknown airman poses with his aircraft. The fate of this aircraft is unknown but it is probable it survived the war. (Everson)

BACK TO THE SACK 42-51288. Heavy damage caused by flak on 9 February 1945 forced the crew to land on the continent. Despite the damage the plane was repaired and returned to the Group. (Everson)

tion continued. As the target loomed nearer, heavy black puffs of exploding flak filled the skies in front of them. Clouds obscured most of the target forcing the bombardiers to use G-H to aid the bombing but bombs were released at 1423 amid a tremendous flak barrage.

Lt. Hugh McFarland's aircraft, LITTLE JO, suffered damage from the barrage but somehow stayed with the formation. Once the formation cleared the enemy coast, he headed his damaged aircraft for the emergency airfield at Woodbridge. They landed without further problems. Almost seven hours after departing, the formation arrived over Seething where low clouds required crews to fly instrument approaches. Despite the low ceilings and icy runways, everyone landed safely. Not all the excitement was in the air. While the bombers braved the elements and the enemy, Military Police captured three Italian POWs on Seething airfield and returned them to their POW camp. They claimed they were on their way to church. The MPs notified the appropriate authorities and they were returned.

A planned mission to Berlin the following day was recalled due to the winter weather. The severe cold made flight operations too hazardous and the planes returned before reaching the enemy coast. Lt. Hershel Hausman's crew dropped their bombs harmlessly into the English Channel. Everyone else did the same.

On 19 February, the 448th participated in a raid on the marshalling yard at Siegen, Germany. Poor weather limited visibility as the formation left the British coast and headed across the North Sea. Thirty-one aircraft attacked the rail yard with a mix of high explosive and incendiary bombs. Again, clouds forced the crews to rely on H2X to identify the target. Flak was sporadic and failed to inflict any damage. When they returned to Seething, crews accomplished yet another instrument letdown into the murky British weather.

The next day, the 2nd Air Division launched a raid against Nurnberg but poor weather conditions over Belgium necessitated a recall. No one from the 448th dropped bombs. The following day, 21 February, the 448th returned to the skies and set out for the same target. Unknown to the crews, this mission marked the beginning of Operation Clarion, a concerted effort to destroy the transportation and communication network of Germany that was previously untouched by the bombing campaign. As the Wehrmacht retreated, rail and road assets were required to move troops and equipment that remained following the Battle of the Bulge. Planners sought to interdict these forces as they were redistributed to fill holes in the German lines. Over the next several days, the entire 8th Air Force flew missions aimed at this vital infrastructure.

Smoke markers drop from a 714th BS aircraft amid the always present flak. Smoke markers were used extensively to identify the release point for the rest of the formation. (LaPoint)

A high-pressure weather system settled across the continent creating clear skies and perfect weather for flying. The bombers took advantage of the nice weather. P-51s filled the sky above the bombers daring the Luftwaffe to fight. They did not. Flak, however, peppered the bombers as they passed near Frankfurt but serious damage was avoided. As the bombers neared the target, weather scouts preceding the B-24s advised the formation to perform a 360-degree turn. B-17s were conducting their own bomb run in the opposite direction. The quick-thinking scouts averted a potential disaster.

Once the B-17s cleared the target area, the Liberators of the 448th led the 20th CBW on the bomb run. Clouds obscured the marshalling yard but the crews still dropped using H2X to identify the target. Very accurate flak filled the sky over the target damaging the lead ship as well as others. Once clear, they turned for home. SLICK CHICK experienced problems leaving the target and the pilot, Lt. Hugh McFarland, elected to land at the emergency airfield at Manston. It was his second such landing in as many missions.

On the second day of Operation Clarion, 22 February, three squadrons of the 448th attacked the marshalling yard at Kreiensen while a fourth squadron augmented the 93rd BG and bombed Northeim as part of a maximum effort directed at targets throughout Germany. In an effort to combat the increasing effectiveness of the flak and minimize collateral damage, the bombers flew below 10,000 feet, much lower than normal. The 448th dropped their bombs from 8,500 feet. The 93rd BG, bombing nearby Northeim, threatened to force the 448th off their planned bomb run with a converging bomb run of their

The most prolific target for the 448th, Magdeburg and its oil refineries, receives more bombs. The Group attacked targets in the city nine times, including five trips in February 1945. During the nine visits they dropped over 550 tons of bombs. (Ray)

own. The formations resolved the conflict and each Group hit their intended targets with good results. Sgt. Ed Chu on WINDHAVEN saw the red flashes as bombs found their mark. Heavy smoke obscured the area as the bombers turned for the rally point. Flying that low, Sgt. Chu saw the roads jammed with military traffic and also watched as P-51s attacked a train. An explosion followed by a large white cloud of smoke marked the end of another German locomotive. Col. Westover commended all four squadrons for the excellent work. "The execu-

BLAKES SNAKES 42-51228. This Liberator completed the war and carried men of the 448th home when they returned to the States on 13 June 1945. The scoreboard under the pilot's window shows its participation in numerous missions. (Everson)

BROADWAY BILL 44-48870. The peeling paint shows the wear and tear this Seething-based B-24 endured. Despite this, she completed the war and returned to the States on 13 June 1945. (LaPoint)

tion of the mission of 22 February 1945, including penetration, attack, and withdrawal, merits high commendation for all those crews who successfully completed the operation. It is an outstanding example of the results that may be achieved through strict observance of air discipline…The targets were actually bombed from 8,500 feet to insure a visual run and the patterns of all four squadrons were so precisely placed that complete destruction of the target resulted."

While the crews attacked targets in Germany, Sgt. Joe Zonyk experienced one of the best treats of the war. He flew as part of a crew that carried Lt. Col. Heber Thompson and Lt. Col. Lester Miller to Paris on official business. During their four hours on the ground, he saw some of the sights including the Cathedral of Notre Dame. He also spent several hundred dollars on numerous gifts: silk stockings, silk handkerchiefs, champagne, perfume and cognac. It all made it to the plane and back to Seething. Although the trip was quick, it was a bright spot in comparison to the dark days of combat.

Crews groaned as the curtain fell on 23 February revealing the target. Gera, Germany, was not a heavily defended target but it lay in eastern Germany, south of Leipzig. The long trip promised unwanted excitement and guaranteed a long day. One aircraft aborted during assembly before the formation headed for the marshalling yard at Gera. The long trip proceeded without problems until they arrived at the target. Clouds foiled visual bombing attempts. Unable to bomb the target with H2X, the formation turned for home. The bombers, usually much lighter after the bomb run, burned fuel at a higher rate and several started running low on fuel. The long trip coupled with the higher than normal fuel consumption forced two aircraft to jettison their bombs in order to complete the return trip. The remaining planes in the formation dropped their bombs on Osnabruk as they passed overhead. Inaccurate flak dirtied the sky over the city but did not cause any damage. Short on fuel, three aircraft landed on the continent to refuel and another refueled at nearby Beccles before returning to Seething.

On 24 February, the 448th attacked an oil refinery near Hannover at Misburg, Germany. The 448th did not fly as an integral Group on this mission, instead two squadrons augmented other Groups. One flew with the 446th BG and the other flew with the 93rd BG. They left Seething at 0930 and joined their respective formations. Lt. Paul Homan and his "Red Caps" flew SONIA as part of the 446th BG formation. They hit their target although the usual clouds prevented seeing the results. No flak opposed the bombers although other Groups were seen to fly through very heavy barrages. As usual, the swarming escorts kept the Luftwaffe out of sight.

GRACE, PETE'S SAKE. It is probably 44-50526 that returned to U.S. after the war. (Patterson)

A crew performs an engine run on UMBREAGO one cold, snowy morning at Seething. The protective covers on the nose turret and top turret remain to protect against the bitter weather. She met an unfortunate demise on 15 February 1945 that tragically claimed the lives of her pilot and copilot. (Bailey)

This 712th BS aircraft fell prey to the icy runways and slid off the end collapsing the nose gear. The winter weather challenged even the most experienced and skilled pilots. (LaPoint)

*How Much Longer?: February 1945*

LILLY'S SISTER 44-50787. Another M model that entered the war in 1945, she also survived combat and returned to the States. (Patterson)

NANCY ANN. The tail number and fate of this 448th aircraft are not known. (Patterson)

On 25 February, the 448th hit a tank repair facility and ordnance depot at Aschaffenburg, Germany. For a change, the weather cooperated and crews bombed the primary target visually. Haze limited visibility slightly but the target was distinctively visible for the bombardiers. Just after noon, bombs fell from thirty-one B-24s of the 448th impacting the target with good results. No flak rose in defense of the target. Not until the Group flew in the vicinity of Strasbourg, France, on return did flak rise in anger. Fortunately, it was not accurate and caused no damage. The relatively uneventful mission was a great 23rd birthday present for Lt. Paul Homan.

Teletypes chattered as orders were printed for the next day's mission. The target: Big B. Berlin! In the briefing room crews groaned with agony as the map was revealed. It was the first trip to Berlin since 21 June 1944. For many, including Sgt. Ed Chu and Lt. Paul Homan, it was their first visit to the Reich capital. Everyone knew the horror stories and expected stiff resistance. The 448th was not alone; the entire 8th Air Force targeted Berlin. The First and Third Air Divisions planned to attack rail stations in central and south Berlin while the Second Air Division, including the 448th, hit rail yards in north Berlin.

Engines coughed to life at 0830 and twenty minutes later the bombers took to flight. Six bombers aborted with various problems before the formation left the assembly area, but otherwise the route to the target was unopposed. Escorting P-47s and P-51s filled the sky above the bombers vainly searching for the Luftwaffe. Clouds completely obscured Berlin and the bombers initiated the bomb run using H2X to identify the Reich capital. Just as expected, intense flak filled the sky over the city in front of the bombers. Crews huddled beneath their flak vests and helmets as a host of prayers asked that the terrifying flak would somehow miss. The bombers waded into the flak and released their bombs over the target. Six, 1,000 pound bombs with time delayed fuses dropped from the bomb racks of MY BUDDIE. A cocktail of high explosive and incendiaries fell from the other bombers. They disappeared into the clouds with unknown results.

Surprisingly, all the bombers arrived at the rally point with no major damage or injuries. A couple of hours later the bombers touched down at Seething. Everyone breathed a tremendous sigh of relief. They had dropped bombs on the very heart of the enemy. Despite Hermann Goering's boast of no enemy aircraft over the Reich capital, the 448th not only attacked the capital but accomplished this feat without a scratch.

The 448th flew its seventh consecutive mission on 27 February. The Group continued to attack the German transportation system by bombing the marshalling yard at Halle in eastern Germany. All the bombers left the English coast on time and attacked the target as briefed. The usual clouds forced the bombers to use H2X once again which they did without a hitch. The problems started after leaving the target. VIRGIN STURGEON lost an engine over the target and Lt. James Guynes pushed the remaining engines to the red line in an effort to stay with the formation. Compounding the problem, strong headwinds from the jet stream buffeted the formation. Their ground speed dropped significantly and navigators figured time remaining while engineers feverishly calculated fuel consumption. Several of the planes landed on the continent to refuel before returning to Seething.

Once clear of enemy territory, Lt. Guynes guided VIRGIN STURGEON, affectionately known as THE STURGEON, out of the formation in search of an airfield. With two engines red-lined and their continued use in question, they found a forward fighter strip in Belgium. THE STURGEON's wheels touched down on the matting but they could not stop the B-24 on the short runway and they ran off the end. No one was injured but the plane suffered heavy damage.

Lt. Ed Anderson and crew, on their first mission, also ran into trouble leaving the target. "On the return trip we ran low on gas. We dropped out of formation, intending to refuel at an airfield in Belgium. I don't recall the name. We were flying low to stay under the clouds. The visibility was limited by fog and we could not see the field we were flying towards. Several engines started to cut out and lose power as they ran out of gas. There was a small farm field ahead and beyond that was woods. I decided to try to put the plane down on the small farm field. I called for the gear to be lowered. The right main gear came down and locked and the left main gear started down but before it could lock we hit the ground. The plane landed on the right main gear, the left wing and the underside of the fuselage and bomb bay. The ground was muddy and the plane slid to a stop. No one was injured." They returned to Seething the following day leaving LITTLE JO on the continent for salvage crews.

While the bombers fought their battles, Pvt. Vincent Padilla of the 1193rd Military Police Company at Seething fought his own battle. He won a clean-cut decision over his opponent for the ETO bantamweight boxing championship held at Rainbow Corner in London. After falling short once before, he was now champion.

Taking a break from the German transportation system, the Group attacked a Me-262 components factory at Meschede, Germany, on the last day of February. Three squadrons of ten aircraft each lifted off the runway at Seething at 1030. Two aborted because of mechanical problems leaving twenty-eight

POOP DECK PAPPY 42-7521. An early entry in the war as denoted by the olive drab paint scheme, this Liberator flew numerous missions with the Group and survived the war. (Everson)

Liberators in the formation. Once again thick winter clouds covered Germany hiding the target from the naked eye. The abundance of bad weather forced crews to become proficient in H2X and G-H bombing procedures. They utilized their skills and dropped their bombs through a solid undercast. The sky remained free of the deadly flak and the Luftwaffe again failed to present a challenge. The six and one half-hour flight concluded uneventfully with all the bombers returning to Seething.

After flying only six missions in the first half of the month, the Group ended the month on a hectic pace. They flew ten missions in the last thirteen days. Amazingly, no aircraft were lost and only one aircraft suffered damage severe enough to require salvage operations. The quick pace of operations did not keep the 715th BS from conducting four practice missions to improve crews' formation and bombing skills. During Feb-

SLICK CHICK 42-50460. Despite an emergency landing at Manston on 21 February 1945, this aircraft survived combat and returned to the U.S. after the war. (Patterson)

LITTLE JO 41-28958. This aircraft crashed on 27 February 1945 while trying to land at an emergency airfield in Belgium. Unable to locate the airfield due to poor weather they crash-landed in a field as their engines quit due to fuel starvation. (Everson)

ruary, crews in the lead squadron, 712th BS, elected to distinguish their aircraft from other B-24s at Seething. They started painting shark's mouths and dragon heads on the noses of the aircraft.

The protracted combat was taking its toll on American forces. A shortage of infantry soldiers brought a recruiting drive to Seething. A draft was instituted although some men volunteered. Numbers were drawn and the selected men were then required to pass a physical. A total of 106 men from Seething transferred to the infantry during March.

One aircraft was lost during the month and two others were salvaged. Most of the missions in February attacked the German transportation system. The heart of the German transportation system, the railways, suffered heavily in the attacks. As a result, the retreating Wehrmacht faced increased confusion and delays in their retreat. Allied ground forces continued to pound them on the ground while the air forces destroyed their infrastructure. How much longer could Germany continue to fight?

eighteen

# LUFTWAFFE RESURGENT: March 1945

The rapid pace of operations did not slow as the calendar turned to March. Even though bad weather continued to cause problems, it steadily improved throughout the month. As the 8th Air Force continued pressuring the battered German war machine, the clearing weather brought a resurgent Luftwaffe. Employing new jet fighters, specifically the Me-262, the Luftwaffe briefly rekindled the fear crews experienced one year earlier. It proved to be too little, too late!

The 448th continued its consecutive mission string flying a mission against the marshalling yards at Augsburg, Germany on the first day of the month. Ironically, the primary target was a components factory for the Me-262. Weather prevented bombing the primary and the formation reverted to its H2X secondary target, marshalling yards at Augsburg. Twenty-eight Liberators from Seething led the 20th CBW on the long route to southern Germany. Only sporadic, inaccurate flak dotted the sky as they dropped their bombs. The real danger was not the flak but the long trip back. Crews fought headwinds and nervously figured fuel rates. Lt. Gordon Brock, flying MY BUDDIE, made it back to Seething after almost nine hours but several others landed on the continent to refuel. The crews of REDDY TEDDY, OLD POP, and EL KORAB elected to land at airfields on the continent instead of risking ditching in the North Sea due to fuel starvation.

For the tenth consecutive day, bombers from Seething attacked targets in Germany. The primary target for 2 March was an oil refinery at Magdeburg but clouds forced the Liberators to hit the secondary target, the Magdeburg rail yards, using H2X. Following an uneventful trip to the target area, crews endured an intense and accurate flak barrage. MOTHER OF TEN, flown by Lt. Karl Augustine, suffered damage but remained with the formation. A B-24 from another group fell to earth. At 1043, ten 500-pound high explosive bombs fell from each aircraft in the two squadrons from Seething. Despite the heavy resistance, the lead squadron dropped with excellent results. Results for the low left squadron were not as good. MOTHER OF TEN, with battle damage, returned safely to Seething as did the other twenty Liberators. That night the Luftwaffe followed returning RAF bombers and strafed airfields throughout England. No one at Seething suffered injuries in the attack. However, nearby airfields were not as lucky.

Ground crews prepared thirty-nine Liberators for another mission on 3 March. Six of the aircraft were designated as a screening force designed to mislead the German air defenses. The six decoys played an important part in the mission but did not drop any bombs. The remaining thirty-three bombers were destined for an oil refinery located at Magdeburg, Germany. The force left Seething at 0630 for the skies of Germany.

Unlike previous missions, German fighters took to the skies. This time however, new jet fighters joined the fray. Real trouble started about forty miles from the target. Me-109s and FW-190s attacked the P-51 escorts, distracting them while two ME-262s attacked the bombers using reckless frontal assaults. The low left squadron endured the brunt of the attack. FEUDEN REBEL, the deputy lead aircraft in the low left squadron, suffered numerous hits from the cannon fire. The pilot, Lt. James Guynes on his eighteenth mission, struggled to maintain position in the formation. Cannon fire raked the cockpit and damaged several of the engines. The aircraft shuddered and slowed as it yawed to the left from the loss of power. FEUDEN REBEL slid to the left and crossed into the path of another aircraft. Lt. Irving Smarinsky, flying GUNG HO II just left of FEUDEN REBEL, saw the damaged aircraft and maneuvered his aircraft to avoid the damaged Liberator. FEUDEN REBEL continued

MY BUDDIE 42-95083. Despite several close calls this Seething-based Liberator survived the war. Here its sits on a hardstand at Seething awaiting its next mission. (Rebeles)

uncontrolled, striking the tail of GUNG HO II. The force of the impact severed the tail of Lt. Smarinsky's aircraft sending it into a loop and then plummeting out of control. Lt. Guynes' aircraft rolled onto its back and then fell to the ground. The centrifugal force pinned Sgt. Don Schleicher to the top of the plane. Smoke filled the aircraft before it was ripped apart by an explosion. It blew Sgt. Schleicher clear of the plane. He pulled the ripcord as debris fell all around him. He passed out again and regained consciousness only after landing and subsequent capture.

Lt. John Rowe, flying SPIRIT OF COOLEY HIGH SCHOOL-DETROIT on the right wing of Lt. Guynes, witnessed the horrible collision. They counted five chutes floating earthward. Only Sgt. Schleicher survived from Lt. Guynes' crew. Although Lt. Guynes initially survived the crash, he died twenty-six days later from an infection in the wounds received during the collision. Two men from Lt. Smarinsky's plane survived the horrible collision. Lt. Arthur Hoffman and Sgt. Gerald Perry miraculously escaped.

Despite the tragedy, the formation continued to the target. The weather cooperated as clouds cleared allowing the bombardiers to identify the target. Intense flak filled the sky in front of the bombers as they started their bomb run. Despite the opposition, general purpose and incendiary bombs arced toward the target at 1031and impacted with good results. Giant columns of smoke billowed from the target area. Thirty minutes later, crews still could see the black smoke as it rose to 20,000 feet.

Leaving the target, Lt. Rowe and crew on SPIRIT OF COOLEY HIGH SCHOOL-DETROIT ran into more trouble. They developed an oxygen leak forcing them to descend and leave the safety of the formation. Leaving the formation this far from home posed serious threats. German fighters, wary of large formations, preferred solitary, damaged aircraft. Answering their distress call, two P-51 Mustangs appeared and escorted them to safety without incident. Debris from the Me-262 attack and ensuing collision damaged Lt. Marcus Horton's aircraft and they were forced to land at Metfield on return. The remainder of the bombers landed at Seething without further loss. That night, the Luftwaffe launched a rare excursion to England. Luftwaffe fighters strafed and bombed Ipswich as well as some of the surrounding airfields sending the crews at Seething scrambling for air raid shelters. They were noticeably shaken by this forceful Luftwaffe resurgence.

In an attempt to combat the jet fighters, planners targeted jet airfields throughout Germany on 4 March. Bad weather over England forced the 2nd Air Division to assemble south of Nancy, France. Once together the formation set out for various targets near Stuttgart. Plans called for the 448th to attack an airfield at Schwabisch Hall, Germany. Bad weather over the continent produced hazardous flying conditions scattering the formation as they crossed the Initial Point. As a result, the lead

BTO 42-100000. This aircraft was one of the veterans of the Group. It flew its first mission on 4 February 1944 and continued flying until the end of the war. (LaPoint)

aircraft abandoned the mission and sent out a recall message. Seven aircraft of the high right squadron joined the 458th BG and bombed the marshalling yards at Stuttgart as a target of opportunity. The remainder of the aircraft answered the Wing recall and returned with the 20th CBW. Once more that night, the Luftwaffe returned. Five times the air alert was sounded. Five times everyone scrambled to the shelters. Lt. Hershel Hausman saw the intense flak fired by the coastal gunners. Their barrage worked as one enemy fighter crashed a few miles from Seething.

The unrelenting pace of operations continued the following day. Twenty-two Liberators left Seething destined for Harburg, Germany in the outskirts of Hamburg. The target was an oil refinery within the city. Two other aircraft from the 448th participated in the mission as part of a screening force but did not conduct bombing operations. Two of the attacking bombers aborted during assembly. An undercast deck totally obscured the target from the attackers requiring them to use H2X. Flak rose to greet the bombers but it was inaccurate and exploded harmlessly away from the B-24s. No fighters opposed the bombers and they all returned to Seething by the middle of the day.

After thirteen consecutive missions, the 448th received a well-needed reprieve when weather grounded all aircraft in England on 6 March. The Group also did not participate in operations on 7 March but did resume flying the following day. The Group's primary target, the marshalling yards at Betzdorf, Germany, escaped some damage as only two of the three attacking Groups dropped their bombs. The third group, the 448th, did not drop after all three squadron leads experienced G-H failure and were unable to bomb. Only one aircraft in the Group dropped its bombs. The pilot of TARFU II tacked onto the 467th BG as they hit the marshalling yards at Dillenburg, just past Betzdorf. All other aircraft returned to Seething with their bombs.

Thirty aircraft attacked the marshalling yards at Rheine, Germany the following day, 9 March. Once again no fighters threatened the Group and flak was only light over the north German town. Clouds necessitated using H2X to aid in the bombing. Results were much the same on 10 March. No flak or fighters opposed the bombers as they attacked the marshalling yards at Paderborn by G-H. One aircraft attacked Arnsberg, Germany, with the 466th BG. Milky clouds made flying difficult but again no damage or injuries were inflicted on the bombers.

The next day, the mission was more demanding. The Group participated in a large Division-size attack on the U-boat pens at Kiel, Germany. Over 350 B-24s of the 2nd Air Division left

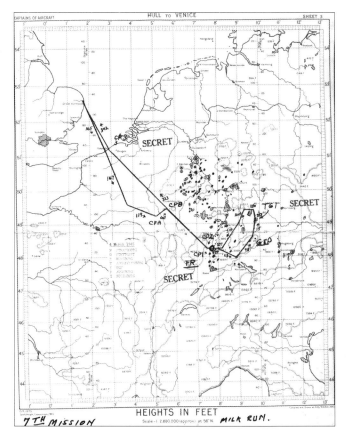

This map shows the route to the target, the return route, and known flak concentrations for the mission to Stuttgart, Germany on 4 March 1945. (Patterson)

East Anglia for assembly over England. For Lt. John Rowe, piloting SPIRIT OF COOLEY HIGH SCHOOL-DETROIT, the mission to Kiel was his eighth in eleven days. This one, however, was special, his thirty-fifth mission in the ETO. According to custom, clouds covered the target forcing the bombers to rely on H2X. The clouds, normally an aid in concealing the bombers from the flak crews, did little to help this day. Numerous thick puffs filled the sky. Surprisingly, only one 448th aircraft suffered damage from the flying shrapnel but it landed safely at Seething.

MY BUDDIE carried Lt. Gordon Brock's crew to the dockyards of Swinemunde, Germany, on 12 March. Three squadrons of eleven aircraft each attacked the German port which was located only twelve miles from the Russian front lines. Clouds concealed the target requiring the use of H2X once more. A moderate flak barrage exploded around the formation but did not cause any damage. Almost eight hours after lifting off Lt. Brock gently sat MY BUDDIE down on the runway at Seething.

While crews attacked Germany, Lt. Charles Platt and crew flew a practice mission in a brand new L model aircraft, tail number 44-50084. Four of these new aircraft, equipped with the latest G-H radar, were assigned to the Group the day prior. Five non-rated enlisted men went along on the practice flight unaware of the pending excitement. Low clouds over Suffolk forced them to fly at 500 feet to stay clear of the clouds. A B-17 flying nearby crossed paths with the B-24. The B-24 pancaked on top of the B-17 ripping the number three engine from the B-24 and tearing away a large portion of the B-17's vertical stabilizer. Miraculously, both aircraft landed safely, however, the tail gunner on the B-17 was killed. Lt. Platt's aircraft was repaired and carried his crew on five more missions before the war ended.

Bad weather on 13 March prevented missions from launching but did not hinder practice flights. Lt. Albert Broadfoot, who recently completed his last combat mission, flew a practice mission with a new crew aboard TARFU. The pilot, Lt. Paul Westrick, and the officers from the crew were using the flight to familiarize themselves with combat procedures as well as local area procedures. The radio operator, Sgt. Selwyn Kaplan, who was only three or four missions away from completing his tour, flew as did Sgt. Earl Jordan who was filing in for the sick regular engineer. Upon return to the airfield, they asked the control tower if they could circle in order to show the local landmarks used to identify the airfield. During this maneuver, a P-47 Thunderbolt, appropriately named GALLOPING CATASTROPHE, from the 5th Emergency Rescue Squadron swooped in on the bomber simulating an attack. Tragically, the pilot, Lt. Thomas Barkett, misjudged and collided with TARFU sending the fighter and bomber crashing to the ground near Langley Park four miles north-northeast of Seething. The pilot of the fighter and the seven men aboard the bomber were killed. It was a tragic end to an otherwise routine training mission.

Ironically, after the crash a poem about death written by Sgt. Selwyn Kaplan was found in his footlocker.

*I came to this land in search of death,*
*But death like life has passed me by,*
*Permitting me only to feel its breath,*
*Not having the grace to let me die,*

*Death like life has played me,*
*Giving and taking as in a game,*
*Refusing to grasp full victory,*
*Not knowing it seems the sport grows tame.*

As Allied forces pushed across the Rhine River at Remagen, the German Wehrmacht reeled under the constant attacks. Their retreating forces put a heavy strain on their transportation infrastructure, especially the train network. Recognizing the vulnerability, Allied air forces continued their punishing assault on the German transportation system. On 14 March, the 448th participated in such a mission. The target was the rail yards at Gutersloh, Germany. Thirty-three bombers in three squadron formations attacked the yards at 1537 hours. Haze and clouds made target acquisition difficult but nevertheless high explosive and incendiary bombs tumbled earthward. Only very light flak dirtied the sky and the Luftwaffe remained out of sight.

The onslaught continued the following day. This time the 2nd Air Division attacked a military headquarters complex south of Berlin at Zossen. Sgt. Ed Chu readied himself in the tail of

OLD GLORY 42-50482. Three unidentified buddies pose with their aircraft. This B-24 survived the war and returned to the States. (Everson)

WHAT DA HELL. This nose art reflects how crews felt about the incessant flak. Until the end of the war, German flak batteries inflicted damage on the bombers. The tail number of this M model Liberator is not known. (Patterson)

MY BUDDIE. This was his 10th mission and first to the Berlin area. The proximity of the target to Berlin promised a strong reaction from the Germans. Heavy fog delayed the takeoff by one hour; despite the delay visibility remained very low. The bombers, carrying delayed-action high explosive bombs as well as 500-pound bundles of incendiary bombs, broke clear of the clouds passing 4,000 feet. Assembly was completed without further complications. Scattered clouds and haze over the target presented some difficulty to the bombardiers but they dropped visually using H2X as a backup. The low left squadron in the 448th formation did not release their bombs due to limited visibility. As the formation left the target and headed west, those bombers still laden with bombs found a target of opportunity and dropped. Their selected target was a bridge two miles south of Parey, Germany. Results were poor. The scattered clouds provided just enough interference to hinder the bombardiers' aim but to everyone's pleasant surprise, the sky remained clear of flak and fighters for the entire trip, a crewmember's dream. The fate of the ground war was clearly reflected in the air. Opposition continued to decrease to everyone's delight.

Weather prevented combat operations on 16 March but did not preclude six crews from the 715th BS from flying a practice mission. The following day, they got their chance in combat. Thick clouds hampered the Liberators' departure and complicated assembly. Once the Group finally completed the assembly they set out for Germany. The bad weather extended over the continent forcing the crews to use H2X once again. They identified the target, industrial works in Hannover. An ineffective flak barrage weakly tried to deter the formation to

A unidentified airman stands in the waist window of aircraft "P" of the 713th BS which endured serious damage from flak as evidenced by the numerous holes in the aluminum skin. (LaPoint)

TARFU II 44-40099. This aircraft was lost to jet fighters on 25 March 1945. The navigator was the only survivor. (LaPoint)

no avail and bombardiers called 'bombs away' at 1400. Immediately the formation turned for the designated rally point and headed home. No Liberators from Seething were damaged.

On 18 March the 448th participated in a mass attack on targets in Berlin. All three Divisions of the 8th Air Force simultaneously hit targets in the city. The main 2nd Air Division attack focused on armored vehicle factories at Tegel, a suburb of Berlin. The massive assault on the Reich capital overwhelmed the German defenses. Heavy contrails from the preceding groups created hazardous formation conditions as they crossed the Initial Point. The trailing contrails also provided cover for enemy fighters to attack from the rear. However, the Luftwaffe failed to rise in opposition but the numerous flak batteries ringing the city threw up an intense curtain of steel.

Clear skies made formation assembly and the route to the target much easier. Here a formation of 448th Liberators head toward Germany. (LaPoint)

Twenty-two aircraft, including MY BUDDIE, avoided the flak and successfully attacked the target.

While crews continued the fight over Germany, Lt. Hershel Hausman and crew enjoyed a trip to London. A sudden terrific explosion shattered their peaceful Sunday morning stroll. A V-2 rocket landed in Hyde Park, just four blocks away. Miraculously no one in the crew was seriously hurt, although they suffered numerous lacerations from the flying glass. Although trips to London and other towns were great R&R, they also provided an insight into the great sacrifices that the British people made. The war disrupted their lives far beyond what the typical "Yank" realized or understood. This incident gave Lt. Hausman a new respect for his host nation:

> "While walking down the street one afternoon, we passed a store front with chocolate candy on display in the window. As I am a confirmed chocoholic, I insisted that we enter the store and purchase some candy. Ahead of us at the counter were a young matron and her eight-year old daughter. The young mother on hearing how verbally ecstatic I was on seeing the chocolate on display turned to the clerk and said 'Please wait on these gentlemen first.' I thanked her and then enthusiastically said to the clerk 'I will take a pound of this and a pound of that – .' With as apologetic look on her face, the clerk said, 'Sir, do you know that the candy is rationed? Do you have ration coupons?' I was chagrined and told her that I did not have coupons and thanked her for letting me look at the chocolates. At this point the young matron turned to me and said, 'Please sir, let me give you my ration coupons.' I was deeply touched. However, the look of relief on her daughter's face was reward enough as I thanked her profusely and refused her offer. I truly felt very humble at her gesture for she was offering her limited allotment to me as thanks for my being a soldier in her country. What made me really feel small was the fact that we had all the candy bars we could possibly eat at our air base. But it was a magnificent gesture on her part to a lonely young man in a foreign country."

The war for the soldiers and civilians continued. Beautiful weather over the continent greeted the bombers on 19 March as they assembled near Brussels, Belgium. Col. Westover flew as command pilot and led the 448th as well as the entire 20th CBW. The target was a power plant for a jet component factory at Baumenheim, Germany. Crews did not takeoff until 1030 for the long trip to southeastern Germany. Flight Officer Harold Dorfman, flying in the nose turret of B-24 4-F as pilotage navigator, had the best seat in the house. All of Europe spread out beneath him as they crossed into Germany near Strasbourg, France. Sgt. Lucian Whipple in REDDY TEDDY enjoyed the view as well. A comforting escort of P-51s circled around the formation. Gunners enjoyed the view of the beautiful Swiss Alps and Lake Constance. The only opposition encountered was contrails. A change of altitude alleviated the problem and the formation commenced the bomb run.

Crews were ecstatic over the absence of flak as they released their bombs. Cameras recorded bomb strikes all over the target area. The formation left the target area and picked up a course for home. As the formation passed the Rhine River, Lt. Dorfman saw fires raging in Frankfurt and Koblenz from previous attacks. Artillery flashes were clearly visible near Remagen where heavy fighting continued over the newly established Rhine bridgehead. The ancient city of Aachen lay shattered as the bombers passed overhead. As Lt. William Voight brought 4-F into the pattern, the navigator, Lt. Solomon Block generously fired flares in celebration of his last mission. He was going home.

Good weather persisted over Germany the next day as well. Eleven aircraft from Seething joined the 446th BG in a mission to northern Germany. They hit an oil refinery at Hemmingstedt, Germany. Light flak did not impede the attack but smoke from previous attacks obscured the target area. Still, the Group hit the target with good results. Everyone landed at Seething late in the afternoon of 20 March.

The growing threat from new jet fighters forced planners to take action. The entire 8th Air Force attacked airfields throughout Germany on 21 March. The 448th took part in two missions this day. They first hit the German airfield at Ahlhorn, Germany. Three squadrons of B-24s took off at 0626 hours. Conditions were great, ceiling and visibility unlimited, a rare treat crews referred to as CAVU. Sgt. Lucian Whipple watched the airfield erupt in smoke and dust as bombs impacted throughout the target area. In a matter of seconds, the target was completely obliterated. The short trip was a milk run for the crews. No flak was encountered and P-51s swarmed over the formation just in case the Luftwaffe attempted to play. Everyone landed back at Seething before noon.

As the first mission landed, another squadron from the 448th prepared for its mission. The target was another airfield, this one located in the Ruhr valley at Mulheim. Plans called for the squadron to form on the 446th BG as the low left squadron. Clear weather that the first mission had enjoyed continued in the afternoon. Heavy flak rose from the heavily defended Ruhr Valley as they started the bomb run, but to no avail. No bombers were damaged and bombs plastered the airfield with

The Ludendorff bridge at Remagen lies in the Rhine River. Before it collapsed on 17 March 1945 U.S. troops were able to establish a bridgehead on the east side of the river. Ironically, the 448th attacked the bridge on 31 December 1944 without success. (Bailey)

Another view of the collapsed bridge at Remagen. Notice the bomb craters to the right of the bridge. They are probably from German aerial attacks following the capture of the bridge. (Hipkins)

very good results, although just like the morning mission, smoke and dust covered the area making aiming difficult. The bombers landed at Seething in the growing darkness of evening.

The 448th continued its attack on jet airfields the following day. The airfield at Kitzingen in south central Germany fell under the sights of the bombers. The attack was unopposed and the bombers hit the target with excellent results. Lt. Hershel Hausman saw four Me-262s but they remained at arm's length and did not attack. "Stars and Stripes" heralded the missions on 21 March:

"In a savage blow aimed at crippling the Luftwaffe, which had appeared to be forming for a comeback in the past few weeks, U.S. heavy bombers and fighters yesterday thundered out to hammer eleven airfields, many of them bases for jet-propelled fighters and fighter-bombers, in northwest Germany, the Ruhr and southern Germany. The 8th and 15th Air Forces combined to deliver this triphammer punch. The 8th had some 2,200 planes out, nearly 2,000 of which figured in the drive on airdromes, while the 15th dispatched a separate force of Liberators to lash at the Neuburg drome 50 miles north of Munich. The bombers and fighters carried out their assault under excellent conditions – ceiling and visibility unlimited. In the greatest blow of the whole operation, approximately 1,100 bombers of the 8th and most of its 800 fighters zoomed in over nine airfields in northwest Germany to wield a three-ply blow. First the bombers came in for their run, followed by fighters who laid fragmentation bombs on runways and other vital spots on the fields. Fighters carried out the third phase of the attack sweeping in to strafe the dromes. In a later operation, approximately 100 Liberators, covered by 100 Mustangs, struck an additional blow, pounding the Mulheim airfield between Duisburg and Essen."

On 23 March, the 448th attacked the marshalling yards at Munster, Germany. Sgt. Donald Zeldin, returning from his weeklong tour at a flak home, rushed to the airplane just in time for takeoff. After his horrible crash, numerous months in the hospital and a rest tour, he was eager to fly and complete his required missions. It was not a good choice. Heavy flak filled the sky over the city. He huddled in the waist of WAZZLE DAZZLE throwing out chaff in an effort to confuse German radar. From the nose of MISS B-HAVIN, Lt. Neal Pettit saw the heavy flak fill the sky in front of them. It destroyed the number one engine and punched numerous holes in the thin skin. An unexploded eighty-eight millimeter shell entered MISS B-HAVIN in the waist area passing from the rear hatch to the leading edge of the horizontal stabilizer. The shell exited the aircraft and exploded above the plane. It severed the rudder control cables and knocked off eighteen inches of the vertical stabilizer but amazingly did little other damage. Despite the heavy flak the Group achieved excellent results with the only damage being to MISS B-HAVIN.

The versatility of the B-24 and its crews were once again utilized on 24 March. British Field Marshall Bernard Montgomery initiated Operation Varsity, an enormous attack across the Rhine River. The attack included a large airborne assault with a simultaneous amphibious crossing of the Rhine. Rumors of the mission circulated around the base during the evening of 22 March after three crews returned from the 20th CBW Headquarters and a briefing detailing the mission. Plans required the B-24s of the 2nd Air Division to resupply the airborne troops. Twenty-six B-24s from Seething were to drop resupply bundles on drop zones located near Wesel, Germany. To ensure the accuracy of the drops, the bombers were to drop from low level, much like the tactics used during Operation Market-Garden. The formation was to fly at 500 feet until the drop zone was acquired, then descend to 300 feet and drop their bundles which were stowed in the bomb bays. Army engineers were scheduled to fly with each crew to oversee the drops and coordinate Army requirements.

Lt. Harold Dorfman was the lead dead-reckoning navigator for the 448th aboard Lt. William Voight's aircraft. "The night of the 23rd we were taken in to study the drop area models and maps. Now we knew it was coming off the following day. The route in and out was over friendly territory all the way. We labeled it a milk run, another sightseeing trip. Except for about five minutes at the drop area we would never be over enemy territory."

After a 0530 briefing, twenty-six aircraft split in three formations left Seething at 0930. Crews passed up the normal heavy flying clothes and donned just Class B uniforms, a light jacket and a back parachute. The formation left England and took a circuitous route across France, Belgium and Holland. As they prepared to cross into enemy territory the formation dropped lower and lower, some as low as seventy-five feet. One aircraft in the lead element accidentally released his bundles on the west side of the Rhine. Two others seeing bundles falling to earth, released theirs on cue, far from the designated drop zone. The formation crossed the Rhine River below 500 feet. Small arms fire immediately erupted around the formation. Sgt. Donald Zeldin, flying on OLD GLORY, gazed out at the ravages of war. Brightly colored chutes littered the ground and dead animals were strewn around the area. The tremendous artillery barrage preceding the attack had devastated the countryside.

All navigation beacons were inoperative and visibility was only about two miles making navigation difficult. Despite the haze and smoke the formation crossed the Rhine River near the American sector just south of their designated course. Since the supplies were earmarked for British forces, they made a 270-degree turn and found the target three miles north of Wesel. Two hundred and twenty-eight bundles of supplies fell from the bombers. However during the turn a hail of gunfire met the bombers.

The fifty-two tons of supplies dropped by the 448th were quickly utilized by the ground troops in exploiting the new

REDDY TEDDY 42-51551. This aircraft was known by two names. The left side of the aircraft carried the nose art of BUGS BUNNY while the right side was adorned as REDDY TEDDY. It survived the war and returned to the U.S. (Everson)

bridgehead, but not without loss to the Liberators. The B-24s crossed the drop zones at low altitude and quite slow. The German soldiers fired all available guns at the vulnerable planes. The hail of bullets took its toll. PICADILLY LILLY, on its 106th mission, suffered numerous hits from the ground fire. Lt. Hugh McFarland struggled with the damaged aircraft but with the control cables shot out, the damage was too great. They crashed into a hill near Wesel. Eight of the crew were killed; only Sgt. Fred Yule and Sgt. Girthel Morrison survived.

The lead aircraft, MISS MUFF, carrying the Group Commander Col. Westover, suffered serious damage. The pilot, Capt. William Wilhelmi, and a waist gunner were injured by ground fire as they crossed the target. The tremendous barrage of small arms fire also damaged the landing gear. They flew the stricken aircraft back to England and because of the damaged landing gear, they chose to crash land at Manston. No one was injured in the landing.

Lt. Harold Dorfman, flying on 42-51504, experienced the lethal fire first hand:

"A large bullet hole the size of a handball appeared in the skin (of the plane) along side of me, probably a twenty-millimeter shell. I dropped to the armored floor between the ammunition cans and landed on Lt. "Shabby" Shabsis the bombardier. Our faces were about eighteen inches apart from the nose window and we watched more holes appear in the glass from machine gun fire. We covered ourselves with our flak jackets. I expected a bullet in the face at any moment. I could hear and feel the machine gun bullets raking the bottom of our armor plate that we were laying on. Suddenly there was a loud whine of high pressure being released behind us. We looked back. The nose wheel had been hit, just about a foot from my rear end. Still the machine guns raked the nose. The hydraulic lines in the wheel housing were hit, red fluid came out over everything, including my chute, which was soaked. A shell exploded between the pilot's legs, but mercifully the direction was away from him. However, the pilot's control cables were shot out. He lost control of the rudder; his controls were crippled. The thing that saved us was their snapping-on the autopilot which continued to work. We were at fifty feet all this time and still to clear 100 foot high electric wires as we came out of the turn to re-cross the Rhine River. By the grace of God did we get over that line of wires and out across to the friendly side of the Rhine."

Following their safe return from Munster on 23 March 1945, the crew of MISS B-HAVIN pose by the tail where an 88-millimeter shell exited the aircraft without exploding. Standing (left to right) Sgt. Jerry Kearney, Sgt. William Kaiser, Sgt. Tyler Tarkington, Sgt. Daniel Graham, Sgt. Anderson Wright. Kneeling (left to right) Lt. Calvin Ellis, Lt. Neal Pettit, Lt. John Paxson, Capt. James Shafter, Sgt. Walter Petrovich. (Pettit)

Arriving over the emergency airfield at Manston, the crew circled for thirty minutes assessing the damage of their plane. The landing gear would not extend and all the tires were blown. The only control the pilots had over the airplane was the elevator. The autopilot kept the plane level but could only make left turns. Finally, Lt. Voight ordered the crew to bail out. Lt. Dorfman with his parachute soaked with hydraulic fluid was the second man out. The radio operator snagged his parachute and Lt. Fred Risinger gave him his and attempted to repack the damaged chute. At 6,000 feet he jumped with some of the parachute still flapping about. He pulled the ripcord with great anticipation and it opened. One person was injured while jumping out of the doomed aircraft. Everyone else, including the 713th BS commander Lt. Col. Heber Thompson, landed without injury. The badly damaged Liberator valiantly continued flying before eventually crashing into the sea.

After landing at various places, the crew eventually reassembled at Manston where they found their Group Commander, Col. Westover, and Capt. Wilhelmi and crew. They too had crash-landed. Capt. Wilhelmi suffered a gunshot wound to the leg and a gunner was shot in the shoulder. That afternoon two B-24s from Seething arrived to take the crews back. Lt. Dorfman's "ride" experienced gear problems and they were advised they might have to bail out. Fortunately, the landing gear eventually extended and they landed without further misfortune. After escaping two potential catastrophes without injury, Lt. Dorfman succumbed to a more humorous accident. While riding his bicycle to the airfield to photograph the damaged aircraft, his brakes jammed and he was thrown from his bike. An ambulance transported him to the hospital where he joined his wounded crewmates from the earlier action.

Damaged aircraft filled the hardstands at Seething after the mission. WAG'S WAGON, flown by Lt. Neil McCluhan, suffered numerous hits and serious damage. "We lost the Gee Box, V.H.F., radio compass, remote compasses, engine instruments for number one and number two, vacuum, and inverters. Engines one and four were shot up so that we could only get about twenty inches of mercury on each of them. We climbed to up to 5,000 feet and came back to the enemy coastline. We then decided to come on across the Channel. Just as we got over England, our number one and four engines went out. The hydraulic fluid was gone and only one good pump on the brakes. Just as we sighted the field, we lost number three engine and so we came in with only number two engine operating. We cranked the gear down. We landed with an airspeed of 150 mph and then we went off to the right side of the runway."

Capt. James Shafter taxied a damaged JOKER'S WILD around the perimeter track toward its designated hardstand. Frayed nerves started to settle as the crew relaxed. Their 24th combat mission was history. As they taxied on the perimeter track, Capt. Shafter experienced difficulties with the nose wheel and stopped to check if it was flat. As he stopped, Lt. McCluhan landed his battle-damaged aircraft. After veering off the runway, they continued across the grass toward the perimeter track directly toward JOKER'S WILD. Seeing the out of control Liberator heading their way, the copilot of JOKER'S WILD,

Lt. John Paxson, gunned the engines in an effort to avoid a collision. The B-24 was sluggish and despite their efforts, the two aircraft collided, right wing tip to right wing tip. The collision sheared the horizontal stabilizer off of JOKER'S WILD. Surprisingly, only one man on Lt. McCluhan's crew, Sgt. Donald Clark, was slightly injured. Lt. McCluhan received the Distinguished Flying Cross for his gallant efforts bringing his crew and badly damaged Liberator home.

Numerous other aircraft endured damage from the ground fire and many brought wounded crew members home. Lt. Francis Piliere landed MISS MINOOKIE on the continent after two men suffered life-threatening injuries. Medics carried Sgt. William Garrett and Pvt. Ronald Burke to the 40th Field Hospital for medical care. Other aircraft, including MOTHER OF TEN and BTO, landed at Seething with wounded on board. It was a busy day for the meat wagons when the bombers entered the pattern. Red flares identifying those aircraft needing medical assistance filled the sky over the airfield at Seething. Medics treated ten men for wounds inflicted by small arms fire over the drop zone.

While the main force of Liberators attempted to resupply the airborne troops, a smaller force attacked the jet airfield at Stormede, Germany. The afternoon mission experienced no opposition as most of the enemy's attention was focused on the Rhine crossing at Wesel. The Liberators bombed the airfield through a heavy haze with excellent results. Unlike the morning mission, everyone returned without damage. The jets got their revenge the next day.

In an unrelated event, another 448th man, Capt. Leonard Monefeldt, died in an aircraft accident on 24 March. On detached service to a fighter group, he was flying a P-51 from a forward airfield on the continent. On return from a combat mission his aircraft crashed at the end of the runway. Although not participating in 448th operations, it was still a loss for the Bomb Group.

On 25 March, crews were awakened early, 0130, for the briefing. John Stanford was one of the many men listening to the briefing that morning. "Gentlemen, your target for today is the underground oil storage depot at Buchen. The secondary target will be the marshalling yards at Osnabruk. You will carry 20 300-pound GP's, have 2,500 gallons of gasoline, and depart with a weight of 63,000 pounds." Lt. Robinson was the Briefing officer and he continued with details of the mission. "He shows us areas where we can expect flak, and says we could come under attack by 100 to 150 single-engine prop fighters and 40 to 50 Me-262 jet fighters. He runs through the order of penetration and other miscellany – code word for the target is 'Hayride'; for the secondary, 'Cornsilk'; the recall word is

PICADILLY LILLY 42-50341. Another of the veterans of the 448th, she was lost on her 106th mission. She fell to ground fire while delivering supplies as part of the Wesel supply drop on 24 March 1945. (Everson)

JOKER'S WILD 42-50772. This aircraft was involved in the taxi accident on 24 March 1945. Its damaged rudder was repaired and it returned to combat. (Everson)

'Coke'; fighter call signs, 'Balance 21 and 22'; weather scout, 'Bootleg Rum'; code word for dropping chaff, 'Black Sheep'; time in the air, 6:40; and time on oxygen, 3:30."

Palm Sunday morning erupted with a roar as the propellers of the Liberators spun to life at Seething. After an uneventful assembly, the formation encountered trouble leaving the English coast. Clouds thickened as they neared the Zuider Zee making formation flying difficult. The lead squadron elected to circle in an effort to climb above the clouds. Lt. Elmer Homelvig, flying OLD POP, struggled to stay with his formation as they entered the clouds. "We did alright for a few seconds. Apparently they increased their turn rate, and it's like snapping a whip, the guy on the tail end always gets the message later, and of course they increased their rate of turn, I lost sight of Tod [Lt. Fred Tod], and the only thing I could do was gradually increase my turn, hoping I'd catch sight of him before I ran into him. It wasn't more than a few seconds later that I see a plane coming at me in the opposite direction, coming right at me. After we got on top of the clouds, Stalland was calling out giving his position, and in a few minutes, we found and caught up with him. Stalland had only two airplanes with him, I assumed they were Ray and Wikander. Tod was not with him, or had taken a different position. Stalland kept flying in a circle, and we kept picking up additional airplanes. We finally ended up with eight airplanes." The turn scattered the formation, leaving the 713th BS perilously out of position and lagging behind the remainder of the Group. Despite their scattered state the Group continued with the 713th vainly trying to make up time. By the time they reached the Wing Initial Point, they were still two minutes behind.

Lt. Calvin Ellis, Lt. John Paxson, Lt. Neal Pettit, and Capt. James Shafter (left to right) hold the sheared off tip of JOKER'S WILD horizontal stabilizer after a taxi collision following the morning mission of 24 March 1945. (Pettit)

DO BUNNY 42-95185. An armor plate under the copilot's window partially covers the nose art. This 448th Liberator fell prey to jet fighters on 25 March 1945. (LaPoint)

Just after 1000, the sky filled with German jet fighters. A large force of Me-262s attacked the 448th with deadly results. Sgt. Clair Rowe, a gunner on SONIA, witnessed the jet attacks. He shot one of the attackers down before his aircraft was seriously damaged. "When he began the attack, I began firing as soon as I thought he was in range. I saw two puffs of smoke when he fired his cannon, which was followed by a loud explosion, and I was blown out of my turret. I immediately got up and put on my chute and opened the escape hatch. We had no communication with anyone up front. Since the waist gunner was lying on the floor, I took over the waist gun. The explosion on the right side of the ship had knocked out our communication system and rudder controls. Our fighter escort then arrived and we got home without further incident." The shrapnel dug into his right leg and foot as well as causing lacerations on his face. With several wounded crewmembers and a severely damaged plane, the crew of SONIA faced a daunting return trip.

Sgt. Ed Chu fought a raging battle in the tail turret of MY BUDDIE. "Before the I.P., Max, our co-pilot, called 'Bandits in the area.' I saw three planes approaching out of the sun at six o'clock level. When they got within range, I recognized them as Me-262s. I opened fire at approximately one thousand

Above: A German guard oversees the wreckage of DO BUNNY which crashed on 25 March 1945 near Soltau, Germany. (Blaney)

Right: Amazingly everyone survived the crash landing of DO BUNNY. German captors took these photos. (Blaney)

yards at the closer of the two, the third in trail. I continued firing until he broke away through the squadron toward two o'clock high. I observed no hits or damage to confirm hits, although my tracers appeared to go right into the jet. P-51s boxed one up in front of the squadron and he exploded. Out of the corner of my eye, I could see the plane on our left wing flying the left element, 'Purple Heart Corner', peel off and in flames. No chutes were observed and the plane was later seen hitting the ground and exploding. We later were told that it was Steffan's crew. Another Me-262 appeared at six o'clock, and this time I opened fire at extreme range. My left gun jammed, the ammo locking up in the booster motor sprockets. P-51s kept this jet from the formation. 'Bombs were away' and they appeared to have landed in an open field. Four more planes approached from six o'clock level. They looked like jets and I opened fire at extreme range with my one remaining gun, but stopped firing when I recognized them as P-51s. There was a P-47 off to the right being shot at by some other gunners, but luckily they recognized it before it was hit. I recall swearing at those gunners under my breath not to shoot, as it was one of ours. P-51s and P-47s dove by our formation after the jets, one P-47 cutting real close to our tail. I saw a B-24 explode and another one spin down in flames, as jets hit a trailing squadron." It was the 713th BS still trying to catch the formation.

The first attacks hit the main part of the formation. Lt. Joseph Steffan and crew in TARFU II fell first at 1017 after Me-262s hit them with cannon fire over Domitz, Germany. The navigator, Lt. Gerald Gottlieb, bailed out of the burning bomber after the interphone was destroyed. "I suspect the plane exploded almost immediately after I left it because I was knocked unconscious for a short while and regained consciousness just before hitting the ground." The last time Lt. Gottlieb heard from his crewmates was during an interphone check two minutes before the attack. Only he survived. The aircraft crashed near Langenhorst, Germany, and was totally destroyed.

After the initial pass, the fighters focused their attention on the straggling low left squadron. They decimated the formation. Cannon fire exploded in the flight deck of the lead aircraft, 42-50646, at 1020 mortally wounding the pilot, Lt. Knute Stalland, and the copilot Lt. Theodore Warner. F/O John Stanford watched from his nearby aircraft. "The bursts move up on Stalland's plane, and suddenly he's on fire – bright red-orange flames sweeping back from the left wing inboard fuel tanks. The plane drops fifty feet or so, recovers, slides off out of formation to the right about 200 feet, in a shallow climb. Then it pauses and starts to swing back towards the squadron. Someone jumps from the stricken bomber, his chute opening immediately, boots flying off from the sudden jar. We are 19,500 feet. It is 1046."

Suddenly, the right wing ripped off the fuselage and the plane exploded. The bombardier, Lt. John McHugh, bailed out through the nose wheel door escape hatch. Flames quickly engulfed the B-24 as it started its death spiral. The pilotage navigator, Lt. William Whitson, intended to follow Lt. McHugh but the plane exploded blowing him clear. Although injured, he pulled his ripcord and was captured. The explosion also saved the radio operator, Sgt. Bobbie Glass. The force of the explosion rendered him unconscious just as he buckled his parachute. He fell ten to fifteen thousand feet before regaining consciousness. Amazingly, his parachute dangled from his chest harness by a single clip. He connected it and pulled the ripcord only to be captured shortly after landing near Schnever-Dingen, Germany. These three men were the only survivors from the crew of twelve. (A normal crew complement consisted of nine or ten men, but lead crews routinely carried twelve due to the specialized PFF equipment.)

DO BUNNY with Lt. Paul Jones and his crew went down next. On the first pass, Me-262's knocked out their number one engine with 20-millimeter cannon fire despite the curtain of lead from the B-24's fifty caliber machine guns. Subsequent passes by the Me-262 from JG 7 and flown by Luftwaffe ace Lt. Rudolf Rademacher further crippled Lt. Jones' Liberator. Two more engines ceased and numerous holes filled the aluminum skin of the plane. An exploding shell knocked the engineer out of the top turret and the plexiglass tail turret exploded in the face of the tail gunner. Somehow both escaped injury. Fuel and hydraulic fluid from ruptured lines filled the inside of the aircraft creating a potentially explosive situation. Also, the electrical system and intercom both failed. To the radio operator, Sgt. Chuck Blaney, the aircraft looked like a sieve from the inside:

"Lt. Jones ordered everyone to prepare to bail out, but with no intercom it was obvious that the word did not get out. Also, we were reluctant to jump because intelligence reports suggested that a crew's chances were amplified if capture as a group were at all possible. Single crewmen in the hands of angry German civilians were a poor risk in these times. Our navigator, Lt. Herman Engel, could see the heavy clouds of smoke caused by our heavy bombing in the Hamburg area. He was able to set a course toward Wesel on the Rhine where British paratroopers had landed just the day before. I guess that we never really expected to make the Rhine, even as we threw everything out of the plane that was not nailed down. Our copilot, Lt. Jim Mucha, kept his eye peeled for a safe place to set DO BUNNY in for a soft landing. With minimum power and controllability, our candidate landing sites were always dead ahead. At 1143 we were at 2,000 feet altitude and sinking fast. One sputtering engine does not provide much power to a B-24 even at minimum loading. Our pilot ordered us to ditching stations and for the inevitable crash landing. The pilots had selected a perfect field to put DO BUNNY down. It was right at the edge of the town of Soltau. We came in wheels-up and all went smooth until one wing dipped and the plane broke up. It was now 1148 and we had covered all of thirty-six of the 180 miles needed to reach the front lines and freedom. The pilot, copilot, tail, and ball gunners were able to get out of the aircraft and were immediately greeted by angry town folks with pitchforks. An SS officer appeared on the scene and arranged to have those crew members already outside of the aircraft run into the town square about 500 yards away. There they were all pinned to a wall across the street from the Mehr Hotel. I was trapped in the wreckage along with the navigator, flight engineer, and nose gunner. We had been pinned there by the top turret that broke away from the aircraft frame and lodged in the flight deck well. The navigator and engineer were unharmed and finally got out after German soldiers axed their way into the wreckage. The nose gunner and I were not so lucky. We were trapped by the top turret and each of us suffered a broken leg. The Soltau Chief of Police joined the German soldiers from the nearby riding academy and after much hard prying and much hacksawing we were freed from the wreckage. They put us on a horse drawn cart and took us to the town hospital where our legs were set and put in soft casts – we then rejoined the other crew members that were not locked up at the riding academy."

Meanwhile, cannon fire from four Me-262s tore into EAGER ONE, flown by Lt. Fredrick Tod. The damage was severe: right side flight control cables severed, right flap shot off, right rudder missing, four foot hole in the left wing, generators out, amplifiers out, main fuel line leaking, upper and tail turrets inoperative, hydraulics gone, radio destroyed, and pilots interphone not working. EAGER ONE immediately fell out of formation and started lagging behind. The B-24 vibrated and shuddered testifying to the tremendous damage the Liberator endured. Despite the terrific pounding, Lt. Tod and the copilot, Lt. Warren Peterson, kept the aircraft flying. With some difficulty, the bombs were jettisoned and the engineer stopped the fuel leak. Shortly afterwards, a fire started in the number four engine but extinguished itself after they feathered the engine.

The navigator, Lt. Herman James, provided a heading and distance to the nearest emergency airfield, Malmo, Sweden. Lt. Tod ordered everyone to prepare to evacuate the aircraft, as continued flight was uncertain. He advised everyone they could bail out over Germany if they did not wish to risk an over water flight. Everyone remained. With control problems and an engine out, the crew threw all non-essential equipment overboard. Still, they descended while a German Ju-88 followed to witness their demise.

Approaching the southern coast of Sweden, the number three engine started running very erratically and another engine became uncontrollable. Lt. Peterson told everyone to prepare to bail out as soon as they reached the coast. From his position in the nose of the aircraft, Lt. Herman James noticed the incredible physical strain on the pilots. Lt. Tod's right leg shook violently from fatigue. They kept the airplane in level flight by sheer strength.

The number three engine abruptly stopped and Lt. Tod issued the bail out order. After Lt. James exited the aircraft, he watched the plane turn away from the village of Falsterbo and head back toward the sea, a selfless act that undoubtedly saved many lives in the village. As the eighth man in the nine-man crew left the airplane, the B-24 entered a spin and crashed into the Baltic Sea just off the coast. Obviously, fatigue caught up with the pilots and as Lt. Peterson attempted to bail out, Lt. Tod was not able to fly the crippled airplane any longer.

On the ground, numerous people witnessed the life or death struggle. Mr. Harald Anderson and Mr. Lennart Ahlstrom were two men who rushed to the waterfront to help in the rescue. As the parachutes floated earthward, the pursuing German Ju-88 finally caught up with its intended victim although too late to inflict more damage. Swedish anti-aircraft fire scared him away. The two Swedish men quickly located a boat and set out for

OLD 75 POP 42-50391. This aircraft survived the harrowing jet attacks on 25 March 1945 and endured several other combat missions before the war ended. (Everson)

the crash site to help anyone in need. Meanwhile, other locals rolled up their pants and waded into the water to help those who landed short of land. One, Sgt. Chester Labus, suffered shrapnel wounds in his leg but managed to make it safely ashore. The wind blew some of the men over land where local residents quickly aided them. Mr. Ahlstrom and Mr. Anderson recovered Lt. Peterson but despite their valiant efforts, Lt. Peterson drowned. Lt. Tod perished in the crash of the B-24. Due to their heroic actions seven of the crew survived. These two gallant pilots were posthumously awarded the Silver Star for their heroic actions.

After the brutal attacks, the low left squadron fell further behind. However, they dropped their bombs on the target at 1034, nine minutes after the rest of the 448th. Realizing they would not catch the formation, the decimated squadron took a more direct route back to Seething instead of the intended route flown by the rest of the bombers. With many damaged planes and wounded men they needed the shortest route home.

While the crew of EAGER ONE fought for their lives, the crew of SONIA held their breath as their plane limped home. A thorough examination of their plane revealed extensive damage. Their hydraulic system was ruptured rendering it inoperative. The pilot, Lt. William Holden, elected to land at the long emergency runway at Manston. They manually lowered the landing gear and prepared to land without brakes and flaps. Despite missing one rudder, they landed on the long runway without any complications. Lt. Douglas Torrance landed short of home. He selected a forward airfield in Belgium to land his shot up Liberator.

Elsewhere, Lt. Ed Anderson struggled to keep his damaged aircraft 42-50590 airborne. With two engines shutdown, the crew dumped everything overboard. They even resorted, although unsuccessfully, to using a crash ax in an attempt to jettison the ball turret. With four P-51s providing escort, they received headings from a homing station to a forward airfield. Using maximum braking they stopped the damaged Liberator on the short runway and followed a jeep to a parking spot. As they parked, the two remaining engines sputtered and shutdown as they ran out of fuel.

After taking the shorter route home, the battered remnants of the 713th BS arrived over Seething thirty minutes before the rest of the formation. Red flares indicating wounded on board shot skyward from numerous planes. It was the first indication to the ground crews of the severe beating the Liberators had endured.

Damage from the jets was tremendous. Four crews were missing and their friends at Seething wondered about their fates. Thirteen B-24s endured damage but still brought their crews home. Still, numerous men suffered injuries ranging from small lacerations to more serious shrapnel injuries. After a long absence, the Luftwaffe struck back with a mighty blow. Only three other missions flown by the 448th during the entire war suffered more losses. All were early in the war except this one. The new jets attacked with near impunity, as the friendly escorts were unable to match their tremendous speed. They added a new dimension to the air war and reinstilled the fear of the Luftwaffe in the aircrews.

On 27 March, operations sent for Lt. Hershel Hausman and crew to test fly a newly arrived B-24, a M model 44-50509, since they were scheduled to fly it on a mission the following day. During the initial climb, the engineer noticed oil leaking from the number one engine so they chose to turn for home rather than risk damaging the new plane. After a normal approach and landing, Lt. Hausman applied the brakes but nothing happened. The copilot quickly attempted to brake but to no avail. Unable to stop, they ran off the runway into the soft mud collapsing the nose gear and cracking the fuselage behind the flight deck. The dejected crew's shiny new airplane was now only good for spare parts and salvage.

Following the tremendous onslaught, the 448th did not return to the skies until 30 March when they attacked the docks at Wilhelmshaven, Germany. Five bombers aborted during assembly and returned to Seething. The rest of the formation crossed the North Sea and attacked the target. Clouds over the target complicated the bombing as did the heavy flak. Using a combination of H2X and visual cues, the bombers released their bombs at 1341 with excellent results. Sgt. Ray Bailey flew the mission on a brand new airplane christened RED BOW. The joy of flying a new aircraft turned to anger when the guns still covered in cosmoline, froze and were unusable. Although the

DON'T FENCE ME IN 42-50525. This admirer no doubt has thoughts of home. This B-24 served as a lead aircraft and completed the war before returning home after the war. (Everson)

flak was heavy, the guns were not required and everyone returned safely to Seething.

The following day, 31 March, crews from Seething returned to northern Germany. As the bombers made landfall, they set course for an ammunition dump at Hassel, Germany. Clouds covered the primary target so they reverted to the secondary target, the rail yards at Brunswick. Nearing the target, Me-262s and Me-163s attacked the formation, but this time the escorts arrived in time to fend off the jets before they inflicted any damage. While the fighters dueled, light flak dotted the sky as the Liberators started the bomb run. Lt. Paul Homan, flying DON'T FENCE ME IN, assumed the lead of the high right squadron just prior to the Initial Point after the lead aircraft experienced problems.

The sudden change of positions caught Lt. Gordon Brock and crew by surprise. Flying the bucket position, Lt. Brock put MY BUDDIE into a sudden dive to avoid hitting the deputy lead aircraft as it descended and moved into the lead position. Sgt. Ed Chu confined in the tail turret and unaware of the other airplane wondered what happened. "I found my head rammed against the top of my tail turret. It was too sudden to think. The dirt in my turret covered everything. In the waist, Deane and Anthony found themselves thrust up against the ceiling of the plane's waist section with everything floating. When Brock pulled out of the dive, everything was strewn about the waist. The ammo flew out of my ammo boxes and wrapped around the control cables running along the sides of the waist section. We had just missed being hit by the deputy lead by a matter of a few feet. Brock thought our tail had hit the other plane as he was having difficulty with the controls. Deane and Anthony finally untangled the ammo from the control cables in the waist, and where they could, visually checked the wings and tail surfaces for damage. Everything appeared to be o.k."

Now out of the formation with most of the guns inoperative, Lt. Brock tried to catch the formation. Nearing the target they discovered the sudden dive had snapped the bands holding the incendiary clusters together. Incendiary bombs lay scattered around the bomb bay. Fearing the fragile bombs may explode, they cracked the bomb bay doors and dropped the single bombs out by hand. Once the stray bombs were disposed of, they opened the doors and jettisoned the remaining bombs. Then an engine started losing oil and they were forced to feather the propeller and shut the down the engine. As they struggled with their problems, the Group passed over the target amidst heavy flak.

Problems continued. Shortly after assuming the lead of the high right squadron, Lt. Homan experienced his own prob-

EAGER ONE 42-50326. Jet fighters severely damaged this aircraft on 25 March 1945 but the crew kept it airborne until they reached the coast of Sweden. The heroic efforts of the pilot and copilot allowed the crew to bail out before the plane crashed killing the two pilots. (Bailey)

lems. The bombsight on DON'T FENCE ME IN failed but the bombardier toggled the release switch anyway sending the high explosive and incendiary bombs tumbling toward earth. They landed short of the target causing significant collateral damage. In spite of all the problems, the bombers were on their hardstands at Seething by noon.

The strong opposition from the Luftwaffe jets rekindled a forgotten fear in the aircrews. While all crews feared the dreaded flak, most of the crews had never experienced fighter attacks because of the overwhelming Allied air superiority. Only a handful of aviators who experienced the swarming fighter attacks during the spring of 1944 remained in the 448th. Now, one year later, the fighter threat re-emerged with the introduction of jets. Crews feared them and rightfully so. The twenty-five missions in March cost the Group seven crews and another was lost on a training mission. Numerous planes were heavily damaged requiring sub-depot repair. It was a painful month just when Germany teetered on the edge of collapse. Yet, for the third consecutive month, the Group retained the "B for Bombing" pennant for the best bombing in the 20th Combat Bomb Wing.

## nineteen

# NOT OVER YET:
# April 1945

In April, the ground war exploded with unprecedented gains by the Allies. American and British forces crossed the Rhine river in several places and the Soviet armies closed on Berlin. Everyone realized the war was won; however, the question remained how much longer could Germany continue to fight. The beginning of April provided unwanted answers for the 448th.

Despite the gains of spring, lingering winter weather prevailed the first two days of April. On April Fool's Day, all offensive operations by the 8th Air Force were grounded by weather. The next day, airfields in Denmark were targeted but again weather stopped the air armada's effort. The 448th planned to attack the airdrome at Albourg, Denmark. High clouds and unfavorable weather conditions along the Danish coast forced the formation to abort the mission and return to Seething.

The 2nd Air Division did not participate in combat operations on 3 April but preparations were set in motion for a mission the following day, 4 April. Forty-five Liberators departed Seething in a maximum effort mission against the jet airfields in northern Germany. The Group's original target, Parchim, was located north of Berlin. The long trip to central Germany evolved quietly. Sgt. Ed Chu, in the tail turret of MY BUDDIE, watched RAF Lancasters pound the German city of Hannover. Passing south of the city, flak was clearly visible around the British bombers. A pall of heavy, black smoke spiraled up over the city. The formation continued to the target without problems. The first hint of trouble arose as the formation approached the initial point. Bandits were reported in the area! Gunners leaned forward in an extra effort to spot the bad guys.

To defend the airfields, the Luftwaffe scrambled fifty fighters, mostly jets and rocket planes. About fifty miles north of Berlin, the first fighters hit. Sgt. Chu first saw two fighters, Me-262s, overtake the squadron, flying parallel and below the Group. He fired a quick burst and watched the tracers fall short as the jets were out of range. Simultaneously, another twin-engine fighter approached the Group from the other side. The left waist gunner of MY BUDDIE opened fire as it approached. Just as Sgt. Chu prepared to fire the fighter tipped its wing revealing it as a British Mosquito. He seemed to escape damage from the friendly fire as he pulled away from the B-24s.

Below the bombers, clouds thickened as the formation crossed the Initial Point and closed on the target. The ground was visible only through scattered holes in the clouds making target acquisition extremely difficult. At the same time, a half-dozen Luftwaffe jets stepped up their attack. Two Me-262s swooped down on the bombers. Sgt. Ed Chu fought for his life in the tail of MY BUDDIE:

> "They were going in a direction opposite to us and then turned toward our squadron. At extreme range, possibly 2,000 yards – at five o'clock low, I recognized them as Me-262s and opened fire at the nearest one. I continued firing to point blank range, and could see my tracers going into him. He then dipped below the tail and broke away to ten o'clock low. As he was coming in, I could see the flashes of his cannon firing from the nose. He was so close that I could see the white of the pilot's face and his flying helmet when he looked up at us as he was passing under our tail. Deane in the waist told me he saw pieces of the fuselage flying off the jet's wing as he flew by. He also opened fire on the same jet. Later, we talked about it, and he thought I may have hit him. Two P-51s were also pursing the same jet at the same time. A moment later, I saw a

piece of B-24 fuselage with the tail section intact and attached, falling and twisting slowly down below my turret. I watched it fall until it was quite a ways down beneath our formation. At first, my mind wouldn't register what it was, and then it hit home. The plane flying left element, in Purple Heart Corner, directly off our left wing, broke in half when it was apparently hit by the jet I'd just been firing at."

The exploding aircraft, 44-50838 piloted by Lt. Robert Mains, broke in half at the waist from canon fire from the Me-262s. The two halves of a once powerful B-24 grotesquely twisted to the ground like falling leaves. Aboard the doomed aircraft the crew fought for their lives. The radio operator, Sgt. Charles Cupp, was the only survivor:

"We had just opened the bomb bay doors. Flak was coming up. I was sitting there watching the bomb bay, which I usually do when we start the bomb run. I always did this to see that the bombs all went out. If they hung up, either Steve or I would go kick them out. (Steve was Frank Merkovich, Engineer/Upper Turret Gunner.) Suddenly there was a big crash, and big balls of fire came into the fuselage on the right side of the plane. We immediately headed for the ground. I began bouncing around the flight deck, with no ability to control any of my motions at all. I reconciled myself to the fact I was going to die, as the plane just kept going down, down, down – . I had no control over my hands, arms, or anything else. All of a sudden things changed. I was sitting there, and suddenly I was able to move a little. I thought, maybe I have a chance to get out. What was going through my mind was, 'is there time enough.' The plane was shot in half at the waist, but I did not know this at the time. I was settled down in one spot, and was sitting on the ceiling of the plane, not on the flight deck. I could see daylight up above me, which must have been the open bomb bay. I'm sitting there, and there is a chute laying in front of me. I grabbed the chute, attached it, and at the same time, stepped over the engineer, who was either dead or unconscious, one or the other. I had to jump up, and grabbed a hold onto what must have been the catwalk, as I can best remember, since the plane was upside down. I pulled myself up, and as quick as I cleared metal, I immediately pulled the ripcord. The chute opened, and I guess I was no more than 2,000 feet up. A moment later, I heard a tremendous explosion, which must have been our plane with the gasoline and the bombs exploding as it hit the ground. I actually looked around to

Arthur Hipkins stands in a drainage ditch behind the 459th Sub Depot Group living site. On several occasions this ditch served as an air raid shelter. (Hipkins)

see if I could tell where it hit. Looking down, I also thought that I'd never make it. I saw all these tall pine trees, and thought, 'oh my God!'"

The aircraft to the left of MY BUDDIE exploded in mid-air and now Sgt. Ed Chu watched as the plane on their right wavered from the fighter attack. MISS B-HAVIN, flown by Lt. James Shafter, suffered serious damage from cannon fire and struggled to stay airborne. The damaged plane peeled out of formation after feathering both engines on the left side. The number two engine spouted flames and Sgt. Chu saw debris falling from the wounded aircraft followed shortly afterward by men jumping from exits.

Cannon fire started a fire in the wheel well below the flight deck. Lt. Neal Pettit, the navigator on MISS B-HAVIN, tried to make contact via interphone with the rest of the crew but was unable to do so. The fire spread quickly and Lt. Shafter ordered the crew to bail out. Lt. Pettit burned his face and hands trying to open the escape hatch. He finally kicked the door open. The nose gunner jumped out the hatch while the bombardier, Lt. Calvin Ellis, scrambled up to the flight deck to retrieve a fire extinguisher. Unfortunately, he left his parachute and upon returning, found the fire had completely consumed the nose of the B-24. The stricken plane continued flying briefly before erupting in a giant fireball. Only seven of the crew survived. Lt. Ellis, the copilot, Lt. Harold Major, the left waist gunner, Sgt. Anderson Wright, and the tail gunner, Sgt. Taylor Tarkington, perished.

TROUBLE N' MIND piloted by Capt. John Ray also fell to the desperate fighter onslaught. A flak round burst in the cockpit of Capt. Ray's aircraft severing oxygen lines and communications cables. Fire, fed by the oxygen lines, consumed the cockpit. The aircraft started a shallow dive that gradually increased until it entered a spin. Although stunned by the attack Lt. Ara Adams, the navigator, helped Sgt. Jack Garrity out of the nose turret and they both jumped out through the nose wheel doors. That was the last time anyone saw Lt. Adams. Most of the crew escaped the inferno and floated toward earth under silk canopies of their parachutes. Six of the men were promptly captured between Kyritz and Havelberg, Germany, but three men, Lt. Adams, Lt. Alden Hershiser, and Sgt. Peter Fager, were killed. Lt. Hershiser was last seen bailing out and his chute opened. The last time anyone saw Sgt. Fager he was pinned inside the aircraft. The aircraft crashed in the small village of Krullenkempe narrowly missing the homes. In a matter of seconds the 448th lost three aircraft at the hands of the new jets.

Intense flak enveloped the formation just as the fighters ceased their attack. The sudden flak burst reminded Sgt. Donald Zeldin to don his flak vest and helmet. The unexpected fighter attacks occupied everyone's attention not giving them time to put on their protective equipment. Clouds hid the target beneath their protective veil and the formation was unable to drop. As the formation arced in a wide circling 360-degree turn in preparation for a second pass, the fighters lined up for another attack. A Me-262 dove on the formation from five o'clock high. Sgt. Ed Chu noticed black smoke trailing the fighter; it was smoke from the cannons in the nose. He opened fire on the jet as it swooped past his turret. Silhouettes of fighters appeared again behind the formation. This time as the fighters approached, tracers from the fifty caliber guns on the B-24s filled they sky. He quickly realized these were P-51s and cursed to himself at the gunners who were firing. Fortunately, the gunners' aim was poor and no P-51s suffered apparent damage.

Fighters of all types swarmed around the bombers. Me-262s flew through the formation firing their cannons with P-51s desperately trying to chase them away. Sgt. Lucian Whipple fired at a Me-163 as it rocketed toward the formation with its cannon belching smoke. The rocket plane raised its belly and dove back toward earth. The bombers crossed over the target for a second pass but once again the clouds failed to cooperate and the target was not visual. The same harrowing flak bracketed the formation as the bombers held steady on their bomb run. Unable to drop their bombs, the formation departed the target area.

Lt. Robert Main's aircraft breaks in half after being hit by cannon fire from attacking Me-262s on 4 April 1945. Amazingly one man was blown clear of the debris and survived. Note the hole in the right wing. (Bailey)

On the ground below, surviving crewmembers faced the cruel reality of war. Sgt. Charles Cupp, the lone survivor of his crew, landed on the edge of a small town and tried to quickly deflate his chute. A young boy promptly arrived on the scene and knocked him to the ground. Villagers surrounded him, kicking him while he lay face down on the ground. A young boy pulled out a pistol and pointed it at his head. The trigger clicked; miraculously the clip was empty. Wehrmacht soldiers arrived shortly afterwards and led him off to captivity. He joined the seven survivors of Lt. Shafter's crew at a nearby airbase. They spent the last month of the war as POWs.

Two squadrons from the 448th bombed an airfield at Wesendorf, Germany, but the other three carried their bombs back to Seething. (The 448th consisted of four squadrons how-

TROUBLE N' MIND 42-95298. Two unknown men pose with the aircraft before it was lost to flak on 4 April 1945. Three of the crew were killed during the encounter. (Everson)

ever, during a mission the formation was sometimes comprised of five squadron formations. It was not uncommon for several squadrons to make up a single squadron formation in the Group formation.) Clouds made visual bombing too difficult. Tensions ran high as crews landed at Seething. Sgt. Ed Chu and the rest of the crew climbed over spent shell casings littering the floor of MY BUDDIE and crawled out their exits. They were called together by their pilot, Lt. Gordon Brock:

"He was mad, and in no uncertain terms said that when we were in flight, we were to stay off the intercom unless we had vital information to relay to the other crewmembers. He referred in particular to the event of the crew conversation relating to Shafters' plane burning, and what he called the blow by blow description of the crew bailing out. He emphasized that tying up the intercom, especially as we were being hit by fighters at the time, could have been fatal. He also said that a vivid description of a plane burning under the conditions of stress we were under at the time was uncalled for. All of the crewmembers involved were silent, as the impact of what he said was obvious to all of us."

The Luftwaffe, supposedly defeated, rattled the bomber crews. The sudden impact of the new jets worried crews and leaders. Suddenly, the hard-won air superiority over Germany was in doubt. Gunners showed a lack of aircraft recognition. Friendly gunners shot down a 446th BG Mosquito with their Group commander on board. Numerous Fighter Groups reported cases of gunners firing on the escorting fighters as they attempted to fend off the jets. As a result, all the gunners in the 448th, when not flying, were subjected to mandatory aircraft recognition classes. Despite the impending defeat of Germany, a war still raged in the skies over the Reich.

Sgt. Donald Zeldin volunteered to fly the following day as a replacement after an engineer broke his arm. Aircraft departed Seething independently in the early morning darkness amid heavy clouds and rain for the assembly area over Merville, France. Due to poor weather in the area, many of the bombers never located the formation and returned to Seething. Four elected to land on the continent instead of returning to Seething. One, 44-50546 piloted by Lt. Forrest Anderson, joined the formation but radio problems and radar navigation difficulties forced them to leave the formation. They elected to land at a fighter base at Mons, Belgium. After having their equipment repaired, they returned to Seething just as the formation arrived back from the mission.

Lt. Kay Flinders, flying ACHTUNG! NOON BALLOON, also experienced radio trouble and the thick clouds and fog forced them to leave the formation. They landed at a fighter strip in Belgium. Only twenty-two of the thirty-nine bombers dispatched continued to the target, an ordnance depot at

MISS B-HAVIN 42-95620. Part of a crew pose with the aircraft. This plane was lost on 4 April 1945 after cannon fire from attacking jet fighters ripped through the aircraft. Four men perished in the crash. (Everson)

MISS B-HAVIN 42-95620. Some aircraft carried nose art on both sides of the aircraft. This one carried the same name on both sides but slightly different nose art. (LaPoint)

Two friends pose atop a B-24 in one of the numerous hardstands scattered around the airfield. Numerous 448th Liberators are visible on other parking spots. (LaPoint)

Smoke from the crash of 44-50572 billows over a living site at Seething on 17 April 1945. (Everson)

Bayreuth, Germany. Although clouds obscured the target requiring the bombers to use H2X to identify the target, no flak or fighters challenged the attackers.

On 6 April, all three divisions of the 8th Air Force bombed targets in the Liepzig area. One 448th squadron flew with the 446th BG against the marshalling yard at Halle, Germany. After an early morning takeoff, ten Seething Liberators formed with the 446th BG. Clouds once more forced bombardiers to use H2X to identify the target. Bombs fell from the B-24s amid inaccurate flak barrages. No aircraft suffered damage and all the crews returned to Seething safely.

The 448th continued its part in the assault on 7 April. Duneberg, Germany, on the outskirts of Hamburg, was the location of an explosives factory targeted by the 448th. Twenty-eight bombers of the 448th received warning via radio that bandits were in the area as they approached the target. Two thousand feet above the bombers a fierce battle raged between the escorting fighters and the Luftwaffe. Lt. Hershel Hausman watched two M-262s head straight for their formation before they abruptly turned and attacked the preceding formation from behind. Two more jets followed. Dogfights filled the skies as the escorts vainly tried to catch the much faster jets. Sgt. Lucian Whipple saw three P-51s fall at the hands of the fighters.

After a brief lapse in the battle, the fighters turned to the 448th. Planes of all types and nationalities swarmed through the bomber formation. P-51s chased jets and conventional Luftwaffe fighters as they attacked the bombers. A Me-262 passed beneath Sgt. Whipple's aircraft and stopped thirty yards off the right wing. He reeled off a quick burst from his waist gun before the jet dove toward the ground with P-51s in pursuit. Suddenly his aircraft, QUEENIE, lost power on the number three engine. The pilot, Lt. Horace Rigel, stayed in the formation until they dropped their bombs. Bombs struck the target squarely setting off a chain of explosions that rocked the aircraft. Sgt. William Meyers thought his aircraft was hit. The shock wave rattled the bombers at 20,000 feet and sent smoke and debris 10,000 feet into the air.

Once clear of the target, QUEENIE fell out of formation in order to save the three remaining engines. Fighter escorts shepherded the vulnerable bomber back to England while the rest of the formation endured twenty-five more minutes of harassing fighter attacks. Despite the concentrated fighter attacks, the 448th did not lose an aircraft, although several were damaged. Everyone landed at Seething following the six and half-hour flight.

The sudden resurgence of the Luftwaffe forced tacticians to continue attacking the German airfields. On 8 April, the 448th attacked a German airfield at Roth, south of Nurnberg. Enroute to the target nerves were tense after the formation received radio calls describing fighter attacks on another Group. Gunners diligently scanned the skies for a flash of sun off a wing or telltale movement of any enemy fighters. Scattered clouds dotted the landscape but otherwise conditions were perfect for bombing. Despite the radio call, no fighters opposed the Group and the flak was absent as well. A milk run! Bombs fell from the lead aircraft at 1222 hitting the target with excellent results. The other two squadrons missed badly.

As Allied ground forces raced across Germany almost unopposed, strategic bombing targets were quickly overrun. Due to these large gains, most of the missions were over friendly territory. Targets for the bombers grew scarce as German-controlled areas grew smaller and smaller. German airfields remained a high priority as the Luftwaffe somehow launched their new jet fighters with good results. The target for the 448th on 9 April was one such airfield. Landsberg, Germany, near Munich, required a long flight for the bombers. Weather for the mission was ideal, no clouds and twenty miles visibility. The bombers assembled over Belgium away from the congested airspace of East Anglia and the route of flight provided a picturesque tour of southern Germany. Crossing the Rhine, Sgt. Lucian Whipple watched as hundreds of bombers shared the sky, all destined for various airfields in Germany. The Alps in nearby Switzerland stretched skyward with their snow-capped peaks. Beautiful Lake Constance on the Swiss-German border shimmered in the bright sunlight. The lack of fighters and flak allowed the crews a rare chance to relax and enjoy the sights.

The bombers dropped their bombs at 1729 from 22,000 feet with excellent results, however, a malfunction forced one aircraft to retain its bombs. Leaving the target, all of southern Germany seemed to be on fire. Smoke billowed into the sky from nearby airfields all hit by different Groups. Munich lay in shambles. Sgt. Lucian Whipple peered down into the pulverized city as they passed south of it. How could anyone survive?

Over the target, BTO, piloted by Lt. Hershel Hausman, experienced engine problems. They elected to head for an airfield on the continent rather than undertake the riskier Channel crossing. They landed at forward airfield B-53 near Merville, France, without incident. The concentrated effort against the German airfields appeared to work; no enemy fighters attacked the formation. There was not even flak. Another milk run for Sgt. Donald Zeldin on EL KORAB.

The following day the 448th returned to attack another German airfield. Sgt. Zeldin flew on his fourth consecutive mission as the bombers hit the jet airfield at Rechlin, Germany, while Sgt. Ed Chu flew his 21st mission, his third consecutive. The Luftwaffe launched fifty-five Me-262s of JG 7 to counter the bombers on 10 April. A large battle ensued between the escorts and the jets, but by the time the 448th arrived at the target, there was no trace of the Luftwaffe. Broken clouds forced the lead bombardier to shift the point of impact but bombs were recorded impacting the hangars and dispersal area with excellent results.

Only scattered flak dotted the sky and the reported fighters never materialized. Leaving the target, LEADING LADY, the lead aircraft of low left squadron, lost the number three engine. The pilot, Lt. Paul Homan, slowly slipped LEADING LADY out of the formation. He tagged onto a B-17 formation and followed them home. The rest of the formation continued to Seething without incident. After debriefing, exhausted men sought sleep. The strain of consecutive missions was weighing heavily on the crews.

On 11 April, the 448th participated in another mission designed to continue the pressure on the Luftwaffe. Crews sat in the morning briefing listening to the details of the mission to Regensburg, Germany. Thirty B-24s comprised the formation launched from Seething. On the way, Sgt. Lucian Whipple on QUEENIE saw the flashes of battle as they approached the front lines. Flying along the southern Rhine River provided the crews with an excellent tour of the front lines. Flashes from artillery duels were clearly visible as were towns burning in the distance. Clear weather prevailed over the target but the smoke and debris from previous bombers clouded the area. Some flak dotted the sky but it was not accurate. The bombers hit the target with excellent results and turned for home. Crews spent only an hour and a half over enemy territory. Everyone wondered how Germany could take such a pounding and continue fighting.

The 2nd Air Division designated the week 8-14 April "Salute the Ground Man Week." The 448th participated by holding mock briefings for those interested and providing aircraft rides. Unfortunately, the death of President Roosevelt marred

ACHTUNG! NOON BALLOON 44-50540. This M model B-24 arrived at Seething late in the war and survived, returning to the States on 13 June 1945. (Everson)

FEARLESS FOSDICK 42-50699. This Liberator also survived the war and returned to home. (Everson)

the events. News of the president's death spread rapidly around the base on the evening of 12 April. The Secretary of War, Henry Stimson, sent a general order to the Army announcing the sad news. It was published in the Seething daily newsletter 'Liberator':

> "It is my duty as Secretary of War to announce to the Army the death of Franklin Delano Roosevelt, President of the United States, which occurred at Warm Springs, Georgia at 1635 hours 12 April 1945. The Army is deeply grieved at the untimely death of our Commander in Chief. He prepared us to meet the savage onslaught of our enemies and he led us through the bitterness of our early reversals. His unwavering courage in the face of overwhelming odds, his abiding faith in the final triumph of democratic ideals and his clear vision of the paths to be followed were a source of constant inspiration. He gave the Army unstintedly of his strength and wisdom, and his unremitting labors hastened his death. Although he leaves us while there is still much hard fighting ahead, the ultimate victory has been fashioned of his heart and spirit. Memorial services shall be held on the day of the funeral 15 April at all posts, camps and stations, war operations permitting, at which time this order will be read. The former Vice President of the United States, Harry Sugar Truman, has taken the oath of office and assumed the duty of President in accordance with the provisions of the Constitution. The national flag will be displayed at half staff at the headquarters of all military commands and vessels under the control of the War Department from 0800 local time 12 April 1945 until sunset 12 May 1945 west longitude dates, war operations permitting. The wearing of mourning bands, the draping of colors and standards and the firing of salutes will be dispensed with because of war conditions."

Due to the tragic events, an enlisted men's dance, the culmination of the "Salute the Ground Man Week," was cancelled for the thirty day mourning period. It was a sad loss when victory was so close.

The number of targets in Germany continued to shrink as ground forces pressed their breath-taking onslaught. On 14 April, the 8th Air Force directed the bombers not against targets in Germany, but stubborn holdouts along the French coast. The rapid breakout from France the previous summer left isolated pockets of German soldiers holding fortified port cities. Hitler intended to deny the Allies use of these port facilities. The Allies chose to bypass these garrisons and attack inland.

William Schneider and Franklin Hastings stand under the tail of their B-24 QUEENIE from the 715th BS. (Whipple)

On 14 April, the 448th attacked a stronghold along the Gironde estuary at Pointe de la Coubre, France. No gunners flew on this mission due to the great distance from enemy air threats. 2,000-pound bombs fell on the enemy gun installations and troop concentrations in an effort to convince them to surrender. Only light flak annoyed the bombers as they pounded the hold outs from 15,000 feet. While some of the 448th took part in the mission, other crews flew to Chevieres, France, to evacuate a fighter unit to England. They filled the bomb bays with equipment and carried three men in the waist area. They ferried them to Watton, England before returning to Seething.

The next day, the Group flew a bombing mission to the same area, this time to Royan, France. Again, no enemy air threat was perceived but gunners flew although without guns or ammunition. The 448th participated in a first for heavy bombers. They dropped new napalm bombs designed for use against strongholds like the forts at Royan. Crews called the odd bombs 'jelly rolls'; they were nothing more than drop tanks filled with jellied gasoline. Some planes carried these "new bombs" while others carried incendiary bombs. The combination was deadly. Beautiful weather graced France as the B-24s undertook the long flight to southern France. Just before noon, the bombers hit the gun emplacements with the napalm. The thick swirling smoke from earlier napalm attacks hid the target and the results were unobserved. Although not a failure, the ballistic characteristics of the napalm proved unsatisfactory and heavy bombers did not use them again.

As the war ground to its inevitable end, much attention was directed toward German efforts to consolidate forces in the south for a last stand. On 16 April, the 448th resumed the offensive against Germany proper after a five-day break. The target was the marshalling yards at Landshut, Germany, near Munich. The calculated attack was designed to deny the Nazi fanatics a chance to reform in the south.

The bombers assembled in clear skies over Belgium on 16 April and set out on the long flight to southern Germany. Heavy smoke and clouds hid the front lines as the Liberators crossed into enemy territory. No flak or fighters attacked the formation prior to the target. The formation released their bombs over the target and turned for the rally point. Leaving the target, undetected flak batteries suddenly erupted surrounding the bombers with heavy, accurate flak. Sgt. Donald Zeldin saw the flak and held tight as the pilot took evasive action. All the aircraft scrambled to shake the unexpected outburst and clear the area. Sgt. Lucian Whipple watched as flak struck JUNIOR, flown by Lt. John McCoy. A direct hit sent the airplane nosing to the ground, landing gear and propellers falling off the mortally wounded plane. Ironically, a crew on their first mission gained infamous notoriety, as the last 448th crew lost in combat, but unfortunately not the last casualties of the war. Four other aircraft suffered damage from the barrage but all returned to Seething.

That evening, Flight Officer Alexander Calomeni, the co-pilot on Lt. Douglas Torrance's crew checked into the infirmary with a high fever and bronchitis. He tried to convince others on his crew who felt ill to do the same, but they elected to see a movie instead. They flew three consecutive missions and expected several days off before their 13th mission. They were wrong!

Tragedy struck early in the mission on 17 April. Just as Lt. Torrance lifted 44-50572 off the runway, a sudden power loss brought the B-24 back down to earth. The tail gunner, Sgt. Edward Paretti, sensed the impending crash as he watched the main wheels return to the runway. Insufficient runway remained to stop the B-24 and it hurtled off the end of the long runway. The nose gear sank into the mud twisting the nose grotesquely ninety degrees. After being tossed around in the back of the

RUGGED BUT RIGHT 42-94953. This H model proudly displays the shark's teeth that identified the lead aircraft of the 712th BS. This aircraft completed the war and returned to the States. (Everson)

aircraft. Sgt. Paretti and Sgt. Larry Caruso struggled with the escape hatch before eventually prying it open. The pilot and fill-in copilot, Lt. Donald DiFrancisco, climbed out of the pilot's window and ran to safety.

Cpl. William Schwinn and another man witnessed the horrible event from the control tower. As part of the alert crew, they jumped into a nearby jeep and headed for the crash site. Arriving at the scene, they heard the screams and pounding of four men trapped in the front of the bomber. A fire in the bomb bay added urgency to the rescue attempt. They struggled for five minutes to free the trapped men before someone began yelling for the men to leave. The fire spread to the number three and four engines and the full fuel tanks were in danger of exploding. Reluctant to leave but fearing for his own life, Cpl. Schwinn started running. "It felt like a big hand that reaches out and pushes you in the back and you had a problem to stay upright because the concussion is so great." Sgt. Paretti dove under an ambulance as the fierce explosion blew debris around the area.

Lying on his cot in the infirmary, Flight Officer Calomeni heard and felt the tremendous explosion that shattered an otherwise sunny, peaceful morning. He immediately felt something had happened to his crew. About fifteen minutes later, his premonition proved correct. Part of his crew, in shock with blackened faces, entered the doors of the infirmary. Medics hurried them through the doors before he could speak to any of them. It was the last time Flight Officer Calomeni saw his crew. Tragically, the four trapped men, Lt. Fredrick McKinley, Lt. Harold Beanland, Sgt. Hairman Merrill, and Sgt. James Gleason, perished in the explosion. The sad events forced the cancellation of the day's mission.

The Group returned to skies of Germany the next day, 18 April, for a mission against the marshalling yards at Passau,

Germany. The bombers fared better on this mission as the sky remained clear of the deadly flak. The attackers achieved excellent results. They returned with no damage.

Following a day off, crews prepared for another mission on 20 April. Marshalling yards in southern Germany were earmarked again for attacks. The 448th hit their assigned target, the marshalling yard at Muhldorf, Germany, with mixed results. Ceiling and visibility were unlimited as the bombers started the bomb run. Just before the target, the bombardier in the lead aircraft succumbed to hypoxia. An oxygen malfunction left him without oxygen and he quickly passed out. The pilot, Lt. James Blank, elected to jettison the bombs near the target. As a result, the lead squadron missed the target, but the other squadron scored excellent results after their separate bomb run.

On 21 April, the 2nd Air Division, including the 448th, left the East Anglian airfields for the marshalling yard at Salzburg, Austria. Bad weather hampered the assembly. Clouds filled the sky up to 33,000 feet making assembly very hazardous. Formations ingeniously assembled between cloud layers. Aircraft climbing through the soup suddenly burst into the clear areas surrounded by other airplanes. On several occasions Sgt.

REDDY TEDDY sits on the tarmac at Seething under a beautiful spring sky. (LaPoint)

Donald Zeldin let out a sigh of relief as another aircraft passed too close for comfort. Pilots struggled to maintain formation, dodging the clouds and trying to keep all the airplanes in sight. Despite their valiant, tiring efforts, clouds obscured the target and radio operators received the recall message over the radios. Everyone turned for home with full bomb bays. Seven hours after taking off Lt. Paul Homan touched his aircraft's wheels to the runway at Seething.

JOKER'S WILD 42-50772. This photo, taken from the waist window of an adjacent aircraft shows just how close formation flying could be. JOKER'S WILD survived the war and returned to the U.S. (LaPoint)

A mission on 23 April was scrubbed at 0418 and it was not until 25 April before the Group received another chance. That chance included a return trip to Salzburg. Lt. Homan flew as lead of the high right squadron. Weather cooperated unlike four days earlier. Ceiling and visibility posed no problems but the distance to the target did. The trip to Salzburg placed the B-24 at its extreme limit. The trip to the target passed uneventfully as most of it was over friendly territory. The route of flight carried the formation within sight of Berchtesgaden and the snow-capped Alps. Both were a breath-taking sight to Sgt. Donald Zeldin. Scattered flak rose to meet the bombers as they closed on the target and struck BLUES IN THE NIGHT, flown by Lt. William Hammes, wounding one man. Bombardiers called "bombs away" at 1058 and the B-24s turned for home one last time. FEARLESS FOSDICK landed at Merville, France, unable to complete the long flight without refueling. Attempting to maximize their visit, they planned to find other 'mechanical' problems with the plane that would delay their takeoff and provide enough time to visit nearby Lille. They scrapped the idea after they were unable to gather enough money for their visit. After refueling, they continued to Seething.

Several of the planes suffered minor flak damage but none were serious. The bombers landed one after the other. Clearing the runway they started their gradual crawl to their assigned hard stands. Pilots shut down the engines and the propellers stopped spinning. Although no one knew it at that moment, this was the last time the B-24s of the 448th flew to Germany with destruction as their mission. Ground crews started repairing damage and preparing the bombers for the next mission. Although plans were made for a mission on 26 April, it never materialized and an era ended.

twenty

# THE END:
# May-June 1945

As the grass turned green and the trees filled with leaves, the bombers sat idle on their hardstands. The near total collapse of Germany left no targets for the strategic bombers of the 8th Air Force. Americans linked up with Soviet troops on the Elbe River cutting Germany in half. Throughout Germany, mop up operations continued while Soviet forces attacked Berlin in fierce fighting. It was only a matter of time until Germany either surrendered or was completely destroyed. Scuttlebutt circulated throughout the base about the Group's fate. The war still raged in the Pacific and many speculated about going to aid in that conflict. Others optimistically hoped to go home for good.

While the B-17s of the 3rd Air Division flew supply missions to Dutch civilians, the 448th with other B-24 Groups flew "Trolley" missions. Starting 1 May, forty aircraft from Seething were scheduled to take part in Operation Trolley. These "Trolley" missions allowed non-flying personnel to observe the destruction they had helped inflict on the German Reich. Twenty-four aircraft flew to Wattisham to ferry ground crews and sixteen men from Seething. The bombers flying a carefully prescribed route in three-ship formations carried the men to Ostend, Mannheim, Aschaffenburg, Frankfurt, Bingen, Koblenz, Bonn, Cologne, Dusseldorf, then back to Ostend and home.

Sgt. Don Zeldin flew on one of these missions on 3 May. They landed at a fighter base and loaded the plane with a group of "sightseers." They flew across Germany at 1,000 feet then along the Rhine. The beautiful scenery ceased as they approached Ludwigshafen. The town lay in complete shambles. They continued up to the Ruhr Valley where the destruction was more complete. Towns in all directions were decimated except for a few lone buildings. Sgt. Zeldin failed to comprehend how people lived through such devastation. As the bomber turned for home a sick feeling overcame him. The low-level tour exposed damage that was not visible from the high altitudes. War was truly a horrible thing.

Although not sudden, the German surrender brought immense relief and anticipation of returning home. On 7 May, units around the base held meetings during which one minute of silent prayer offered thanks for the victory and paid homage to all those who gave their lives in the struggle. Lt. John Stanford wrote home describing the anticipation of the surrender. "The inactivity has been great. It's the first real vacation we have had in some time. I have used a great deal of this time just catching up on sleep, and most of the rest of it is writing up missions we have been on, and taking notes from intelligence summaries, etc. They've been expecting VE Day to be announced all day long but so far (1900) it has not been."

Word came later that night announcing the end of the war, although officially VE Day would be the following day. Lt. Stanford watched the unauthorized celebration. "It had been expected for so long that it was greeted at first with relatively mild displays of excitement. By late evening, with a few celebratory drinks under their belts, the men got more into the spirit of the occasion. Flares and fireworks lit up the sky, and the sharp report of .45s (automatic revolvers) added to the noise level. For me, the simple act of moving the blackout curtains back off the windows to let the light shine out was ceremony enough. But all around outside it remained very dark, except for the bursts of colored flares over the field. Hal, Ed and I stood beside our barracks watching. Someone came up next to Ed and asked, 'Who's shooting off those .45s?' Ed said he didn't know and the figure disappeared into the darkness. 'That,' said Ed, 'was Col. Westover.'"

The men continued using their sidearms as unauthorized noisemakers firing them randomly into the night sky unaware Col. Westover was searching for the culprits. During the merrymaking, one man, Flight Officer Frank Barilla of the 715th BS, landed in the base hospital after a .45 slug penetrated his hand. Lt. Donald Ford took part in the revelry. "On the evening when V-E Day was announced, many of the lads took out their sidearms and began shooting at the sky in celebration. My bombardier had a revolver and fired it. He reloaded it and handed it to me. Just as I started to raise it up I saw something flash, along came the Group CO. He told several lads to finish firing their weapons. He then ordered two of the lads to report to his office on the double. They spent the night walking back and forth in front of the tower. The next morning the CO had all the troops out doing calisthenics in retribution, all were very irate at such treatment. As soon as he had all the officers lined up in front of the tower he gave the squadron COs orders to make sure we all exercised thoroughly. Lt. Col. Thomas asked one man to give exercises. As soon as Col. Westover was out of sight, Lt. Col. Thomas told the one lad he was too energetic, another man was assigned who then gave us simple finger exercises. One man was posted to keep watch out for Col. Westover."

VE Day dawned with men assembled at 0500 performing "calisthenics" for the previous night's fun. When Germany signed the surrender papers in a small schoolhouse in Reims, France, on 8 May, the war officially ended in Europe. Sgt. George DuPont learned of the surrender while flying a Trolley mission. The war was over! Everyone was given the day off except for necessary details, however, everyone was restricted to base. At 1500 hours, all radios were turned on to listen to Winston Churchill and the King deliver their victory speeches. A previously scheduled officer's dance that night went as planned while a large fireworks display, this time authorized, was held at the control tower. The night sky erupted in a ceremonial display of pyrotechnics. Sgt. William Myers and other men from Lt. Kay Flinder's crew watched the show from atop the base water tower. The water tower standing near their home, hut 21, provided the perfect vantage point. Sgt. Noble Germany and his crew missed the celebration at Seething as they were at a Flak House on R&R. However, they enjoyed their own celebration.

On 13 May, the 448th participated in the 8th Air Force's aerial review. Dubbed Operation "Show Off", the entire 8th Air Force filled the skies over Europe celebrating the end of the war and commemorating the vital part they had played in the victory. Thirty-six B-24s of the 448th assembled one last time in a combat formation on STRIPED APE. This time however, they flew in peace.

The end of war brought a return of many duties forgotten since arriving in Europe. Without scheduled combat missions, duty rosters, a relic of pre-combat days, reappeared for crewmembers who had grown accustomed to their absence. Reveille was reinstated at 0600 every morning followed by retreat every afternoon at 1700 with the men attired in full dress uniform. On a bright note, censorship rules were relaxed and men were allowed to disclose the location of their base for the first time. Sgt. Harry Holmberg wrote to his parents describing the post-war changes. "We thought things would get easier now that it's over, but we have to get our ships ready to leave here so that's a lot more work to do. I've got a couple of engines to change and lots of flak holes to repair."

While some facets of life improved others worsened as the Group's official records indicated. "One unpleasant feature of post-war era concerned the mess halls, with a reduction in the quantity and more men eating 'at home' since they weren't allowed out on a pass. When the reduced quantity of meat ra-

SWEAT'ER OUT. This Liberator displays a common phrase spoken by many combat veterans. She sweated through the missions and returned to the States after the war. (Everson)

Fireworks and flares illuminate the night sky around the perimeter track on the night of 8 May 1945. Despite orders to the contrary, some men participated in unauthorized celebrations. (LaPoint)

tion was exhausted, often there was not even Spam or corned beef to replace it and cheese was substituted while the chow line was still quite long, many breadless meals, no fresh eggs and lack of variety meant more reason for everyone wanting to get home in a hurry." Sgt. Benjamin Everett spent much of the time hungry. "We were told that now that we had defeated those people we had to feed them, so our rations were to be cut 25 percent…After the war everyone started working days, so the chow lines were very long. To stand in line for 30 minutes and only get one thin slice of meat, a very small potato, a spoon of peas, and one slice of bread was very bad on morale. Everyone was hungry most of the time and were about ready to start fighting again in order to get more food."

Finally, the 448th received orders to return the States. Once home they were to retrain in the new B-29 Superfortress in preparation for operations in the Pacific. The newly appointed Group Operations officer, Maj. Leroy "Jack" Smith, prepared a detailed plan of the return trip. It called for each aircraft to carry its crew of ten plus ten passengers and luggage. The exact time of the Group's departure from England remained a secret but everyone knew it would be soon. Ground crews prepared the B-24s for the long trip. Those aircraft not capable of making the trip were salvaged.

On 27 May, Lt. Col. Lester Miller replaced Col. Westover as the Group Commander. Col. Westover returned to the States. Before leaving, Division ordered Col. Westover to select one flying officer to remain behind and supervise the base closure and return to British control. Maj. Smith was devastated when Col. Westover told him, he was his choice. It turned out to be a fateful choice. When Maj. Smith told his roommate, Lt. Col. Heber Thompson the 713th BS commander, the news, he quickly interjected, "Hey that's great! I'll take your place - that's a good crew."

Two Liberators from the 448th fly low over the countryside during a low-level "Trolley" mission. These missions ferried ground personnel on trips across Europe to witness the damage caused in the war. (Hipkins)

A destroyed factory is clearly visible in this photo. (Ray)

One of the hardest hit areas was the German transportation system, chiefly the railroads. This rail yard is devastated as its surroundings. (Ray)

Everyone spent the remainder of the month preparing for the much awaited trip home. On 5 June, Group Headquarters received orders directing the air echelon to depart Seething for Valley, Wales. From there, the bombers were ordered to proceed to Bradley Field, Connecticut, where the aircraft would be left. Crews were ordered to continue to Camp Miles Standish in Boston for in-processing.

The first bombers left Seething on 11 June amidst great fanfare. Some flew to Valley, Wales then to the Azores and Gander, Newfoundland, before returning to the States while others utilized a route taking to them to Prestwick, Scotland, then to Gander before reaching Bradley Field, Connecticut. Tragedy struck the next day as more bombers departed Seething destined for Prestwick, Scotland. Capt. James Blank's aircraft took off followed several minutes later by Lt. John Caldwell's. As they neared Prestwick, Lt. Delmo Pearce, the bombardier on Lt. Caldwell's B-24 nervously listened to rain battering against the aluminum skin, Thick, dark clouds enveloped the plane. Weather conditions at Prestwick deteriorated as the B-24s approached. Lt. Caldwell elected to descend over the water until they cleared the clouds. They landed safely surprised they had arrived before Capt. Blank's crew.

Tragically, Capt. James Blank and crew, including Lt. Col. Heber Thompson, started their descent while still over land. At 1000 their Liberator, 44-50695, hit the 400-foot high Pildinny Hill, five miles south of Ballantrae. Ironically, the hill sits overlooking the ocean just south of Prestwick. That night, Maj. Smith received a phone call from Lt. Col. Miller asking if anyone had heard from Capt. Blank or Lt. Col. Thompson. Their plane was overdue at Prestwick.

The next morning, there was still no sign of the missing plane. That afternoon, Maj. Smith received the bad news. Wreckage from the plane was found with only three survivors. He immediately started a search for transportation to Scotland. There were no planes at Seething and Division no longer had any. The motor pool was decimated as well. Finally, Maj. Smith, the Group Flight Surgeon, Maj. Patrick Hoey, and a driver selected an ambulance and set out for the crash site. After a sixteen-hour trip, they arrived. Wreckage littered the top of the hill for over a quarter of a mile. Even for combat-hardened aviators, it was a sickening sight.

It was not until they visited the hospital did they learn the full story. One of the survivors, Sgt. Robert May, described the horrible events. There was no indication of trouble before the airplane hit the hill. Sgt. May, knocked unconscious in the crash, regained consciousness and crawled from the back of the airplane. Three other men, all riding in the back of the airplane,

Most rail yards were located in the center of large cities, much like this one. The air attacks destroyed these railroads as well as many city centers. (Ray)

Above: SPIRIT OF NOTRE DAME sits on a rain-drenched hardstand as it is prepared for the return trip to the U.S. (Whipple)

Right: SPIRIT OF NOTRE DAME 42-95102. With the end of hostilities, there was large surplus of B-24s. Those aircraft deemed war weary or too costly to repair were salvaged. Those aircraft still fit for service flew to the States where most were parked and later destroyed. This H model was one such aircraft. (Everson)

survived and crawled from the wreckage. All were badly injured, one seriously. Despite a broken back, broken ankle, and broken wrist, Sgt. May scavenged a couple of canteens and parachutes from the wreckage. They used the parachutes to stave off the cold night air.

The following day, the seriously injured man, Pfc. George Gaffney, died. The three survivors, Sgt. May, Sgt. Kenneth Nelson and Sgt. Richard Pokorny, waited desperately hoping help would come. After another night on the hill without signs of help, Sgt. May summoned the strength to carry his badly injured body down the hill. After two tortuous miles, he attracted the attention of a gamekeeper, Mr. James Wright, who helped him to his house. In newspaper interviews, Mr. Wright described finding Sgt. May. "I was going along the side of a wood with my axe over my shoulder to gather pea-sticks when I saw a man in uniform staggering along, dazed and covered with blood. He waved in a tired kind of way and said, 'Please, please.' I ran to him and he said 'Never mind me. See what you can do for my comrades.' I helped him to my home, phoned the police and RAF and was asked to wait and guide searchers to the spot. When they came we began climbing the hill. Meanwhile an Anson had been sent out and we could see it circling and dipping its wings to indicate where the wreckage was." The injured men were taken to a nearby hospital in serious condition. For Sgt. May's heroic efforts, he received the Soldier's Medal.

The bombers continued their homeward journey. Strong head winds over the Atlantic delayed departures for several days for those stopping at Keflavik, Iceland for gas. During this delay, Sgt. Noble Germany learned the unfortunate story of the tragedy involving Capt. Blank's crew. Finally, the strong winds abated and the Liberators continued westbound. The next fuel stop was at Goose Bay, Labrador, and finally they landed at Bradley Field, near Hartford, Connecticut, starting 14 June. Pilots parked their aircraft one last time. As the propellers stopped spinning, ecstatic crewmembers spilled out of the bombers, eager to touch American soil. Crews took a train to Camp Miles Standish near Boston where they in-processed and

## The End: May-June 1945

This awe-inspiring aerial view of the Cambridge military cemetery in 1987 is a small reminder of the sacrifices made by these men. This is the final resting place for thirty-two men from the 448th, as well as seventy-nine men listed on the Wall of the Missing. (Everson)

This aerial photo of Seething taken in 2000 shows the remains of the airfield. The control tower is in the top right and half of runway 25 (now 24) is still active. Portions of the perimeter track and one of the crossing runways are still visible. (Everson)

were promptly given leave. As men scattered to the far reaches of their great country, no one knew or cared whether the 448th would ever reform. They had one thing on their minds: wives, girlfriends, mothers, fathers, and family. They were going home!

Back at Seething, the ground echelon continued their preparations for their move. They left England aboard the ocean liner QUEEN MARY on 11 July. After a six-day trip, the Statue of Liberty appeared on the horizon, welcoming her troops home. What a sight! They in-processed at Fort Dix, New Jersey, and were given thirty days of leave with orders directing them to report to Sioux Falls Airbase, South Dakota, for B-29 training following their R&R. Filled with anticipation, the men headed home for thirty glorious days. In the interim, the war in Japan ended with the cataclysmic atomic bombings. After leave and upon arrival at Sioux Falls, men were put in a casual status and the 448th Bombardment Group (Heavy) designation entered history.

While the success or failure of air power still draws debate among air power theorists, it is much clearer on a personal level, soldier to soldier or airman to airman. The best measure of success is summed up in a letter written by an infantryman to "Stars and Stripes." Published in the 25 April 1945 edition it states:

"I got to thinking that plenty of cracks that we in the infantry make about the Air Force, while seemingly ignored, may be taken more to heart than we think. I feel

The abandoned control at Seething in 1985 before restoration efforts were initiated. (Everson)

A fully restored control tower in 2000! (Everson)

that I could never express the feeling I have toward the Air Force. I believe that most infantrymen feel the same way. The Air Force is as necessary to the infantry soldier as the weapon he holds in his hands. When you are up on the front, and the heavy artillery and mortars come crashing down on you from behind the enemy lines, you have a peculiar feeling in your stomach because you can do nothing about it. It has been the same since D-Day. But let the rain stop momentarily or a few breaks appear in the clouds. The Air Force needs no more. You see flight after flight of our planes – on their way to return death and destruction to the enemy. You feel strangely elated. The enemy mortars, the rockets, the artillery have ceased firing. The flak starts mushrooming into life in the sky. But the planes, as far as you can see, are still coming. Your eyes grow moist as you say to yourself simply, Thank God."

# APPENDICES

*Appendix 1*
# HONOR ROLL

| Name | Rank | Serial Number | Date | MACR | Burial Location |
|---|---|---|---|---|---|
| Abraham, Adolph T. | 2Lt. | O-810982 | 31 Dec 43 | 3605 | |
| Acido, Norman F. | SSgt. | 19174760 | 5 Jan 44 | 2515 | Cambridge |
| Adams, Ara J. | 1Lt. | O-2000376 | 4 Apr 45 | 13731 | Netherlands |
| Adelizzi, Angelo A. | SSgt. | 16036369 | 7 Jan 45 | 11584 | Cambridge |
| Alexander, George S. | 2Lt. | O-869037 | 4 Apr 45 | 13730 | Netherlands |
| Allen, George C. | 2Lt. | O-684457 | 8 Mar 44 | 3189 | Cambridge |
| Allen, George L. | SSgt. | 34580133 | 23 Nov 44 | | |
| Allen, Harry J. | SSgt. | 39621301 | 4 Apr 45 | 13730 | Netherlands |
| Allen, Henry W., Jr. | SSgt. | 14187838 | 5 Jan 44 | 2515 | Cambridge |
| Allen, Luther E., Jr. | 2Lt. | O-805711 | 19 Apr 44 | 4303 | Ardennes |
| Alvis, Russel B. | Sgt. | 13094122 | 24 Dec 44 | 11120 | |
| Ambrosini, Harry J. | TSgt. | 19003327 | 29 Apr 44 | 4488 | |
| Anderson, Bernard W. | SSgt. | 16069755 | 22 Dec 43 | 3313 | Netherlands |
| Anderson, Carl | SSgt. | 19185853 | 22 Dec 43 | 2442 | Netherlands |
| Anderson, Truman K. | TSgt. | 18167649 | 24 Aug 44 | 8292 | Netherlands |
| Angelo, Arthur | TSgt. | 11100398 | 22 Apr 44 | 4301 | Cambridge |
| Anthony, Alden P. | 2Lt. | O-808492 | 11 Jan 44 | 2518 | |
| Arluck, Jack | Sgt. | 33426161 | 29 Apr 44 | 4490 | Netherlands |
| Armstrong, Bernard F. | 2Lt. | O-745028 | 5 May 44 | | Ardennes |
| Arrons, Sidney G. | SSgt. | 31204878 | 12 Jun 45 | | |
| Askew, Josh K. | Sgt. | 34503344 | 5 Jan 44 | 2517 | Ardennes |
| Aucker, Harold L. | Sgt. | 37115617 | 10 Feb 44 | | Cambridge |
| Ausfresser, Seymour | 1Lt. | O-669946 | 21 Apr 44 | 4343 | Cambridge |
| Ayrest, Robert C. | 2Lt. | O-680984 | 10 Feb 44 | | Cambridge |
| Babini, Louis D. | Cpl. | 31420626 | 16 Aug 44 | 8426 | |
| Backhaus, Albert H. | SSgt. | 38273742 | 5 Jan 44 | 2516 | Cambridge |
| Bair, Parke J. | Sgt. | 13095286 | 23 Nov 43 | 1247 | |
| Ball, Roy E. | SSgt. | 35595557 | 31 Dec 43 | 3605 | Epinal |
| Banas, Bernard J. | F/O | T-065103 | 16 Apr 45 | 14185 | Lorraine |
| Barbarito, William N., Jr. | Sgt. | 31280111 | 8 Dec 43 | 3907/4135 | |
| Barber, Roy D. | TSgt. | 14040228 | 11 Jan 44 | 2518 | |
| Barney, Hulick H. | SSgt. | 39857402 | 21 Jun 44 | 6231 | |
| Barneycastle, L.C. | 1Lt. | O-685258 | 24 Dec 44 | 11120 | Henri-Chapelle |
| Baron, Walter S. | SSgt. | 31133831 | 31 Dec 43 | 3605 | |
| Bass, Robert V. | 2Lt. | O-688162 | 10 Feb 44 | | |
| Beanland, Harold H. | 2Lt. | O-2063146 | 17 Apr 45 | | Cambridge |
| Bell, James T. | 2Lt. | O-769089 | 16 Aug 44 | 8467 | |
| Bender, Clark S. | 2Lt. | O-736682 | 23 Nov 43 | 1247 | |
| Benoit, Adley V. | Sgt. | 38486715 | 26 Aug 44 | 8467 | Ardennes |
| Bergum, Arne O. | 2Lt. | O-747568 | 22 Dec 43 | 3313 | Netherlands |
| Beverlin, Dellas D. | Sgt. | 36707999 | 5 Jan 44 | 2516 | Cambridge |
| Bienapfl, John A. | 2Lt. | O-744233 | 31 Dec 43 | 3605 | |
| Biggerstaff, Jim | 2Lt. | O-676535 | 11 Jan 44 | 2519 | |
| Bilyk, John | 2Lt. | O-873020 | 8 Dec 43 | 3907/4135 | |
| Binkley, Raymond T. | 1Lt. | O-706675 | 13 Jan 45 | 11583 | Lorraine |
| Blank, James G. | Capt. | O-567334 | 12 Jun 45 | | Cambridge |
| Blanton, Charlie L. | SSgt. | 39394021 | 19 May 44 | 5029 | |

*Appendices*

| | | | | | |
|---|---|---|---|---|---|
| Blashe, Oryn M. | Sgt. | 36841206 | 25 Mar 45 | 13548 | |
| Bookamer, Walter D. | SSgt. | 13109114 | 27 Jun 44 | 6728 | Cambridge |
| Boss, Edward F. | Sgt. | 33227643 | 5 Jan 44 | 2516 | Cambridge |
| Boula, Frank E. | SSgt. | 36619461 | 10 Feb 44 | | Cambridge |
| Bowers, Joe P. | 2Lt. | 0-711120 | 21 Nov 44 | 10407 | Ardennes |
| Bradley, Ulmer E., Jr. | SSgt. | 34333520 | 5 Jan 44 | 2517 | |
| Bramhall, Aaron R. | 2Lt. | 0-683897 | 8 Mar 44 | 3189 | |
| Brant, James M. | TSgt. | 35741985 | 30 Dec 44 | 4178 | Epinal |
| Brenner, Fred | 2Lt. | 0-687928 | 11 Jan 44 | 2518 | |
| Brewer, Gilman P., Jr. | SSgt. | 20102011 | 19 Apr 44 | 4303 | Ardennes |
| Briquelet, Gordon F. | SSgt. | 36826978 | 24 Mar 45 | 13546 | |
| Brittingham, Donald | SSgt. | 13141526 | 7 Jan 45 | 11584 | Cambridge |
| Brizzolara, Leroy J. | Sgt. | 32771334 | 26 Aug 44 | 8467 | |
| Broadfoot, Albert S., Jr. | 1Lt. | 0-719548 | 13 Mar 45 | | |
| Brown, Charles E. | Sgt. | 16076433 | 19 Apr 44 | 4303 | Ardennes |
| Brown, Earl T. | 2Lt. | 0-722813 | 16 Aug 44 | 8426 | |
| Brown, Millard L. | SSgt. | 15171462 | 29 Jun 44 | 7090 | |
| Brown, Paul E. | SSgt. | 33380722 | 20 Mar 44 | 3549 | |
| Bullard, Sam | 2Lt. | 0-692604 | 24 Feb 44 | 2933 | |
| Bunday, Raymond E. | 2Lt. | 0-721341 | 18 Jan 45 | | |
| Burke, Raymond G. | Sgt. | 36730017 | 21 Nov 44 | 10407 | |
| Burkhead, John L. | Cpl. | 36770874 | 16 Aug 44 | 8426 | |
| Butler, Robert D. | 2Lt. | 0-772937 | 16 Aug 44 | 8426 | |
| Buxton, John M. | 2Lt. | 0-705311 | 16 Aug 44 | 8426 | |
| Callahan, Woodrow | TSgt. | 6964020 | 5 Jan 44 | 2517 | |
| Campbell, Rayland | SSgt. | 19169499 | | | |
| Campbell, Robert L.J. | Maj. | 0-406707 | 20 Mar 44 | 3549 | Epinal |
| Canning, Leroy J. | SSgt. | 13151539 | 26 Aug 44 | 8467 | |
| Carcelli, William | 2Lt. | 0-692476 | 22 Apr 44 | | Cambridge |
| Carlson, Carl M. | F/O | T-1686 | 29 Apr 44 | 4490 | Netherlands |
| Carney, Lawrence E., Jr. | 2Lt. | 0-697988 | 27 Jun 44 | 6727 | |
| Carr, Floyd D. | SSgt. | 34132932 | 27 Jun 44 | 6726 | Cambridge |
| Cashin, Robert J. | SSgt. | 37553432 | 29 Jun 44 | 7090 | |
| Casto, Orvie O., Jr. | 2Lt. | 0-722944 | 2 Jan 45 | | |
| Chaffin, Joseph C. | 1Lt. | 0-797265 | 25 Feb 44 | 2956 | Cambridge |
| Charette, Albee G. | 1Lt. | 0-687515 | 19 Apr 44 | 4303 | |
| Chidester, Ned A. | Sgt. | 17014458 | 5 Aug 44 | 8669 | |
| Chioda, Joseph A. | SSgt. | 11102704 | 10 Feb 44 | | |
| Chisolm, John L. | Cpl. | 38395834 | 29 Apr 44 | 4486 | |
| Ciolek, Kenneth | Sgt. | 36180354 | 5 Jan 44 | 2517 | Ardennes |
| Clark, Charles V. | Sgt. | 35233954 | 18 Jan 45 | | Normandy |
| Clark, James G. | 1Lt. | 0-796780 | 29 Apr 44 | 4486 | |
| Cockings, Rector B. | Sgt. | 18053195 | 8 Dec 43 | 3907/4135 | |
| Cole, Albert P. | SSgt. | 12228133 | 6 Nov 44 | 10348 | |
| Concepcion, Ernesto | Sgt. | 32886753 | 26 Aug 44 | 8467 | |
| Conway, Theodore J., Jr. | 2Lt. | 0-703454 | 27 Jun 44 | 6728 | Cambridge |
| Cook, Merwyn G. | TSgt. | 18162136 | 7 Dec 44 | | |
| Coolman, Dean E. | Sgt. | 16159504 | 7 Jan 45 | 11584 | Cambridge |
| Corziatti, Joe K. | TSgt. | 18194485 | 27 Apr 44 | | Cambridge |
| Crow, Malcolm W. | SSgt. | 38446782 | 5 Jan 44 | 2514 | |
| Crumbley, Calvin C. | TSgt. | 34686327 | 12 Jul 44 | 7562 | Epinal |
| Curio, Louis W. | SSgt. | 12120417 | 29 Apr 44 | 4486 | |
| Cuthbert, William B. | 2Lt. | 0-687930 | 20 Apr 44 | 4302 | Normandy |
| Daidone, Anthony, J. | SSgt. | 32306952 | 31 Dec 43 | 3605 | |

| | | | | | |
|---|---|---|---|---|---|
| Dailey, James W., Jr. | TSgt. | 38236448 | 25 Feb 44 | 2956 | Cambridge |
| Dailey, Paul V. | 2Lt. | 0-736627 | 30 Dec 43 | 4178 | |
| Daley, John N. | 2Lt. | 0-801257 | 8 Mar 44 | 3189 | Cambridge |
| Daman, Charles H. | Sgt. | 39463945 | 4 Apr 45 | 13730 | Netherlands |
| Daneau, George A. | SSgt. | 11021582 | 29 Apr 44 | 4489 | Ardennes |
| Dansereau, Armand J. | Sgt. | 31439980 | 26 Aug 44 | 8469 | Ardennes |
| Davidson, Lester E. | 1Lt. | 0-739076 | 5 Jan 44 | 2516 | Cambridge |
| Davis, Roy W. | 2Lt. | 0-681347 | 19 Apr 44 | 4303 | Ardennes |
| Davis, William S. | Sgt. | 14135368 | 22 Apr 44 | | Cambridge |
| Deal, Philip A. | Cpl. | 61915668 | | | |
| Dean, William C., Jr. | SSgt. | 39555534 | 19 Apr 44 | 4303 | Ardennes |
| Deffner, Joseph B. | SSgt. | 32675587 | 11 Jan 44 | 2518 | |
| Delay, Harold L. | SSgt. | 19188393 | 25 Feb 44 | 2956 | Cambridge |
| Denning, David H. | Sgt. | 34870892 | 25 Mar 45 | 13547 | |
| DeSoto, Kenneth H. | Sgt. | 19186251 | 30 Dec 43 | 4177 | |
| Dickey, Harvey L., Jr. | Sgt. | 18226840 | 1 Apr 44 | 3567 | |
| Dolecek, Victor D. | 2Lt. | 0-697998 | 21 Jun 44 | 6264 | Netherlands |
| Dougherty, Daniel B. | Sgt. | 12164002 | 10 Feb 44 | | Cambridge |
| Drowne, Richard E. | 2Lt. | 0-717044 | 21 Nov 44 | 10407 | |
| Dunham, William D. | SSgt. | 34303305 | 31 Dec 43 | 3093 | Rhone |
| Durant, Alexander M., Jr. | 2Lt. | 0-834771 | 24 Mar 45 | 13546 | |
| Durant, William H. | Sgt. | 35623918 | 22 Apr 44 | | Cambridge |
| Duval, Albert F. | 2Lt. | 0-716390 | 5 Aug 44 | 8669 | Cambridge |
| Earhart, Harold L. | Capt. | 0-1551003 | 12 Jun 45 | | |
| Easterling, Ernest R. | SSgt. | 14130012 | 6 Aug 44 | | |
| Eckrosh, Floyd D. | SSgt. | 17097350 | 24 Aug 44 | 8292 | Ardennes |
| Edgerton, Henry F., Jr. | 2Lt. | 0-2061656 | 25 Mar 45 | 13547 | Netherlands |
| Edman, Lawrence M. | 1Lt. | 0-740775 | 25 Feb 44 | 2956 | Cambridge |
| Edson, Harlan, R. | 2Lt. | 0-2065971 | 7 Dec 44 | | |
| Edwards, Jacob E. | TSgt. | 18000037 | 31 Dec 43 | 3605 | |
| Edwards, Samuel | Sgt. | 32738235 | 8 Dec 43 | 3907/4135 | |
| Ellis, Calvin J., Jr. | 1Lt. | 0-689455 | 4 Apr 45 | 13732 | Ardennes |
| End, Calvin H. | Cpl. | 39453465 | 23 Nov 43 | 1247 | |
| Fager, Peter, J. | SSgt. | 39617321 | 4 Apr 45 | 13731 | Netherlands |
| Fahr, George S. | 2Lt. | 0-697831 | 22 Apr 44 | | Brittany |
| Falconi, Allessandro L. | SSgt. | 31231533 | 26 Aug 44 | 8467 | Lorraine |
| Fallert, Emmett E. | Sgt. | 16144615 | 10 Feb 44 | | |
| Faris, Eber D.O. | 2Lt. | 0-751289 | 8 Dec 43 | 3907/4135 | |
| Fay, Jerome C. | 2Lt. | 0-817411 | 27 Jun 44 | 6726 | Cambridge |
| Feingold, Leonard H. | 2Lt. | 0-688744 | 5 Jan 44 | 2514 | Ardennes |
| Fena, William G. | 2Lt. | 0-679487 | 8 Mar 44 | 3189 | Cambridge |
| Ferguson, William F. | 2Lt. | 0-681366 | 5 Jan 44 | 2515 | Cambridge |
| Ferrari, Bernard X. | 2Lt. | 0-717048 | 24 Dec 44 | 11120 | |
| Feyti, John J. | SSgt. | 20248174 | 24 Feb 44 | 2955 | |
| Fiego, Nicholas A. | SSgt. | 33084134 | 8 Mar 44 | 3189 | Cambridge |
| Fields, James E. | 1Lt. | 0-769111 | 2 Jan 45 | | |
| Finberg, Maurice M. | Sgt. | 37573487 | 11 Sep 44 | 8926 | |
| Fink, Thomas M. | SSgt. | 32413485 | 8 Dec 43 | 3907/4135 | |
| Fitch, William D., Jr. | 2Lt. | 0-717419 | | | |
| Ford, Walter A. | 2Lt. | 0-717420 | 6 Nov 44 | 10348 | |
| Fortin, Edmond C. | Cpl. | 31160104 | 12 Jun 44 | | Cambridge |
| Foss, John R. | Sgt. | 34707354 | 19 May 44 | 5029 | Ardennes |
| Foster, Thomas A.E. | 2Lt. | 0-714894 | 30 Dec 43 | 4178 | |
| Foster, Thomas K. | 2Lt. | 0-744267 | 10 Jun 44 | 5630 | |

| | | | | | |
|---|---|---|---|---|---|
| Fourneyron, Matthew F., Jr. | SSgt. | 12162824 | 21 Nov 44 | 10407 | |
| Fowler, Harold E. | Sgt. | 34677771 | 11 Jul 44 | | |
| Fox, Edward A. | 2Lt. | 0-808980 | 30 Dec 43 | 4177 | Ardennes |
| Freeman, Francis E. | SSgt. | 37236298 | 19 Apr 44 | 4303 | Ardennes |
| French, Frank L. | 2Lt. | 0-689147 | 24 Feb 44 | 2955 | |
| Friedman, Stanley | 2Lt. | 0-688184 | 11 Jan 44 | 2518 | |
| Gaffney, George T. | Pfc. | 17132724 | 12 Jun 45 | | |
| Gartman, Woodrow W. | SSgt. | 14040410 | 19 Apr 44 | 4303 | Ardennes |
| Gauthier, Henry I. | SSgt. | 31147965 | 5 Jan 44 | 2514 | |
| Genarlsky, Frank R. | 1Lt. | 0-886510 | 6 Nov 44 | 10348 | |
| George, Patrick H. | TSgt. | 19022087 | 24 Dec 44 | 11120 | |
| Gerber, John R. | Sgt. | 35225207 | 7 Jan 44 | 11584 | Cambridge |
| Getz, Richard | TSgt. | 12034631 | 20 Mar 44 | 3549 | |
| Ghormley, Jack A. | TSgt. | 32626123 | 25 Feb 44 | 2956 | Normandy |
| Gikas, Miltiades C. | SSgt. | 31258989 | 20 Mar 44 | 3549 | Ardennes |
| Gilmore, James H., Jr. | Sgt. | 14185681 | 25 Mar 45 | 13547 | |
| Ginevan, Donald G. | 2Lt. | 0-817832 | 5 Aug 44 | 8669 | Cambridge |
| Gleason, James E., Jr. | Sgt. | 31302822 | 17 Apr 45 | | |
| Good, Allen D. | SSgt. | 33479219 | 12 Jun 45 | | Cambridge |
| Goodpasture, Morgan | F/O | T-61188 | 24 Feb 44 | 2955 | |
| Greene, George C. | Sgt. | 31333352 | 29 Jan 45 | | |
| Greenwade, Billy R. | 2Lt. | 0-820260 | 5 Aug 44 | 8669 | Cambridge |
| Grogan, Frank E., Jr. | TSgt. | 14124065 | 25 Mar 45 | 13548 | Ardennes |
| Grossman, Edward J. | SSgt. | 32931086 | 3 Mar 45 | 12884 | |
| Grubbs, Roland L. | Sgt. | 37488400 | 24 Dec 44 | 11120 | |
| Grubisa, George J. | Sgt. | 32916135 | 21 Jun 44 | 6264 | Ardennes |
| Guynes, James W., Jr. | 2Lt. | 0-831159 | 3 Mar 45 | 12885 | |
| Guyton, Graham G. | 1Lt. | 0-390541 | 5 Jan 44 | 2517 | |
| Hamblin, Herschel O. | SSgt. | 13062783 | 21 Jun 44 | 6264 | Netherlands |
| Hammond, James M., Jr. | Sgt. | 34816783 | 3 Mar 45 | 12885 | Netherlands |
| Handzlik, Edwin F. | Sgt. | 36585389 | 21 Nov 44 | 10407 | Ardennes |
| Haney, Edward E. | Cpl. | 33294051 | 10 Dec 43 | | |
| Hanson, Edward E. | 1Lt. | 0-820006 | 24 Dec 44 | 11120 | Ardennes |
| Hardin, James R. | SSgt. | 17077107 | 22 Apr 44 | | Cambridge |
| Harriman, William T. | SSgt. | 17111211 | 12 Jun 45 | | Cambridge |
| Harwood, Thomas H. | SSgt. | 34374445 | 22 Apr 44 | 4301 | Cambridge |
| Hasiak, Henry S. | 2Lt. | 0-702043 | 29 Jun 44 | 7090 | |
| Hauver, Roland T. | 2Lt. | 0-2008958 | 25 Mar 45 | 13548 | |
| Haynes, Harold C. | Sgt. | 39712761 | 3 Mar 45 | 12884 | |
| Heard, John M. | 1Lt. | 0-2069015 | 25 Mar 45 | 13548 | |
| Heaton, Charles A., Jr. | Sgt. | 18226363 | 8 Dec 43 | 3907/4135 | |
| Helm, George H. | 2Lt. | 0-674042 | 18 Jan 45 | | |
| Helvey, Wesley V. | 2Lt. | 0-805916 | 21 Feb 44 | 2468 | Netherlands |
| Hershiser, Alden J. | 1Lt. | 0-768545 | 4 Apr 45 | 13731 | Netherlands |
| Higgins, Eugene E. | Sgt. | 36273563 | 5 Jan 44 | 2514 | Ardennes |
| Hill, Charles D. | 1Lt. | 0-379552 | 22 Dec 43 | 2442 | Netherlands |
| Hill, Glenn D. | Sgt. | 17147109 | 18 Jan 45 | | |
| Hill, John A. | Sgt. | 36576975 | 29 Apr 44 | 4491 | Ardennes |
| Hinckley, Howard D. | 2Lt. | 0-771716 | 16 Aug 44 | 8426 | |
| Hines, Thomas H. | Sgt. | 335603093 | 29 Apr 44 | 4491 | |
| Hipps, Charles F. | SSgt. | 3325398 | 30 Dec 43 | 4178 | |
| Hogan, Thomas R. | Sgt. | 13116127 | 20 Mar 44 | 3549 | |
| Holesa, John J. | TSgt. | 15072502 | 31 Dec 43 | 3605 | Epinal |
| Holland, Robert F. | Sgt. | 12171282 | 10 Feb 44 | | |

| Name | Rank | Serial | Date | | Location |
|---|---|---|---|---|---|
| Huber, John K. | 1Lt. | O-831942 | 12 Jun 45 | | |
| Hudson, Robert E. | Sgt. | 16162433 | 10 Jan 45 | | |
| Hudson, Robert P., Jr. | TSgt. | 32176645 | 5 Jan 44 | 2514 | Ardennes |
| Hughey, Edward, D., Jr. | 2Lt. | O-677748 | 22 Dec 43 | 2442 | Netherlands |
| Hugley, Ned R. | Sgt. | 14152714 | 23 Nov 43 | 1247 | |
| Hurton, Paul J. | Sgt. | 31427378 | 18 Jan 45 | | Normandy |
| Indorf, Archie B. | SSgt. | 35595495 | 31 Dec 43 | 3605 | Epinal |
| Jackson, William E., Jr. | SSgt. | 15102452 | 9 May 44 | | Cambridge |
| Jackson, Zachariah F. | 2Lt. | O-681101 | 10 Feb 44 | | Cambridge |
| Jacobsen, Edwin S. | 2Lt. | O-684114 | 22 Dec 43 | 2442 | Netherlands |
| Jarol, Seymour | 1Lt. | O-750995 | 27 Jun 44 | 6726 | Cambridge |
| Jasura, Peter, Jr. | Pfc. | | 18 Sep 44 | | |
| Jennings, Wyllys B. | 2Lt. | O-715414 | 26 Aug 44 | 8467 | |
| Johnson, Bertil S. | SSgt. | 35099877 | 21 Jun 44 | 6264 | Netherlands |
| Johnson, Kenneth D. | Sgt. | 36696966 | 26 Aug 44 | 8469 | |
| Johnson, Mabron P. | TSgt. | 18007370 | 10 Jun 44 | 5630 | |
| Johnson, Norman H. | SSgt. | 36740128 | 29 Jun 44 | 7090 | Lorraine |
| Jones, Ernest E. | SSgt. | 13062713 | 23 Nov 43 | 1247 | |
| Jones, Robert J. | Sgt. | 12093034 | 5 Aug 44 | 8669 | |
| Jordan, Earl B. | Sgt. | 34782003 | 13 Mar 45 | | |
| Jordan, Max R. | 1Lt. | O-725740 | 31 Dec 43 | 3605 | Epinal |
| Juliano, Saverio J. | F/O | T-134418 | 16 Apr 45 | 14185 | Lorraine |
| Kadehjian, Aram G. | F/O | T-132946 | 3 Mar 45 | 12885 | |
| Kanarek, Morris L. | TSgt. | 13133595 | 12 Jun 45 | | |
| Kaplan, Selwyn | TSgt. | 12088705 | 13 Mar 45 | | |
| Kardys, Bronislaw J. | Sgt. | 31252333 | 16 Apr 45 | 14185 | |
| Kary, John V. | SSgt. | 19071394 | 5 Aug 44 | 8669 | Cambridge |
| Kasbarian, Harry | 2Lt. | O-688359 | 5 Jan 44 | 2515 | Cambridge |
| Kasprzak, Joseph F. | SSgt. | 13155970 | 22 Dec 43 | 3313 | Netherlands |
| Keegstra, Donald | 2Lt. | O-679887 | 10 Feb 44 | | Lorraine |
| Kelley, Francis J., Jr. | SSgt. | 11103783 | 3 Mar 45 | 12884 | |
| Kelly, John J. | SSgt. | 32782047 | 11 Jan 44 | 2518 | |
| Kelly, Leon E. | SSgt. | 34671680 | 21 Nov 44 | 10407 | |
| Kelly, Walter T. | TSgt. | 14040497 | 5 Jan 44 | 2515 | Cambridge |
| Kiehn, Donald R. | Sgt. | 35898639 | 24 Mar 45 | 13546 | Netherlands |
| Kindt, Harold L. | SSgt. | 13094588 | 5 Jan 44 | 2517 | |
| King, Christopher C. | SSgt. | 32747076 | 12 Jun 45 | | Cambridge |
| Kittredge, Abraham J. | 2Lt. | O-745907 | 30 Dec 43 | 4177 | |
| Klum, Philip H. | SSgt. | 12186879 | 22 Dec 43 | 3313 | Netherlands |
| Knight, Raymond F. | 2Lt. | O-744532 | 24 Aug 44 | 8292 | |
| Kocheran, Alex | Cpl. | 33414605 | 16 Aug 44 | 8426 | |
| Kolokow, Irving | Sgt. | 12189812 | 5 Jan 44 | 2516 | Cambridge |
| Koon, Curtis L. | TSgt. | 33252697 | 22 Dec 43 | 2442 | Netherlands |
| Korte, Jerome A. | Sgt. | 36477450 | 7 Dec 44 | | Cambridge |
| Kracyla, Henry J. | Sgt. | 32488180 | 25 Feb 44 | 2956 | Cambridge |
| Kraemer, Daniel J. | TSgt. | 13028896 | 24 Aug 44 | 8292 | Netherlands |
| Kraft, Alton L. | 1Lt. | O-702055 | 6 Nov 44 | 10348 | |
| Kropp, John M. | Sgt. | 36819181 | 25 Mar 45 | 13548 | Ardennes |
| Krueger, Ardell F. | SSgt. | 36292349 | 27 Jun 44 | 6726 | Cambridge |
| Kubinski, Henry | SSgt. | 36555495 | 21 Feb 44 | | Netherlands |
| Kuzminski, Anthony J. | SSgt. | 11101473 | 20 Apr 44 | 4302 | Normandy |
| Lackey, David E. | SSgt. | 15082537 | 22 Dec 43 | 3313 | Netherlands |
| Laing, Richard B. | Sgt. | 19164878 | 22 Dec 43 | 2442 | Netherlands |
| Lajoie, Leonard G. | SSgt. | 16023435 | 23 Nov 43 | 1247 | |

| | | | | | |
|---|---|---|---|---|---|
| Lake, Allan L. | 1Lt. | 0-2060318 | 4 Apr 45 | 13730 | Netherlands |
| Lane, Harold J. | Sgt. | 15383034 | 30 Dec 43 | 4176 | Epinal |
| Lane, Horace B. | 2Lt. | 0-771451 | 3 Mar 45 | 12884 | |
| Lang, Harold L. | Sgt. | 39704462 | 7 Jan 45 | 11584 | Cambridge |
| Lanphear, Byron E. | 2Lt. | 0-432610 | 22 Dec 43 | 3313 | Netherlands |
| LaRiviere, John E. | 1Lt. | 0-886809 | 4 Apr 45 | 13730 | Netherlands |
| Larson, James M. | 2Lt. | 0-2061983 | 3 Mar 45 | 12885 | |
| Lazarus, Arthur C. | Cpl. | 20480080 | 10 Dec 43 | | |
| Leis, William E. | TSgt. | 18192905 | 8 Dec 43 | 3907/4135 | |
| Lepley, Howard P. | TSgt. | 13137364 | 10 Jun 44 | 5630 | |
| Lerner, Herbert M. | 2Lt. | 0-771943 | 7 Dec 44 | | Cambridge |
| Levey, Paul D. | Sgt. | 36654552 | 21 Nov 44 | 10407 | Netherlands |
| Lindsey, Aubrey W. | Sgt. | 38444071 | 12 Jun 45 | | Cambridge |
| Long, Andrew P. | SSgt. | 32186301 | 9 May 44 | 4775 | |
| Loprete, William A. | 1Lt. | 0-807004 | 29 Jun 44 | 7090 | Lorraine |
| Loughlin, Thomas A., Jr. | Sgt. | 36851964 | 27 Jun 44 | 6728 | Cambridge |
| Loyd, Ira H. | SSgt. | 36431449 | 21 Feb 44 | 2469 | Ardennes |
| Lunt, James C. | TSgt. | 19064266 | 24 Dec 44 | 11120 | Henri-Chapelle |
| Lyons, Edgar E. | SSgt. | 16033854 | 2 Jan 45 | | Cambridge |
| Mains, Robert L. | 1Lt. | 0-680467 | 4 Apr 45 | 13730 | Netherlands |
| Majestic, Arthur B. | 2Lt. | 0-701940 | 21 Jun 44 | 6264 | Netherlands |
| Major, Harold, Jr. | 2Lt. | 0-2067316 | 4 Apr 45 | 13732 | |
| Malwitz, Willard R. | SSgt. | 19175397 | 11 Jan 44 | 2519 | |
| Manning, David E. | 2Lt. | 0-745722 | 22 Dec 43 | 3313 | Netherlands |
| Markewicz, Edward A. | 2Lt. | 0-677543 | 10 Feb 44 | | |
| Marsh, David T. | SSgt. | 33300005 | 10 Feb 44 | | |
| Marshall, Clarence R. | SSgt. | 33231924 | 20 Mar 44 | 3549 | |
| Martin, Leon H. | 2Lt. | 0-2074515 | 13 Mar 45 | | |
| Martin, Raymond L. | Sgt. | 39453037 | 23 Nov 43 | 1247 | |
| Masters, John D. | 2Lt. | 0-758712 | 3 Jul 44 | 6231 | |
| Matthews, James P., Jr. | Sgt. | 34829996 | 18 Jan 45 | | |
| Matula, Frank J. | Sgt. | 12100312 | 16 Apr 45 | 14185 | Lorraine |
| Maxton, Marion C. | SSgt. | 37205700 | 5 Jan 44 | 2514 | |
| Mazur, Irving | SSgt. | 12036294 | 22 Dec 43 | 3313 | Netherlands |
| Mazzagatti, Philip | TSgt. | 38420922 | 2 Jan 45 | | |
| McCleary, Donald C. | 2Lt. | 0-1314163 | 24 Mar 45 | 13546 | |
| McClellan, Wendell R. | SSgt. | 38285507 | 11 Jan 44 | 2518 | |
| McCoy, Donald M. | Cpl. | 35763351 | 16 Aug 44 | 8426 | |
| McCoy, John C. | 2Lt. | 0-2057337 | 16 Apr 45 | 14185 | |
| McFarland, Hugh, Jr. | 2Lt. | 0-2059558 | 24 Mar 45 | 13546 | |
| McGinnis, Edward E. | Sgt. | 39208597 | 16 Jul 44 | | |
| McGlone, John E., Jr. | SSgt. | 32202248 | 5 Jan 44 | 2514 | |
| McKinley, Fredrick T. | 2Lt. | 0-2074504 | 17 Apr 45 | | |
| McLaughlin, James W. | TSgt. | 32440514 | 6 Nov 44 | 10348 | |
| McMahan, Lewis A. | SSgt. | 18182755 | 6 Nov 44 | 10348 | |
| McNamara, Thomas M. | SSgt. | 16088447 | 31 Dec 43 | 3093 | |
| Meents, Edward P., Jr. | 2Lt. | 0-750812 | 25 Feb 44 | 2956 | Cambridge |
| Meier, Elmer P. | 2Lt. | 0-815366 | 22 Apr 44 | | |
| Menrad, Louis F. | SSgt. | 36605687 | 12 Jun 45 | | Cambridge |
| Merkling, John L. | 2Lt. | 0-683045 | 22 Apr 44 | 4301 | |
| Merkovich, Frank S. | TSgt. | 3684180 | 4 Apr 45 | 13730 | Netherlands |
| Merrill, Hairman M. | Sgt. | 23498186 | 17 Apr 45 | | |
| Merrow, Derward E. | TSgt. | 11043372 | 12 Jun 45 | | |
| Mezzetti, Peter A. | Sgt. | 12034349 | 11 Sep 44 | 8926 | Netherlands |

| | | | | | |
|---|---|---|---|---|---|
| Michels, Carroll A. | 1Lt. | O-1844474 | 3 Mar 45 | 12885 | Netherlands |
| Mied, Arthur F. | TSgt. | 16000220 | 5 Mar 44 | 2913 | Epinal |
| Mielke, Henry E. | 2Lt. | O-7754750 | 7 Dec 44 | | |
| Miltner, Robert F. | SSgt. | 30927755 | 18 Mar 44 | 3341 | |
| Mitchell, John H. | Sgt. | 32910987 | 5 Aug 44 | 8669 | Cambridge |
| Mize, Thomas N. | 2Lt. | O-773427 | 26 Aug 44 | 8469 | Ardennes |
| Monefeldt, Leonard H. | Capt. | O-681460 | 24 Mar 45 | | Cambridge |
| Moran, Joseph E. | Sgt. | 15337063 | 29 Apr 44 | 4491 | Ardennes |
| Morgan Abe L. | Sgt. | 34738387 | 16 Apr 45 | 14185 | |
| Morse, Douglas C. | 1Lt. | O-686267 | 24 Aug 44 | 8292 | |
| Morrison, Girthel R. | SSgt. | 16065555 | 24 Mar 45 | 13546 | |
| Mroczek, Elmer J. | Sgt. | 35065303 | 24 Mar 45 | 13546 | |
| Mufson, Philip | 2Lt. | O-684181 | 25 Feb 44 | 2956 | Cambridge |
| Mull, Joseph H., Jr. | Pvt. | 14174063 | 3 Mar 45 | 12885 | |
| Murphy, Ebonezer J. | 2Lt. | O-7724466 | 21 Nov 44 | 10407 | |
| Murphy, Edward C., Jr. | SSgt. | 35370754 | 25 Mar 45 | 13547 | Netherlands |
| Murphy, Kenneth J. | 2Lt. | O-687958 | 8 Dec 43 | 3907/4135 | |
| Myers, Charles H. | SSgt. | 33257178 | 29 Apr 44 | 4486 | |
| Neidig, Harold S. | 2Lt. | O-684182 | 29 Apr 44 | 4491 | Henri-Chapelle |
| Nelson, John B. | SSgt. | 39275415 | 25 Feb 44 | 2956 | Cambridge |
| Newton, Lloyd E. | SSgt | 18110354 | 3 Mar 45 | 12885 | |
| Nichols, Elton L. | Sgt. | 38518612 | 3 Mar 45 | 12884 | Netherlands |
| Nickerson, Joseph F. | Sgt. | 35572323 | 21 Feb 44 | 2468 | |
| Nissen, Charles C. | TSgt. | 32589610 | 1 Apr 44 | 5030 | |
| Novichenk, Paul | TSgt. | 15323363 | 6 Nov 44 | 10348 | |
| Nystrom, Robert J. | 2Lt. | O-715412 | 27 Jun 44 | 6728 | Cambridge |
| O'Brien, John R. | SSgt. | 12187452 | 29 Apr 44 | 4491 | |
| Oden, Howard M. | 2Lt. | O-735439 | 8 Dec 43 | 3907/4135 | |
| Odiorne, Edward A. | SSgt. | 35582659 | 10 Feb 44 | | Cambridge |
| Olhaber, John H. | 2Lt. | O-873363 | 23 Nov 43 | 1247 | North Africa |
| O'Neil, Paul G. | 2Lt. | O-771113 | 7 Jan 45 | 11584 | Cambridge |
| O'Neil, Ralph F. | 2Lt. | O-886634 | 6 Nov 44 | 10348 | |
| Oppelt, Harry J. | 1Lt. | O-801356 | 10 Feb 44 | | |
| Opper, Henry J. | SSgt. | 12125979 | 10 Feb 44 | | |
| Ostarello, Umberto F. | Sgt. | 36604398 | 30 Dec 43 | 4178 | |
| Overy, Dale W. | Sgt. | 35296070 | 25 Mar 45 | 13548 | Ardennes |
| Packer, Cyrus | SSgt. | 18130544 | 10 Jun 44 | 5630 | |
| Palicki, Robert F. | 2Lt. | O-808887 | 22 Dec 43 | 3313 | |
| Pargh, Bernard | 1Lt. | O-2085215 | 12 Jun 45 | | |
| Parks, James W. | Cpl. | 18217836 | 5 Jan 44 | 2515 | Cambridge |
| Parks, Joseph F. | Sgt. | 38579450 | 25 Mar 45 | 13548 | |
| Parsons, Jerry M. | TSgt. | 16035535 | 29 Jun 44 | 7090 | |
| Payne, Charles C. | Sgt. | 34737456 | 26 Aug 44 | 8467 | |
| Payne, Donald A. | SSgt. | 13178062 | 25 Mar 45 | 13547 | |
| Pearce, Edward J. | 1Lt. | O-393388 | 23 Nov 43 | 1247 | |
| Pempek, Albert A. | 2Lt. | O-699905 | 6 Nov 44 | 10348 | |
| Pennypacker, William S. | SSgt. | 13137693 | 22 Dec 43 | 3313 | |
| Perkowski, Michael | SSgt. | 32836132 | 21 Nov 44 | | |
| Peterson, Irria J. | Sgt. | 38174375 | 16 Apr 45 | 14185 | |
| Peterson, Warren N., Jr. | 1Lt. | O-928144 | 25 Mar 45 | 14179 | |
| Petula, George | SSgt. | 12183208 | 11 Jan 44 | 2518 | Netherlands |
| Phillips, Frank S. | Capt. | O-2043738 | 20 Mar 44 | 3549 | |
| Phillips, John S. | Sgt. | 17057058 | 11 Sep 44 | 8927 | Netherlands |
| Pickering, Everett R. | 2Lt. | O-2067346 | 13 Mar 45 | | Cambridge |

*Appendices*

| | | | | | |
|---|---|---|---|---|---|
| Pinkus, Robert F. | 1Lt. | 0-661557 | 23 Mar 44 | 3547 | Netherlands |
| Pitts, Cherry C. | 1Lt. | 0-679115 | 22 Apr 44 | 4301 | Netherlands |
| Pogge, George H. | 1Lt. | 0-718974 | 26 Aug 44 | 8469 | |
| Pollard, Billie C. | SSgt. | 38425993 | 30 Dec 43 | 4178 | Epinal |
| Pollio, Francis X. | 1Lt. | 0-505988 | 12 Jun 45 | | |
| Pomfret, Joseph | 2Lt. | 0-810214 | 1 Apr 44 | | Ardennes |
| Ponge, William F. | 2Lt. | 0-807869 | 29 Apr 44 | 4491 | |
| Postemsky, Edmond G. | 2Lt. | 0-821075 | 26 Aug 44 | 8467 | |
| Prabucki, Bernard D. | SSgt. | 32731039 | 5 Jan 44 | 2515 | Cambridge |
| Prieb, Kenneth W. | SSgt. | 17131824 | 11 Jul 44 | | Cambridge |
| Prior, Virgle V. | SSgt. | 37653038 | 27 Jun 44 | 6726 | Cambridge |
| Pulcipher, Eugene V. | 2Lt. | 0-750685 | 22 Apr 44 | | Cambridge |
| Quinlan, Dennis C. | SSgt. | 32456595 | 5 Jan 44 | 2517 | |
| Raber, William D. | Sgt. | 39118166 | 23 Nov 43 | 1247 | |
| Resnikoff, Harold | SSgt. | 18231595 | 25 Mar 45 | 13547 | |
| Rhodes, John P. | 2Lt. | 0-745180 | 8 Dec 43 | 3907/4135 | |
| Rigg, Robert W. | SSgt. | 39549975 | 20 Mar 44 | 3549 | Normandy |
| Rikard, Robert F. | 2Lt. | 0-706903 | 11 Sep 44 | 8926 | |
| Risner, Homer C. | Sgt. | 38333496 | 22 Dec 43 | 2442 | Cambridge |
| Robinson, Ernest W., Jr. | TSgt. | 14055978 | 22 Apr 44 | 4301 | |
| Robinson, James R. | Cpl. | 35756585 | 16 Aug 44 | 8426 | |
| Rogers, Francis G. | 2Lt. | 0-687768 | 30 Dec 43 | 4178 | Epinal |
| Rogers, William W. | 2Lt. | 0-2007843 | 29 Apr 44 | 4487 | Ardennes |
| Romanosky, Chester J. | SSgt. | 13056696 | 22 Apr 44 | | Cambridge |
| Roorda, Delwin D. | 2Lt. | 0-2062050 | 15 Feb 45 | | |
| Rose, Jerome S. | Pfc. | 35055269 | 8 Dec 43 | 3907/4135 | |
| Rosenthal, Myron | Sgt. | 36869665 | 7 Jan 45 | 11584 | Cambridge |
| Savo, Enrico | Sgt. | 32805787 | 27 Jun 44 | 6726 | Cambridge |
| Scanlon, Edgar C., Jr. | 2Lt. | 0-701312 | 29 Jun 44 | 7090 | |
| Schierbrock, James J. | 2Lt. | 0-699776 | 7 Jul 44 | | Cambridge |
| Schierenbeck, Edmund A. | Pvt. | 37547985 | 3 Mar 45 | 12885 | |
| Schilling, John E. | Sgt. | 36841583 | 3 Mar 45 | 12885 | |
| Schlorman, Louis J. | Sgt. | 35622664 | 11 Sep 44 | 8926 | Netherlands |
| Schmidt, Charles R. | 2Lt. | 0-750861 | 5 Jan 44 | 2516 | Cambridge |
| Schoonmaker, Fredrick A. | SSgt. | 32734039 | 21 Jun 44 | 6231 | Ardennes |
| Schroeder, Harlyn H. | 1Lt. | 0-826014 | 15 Feb 45 | | |
| Schroeder, Russell F. | 2Lt. | 0-704067 | 27 Jun 44 | 6726 | Cambridge |
| Schrom, Clifford R. | 2Lt. | 0-679664 | 22 Apr 44 | 4301 | Cambridge |
| Scott, Thomas V. | 2Lt. | 0-785138 | 13 Mar 45 | | |
| Seiders, Pinkney W. | 2Lt. | 0-751847 | 5 Jan 44 | 2517 | |
| Shank, Joseph W., Jr. | 2Lt. | 0-743108 | 23 Nov 43 | 1247 | |
| Shipp, Charles A. | TSgt. | 18025053 | 26 Aug 44 | 8469 | Ardennes |
| Skledar, Joseph, Jr. | Sgt. | 33431803 | 27 Jun 44 | 6728 | Cambridge |
| Slack, Robert W. | Sgt. | 39912946 | 27 Jun 44 | 6727 | Epinal |
| Slepin, Jerome | 2Lt. | 0-689676 | 22 Dec 43 | 3313 | Netherlands |
| Sloan, Charles E. | 2Lt. | 0-689677 | 5 Jan 44 | 2516 | Cambridge |
| Smarinsky, Irving | 1Lt. | 0-710613 | 3 Mar 45 | 12884 | |
| Smidy, James E. | Sgt. | 33667892 | 27 Jun 44 | 6728 | Cambridge |
| Smith, Castleton D. | 1Lt. | 0-750225 | 20 Jul 44 | 3341 | |
| Smith, Harvey E. | SSgt. | 34516429 | 11 Jan 44 | 2519 | |
| Smith, Howard M. | SSgt. | 17038532 | 11 Jan 44 | 2518 | |
| Smith, Salem A., Jr. | 2Lt. | 0-688978 | 8 Dec 43 | 3907/4135 | |
| Sowell, Turner A., Jr. | 2Lt. | 0-747607 | 23 Nov 43 | 1247 | |
| Spadafore, Albert N. | SSgt. | 11090413 | 8 Mar 44 | 3189 | Cambridge |

| Name | Rank | Serial | Date | | Cemetery |
|---|---|---|---|---|---|
| Spellman, Carl E. | Sgt. | 37368208 | 22 Apr 44 | | |
| Spell, Ira L. | Sgt. | 14188306 | 19 Apr 44 | 4303 | Ardennes |
| Spruill, Robert B. | Sgt. | 14080837 | 19 May 44 | 5029 | Ardennes |
| Squyres, Kenneth D. | Maj. | 0-406033 | 5 Jan 44 | 2515 | Cambridge |
| Stalland, Knute P. | 1Lt. | 0-721875 | 25 Mar 45 | 13548 | |
| Steffan, Joseph F. | 2Lt. | 0-719786 | 25 Mar 45 | 13547 | |
| Stennes, John J. | 2Lt. | 0-694456 | 27 Jun 44 | 6726 | Cambridge |
| Stevens, Sumner W. | 2Lt. | 0-670285 | 22 Dec 43 | 2442 | Netherlands |
| Stine, Everett F. | 2Lt. | 0-689683 | 20 Mar 44 | 3549 | |
| Stone, Robert B. | 2Lt. | 0-689684 | 31 Dec 43 | 3605 | |
| Strawn, Harold E. | Sgt. | 13029830 | 30 Dec 43 | 4178 | Epinal |
| Sullivan, John J. | 2Lt. | 0-696718 | 24 Aug 44 | 8292 | Netherlands |
| Swanson, Charles P. | 2Lt. | 0-717118 | 5 Aug 44 | 8669 | |
| Szudarek, Severyn G. | 2Lt. | 0-1998528 | 7 Dec 44 | | Cambridge |
| Tarkington, Taylor L. | Sgt. | 18177172 | 4 Apr 45 | 13732 | Ardennes |
| Taylor, Carl E. | SSgt. | 39835597 | 8 Mar 44 | 3189 | Cambridge |
| Taylor, Earl L. | SSgt. | 37498287 | 23 Jun 44 | 5630 | Normandy |
| Taylor, James E. | Sgt. | 36446039 | 23 Nov 43 | 1247 | |
| Tennant, Marion A. | SSgt. | 35382677 | 29 Apr 44 | 4486 | |
| Thieleu, Charles M. | 2Lt. | 0-701346 | 11 Jul 44 | | Normandy |
| Thompson, Heber H. | Lt. Col. | 0-1699234 | 12 Jun 45 | | |
| Thompson, James McKay | Col. | 0-017992 | 1 Apr 44 | | Normandy |
| Thompson, Robert W. | SSgt. | 17066011 | 29 Jun 44 | 7090 | |
| Thurber, Raymond L. | 2Lt. | 0-689693 | 11 Jan 44 | 2519 | |
| Tod, Fredrick W. | 1Lt. | 0-776133 | 25 Mar 45 | 14179 | Luxembourg |
| Totman, Edward F. | SSgt. | 35596301 | 8 Mar 44 | 3189 | Cambridge |
| Towles, Raymond S. | 1Lt. | 0-2045237 | 10 Jun 44 | 5630 | |
| Turner, Arlin W. | SSgt. | 18140043 | 21 Jun 44 | 6231 | |
| Turner, Robert Jr. | Sgt. | 38519135 | 3 Mar 45 | 12884 | |
| Turpin, Harold C. | 1Lt. | 0-807541 | 27 Jun 44 | 6728 | Cambridge |
| Urban, James E. | 2Lt. | 0-676878 | 11 Jan 44 | 2518 | |
| Vajgyl, James L. | Sgt. | 35527301 | 21 Jun 44 | 6264 | Ardennes |
| Van Deventer, Stuart D. | Sgt. | 38588932 | 4 Apr 45 | 13730 | Netherlands |
| VanVorst, Leo A., Jr. | Sgt. | 32813059 | 27 Jun 44 | 6728 | Cambridge |
| Vetter, Harry F. | SSgt. | 16088095 | 22 Apr 44 | 4301 | Cambridge |
| Villari, Anthony C. | Sgt. | 35919619 | 4 Apr 45 | 13730 | Netherlands |
| Vinson, Sammie D. | Sgt. | 14177911 | 21 Jun 44 | 6264 | Netherlands |
| Wais, Dan A. | Sgt. | 35788495 | 18 Jun 44 | 5929 | Ardennes |
| Walasik, Anthony | SSgt. | 33034529 | 26 Apr 44 | | |
| Wallenda, John, Jr. | SSgt. | 35514944 | 25 Feb 44 | 2956 | Cambridge |
| Wann, Keith D. | SSgt. | 39117819 | 27 Jun 44 | 6728 | Cambridge |
| Ward, Leon A., Jr. | SSgt. | 19203253 | 5 Jan 44 | 2515 | Cambridge |
| Warke, William A. | 1Lt. | 0-693082 | 29 Jun 44 | 7090 | |
| Warner, Theodore, Jr. | 1Lt. | 0-928014 | 25 Mar 45 | 13548 | |
| Warnock, Roland D. | 2Lt. | 0-731707 | 24 Feb 44 | 2955 | |
| Warren, William A. | TSgt. | 18028026 | 1 Apr 44 | 3566 | Epinal |
| Wasara, Tolva A. | Sgt. | 15133297 | 27 Jun 44 | 6726 | Cambridge |
| Waters, Herman L. | Sgt. | 34587603 | 24 Mar 45 | 13546 | Netherlands |
| Weisenburgh, Charles P. | 2Lt. | 0-689706 | 5 Jan 44 | 2515 | Cambridge |
| Weishaar, Eugene C. | SSgt. | 36704194 | 23 Nov 43 | 1247 | |
| Werner, Samuel | TSgt. | 12156211 | 22 Apr 44 | 4301 | Cambridge |
| Westrick, Paul E. | 2Lt. | 0-2062099 | 13 Mar 45 | | Cambridge |
| White, Harold E. | 2Lt. | 0-747535 | 11 Jan 44 | 2519 | Netherlands |
| Wilder, Charles W., Jr. | 2Lt. | 0-683885 | 22 Apr 44 | 4301 | Cambridge |

| | | | | | |
|---|---|---|---|---|---|
| Wildman, John A. | SSgt. | 35298840 | 12 Jun 45 | | |
| Wilhelm, Lawrence M. | SSgt. | 37604384 | 26 Aug 44 | 8469 | Ardennes |
| Wilkins, Robert J. | 2Lt. | 0-2066088 | 7 Jan 45 | 11584 | Cambridge |
| Wilson, Charles W. | 2Lt. | 0-704863 | 27 Jun 44 | 6728 | Cambridge |
| Wilson, Howard L. | SSgt. | 33215816 | 5 Jan 44 | 2516 | Cambridge |
| Wilson, Jap R., Jr. | SSgt. | 18076005 | 21 Jun 44 | 6231 | |
| Wilson, Stanley L. | SSgt. | 39550446 | 22 Apr 44 | 4301 | Cambridge |
| Wittenberg, Herman | Sgt. | 13124383 | 5 Jan 44 | 2516 | Cambridge |
| Wright, Anderson C. | SSgt. | 33836022 | 4 Apr 45 | 13732 | |
| Wright, Gordon W. | SSgt. | 12072510 | 29 Jun 44 | 7090 | Lorraine |
| Wright, William A., Jr. | Sgt. | 32858722 | 25 Mar 45 | 13547 | |
| Yarnell, Robert H. | SSgt. | 35668792 | 21 Feb 44 | 2469 | Netherlands |
| Yates, Joseph A., Jr. | SSgt. | 13184568 | 6 Nov 44 | 10348 | |
| Young, John B. | F/O | T-134801 | 24 Mar 45 | 13546 | |
| Young, Maynard H. | Sgt. | 31152846 | 22 Apr 44 | | Cambridge |
| Yuengert, Walter A. | 2Lt. | 0-742103 | 5 Jan 44 | 2516 | Cambridge |
| Zemba, Joseph M. | Sgt. | 33427181 | 27 Jun 44 | 6726 | Cambridge |
| Zierdt, Kenneth A. | Sgt. | 17064455 | 12 Jun 44 | 5803 | |
| Zimmerman, Theodore R. | 2Lt. | 0-2058344 | 7 Jan 44 | 11584 | Cambridge |

## Appendix 2
# AIRCRAFT

This list does not include the numerous aircraft that flew as lead PFF aircraft for the 448th while assigned to other Groups. This list identifies only those aircraft that are believed to have been assigned specifically to the 448th BG at one time.

| AIRCRAFT NAME | S/N | MODEL | FATE | DATE | MACR |
|---|---|---|---|---|---|
| YOU CAWN'T MISS IT | 41-23809 | D-5-CO | Salvaged | 19 Jan 45 | |
| LADY LUCK | 41-28578 | H-1-DT | Returned | 7 Mar 45 | |
| RUM RUNNER | 41-28583 | H-1-DT | Interned | 25 Apr 44 | 4300 |
| SAD SACK | 41-28588 | H-1-DT | Missing | 30 Dec 43 | 4176 |
| OLD IRONSIDES | 41-28589 | H-1-DT | Transferred | | |
| | 41-28590 | H-1-DT | Transferred | | |
| PRODIGAL SON | 41-28593 | H-1-DT | Missing | 11 Jan 44 | 2519 |
| ICE COLD KATIE | 41-28595 | H-1-DT | Crashed | 22 Apr 44 | |
| | 41-28599 | H-1-DT | Missing | 30 Dec 43 | 4177 |
| CRAZY MARY | 41-28601 | H-1-DT | Crashed | 31 Dec 43 | |
| COMMANDO | 41-28602 | H-1-DT | Interned | 25 Apr 44 | 4364 |
| | 41-28609 | H-1-DT | Missing | 22 Dec 43 | 2442 |
| BABY SHOES | 41-28611 | H-1-DT | Ditched | 9 Mar 44 | 3339 |
| | 41-28626 | H-1-DT | Transferred | | |
| | 41-28648 | H-5-DT | Crashed | 10 Aug 44 | |
| SQUAT-N-DROPPIT | 41-28710 | H-10-DT | Missing | 12 Jun 44 | 5803 |
| LITTLE SHEPPARD | 41-28711 | H-10-DT | Interned | 21 Jul 44 | 7251 |
| ROSIE RIVETS | 41-28819 | H-15-DT | Crashed | 16 Jan 45 | |
| REPULSER | 41-28843 | H-15-DT | Crashed | 22 Apr 44 | |
| | 41-28924 | H-15-DT | Missing | 26 Aug 44 | 8467 |
| LITTLE JOE | 41-28925 | H-15-DT | Missing | 21 Nov 44 | 10407 |
| LITTLE IODINE | 41-28941 | H-15-DT | Crashed | 16 Jan 45 | |
| NO NOTHING | 41-28945 | H-20-DT | Interned | 6 Aug 44 | 8217 |
| LITTLE JO | 41-28958 | H-20-DT | Crashed | 27 Feb 45 | |
| | 41-28983 | H-20-DT | | | |
| | 41-29004 | H-20-DT | Missing | 27 Jun 44 | 6728 |
| HELLO NATURAL | 41-29191 | H-5-CF | Interned | 6 Mar 44 | 3548 |
| | 41-29192 | H-5-CF | Transferred | | |
| SHOO SHOO BABY | 41-29208 | H-5-CF | Transferred | | |
| NO NAME JIVE | 41-29230 | H-5-CF | Returned | | |
| IMPATIENT VIRGIN, THE | 41-29231 | H-5-CF | Transferred | | |
| MENACE, THE | 41-29232 | H-5-CF | Crashed | 23 Jul 44 | |
| | 41-29234 | H-5-CF | Missing | 5 Jan 44 | 2517 |
| | 41-29235 | H-5-CF | Crashed | 25 Feb 44 | |
| HELL'S KITTEN | 41-29236 | H-5-CF | Transferred | | |
| TONDELAYO | 41-29240 | H-5-CF | Transferred | | |
| COLD TURKEY | 41-29248 | H-5-CF | Missing | 31 Dec 43 | 3605 |
| LONESOME LOU | 41-29465 | H-15-CF | Ditched | 5 Aug 44 | 8669 |
| BIG BAD WOLF | 41-29479 | H-15-CF | Missing | 29 Apr 44 | 4488 |
| | 41-29489 | H-15-CF | Salvaged | 26 May 45 | |
| EL KORAB | 41-29521 | H-15-CF | Returned | | |
| MISS HAPP | 41-29523 | H-15-CF | Missing | 29 Apr 44 | 4489 |
| MAID OF ORLEANS II | 41-29565 | H-15-CF | Ditched | 19 Apr 44 | 4303 |
| RUTH E.K., THE (ALLAH HASSID) | 41-29575 | H-15-CF | Crashed | 6 Jan 45 | |

*Appendices*

|  |  |  |  |  |  |
|---|---|---|---|---|---|
|  | 42-50289 | H-20-CF | Salvaged | 29 May 45 |  |
| FOUL BALL | 42-50290 | H-20-CF | Salvaged | 26 May 45 |  |
|  | 42-50300 | H-20-CF | Missing | 29 Jun 44 | 7090 |
| OUR HONEY | 42-50302 | H-20-CF | Returned |  |  |
| EAGER ONE | 42-50326 | H-20-CF | Crashed | 15 Mar 45 |  |
| PICCADILLY LILLY | 42-50341 | H-20-CF | Missing | 24 Mar 45 | 13546 |
| RED SOX | 42-50344 | H-20-CF | Missing | 27 Jun 44 | 6727 |
| QUEENIE | 42-50348 | H-20-CF | Returned |  |  |
|  | 42-50357 | H-25-CF | Crashed | 19 Dec 44 |  |
| TARFU | 42-50359 | H-25-CF | Crashed | 13 Mar 45 |  |
|  | 42-50369 | H-25-CF | Missing | 27 Jun 44 | 6726 |
| OLD 75 POP | 42-50391 | H-25-CF | Returned |  |  |
|  | 42-50443 | H-30-CF | Missing | 26 Aug 44 | 8468 |
| WAG'S WAGON | 42-50455 | J-40-CF | Transferred |  |  |
| DRAGON LADY | 42-50457 | J-40-CF | Crashed | 7 Dec 44 |  |
| ST LOUIS WOMAN | 42-50459 | J-40-CF | Missing | 16 Aug 44 | 8426 |
| SLICK CHICK | 42-50460 | J-40-CF | Returned |  |  |
| GUNG HO II | 42-50463 | J-40-CF | Missing | 3 Mar 45 | 12884 |
|  | 42-50468 | J-40-CF | Returned |  |  |
| OUR BABY | 42-50469 | J-40-CF | Crashed | 28 Jan 45 |  |
| PICCADILLY PAT | 42-50470 | J-40-CF | Crashed | 21 Jul 44 |  |
|  | 42-50475 | J-40-CF | Crashed | 12 Jul 44 |  |
| OLD GLORY | 42-50482 | J-40-CF | Returned |  |  |
|  | 42-50493 | J-40-CF | Returned |  |  |
|  | 42-50522 | J-1-FO | Returned |  |  |
| DON'T FENCE ME IN | 42-50525 | J-1-FO | Returned |  |  |
|  | 42-50566 | J-1-FO | Crashed | 2 Aug 44 |  |
|  | 42-50587 | J-1-FO | Salvaged | 11 Dec 44 |  |
|  | 42-50590 | J-1-FO | Returned |  |  |
|  | 42-50646 | J-1-FO | Missing | 25 Mar 45 | 13548 |
|  | 42-50648 | J-1-FO | Interned | 4 Aug 44 | 7386 |
|  | 42-50661 | J-1-FO | Missing | 13 Jan 45 | 11583 |
|  | 42-50676 | J-1-FO | Crashed | 28 Jan 45 |  |
| BLUES IN THE NITE | 42-50677 | J-1-FO | Returned |  |  |
| UMBREOGO | 42-50678 | J-1-FO | Crashed | 15 Feb 45 |  |
| FEARLESS FOSDICK | 42-50699 | J-1-FO | Returned |  |  |
| BROWNIE | 42-50727 | J-1-FO | Returned |  |  |
| FLEXIBLE FLYER | 42-50745 | J-1-FO | Crashed | 13 Jan 45 |  |
| SHADY LADY | 42-50759 | J-1-FO | Crashed | 16 Nov 44 |  |
| WAZZLE DAZZLE | 42-50767 | J-5-FO | Returned |  |  |
| JOKER'S WILD | 42-50772 | J-5-FO | Returned |  |  |
| SPIRIT OF COOLEY HS-DETROIT | 42-50777 | J-5-FO | Returned |  |  |
|  | 42-50788 | J-5-FO | Missing | 26 Aug 44 | 8608 |
| PATRICK DEMPSEY | 42-50799 | J-5-FO | Missing | 24 Dec 44 | 11120 |
| BROWNIE | 42-50809 | J-5-FO | Returned | 13 Jun 45 |  |
|  | 42-50820 | J-5-FO | Missing | 6 Nov 44 | 10348 |
| MISS MUFF | 42-50902 | J-5-FO | Returned | 13 Jun 45 |  |
|  | 42-51030 | J-5-FO | Salvaged | 4 Nov 44 |  |
| LINDA MAE | 42-51075 | J-5-FO | Returned | 13 Jun 45 |  |
|  | 42-51079 | H-20-DT | Interned | 20 Jun 44 | 6163 |
|  | 42-51104 | H-25-DT | Missing | 6 Aug 44 | 8291 |
|  | 42-51119 | H-25-DT | Missing | 18 Jun 44 | 5929 |
|  | 42-51221 | H-30-DT | Salvaged | 16 Jul 44 |  |
| BLAKE'S SNAKES | 42-51228 | J-1-DT | Returned | 13 Jun 45 |  |

| Name | Serial | Model | Status | Date | MACR |
|---|---|---|---|---|---|
| FRISCO FRISKY | 42-51246 | J-1-DT | Returned | | |
| FEUDEN REBEL | 42-51247 | J-1-DT | Missing | 3 Mar 45 | 12885 |
| | 42-51252 | J-1-DT | Returned | | |
| BACK TO THE SACK | 42-51288 | J-1-DT | Returned | | |
| OUR JOY | 42-51291 | J-1-DT | Crashed | 24 Aug 44 | |
| | 42-51349 | J-5-DT | Crashed | 7 Jan 45 | |
| | 42-51358 | J-5-DT | | | |
| | 42-51443 | J-5-FO | | | |
| SWEET ROSE MARIE | 42-51489 | J-5-FO | Returned | | |
| | 42-51497 | J-5-FO | Salvaged | 20 Dec 44 | |
| | 42-51504 | J-5-FO | Crashed | 24 Mar 45 | |
| | 42-51505 | J-5-FO | Crashed | 27 Dec 44 | |
| BUGS BUNNY (REDDY TEDDY) | 42-51551 | J-5-FO | Returned | | |
| MILK RUN | 42-51589 | J-5-FO | Transferred | 8 Apr 45 | |
| | 42-51666 | J-10-FO | Crashed | 3 Jan 45 | |
| | 42-51745 | J-10-FO | Returned | | |
| | 42-52008 | J-15-FO | | | |
| OUTHOUSE MOUSE | 42-52083 | H-5-FO | Transferred | | |
| LONESOME POLECAT | 42-52097 | H-5-FO | Transferred | | |
| CRUD WAGON | 42-52098 | H-5-FO | Missing | 1 Apr 44 | 3565 |
| LADY FROM BRISTOL | 42-52100 | H-5-FO | Missing | 25 Feb 44 | 2956 |
| | 42-52105 | H-5-FO | Missing | 22 Dec 43 | 3313 |
| | 42-52108 | H-5-FO | Crashed | 23 Nov 43 | 1247 |
| HELL'S BELLE | 42-52115 | H-10-FO | Crashed | 10 Feb 44 | |
| | 42-52116 | H-10-FO | Salvaged | 29 May 45 | |
| PICCADILLY PETE | 42-52118 | H-10-FO | Interned | 9 Apr 44 | 3659 |
| BOMB BOOGIE | 42-52120 | H-10-FO | Crashed | 31 Dec 43 | |
| WOLF PACK | 42-52121 | H-10-FO | Salvaged | 9 Sep 44 | |
| THIRTY DAY FURLOUGH | 42-52123 | H-10-FO | Missing | 11 Jan 44 | 2518 |
| | 42-52128 | H-10-FO | Crashed | 19 Nov 43 | |
| BOOMERANG, THE | 42-52132 | H-10-FO | Crashed | 10 Feb 44 | |
| | 42-52135 | H-10-FO | Crashed | | |
| | 42-52145 | H-10-FO | Salvaged | 29 May 45 | |
| | 42-52435 | H-15-FO | Missing | 29 Apr 44 | 4490 |
| | 42-52496 | H-15-FO | Returned | | |
| HELLO NATURAL II | 42-52606 | H-15-FO | Salvaged | 16 Jul 44 | |
| | 42-52608 | H-15-FO | Ditched | 22 Apr 44 | 4301 |
| SKEETER II | 42-52638 | H-15-FO | Missing | 19 May 44 | 5029 |
| CAROL-N-CHICK | 42-63981 | D-20-CF | Salvaged | 13 Feb 45 | |
| BIG ASS BURD | 42-64441 | H-5-CF | Salvaged | | |
| CONSOLIDATED MESS | 42-64444 | H-5-CF | Missing | 30 Dec 43 | 4178 |
| COMANCHE, THE | 42-64447 | H-5-CF | Missing | 20 Mar 44 | 3340 |
| SEQUOIA GAL | 42-64451 | H-5-CF | Crashed | 5 Jan 44 | |
| FASCINATING LADY | 42-72981 | J-1-CO | Missing | 20 Apr 44 | 4302 |
| FEATHER MERCHANT | 42-73477 | J-50-CO | Transferred | | |
| VADIE RAYE | 42-73497 | J-50-CO | Crashed | 22 Apr 44 | |
| PROBLEM CHILD | 42-73512 | J-50-CO | Transferred | | |
| POOP DECK PAPPY | 42-7521 | H-1-FO | Returned | | |
| FAT STUFF II | 42-7591 | H-1-FO | Interned | 12 Jul 44 | 7559 |
| ABIE'S IRISH ROSE | 42-7606 | H-1-FO | Missing | 24 Feb 44 | 2955 |
| CHUBBY CHAMP | 42-7655 | H-1-FO | Missing | 29 Apr 44 | 4486 |
| FINK'S JINX | 42-7681 | H-1-FO | Crashed | 8 Dec 43 | 3907 |
| SWEET SIOUX | 42-7683 | H-1-FO | Missing | 29 Apr 44 | 4487 |
| MAID OF TIN | 42-7709 | H-1-FO | Missing | 5 Jan 44 | 2515 |

*Appendices*

|  | 42-7712 | H-1-FO | Missing | 5 Jan 44 | 2514 |
| --- | --- | --- | --- | --- | --- |
| MERRY MAX, THE | 42-7713 | H-1-FO | Crashed | 26 Nov 43 |  |
| EXTERMINATOR, THE | 42-7717 | H-1-FO | Returned | 25 Apr 45 |  |
|  | 42-7722 | H-5-FO | Missing | 5 Jan 44 | 2516 |
| LAKI-NUKI | 42-7733 | H-5-FO | Crashed | 14 Dec 43 |  |
| MAID OF ORLEANS | 42-7739 | H-5-FO | Missing | 23 Mar 44 | 3547 |
| HARMFUL LIL ARMFUL | 42-7754 | H-5-FO | Missing | 31 Dec 43 | 3093 |
| HARD TIMES | 42-7755 | H-5-FO | Salvaged | 8 Nov 44 |  |
| DOWN AND GO | 42-7758 | H-5-FO | Crashed | 22 Jun 44 |  |
| BAG O' BOLTS | 42-7764 | H-5-FO | Salvaged | 21 Feb 44 |  |
|  | 42-7766 | H-5-FO | Transferred |  |  |
| SHACK RABBIT | 42-7767 | H-5-FO | Transferred |  |  |
| PROUD WANDERLOST, THE | 42-7768 | H-5-FO | Missing | 21 Feb 44 | 2469 |
| BACHELOR'S DELIGHT | 42-78481 | J-1-NT | Crashed | 2 Oct 44 |  |
|  | 42-78491 | J-1-NT | Returned |  |  |
| BIM BAM BOLA | 42-94735 | H-15-FO | Interned | 12 Jul 44 | 7560 |
| PEGGY JO | 42-94744 | H-15-FO | Crashed | 22 Apr 44 |  |
| UNINVITED | 42-94762 | H-15-FO |  |  |  |
| OL' BUDDY | 42-94774 | H-15-FO | Crashed | 26 Aug 44 |  |
| MISS MINOOKIE (BOTTLE BOYS) | 42-94798 | H-20-FO | Returned |  |  |
|  | 42-94809 | H-20-FO | Crashed | 5 Aug 44 |  |
| YOU'RE SAFE AT HOME | 42-94828 | H-20-FO | Returned | 13 Jun 45 |  |
| WINDHAVEN | 42-94853 | H-20-FO | Returned |  |  |
| HAPPY WARRIOR | 42-94860 | H-20-FO | Returned |  |  |
| MISS CARRIAGE | 42-94880 | H-20-FO |  |  |  |
| RUGGED BUT RIGHT | 42-94953 | H-20-FO | Returned |  |  |
| DAISY MAE | 42-94972 | H-20-FO | Missing | 7 Jan 45 | 11584 |
|  | 42-94989 | H-20-FO | Interned | 13 Jul 44 | 7561 |
| DEAD END KIDS | 42-94992 | H-20-FO | Returned |  |  |
| LIBERTY BELLE | 42-94996 | H-20-FO | Salvaged | 26 May 45 |  |
|  | 42-95006 | H-20-FO | Crashed | 3 Oct 44 |  |
| SWEET SIOUX II | 42-95013 | H-20-FO | Interned | 20 Jun 44 | 6267 |
| TANGERINE | 42-95022 | H-20-FO | Missing | 24 Jun 44 | 6266 |
| BAR FLY | 42-95055 | H-25-FO | Salvaged | 25 Oct 44 |  |
| HAPPY HANGOVER | 42-95075 | H-25-FO | Missing | 21 Jun 44 | 6231 |
| MY BUDDIE | 42-95083 | H-25-FO | Returned |  |  |
| DUAL SACK | 42-95089 | H-25-FO | Interned | 21 Jun 44 | 6265 |
| SPIRIT OF NOTRE DAME | 42-95102 | H-25-FO | Returned |  |  |
| LADY MARGARET | 42-95134 | H-25-FO | Missing | 26 Aug 44 | 8469 |
| NO LOVE NO NOTHING | 42-95138 | H-25-FO | Missing | 11 Sep 44 | 8927 |
| MONOTONOUS MAGGIE | 42-95151 | H-25-FO | Returned |  |  |
| INCENDIARY BLONDE, THE | 42-95158 | H-25-FO | Crashed | 11 Jul 44 |  |
| BETSY JAY | 42-95169 | H-25-FO | Crashed | 16 Jul 44 |  |
|  | 42-95182 | H-25-FO | Missing | 24 Aug 44 | 8292 |
| DO BUNNY | 42-95185 | H-25-FO | Missing | 25 Mar 45 | 13549 |
|  | 42-95186 | H-25-FO | Missing | 21 Jun 44 | 6264 |
| FLAK JACK | 42-95200 | H-25-FO | Interned | 20 Jun 44 | 6240 |
| SONIA | 42-95270 | H-25-FO | Returned | 12 Jun 45 |  |
| TROUBLE N' MIND | 42-95298 | H-30-FO | Missing | 4 Apr 45 | 13731 |
| MY BABY | 42-95305 | H-30-FO | Returned |  |  |
|  | 42-95326 | H-30-FO | Missing | 29 Jun 44 | 7091 |
| 4-F | 42-95527 | J-1-FO | Returned | 13 Jun 45 |  |
| UNHOLY VIRGIN | 42-95544 | J-1-FO | Missing | 16 Jan 45 | 11727 |
| MISS B-HAVIN | 42-95620 | J-1-FO | Missing | 4 Apr 45 | 13732 |

| Name | Serial | Block | Fate | Date | MACR |
|---|---|---|---|---|---|
| BATTLIN BABY | 42-99971 | J-55-CO | Transferred | | |
| SAD SACK, THE | 42-99988 | J-60-CO | Missing | 29 Apr 44 | 4491 |
| MARY MICHELE | 42-99993 | J-60-CO | Salvaged | | |
| BTO | 42-100000 | J-60-CO | Returned | | |
| EASTERN QUEEN | 42-100109 | J-70-CO | Missing | 1 Apr 44 | 3567 |
| TWIN TAILS | 42-100122 | J-70-CO | Missing | 8 Mar 44 | 3189 |
| CARRY ME BACK | 42-100178 | J-75-CO | Missing | 21 Feb 44 | 2468 |
| | 42-100284 | J-85-CO | Interned | 18 Mar 44 | 3341 |
| MARGARET L (BUSTED FLUSH) | 42-100287 | J-90-CO | Missing | 9 May 44 | 4775 |
| COME ALONG BOYS | 42-100322 | J-90-CO | Crashed | 2 Jan 45 | |
| | 42-100342 | J-95-CO | Missing | 8 Mar 44 | 2967 |
| BLACK WIDOW | 42-100356 | J-95-CO | Ditched | 1 Apr 44 | 5030 |
| | 42-100414 | J-100-CO | Missing | 5 Mar 44 | 2912 |
| | 42-100430 | J-100-CO | Missing | 5 Mar 44 | 2913 |
| RAB DUCKIT | 42-100435 | J-100-CO | Returned | | |
| | 42-109793 | J-105-CO | Crashed | 2 Jul 44 | |
| | 42-109808 | J-105-CO | Missing | 20 Mar 44 | 3549 |
| SKY QUEEN | 42-110026 | J-125-CO | Crashed | 8 Jun 44 | |
| IMPATIENT VIRGIN, THE | 42-110040 | J-130-CO | Interned | 9 Apr 44 | 3660 |
| SLEEPLESS KNIGHTS | 42-110044 | J-130-CO | Missing | 12 Jul 44 | 7562 |
| STURGEON, THE | 42-110066 | J-130-CO | Returned | | |
| | 42-110069 | J-130-CO | Crashed | 9 Apr 44 | |
| | 42-110079 | J-130-CO | Crashed | 9 Apr 44 | |
| | 42-110087 | J-130-CO | Missing | 1 Apr 44 | 3566 |
| FLYING SAC, THE | 42-110098 | J-135-CO | Interned | 24 Apr 44 | 4298 |
| | 44-10486 | J-55-CF | Crashed | 17 Sep 44 | |
| BUFFALO GAL | 44-10498 | J-55-CF | Returned | | |
| GUNG HO | 44-10505 | J-60-CF | Missing | 11 Sep 44 | 8926 |
| | 44-10512 | J-60-CF | Returned | | |
| LITTLE IODINE | 44-10516 | J-60-CF | Returned | | |
| EAGER ONE | 44-10517 | J-60-CF | Interned | 25 Mar 45 | 14179 |
| | 44-10520 | J-60-CF | Crashed | 12 Dec 44 | |
| JUNIOR | 44-10536 | J-60-CF | Missing | 16 Apr 45 | 14185 |
| SWEETHEART OF THE ROCKIES | 44-10544 | J-60-CF | Returned | | |
| | 44-10547 | J-60-CF | Returned | | |
| MISS AMERICA | 44-10554 | J-65-CF | Returned | 13 Jun 45 | |
| MOTHER OF TEN | 44-10556 | J-65-CF | Returned | | |
| WINDY WINNIE | 44-10599 | J-65-CF | Crashed | 28 Jan 45 | |
| TARFU II | 44-40099 | J-145-CO | Missing | 25 Mar 45 | 13547 |
| | 44-40107 | J-145-CO | Missing | 10 Jun 44 | 5630 |
| SKY LARK | 44-40224 | J-150-CO | Crashed | 15 Jan 45 | |
| HEAVEN CAN WAIT | 44-40875 | J-185-CO | Salvaged | 17 Jul 44 | |
| | 44-40879 | J-185-CO | Transferred | | |
| | 44-48787 | J-20-FO | | | |
| LEADING LADY | 44-48805 | J-20-FO | Returned | 13 Jun 45 | |
| BROADWAY BILL | 44-48870 | J-20-FO | Returned | 13 Jun 45 | |
| | 44-49516 | L-10-FO | Returned | | |
| | 44-49593 | L-10-FO | Returned | 13 Jun 45 | |
| | 44-49995 | L-15-FO | | | |
| | 44-50084 | L-20-FO | | | |
| | 44-50509 | M-5-FO | Salvaged | 27 Mar 45 | |
| FER PETE'S SAKE | 44-50526 | M-5-FO | Returned | | |
| | 44-50531 | M-5-FO | Returned | | |
| ACHTUNG NOON BALLOON | 44-50540 | M-5-FO | Returned | 13 Jun 45 | |

*Appendices*

| | | | | | |
|---|---|---|---|---|---|
| | 44-50546 | M-5-FO | Returned | | |
| | 44-50551 | M-5-FO | Returned | | |
| | 44-50572 | M-5-FO | Crashed | 17 Apr 45 | |
| | 44-50658 | M-10-FO | | | |
| | 44-50676 | M-10-FO | Returned | 13 Jun 45 | |
| BTO | 44-50678 | M-10-FO | Transferred | | |
| | 44-50695 | M-10-FO | Crashed | 12 Jun 45 | |
| | 44-50701 | M-10-FO | Returned | | |
| | 44-50752 | M-10-FO | Returned | | |
| LIILY'S SISTER & ALL FOR AL | 44-50787 | M-10-FO | Returned | | |
| | 44-50819 | M-10-FO | Returned | | |
| RED BOW | 44-50838 | M-10-FO | Missing | 4 Apr 45 | 13730 |
| | 44-50846 | M-10-FO | | | |
| | 44-50859 | M-15-FO | Returned | | |
| | 44-50872 | M-15-FO | Returned | | |
| SNUFFY'S DELIGHT | 44-50976 | M-15-FO | | | |

# Appendix 3
# AIRCREWS

Listed below are Headquarters and Squadron staff who flew with the crews and are listed with the crews as passengers. Ground echelon that traveled via HMS Queen Elizabeth was not listed in the microfilm. Lt. Col. Elver, Captain Kramer & Major Arnold were transported to England via the ATC through the northern route. Of the original 70 crews, only the 62 crews listed below flew their own aircraft overseas as a group. (This list was compiled from original orders. Some of them were hard to read, therefore, ?'s inidicate where the records were unreadable. Unless confirmed elsewhere, this list reflects the original orders.)

Original 448th Bombardment Group Staff And Crews

Col. James McK. Thompson
Major Patrick H. Hoey
Mjor Hubert S. Judy, Jr.
Major James R. Patterson
Captain John S. Laws
Captain William R. Reid
1st Lt. George O. Capp
1st Lt. John B. D. Grunow
1st Lt. Arthur S. Hunt
1st Lt. Minor L. Morgan
2nd Lt. Robert L. Harper
2nd Lt. William C. Cates, Jr.
MSgt. Wilfred Carroll
MSgt. James D. McIntyre
TSgt. Frank M. Schultz
SSgt. Ellis L. Copeland

712th Bombardment Squadron Staff

Major Robert L. Campbell
1st. Lt. William G. Blum
1st. Lt. Robert Lewis
1st. Lt. Clifford C. Gaither
1st. Robert W. McDonough
1st. Lt. Earl M. Parks
1st. Lt. Harold S. Podolsky
2nd Lt. John Bilyk
2nd. Lt. Howard M. Oden
MSgt. Walter C. Brown
MSgt. Frank J. Naglak
TSgt. Henry C. Corbin
TSgt. Michael P.Corce

TSgt. Frederick J. Fennewald
TSgt. Thomas J. Flynn
SSgt. Delmer A. Akerley
SSgt. John J. Bernhard
SSgt. Allen L. Bowman
SSgt. Thomas N. Fink
Sgt. Benjamin Fitzpatrick
SSgt. Edward J. Goodman
SSgt. Russell M. Jenkins
SSgt. Walter J. Kruzich
SSgt. Burton D. Lane
SSgt. John Truscott
SSgt. Robert C. Waddell
SSgt. Henry J. Weigel
Sgt. Nathaniel L. Breunig
Sgt. Robert M. Dowell
Sgt. Paul Dukas
Sgt. Joseph Ekasla
Sgt. Walter C. Engel
Sgt. Benjamin Fitzpatrick
Sgt. Norman B. Jacobsen
Sgt. Lane E. McPhee
Sgt. George B. Palmer
Sgt. Cecil Patton
Sgt. Bernard F. Seufort
Sgt. Eugene S. Teter, Jr.
Cpl. Albert N. Alexander
Cpl. Leonard Aronson
Cpl. Albert P. Boyle
Cpl. Vincent M. Burke
Cpl. Edward A. Butler
Cpl. Robert L. Cohen
Cpl. Carl R. Dolmotsch, Jr.
Cpl. Edward A. Gardiner
Cpl. Samuel Katz
Cpl. George H. LeRoy
Cpl. Harold R. Lewis
Cpl. Balke L. Manler
Cpl. Robert L. McCoy
Cpl. Ralph E. Reeder
Cpl. Harold D. Stroud
Cpl. Thomas J. Towle
Pfc. Lewis A. Noce
Pfc. Jerome S. Rose
Pvt. Marion E. Carter
Pvt. Theodore Wladyka

Crew #1 – Aircraft #42-7591
| | | |
|---|---|---|
| 2nd Lt. Jack W. O'Brien | P | 0741929 |
| 2nd Lt. Raymond L. Boll | CP | 0806394 |
| 2nd Lt. Seymour D. Ausfresser | N | 0669946 |
| 2nd Lt. Arthur D. Steele | B | 0678486 |
| SSgt. Joseph J. Buschek | E | 19056598 |
| SSgt. Jerome R. Haas | R | 36261526 |
| Sgt. Charles L. Hutton | AE | 35661741 |
| Sgt. Walter D. Garland | AR | 36556774 |
| Sgt. Jay R. Dempsey | G | 13145844 |
| SSgt. Thomas W. Abbott | AG | 33162501 |

Passengers
Captain William G. Blum
Captain Harold S. Podolsky
MSgt. Walter G. Brown
SSgt. Edward J.G. Brown, Jr.

Crew #2 – Aircraft #41-29191
| | | |
|---|---|---|
| 2nd Lt. Jack Parker | P | 0742771 |
| 2nd Lt. John P. Shaw | CP | 0805990 |
| 2nd Lt. Joseph T. Myer | N | 0738713 |
| 2nd Lt. Dominic W. Maineri | B | 0684177 |
| Sgt. James V. Nobe | E | 37218528 |
| SSgt. Albert B. Foreman | R | 13176763 |
| Sgt. Kenneth W. Ebaugh | AE | 13136468 |
| Sgt. Robert S. Sale | AR | 38283629 |
| Sgt. James J. Plazio | G | 13012943 |
| Sgt. William V. Biles | AG | 17129606 |

Passengers
| | |
|---|---|
| SSgt. Burton D. Dane | 37195327 |
| SSgt. Theodore Wladyka | 6994020 |
| Sgt. Robert L. Cohen | 31308972 |
| Sgt. Thomas J. Towle | 17142410 |

Crew #3 – Aircraft #42-52083
| | | |
|---|---|---|
| 2nd Lt. William B. Brown | P | 0742841 |
| 2nd Lt. Kenneth W. Barnet | CP | 0808934 |
| 2nd Lt. Frederick W. Saltus | N | 0736658 |
| 2nd Lt. William L. LaBonte | B | 0684174 |
| Pvt. Earl R. Myrick | E | 31152194 |
| SSgt. Oliver L. Bidne | R | 39452901 |
| Sgt. Clarence W. Schrader | AE | 36069204 |
| Sgt. Jacob M. Lebovitz | AR | 11072772 |
| Sgt. Herman B. Johns | G | 35588210 |
| Sgt. Russell E. Flamion | AG | 35715492 |

Passengers
| | |
|---|---|
| 1st Lt. Robert Lewis | 01702816 |
| 1st Lt. William C. Cates, Jr. | 0649608 |
| Major Patrick H. Hoey | 0405121 |

Crew #4 – Aircraft #42-7767
| | | |
|---|---|---|
| 2nd Lt. Leroy E Middleworth, Jr. | P | 0797161 |
| 2nd Lt. Thomas E. Winslett | CP | 0751783 |
| 2nd Lt. William V. Voorhees, Jr. | N | 0750230 |
| 2nd Lt. John D. McGarry | B | 0738564 |
| TSgt. Clarence L. Campbell | E | 16035515 |
| Sgt. James J. Gregan | R | 33193048 |
| Sgt. Walter K. Bickle | AE | 35611165 |
| SSgt. Paul Krasney | AR | 36324023 |
| Pvt. George Henderson | G | 7005828 |
| Sgt. Harold R. Mattice | AG | 12138414 |

Passengers
| | |
|---|---|
| Sgt. Joseph Ekasela | 31082697 |
| Pvt. Martin E. Carter | 37375810 |
| SSgt. Ellis L. Copeland | 34585717 |
| SSgt. John R. Truscott | 17030411 |

Crew #5 – Aircraft #42-7733
| | | |
|---|---|---|
| 2nd Lt. Robert C. Ayrest | P | 0680984 |
| 2nd Lt. Irwin Litman | CP | 0797567 |
| 2nd Lt. Robert F. Boberg | N | 0687980 |
| 2nd Lt. Robert V. Hess | B | 0688162 |
| SSgt. Frank E. Boula | E | 36619461 |
| Sgt. Joseph R. Chioda | R | 11103704 |
| Sgt. Edward N. Schroder | AE | 36285186 |
| Cpl. Harold L. Auker | AR | 37115617 |
| Sgt. Leonard J. Snell | G | 16021987 |
| Sgt. Edward A. Odiorne | AG | 35582869 |

Passengers
| | |
|---|---|
| SSgt. Henry J. Weigel | 39677808 |
| SSgt. Walter J. Kruzich | 36304202 |
| TSgt. Harry C. Corbin | 32260926 |
| TSgt. Frank M. Schultz | R-18697 |

Crew #7 – Aircraft # 42-52908
| | | |
|---|---|---|
| 2nd Lt. Charles Knorr | P | 0745700 |
| 2nd Lt. Herbert J. Bunde | CP | 0751985 |
| 2nd Lt. Stanley Baranofsky | N | 0687922 |
| 2nd Lt. Charles C. McBride | B | 0741207 |
| Sgt. William L. Quigley | E | 12159196 |
| Sgt. Ralph S. Callahan | R | 12207911 |
| Sgt. Ernest J. Schultz | AE | 32405166 |
| Sgt. Stanley J. Sarna | AR | 16147485 |
| Sgt. Jack L. Cooper | G | 35595779 |
| Sgt. Albert C. Padilla | AG | 39117760 |

Passengers
| | |
|---|---|
| 1st Lt. Robert W. McDonough | 02043766 |
| 2nd Lt. Robert L. Harper | 0667402 |
| MSgt. Thomas J. Flynn | 38047787 |
| Sgt. Balke L. Mahler | 18118003 |

Crew #8 – Aircraft #41-29232
| | | |
|---|---|---|
| 2nd Lt. Paul R. Harrison | P | 0731978 |
| 2nd Lt. James E. Berry | CP | 0687977 |
| 2nd Lt. Floyd L. Drake | N | 0750354 |
| 2nd Lt. Frank H. Dial | B | 0751816 |
| Sgt. George H. Chapman, Jr. | E | 32539122 |
| Sgt. Edward J. Lies | R | 33308033 |

| | | |
|---|---|---|
| Sgt. Howard I. Patchell | AE | 31228127 |
| Sgt. Howard M. Donley | AR | 17077488 |
| Sgt. Robert J. McCormick | G | 36421148 |
| Sgt. Donald F. Ransom | AG | 12194131 |
| Passengers | | |
| 1st Lt. George O. Capp | | 0571189 |
| MSgt. Frank J. Naglak | | 37285267 |
| Cpl. Jerome Fisher | | 34585717 |
| Cpl Albert P. Royle | | 11101361 |

Crew #9 – Aircraft #42-7681

| | | |
|---|---|---|
| 2nd Lt. John P. Rhodes | P | 0745180 |
| 2nd Lt. Eber D. O'Faris | CP | 0751289 |
| 2nd Lt. Kenneth J. Murphy | N | 0687958 |
| 2nd Lt. Salem S. Smith, Jr. | B | 0688978 |
| Sgt. Rector R. Cockings | E | 18053195 |
| Pvt. Dominic Rosas | R | 38367454 |
| Sgt. William E. Leis | AE | 18192905 |
| Sgt. Charles A. Heaton, Jr. | AR | 18226363 |
| Sgt. William H. Barbarito, Jr. | G | 31280111 |
| Pvt. Samuel Edwards | AG | 32738235 |
| Passengers | | |
| 2nd Lt. John Bilyk | | 0873020 |
| 2nd Lt. Howard M. Oden | | 0735429 |
| SSgt. Thomas M. Fink | | 32413485 |
| Pvt. Jerome S. Rose | | 35055269 |

Crew #10 – Aircraft #42-52128

| | | |
|---|---|---|
| 2nd Lt. Carroll C. Key | P | 0749479 |
| 2nd Lt. James M. Susoeff | CP | 0752026 |
| 2nd Lt. John W. Brown | N | 0809315 |
| 2nd Lt. Harry Fisher | B | 0688181 |
| SSgt. Clyde L. Baird | E | 14131280 |
| SSgt. Robert W. McKinney | R | 39454113 |
| Pvt. James E. Anderson | AE | 35639566 |
| SSgt. Jack C. Williamson | AR | 34474011 |
| Sgt. Richard L. Auer | G | 31278064 |
| Sgt. Douglas B. Dann | AG | 34599839 |
| Passengers | | |
| Sgt. Allen L. Bowman | | 34289531 |
| Cpl. Ralph F. Reeder | | 36445953 |
| Cpl. Harold D. Stroud | | 37506477 |
| Cpl. Robert L. McCoy | | 39555504 |

Crew #11 – Aircraft #42-52132

| | | |
|---|---|---|
| 2nd Lt. Paul R. Helander | P | 0804331 |
| 2nd Lt. John J. Schneider | CP | 0652243 |
| 2nd Lt. Alfred E. Cannon | N | 0809318 |
| 2nd Lt. Henry M. Snyder | B | 0752588 |
| Sgt. Gerald E. Carroll | E | 19055027 |
| Sgt. Oclotan U. Richmond | R | 39263863 |
| Sgt. Norbert F. Duginski | AE | 16008632 |
| Sgt. Raymond G. Giwojna | AR | 16133962 |
| Sgt. Bill J. McCullah | G | 37413477 |

| | | |
|---|---|---|
| Sgt. Benjamin Z. Means | AG | 33424645 |
| Passengers | | |
| Sgt. Bernard F. Fitzpatrick | | 32260948 |
| Sgt. Norman R. Jacobson | | 12172778 |
| Sgt. George R. Palmer | | 32707922 |
| Sgt. Albert N. Alexander | | 33999071 |

Crew #12 – Aircraft #41-28593

| | | |
|---|---|---|
| 2nd Lt. Donald C.G. Schumann | P | 0672220 |
| 2nd Lt. Harold E. White | CP | 0747535 |
| 2nd Lt. Raymond L. Thurber | N | 0689693 |
| 2nd Lt. Jim Biggerstaff | B | 0676535 |
| Sgt. Harvery E. Smith | E | 34516429 |
| Sgt. John M. Milton | R | 36281759 |
| SSgt. Conrad Holzgraf | AE | 18015585 |
| SSgt. James V. Newton | AR | 16034261 |
| Sgt. Millard R. Malwitz | G | 19175397 |
| Sgt. Isaac H. Odell | AG | 39278598 |
| Passengers | | |
| SSgt. Delmer A. Akerley | | 31300616 |
| Sgt. Cecil Patton | | 34433802 |
| Sgt. Carl R. Dolmetsch, Jr. | | 35654949 |
| Sgt. Paul L. Doukas | | 37448346 |

Crew #14 – Aircraft #42-64451

| | | |
|---|---|---|
| 2nd Lt. Irvin E. Toler | P | 0746484 |
| 2nd Lt. Edward C. O'Hare | CP | 0681173 |
| 2nd Lt. John E. Silvia | N | 0689674 |
| 2nd Lt. Fred J. Bittner | B | 0688503 |
| Sgt. Thurston E. Johnson | E | 13135079 |
| Sgt. Herbert C. Bloom | R | 39101187 |
| Sgt. Lyle E. Steinberg | AE | 32579409 |
| Sgt. Ray L. Jeffers | AR | 32369841 |
| Sgt. David A. Gustafson | G | 16088710 |
| Sgt. Paul E. Crewe | AG | 35679564 |
| Passengers | | |
| TSgt. Frederick J. Fenneweld | | 37184855 |
| Sgt. Bernard F. Soufert | | 13104652 |
| Sgt. Vincent M. Burke | | 32882318 |
| Sgt. Edward A. Butler | | 31138999 |

Crew #15 – Aircraft #41-28611

| | | |
|---|---|---|
| 2nd Lt. Robert C. Voight | P | 0737678 |
| 2nd Lt. William C. Edwards | CP | 0811041 |
| 2nd Lt. Everett F. Stine | N | 0689683 |
| 2nd Lt. Robert T. Ash | B | 0688492 |
| SSgt. Richard Getz | E | 12034631 |
| Sgt. Robert W. Rigg | R | 39549975 |
| Sgt. Clarence R. Marshall | AE | 33231924 |
| Sgt. Paul E. Brown | AR | 33380722 |
| Sgt. Miltiades C. Cikes | G | 31258989 |
| SSgt. Thomas R. Hogan | AG | 3116127 |
| Passengers | | |
| Sgt. Walter C. Engel | | 37292097 |

| | | |
|---|---|---|
| Cpl. Edward R. Gerliner | | 33301403 |
| Cpl. Samuel Katz | | 33311023 |
| Sgt. George H. Leroy | | 37609476 |

Crew #16 – Aircraft #41-29208

| | | | |
|---|---|---|---|
| 2nd Lt. Earle P. Durley, Jr. | P | 0447213 |
| 2nd Lt. Robert B. Haloran | CP | 0808990 |
| 2nd Lt. Naseeb S. Tweel | N | 0689696 |
| 2nd Lt. Lester Bise | B | 0688502 |
| SSgt. Clarence H. Stark | E | 16131107 |
| Sgt. Eamond A. Rock | R | 32635184 |
| Sgt. John Stemmerman | AE | 12155113 |
| Sgt. William D. Hackney | AR | 36539204 |
| Sgt. Joe P. Ford | G | 39407846 |
| Sgt. William C. Walker | AG | 18200358 |

Passengers

| | |
|---|---|
| Major Robert Campbell, Jr. | 0406707 |
| 1st Lt. Earl H. Parks | 0725909 |
| 1st Lt. Clifford C. Gaither | 0857093 |
| MSgt. Michael P. Corce | 6949840 |

Crew #17 – Aircraft #42-52145

| | | | |
|---|---|---|---|
| 1st Lt. Robert K. Winn | P | 0724661 |
| 2nd Lt. James H. Harmon | CP | 0811199 |
| 2nd Lt. David T. Tobin | N | 0690738 |
| 2nd Lt. Richard J. Brady | B | 0688504 |
| SSgt. Charlie D. Lugosh | E | 18201733 |
| Sgt. Kenneth L. Dyer | R | 35568310 |
| Sgt. Edwin H. Pixley | AE | 17107052 |
| Sgt. Billy J. Espich | AR | 15354024 |
| Sgt. Robert R. Cook | G | 15230812 |
| Sgt. Jack W. Porter | AG | 36181194 |

Passengers

| | |
|---|---|
| SSgt. Nathaniel L. Bruning | 35614932 |
| Sgt. Robert M. Dowell | 19114490 |
| Sgt. Eugene S. Teter, Jr. | 13116301 |
| Cpl. Merle E. Davis | 39536529 |

Crew #18 – Aircraft #42-52118

| | | | |
|---|---|---|---|
| 1st Lt. Alan J. Teague | P | 0661892 |
| 2nd Lt. Jesse M. Hamby | CP | 0811196 |
| 2nd Lt. Bruce A. Vaughn | N | 0690742 |
| 2nd Lt. Roy E. Anderson | B | 0688300 |
| SSgt. Edmund J. Rudnicki | E | 35300453 |
| Sgt. Simon Cohen | R | 11130339 |
| Sgt. Joseph M. Redditt | AE | 34427597 |
| Sgt. Kazmierz Pochopin | AR | 32591233 |
| Sgt. John A. Duka | G | 31277387 |
| Sgt. Harvey R. Davis | AG | 33340771 |

Passengers

| | |
|---|---|
| SSgt. John J. Bernard | 35474117 |
| SSgt. Russell A. Jenkins | 17074757 |
| SSgt. Robert G. Waddell | 14121340 |
| SSgt. Lane E. McPhee | 17027206 |

Crew #21 – Aircraft #41-28595

| | | | |
|---|---|---|---|
| 2nd Lt. James J. Bell | P | 0742831 |
| 2nd Lt. David E. Mellott | CP | 0749286 |
| 2nd Lt. Hugh X. Cullinan | N | 0736688 |
| 2nd Lt. Marvin Joseph | B | 0678418 |
| SSgt. James R. Bricker | E | 35588078 |
| SSgt. William E. Ruck, Jr. | R | 13019337 |
| Sgt. Kenneth L. Hess | AE | 36068383 |
| Sgt. Lloyd T. Williams | AR | 19122941 |
| SSgt. Roger O. Vance | G | 39530101 |
| Sgt. Daner E. Anderson | AG | 13033362 |

Passengers

| | |
|---|---|
| 1st Lt. James L. Ferguson | 0568113 |
| 2nd Lt. George E. Cohen | 0861048 |
| 1st Lt. Raymond S. McKeeby | 0500564 |
| MSgt. Robert M. McDowell, Jr. | 14037254 |

Crew #22 – Aircraft #41-28590

| | | | |
|---|---|---|---|
| 2nd Lt. Robert E. Kraus | P | 0732021 |
| 2nd Lt. Clive O. Stevens | CP | 0809059 |
| 2nd Lt. Mathew L. Szydlowski | N | 0690736 |
| 2nd Lt. Harry A. Kohn | B | 0684171 |
| SSgt. Azizes F. Erban | E | 11057792 |
| SSgt. William H. Pehle | R | 37308259 |
| SSgt. Earl Ellis, Jr. | AE | 15335103 |
| SSgt. Junior W. Klug | AR | 13167639 |
| SSgt. Arnold H. Radde | G | 16085975 |
| SSgt. Leon Hawkersmith | AG | 34149089 |

Passengers

| | |
|---|---|
| SSgt. Richard G. Pokorny | 12095697 |
| TSgt. James F. Doherty | 32290419 |
| SSgt. Max M. Myers | 6919517 |

Crew #23 – Aircraft #41-29192

| | | | |
|---|---|---|---|
| 2nd Lt. James P. Sullivan | P | 0675950 |
| 2nd Lt. Evans J. Evans | CP | 0752188 |
| 2nd Lt. Kenneth O. Reed | N | 0416197 |
| 2nd Lt. Leonard B. Harmon | B | 0676384 |
| SSgt. Lawrence H. Vogtmann | E | 37216961 |
| TSgt. William C. Maxwell | R | 33249071 |
| SSgt. William G. Senville | AE | 31160560 |
| SSgt. Clifford W. Harris | AR | 35662194 |
| SSgt. Michael J. Fuller | G | 11038363 |
| SSgt. George A. Herpoulos | AG | 15377097 |

Passengers

| | |
|---|---|
| TSgt. Alva B. Hampton | 37254291 |
| Sgt. Charles E. Gayle | 39404396 |
| Cpl. Vincent E. Lang | 32478867 |
| Cpl. James J. Kveten | 36647841 |

Crew #24 – Aircraft #42-72981

| | | | |
|---|---|---|---|
| 2nd Lt. Thomas R. Apple | P | 0680377 |
| 2nd Lt. Richard L. Henderson | CP | 0751323 |
| 2nd Lt. Bruce B. Winter | N | 0705260 |

| | | | | | |
|---|---|---|---|---|---|
| 2nd Lt. Reese C. Lee | B | 0671260 | Sgt. Van B. Scott | AR | 18109376 |
| TSgt. James A. Pegher | E | 33268724 | Sgt. Bertrand B. Lutz | G | 32605895 |
| SSgt. John F. Decker | R | 13113895 | SSgt. Roy E. Maker | AG | 11116363 |
| TSgt. Furman A. Powell | AE | 34381553 | Passengers | | |
| SSgt. George H. Jepson | AR | 12169791 | SSgt. Veikko Hirviivara | | 37178164 |
| SSgt. Richard L. Maze | G | 6940927 | Sgt. George M. Mathews | | 15089461 |
| Sgt. Roy E. Lewis | AG | 14072402 | TSgt. James F. Ingelsby | | 11037040 |
| Passengers | | | Pvt. Francis G. Koh, Jr. | | 36123974 |
| Major Hubert S. Judy, Jr. | | 0406697 | | | |
| 1st Lt. Arthur S. Hunt | | 0729755 | Crew #28 – Aircraft #42-52123 | | |
| Sgt. Alfred E. Candelaria | | 39258873 | 2nd Lt. James E. Urban | P | 0676878 |
| Sgt. Henry J. Myslinski | | 32422730 | 2nd Lt. Alden P. Anthony | CP | 0808482 |
| | | | 2nd Lt. Fred Brenner | N | 0687928 |
| Crew #25 – Aircraft #42-52120 | | | 2nd Lt. Stanley Friedman | B | 0688184 |
| 2nd Lt. Henry B. Schroeder | P | 0680731 | Sgt. Howard M. Smith | E | 17038532 |
| 2nd Lt. Lewis M. Sarkovich | CP | 0686600 | Sgt. Wendell R. McClellan | R | 38285507 |
| 2nd Lt. Bruce E. Crane | N | 0687815 | SSgt. Roy D. Barber | AE | 14040228 |
| 2nd Lt. Jack R. Smith | B | 0751850 | Sgt. George Petula | AR | 12183208 |
| SSgt. Hugh O. Riley | E | 18052449 | Sgt. Joseph B. Deffner | G | 32675587 |
| Sgt. Moe Liebman | R | 32411484 | Sgt. John J. Kelly | AG | 32782047 |
| Sgt. William E. Seidel | AE | 36058850 | Passengers | | |
| Sgt. Francis W. Scarbrough | AR | 16030978 | Cpl. Clarence E. Gibson | | 16057749 |
| Sgt. Alfred B. Maine | G | 31259024 | TSgt. Curry Dial | | 15041120 |
| Sgt. Robert D. Hiller | AG | 19170207 | Cpl. Harvey D. Casner | | 36581785 |
| Passengers | | | Pvt. Guion J. Allen | | 34609898 |
| Cpl. Robert A. Prouty | | 34248906 | | | |
| TSgt. Ernest H. Lepke | | 39606393 | 713th Bombardment Squadron Staff | | |
| TSgt. Markus A. Schreacke | | 36027064 | | | |
| Sgt. Ernest H. Stark | | 32816868 | 1st Lt. James L. Ferguson | | |
| | | | 1st Lt. Chester B. Hackett, Jr. | | |
| Crew #26 – Aircraft #42-7766 | | | 1st Lt. Raymond S. McKeeby | | |
| 2nd Lt. William M. Martin | P | 0805949 | 1st Lt. William H. Smelter | | |
| 2nd Lt. William A. Bond | CP | 0808938 | 1st Lt. Heber H. Thompson | | |
| 2nd Lt. William S. Cuthbert | N | 0687930 | 2nd Lt. George E. Cone | | |
| 2nd Lt. William E. Wallace | B | 0741384 | 2nd Lt. Arthur Klein | | |
| Sgt. William E. Sherratt | E | 18046310 | 2nd Lt. Francis L. Martin | | |
| Sgt. Peter Edgar | R | 12187861 | 2nd Lt. Bruce B. McCleary | | |
| SSgt. Charles F. Voge | AE | 18046310 | MSgt. Francis Berrigan | | |
| Sgt. Anthony J. Kuzminski | AR | 11101473 | MSgt. Willie G. Brantley | | |
| Sgt. David D. Culp | G | 35538126 | MSgt. Anthony P. Dennis | | |
| Sgt. Ike J. Beasley | AG | 14058704 | MSgt. Robert M. McDowell | | |
| Passengers | | | MSgt. Frank A. Miller | | |
| SSgt. John R. Moore | | 35385158 | MSgt. Seamon A. Pledger | | |
| TSgt. Vernon L. Poppe | | 36175912 | TSgt. Curry Dial | | |
| Pvt. Alvin M. Jampol | | 32883696 | TSgt. James F. Doherty | | |
| | | | TSgt. Ralph C. Flees | | |
| Crew #27 – Aircraft #42-52121 | | | TSgt. Alva B. Hampton | | |
| 2nd Lt. Jack L. Barak | P | 0806776 | TSgt. James C. Ingelsby | | |
| 2nd Lt. Anthony Witzkowski | CP | 0686297 | TSgt. Ernest A. Lapko | | |
| 2nd Lt. Thaine A. Clark | N | 0687297 | TSgt. Hunter W. Martin | | |
| 2nd Lt. Martin F. Survoy | B | 0751854 | TSgt. Vernon L. Poppo | | |
| Sgt. Jesse R. Kain | E | 15337846 | TSgt. Markus A. Schreacke | | |
| Sgt. Joseph S. Kasacjak | R | 33417119 | SSgt. Earl A. Eggleston | | |
| Sgt. Floyd E. Sand | AE | 35347119 | SSgt. John L. Fluke | | |

SSgt. James T. Gwaltney
SSgt. Herbert C. Heidrich
SSgt. Veikko Hirvivara
SSgt. Harley A. Kelly
SSgt. S. J. Laney
SSgt. John R. Moore
SSgt. Merle S. Morris
SSgt. Joseph N. Musuraca
SSgt. Richard G. Pokorny
Sgt. Dale L. Black
Sgt. Glade H. Butterfield, Jr.
Sgt. Charles E. Coyle
Sgt. George M. Mathews
Sgt. Obest B. Rood
Sgt. Thomas C. Schnoor
Sgt. Ernest H. Stark

Crew #29 – Aircraft # 41-28583
| | | |
|---|---|---|
| 1st Lt. Robert T. Lambertson | P | 0732023 |
| 2nd Lt. James L. Thomas | CP | 0752088 |
| 2nd Lt. Ralph E. Brown | N | 0809316 |
| 2nd Lt. Robert D. Larew | B | 0688672 |
| SSgt. Perry L. Davenport | E | 39091485 |
| Sgt. William Sidoruk | R | 32623822 |
| Sgt. Ray K. Littlejohn | AE | 34169612 |
| Sgt. Walter D. Petts | AR | 15382788 |
| Sgt. Andrew P. Long | G | 32186301 |
| Sgt. Edward D. Roser | AG | 32678470 |

Passengers
| | |
|---|---|
| Captain Heber H. Thompson | 01699234 |
| 1st Lt. William D. Smelter | 02043773 |
| MSgt. Willie G. Brantley | 15098971 |
| Sgt. John R. Ray | 18168514 |

Crew #30 – Aircraft # 42-52097
| | | |
|---|---|---|
| 2nd Lt. Richard C. Harris | P | 0494504 |
| 2nd Lt. William C. Moore | CP | 0808576 |
| 2nd Lt. Warren R. Auch | N | 0687803 |
| 1st Lt. Frank S. Phillips | B | 02043738 |
| SSgt. Robert E. Whiteside | E | 15070206 |
| Sgt. Paul T. Dempsey | R | 12135037 |
| SSgt. Alfred F. Massey | AE | 31124673 |
| SSgt. Grady W. McLaughlen | AR | 19175031 |
| SSgt. Harry T. Rummel | G | 37272148 |
| SSgt. George J. Schibler | AG | 13155700 |

Passengers
| | |
|---|---|
| 2nd Lt. Bruce B. McCleary | 0799815 |
| MSgt. Francis Berrigan | 38089023 |
| SSgt. Harley A. Kelley | 19138718 |
| Sgt. Mervin Koffman | 31161869 |

Crew #31 – Aircraft #41-29236
| | | |
|---|---|---|
| 2nd Lt. Stewart F. Chase | P | 0746293 |
| 2nd Lt. Lorenz G. Johnson | CP | 0752416 |
| 2nd Lt. Philip Baskin | N | 0687806 |
| 2nd Lt. John A. Bienapfl | B | 0744233 |
| SSgt. Luther D. Rummage, Jr. | E | 14196490 |
| Sgt. Thomas R. Lee | R | 18171694 |
| Sgt. Demo D. Saranti | AE | 12140760 |
| Sgt. Raymond W. Wood | AR | 19179637 |
| SSgt. Harry D. Quillen | G | 12012673 |
| Sgt. Harry W. Klober | AG | 33035628 |

Passengers
| | |
|---|---|
| TSgt. Hunter W. Martin | 33221158 |
| TSgt. S. J. Laney | 38120892 |
| SSgt. John L. Fluke | 33252139 |
| Pvt. Elmer C. Madsen | 39695963 |

Crew #32 – Aircraft # 41-29248
| | | |
|---|---|---|
| 2nd Lt. George W. Elkins | P | 0804312 |
| 2nd Lt. Eugene J. Coffey | CP | 0748576 |
| 2nd Lt. Edward T. Card | N | 0755172 |
| 2nd Lt. Richard P. Casterline | B | 0688174 |
| Sgt. Morris F. Cooper | E | 34570736 |
| Sgt. Thomas V. Tornillo | R | 32562276 |
| Sgt. George H. Finch | AE | 19171540 |
| Sgt. Oren B. Casto | AR | 15338524 |
| Sgt. Mattie A. Laurelli | G | 32836621 |
| Sgt. Walter Mishanic | AG | 32830147 |

Passengers
| | |
|---|---|
| SSgt. James T. Gwaltney | 33158448 |
| SSgt. Herbert C. Heidrich | 37264787 |
| Cpl. William P. Miller | 16121346 |
| Cpl. Joseph S. Blouin | 39553152 |

Crew # 35 – Aircraft #42-52135
| | | |
|---|---|---|
| 2nd Lt. Lawerence T. Crepeau | P | 0747219 |
| 2nd Lt. Robert E. Lehman | CP | 0750630 |
| 2nd Lt. William F. New | N | 0690491 |
| 2nd Lt. Otto Ciavardoni | B | 0688580 |
| Sgt. Dearl Whittaker | E | 15117053 |
| Sgt. Joseph R. Morrision | R | 36629920 |
| Sgt. Johnny W. Jones | AE | 14149210 |
| Sgt. Bashem B. Weide | AR | 37229920 |
| Sgt. Raymond M. Arnold | G | 39550987 |
| Sgt. Jesse W. Carroll | AG | 35369438 |

Passengers
| | |
|---|---|
| SSgt. Merle S. Morris | 17026290 |
| SSgt. Charles W. Kruse | 37446869 |
| SSgt. Dale L. Black | 14134595 |
| Sgt. Glade H. Butterfield, Jr. | 17071649 |

Crew # 36 – Aircraft # 41-28602
| | | |
|---|---|---|
| 1st Lt. Max R. Jordon | P | 0725740 |
| 2nd Lt. Adolph T. Abraham | CP | 0810982 |
| 2nd Lt. Robert B. Stone | N | 0689684 |
| 2nd Lt. Searl J. Collins, Jr. | B | 0749611 |
| SSgt. John J. Holesa | E | 15072502 |

| | | |
|---|---|---|
| SSgt. Jacob E. Edwards | R | 18000037 |
| Sgt. Anthony J. Daidone | AE | 32306952 |
| Sgt. Walter S. Baron | AR | 31133831 |
| Sgt. Archie B. Indorf | G | 35595495 |
| Sgt. Roy E. Ball | AG | 35595557 |
| Passengers | | |
| Pvt. Glenn Horner | | 38452559 |
| Sgt. Thomas C. Schnoor | | 39082176 |
| MSgt. Anthony P. Dennis | | 6788860 |
| SSgt. Joseph N. Musuraca | | 35340477 |

Crew #37 – Aircraft #41-29240

| | | |
|---|---|---|
| 1st Lt. Philip B. Thompson | P | 0725377 |
| 2nd Lt. Robert C. Beirke | CP | 0811329 |
| 2nd Lt. Bernard C. McGunn | N | 0690467 |
| 2nd Lt. Lloyd C. Drury | B | 0747653 |
| Sgt. Arnie R. Gunderson | E | 17154143 |
| Cpl. Isidore Pertzman | R | 11057888 |
| Sgt. Robert H. Hale | AE | 17159570 |
| Sgt. Joseph T. Michalczyk | AR | 31284067 |
| Cpl. Jerry Mejeur, Jr. | G | 16014523 |
| SSgt. John P. Moran | AG | 31097317 |
| Passengers | | |
| Sgt. Obest B. Reed | | 18043424 |
| MSgt. Frank A. Miller | | 37113148 |
| TSgt. Ralph C. Flees | | 16046785 |
| Cpl. Frank A. Carpenter | | 38379028 |

Crew # 38 – Aircraft # 42-2712

| | | |
|---|---|---|
| 1st Lt. James E. Curtis | P | 0725688 |
| 2nd Lt. Donald Clift | CP | 0811338 |
| 2nd Lt. Emmett J. Moore | N | 0690683 |
| 2nd Lt. Leonard H. Feingold | B | 0688744 |
| Sgt. Marion C. Maxton | E | 37205700 |
| Pvt. Eugene E. Higgins | R | 36273563 |
| Sgt. Henry I. Gouthier | AE | 31147965 |
| Sgt. John E. McClone | AR | 32202248 |
| Sgt. Malcolm W. Crow | G | 38446782 |
| SSgt. Robert P. Hudson, Jr. | AG | 32176645 |
| Passengers | | |
| Captain Chester P. Hackett, Jr. | | 0406683 |
| 1st Lt. Arthur Klein | | 0796549 |
| 1st Lt. Francis L. Martin | | 0857096 |
| MSgt. Seamon A. Pledger | | 14057081 |

Crew #41 – Aircraft # 41-28588

| | | |
|---|---|---|
| 2nd Lt. Edward D. Hughey, Jr. | P | 0677748 |
| 1st Lt. Charles D. Hill | CP | 0397552 |
| 2nd Lt. Sumner W. Stevens | N | 0670285 |
| 2nd Lt. Edwin S. Jacobson | B | 0684114 |
| SSgt. Curtis L. Koon | E | 33252697 |
| Sgt. Carl Anderson | R | 19185853 |
| Sgt. James C. Hicks | AE | 17129097 |
| Sgt. Homer C. Risner | AR | 38333496 |
| Sgt. John L. Reim | G | 13110217 |
| Sgt. Richard B. Laing | AG | 19164878 |
| Passengers | | |
| SSgt. William P. O'Reilly | | 32139452 |
| Cpl. Edgar S. West, Jr. | | 18124229 |
| Cpl. Robert G. Swin | | 38504881 |
| Pvt. Albert L. Dickerson | | 35867339 |

Crew #42 – Aircraft # 42-52108

| | | |
|---|---|---|
| 2nd Lt. Joseph W. Shank, Jr. | P | 0743108 |
| 1st Lt. Edward J. Pearce | CP | 0393388 |
| 2nd Lt. Clark S. Bender | N | 0736682 |
| 2nd Lt. Turner A. Sowell, Jr. | B | 0747607 |
| SSgt. Leonard G. LaJoie | E | 16023435 |
| SSgt. Eugene C. Weishaar | R | 36704194 |
| Sgt. Parke J. Bair | AE | 13095286 |
| Sgt. Raymond E. Martin | AR | 39453037 |
| Sgt. William D. Rabor | G | 39118166 |
| Cpl. Calvin H. End | AG | 39453465 |
| Passengers | | |
| 2nd Lt. John H. Olhaber | | 0873363 |
| SSgt. Ernest E. Jones | | 13062713 |
| Sgt. Ned R. Hugley | | 14152714 |
| Sgt. James E. Taylor | | 36446039 |

Crew #43 – Aircraft #41-29230

| | | |
|---|---|---|
| 2nd Lt. Lawrence M. Edman | P | 0740775 |
| 2nd Lt. Lawson D. Campbell | CP | 0687543 |
| 2nd Lt. Joseph C. Chaffin | N | 0797265 |
| 2nd Lt. Philip Mufsen | B | 0684181 |
| SSgt. James W. Dailey, Jr. | E | 38236448 |
| SSgt. Jack A. Ghormley | R | 32626123 |
| Sgt. John Wallends, Jr. | AE | 35514944 |
| Sgt. Harold W. Beaver | AR | 39269212 |
| Sgt. Tony Gomondo | G | 15377489 |
| Sgt. Harold L. Delay | AG | 19188393 |
| Passengers | | |
| 2nd Lt. Lester E. Davison | | 0739076 |
| TSgt. Edward G. Finnegan | | 32385610 |
| TSgt. Ralph Schwartzkofp | | 19094390 |
| Sgt. Frank Macefe | | 33300883 |

Crew #44 – Aircraft # 42-7606

| | | |
|---|---|---|
| 2nd Lt. Robert W. Carroll | P | 0680173 |
| 1st Lt. Wirt D. Walker | CP | 0661573 |
| 2nd Lt. Castleton D. Smith | N | 0750225 |
| 2nd Lt. John E. Hennessy | B | 0673163 |
| SSgt. Earl F. Brown | E | 35368548 |
| SSgt. Milfred K. Hathaway, Jr. | R | 31157119 |
| Cpl. Albert E. Childs | AE | 34266654 |
| Sgt. Randall C. Laing | AR | 36365095 |
| Cpl. Robert F. Miltner | G | 39827755 |
| SSgt. Melvin F. Schiefelbein | AG | 35307779 |
| Passengers | | |

*Appendices*

| | | |
|---|---|---|
| MSgt. Julius E. Ryan, Jr. | | 18030400 |
| SSgt. Clayton R. Harvey | | 37276174 |
| Sgt. Walter D. Trinder | | 32383444 |
| Cpl. Louis J. Toia | | 12136206 |

Crew #45 – Aircraft #41-28599

| | | |
|---|---|---|
| 1st Lt. James D. Conrad | P | 0365492 |
| 2nd Lt. Joseph R. Gonzales, Jr. | CP | 0752196 |
| 2nd Lt. Royal D. Goldenberg | N | 0687939 |
| 2nd Lt. Benny E. Roark | B | 0751844 |
| SSgt. Paul F. Bland | E | 35370902 |
| Cpl. Julius S. Kopes | R | 35746618 |
| Sgt. Bernard W. Anderson | AE | 16069755 |
| Pvt. Russel E. Towsley | AR | 35377602 |
| Sgt. Byron V. Bacon | G | 17035965 |
| Sgt. Jasper W. Early | AG | 13075590 |

Passengers

| | | |
|---|---|---|
| 2nd Lt. Francis C. Doherty | | 0861158 |
| SSgt. Lawrence E. Voorheis | | 39608515 |
| Cpl. Francis C. Schade | | 17035965 |
| Cpl. Lawrence J. Jablonski | | 13075590 |

Crew #46 – Aircraft #41-29231

| | | |
|---|---|---|
| 2nd Lt. Donald R. Coleman | P | 0805892 |
| 2nd Lt. Edward P. Meents, Jr. | CP | 0750812 |
| 2nd Lt. Albert DiLorenzo | N | 0687932 |
| 2nd Lt. Donald H. James | B | 0752571 |
| SSgt. Stanley Malamut | E | 13154151 |
| Sgt. Royal V. Donihee | R | 38426739 |
| Sgt. George J. Robichau | AE | 11116192 |
| Sgt. Willis Mills | AR | 37419291 |
| Sgt. Ralph Meigs | G | 14135765 |
| Sgt. Bordie S. Haynes | AG | 34490741 |

Passengers

| | | |
|---|---|---|
| 2nd Lt. Marvin C. Onks | | 0799842 |
| TSgt. Henry A. Anderson | | 37164217 |
| SSgt. Roy N. Stroop | | 3312689 |
| Sgt. Mitchell J. Biernat | | 36310611 |

714th Bombardment Squadron Staff

Captain Glassell S. Stringfellow
1st Lt. Lester F. Miller
1st Lt. Thomas J. Maye
1st Lt. Carl T. Yast
2nd Lt. Lester E. Davidson
2nd Lt. Francis C. Doherty
2nd Lt. Robert C. Klein, Jr.
2nd Lt. John E. Olhaber
2nd Lt. Marvin C. Onks
2nd Lt. Harry J. Oppelt
2nd Lt. James E. Smithson
MSgt. Darwin T. Hall
MSgt. Julius E. Ryan, Jr.

TSgt. Harry A. Anderson
TSgt. James E. Cackett
TSgt. Herbert S. Chrzan
TSgt. Manlie A. DePaoli
TSgt. Alfio A. Fontanella
TSgt. Edward G. Finnegan
TSgt. Ralph Swartzkopf
SSgt. Lloyd S. Brewer
SSgt. William M. Camp
SSgt. Ernest E. Jones
SSgt. Robert M. Kellner
SSgt. Michael E. Kruzinski
SSgt. Edgar E. Lyon
SSgt. William P. O'Reilly
SSgt. Roy F. Stroop
Sgt. Mitchell J. Biernat
Sgt. Stephen E. Burzenski
Sgt. Clayton R. Harvey
Sgt. Harry G. Holmberg
Sgt. Will T. Lee
Sgt. Frank Masef
Sgt. John A. Sager
Sgt. Billy C. Smith
Sgt. James E. Taylor
Sgt. William Thorne
Sgt. Walter D. Trinder
Sgt. Lawrence E. Voorheis
Sgt. Ronald E. Weaver
Cpl. Ralph C. Erskine, Jr.
Cpl. Ned R. Hugley
Cpl. Lawrence R. Jablonski
Cpl. John F. Muller, Jr.
Cpl. Anthony Rouscotti
Cpl. Francis C. Schade
Cpl. John R. Shea
Cpl. Joseph Sulin
Cpl. Robert C. Swin
Cpl. Louis J. Tora
Cpl. Edgar S. West, Jr.
Pfc. Walter E. Andress
Pfc. David Frank
Pfc. Dennis W. McLaughlin
Pfc. Milton A. Solomen
Pfc. Emilie E. Tafoya
Pvt. Albert L. Dickerson

Crew #47 – Aircraft #42-7683

| | | |
|---|---|---|
| F/O Karl M. Schlund | P | T-61138 |
| F/O William W. Rogers | CP | T-122253 |
| 2nd Lt. Robert W. Rogers | N | 0687944 |
| 2nd Lt. Raymond L. Cohee | B | 0752551 |
| Sgt. Robert G. Woolweaver | E | 13145931 |
| Cpl. Frank J. Gardner | R | 13048050 |
| Sgt. Carlton McIntosh | AE | 14140523 |

| Sgt. William E. Ervin | AR | 19170231 |
| Sgt. Arthur L. Torness | G | 17036948 |
| Sgt. James E. Jefferson | AG | 33486265 |

Passengers
| Sgt. Harry G. Holmberg | | 36331430 |
| Cpl. Joseph Sulin, Jr. | | 35599259 |
| Cpl. Walter E. Andress | | 32473368 |
| Cpl. Dennis W. McLaughlin | | 38439767 |

Crew #49 – Aircraft #42-7768
| 2nd Lt. Clair W. Cline | P | 0746302 |
| 2nd Lt. Robert O. Brockman | CP | 0687671 |
| 1st Lt. Adin S. Batson | N | 0687924 |
| 2nd Lt. Ted Strain | B | 0686204 |
| Sgt. Lin L. Teing | E | 3660326 |
| Cpl. Maurice M. Taylor | R | 35724736 |
| Sgt. Ira H. Loyd | AE | 36431449 |
| Sgt. Hubert A. Hunt | AR | 38321373 |
| Sgt. Robert H. Yarnell | G | 35668792 |
| Sgt. Albert Giliotti | AG | 32717371 |

Passengers
| 1st Lt. Carl T. Yast | | 0570876 |
| TSgt. Herbert S. Chrzan | | 16027847 |
| SSgt. Lloyd S. Brewer | | 37263834 |
| Cpl. Ralph C. Erskine, Jr. | | 34770116 |

Crew #51 – Aircraft # 42-52115
| 2nd Lt. Robert A. Martin | P | 0732039 |
| 2nd Lt. Joseph B. S. Johnson, Jr. | CP | 0751708 |
| 2nd Lt. Joseph E. Sutphin | N | 0741719 |
| 2nd Lt. John E. Johnston | B | 0752572 |
| SSgt. George C. Hunt | E | 33392182 |
| SSgt. Richard C. Thalhamer | R | 15335752 |
| Sgt. Melvin Porter | AE | 15325318 |
| Sgt. Charlie Flukinger, Jr. | AR | 18191275 |
| Sgt. Robert N. Metcalf | G | 35619527 |
| Sgt. Earl D. Hostetter | AG | 37499896 |

Passengers
| 2nd Lt. Robert C. Klein | | 0864058 |
| MSgt. Wilfred Carroll | | 14040294 |
| SSgt. Robert M. Kellner | | 36262680 |
| Cpl. Milton A. Solomon | | 32573476 |

Crew #52 – Aircraft #42-52105
| 2nd Lt. David E. Manning | P | 0745568 |
| 2nd Lt. Robert F. Palicki | CP | 0808887 |
| 2nd Lt. Jerome Slepin | N | 0689676 |
| 2nd Lt. Arne O. Bergum | B | 0747568 |
| Sgt. David E. Lackney | E | 15082537 |
| Sgt. Philip H. Klum | R | 12186879 |
| Sgt. Irving Mazur | AE | 12036294 |
| Sgt. William S. Pennypacker | AR | 13137693 |
| Sgt. Joseph F. Kasprzak | G | 13155970 |
| Sgt. John B. Nelson | AG | 39275415 |

Passengers
| 2nd Lt. James E. Smithson | | 0861750 |
| MSgt. James D. McIntyre | | 31118322 |
| SSgt. Stephen H. Burzenski | | 32229340 |
| Cpl. David Frank | | 32881627 |

Crew #53 – Aircraft #42-7755
| 2nd Lt. Walter A. Yuengert | P | 0742103 |
| 2nd Lt. Edward A. Fox | CP | 0808980 |
| 2nd Lt. Charles E. Sloan | N | 0689677 |
| 2nd Lt. Rudolf F. Gabrys | B | 0751820 |
| Sgt. Howard L. Wilson | E | 33215816 |
| Sgt. Irving Kolokow | R | 12189812 |
| Sgt. Albert H. Backhus | AE | 38273742 |
| Sgt. Dallas D. Beverlin | AR | 36707999 |
| Sgt. Edward F. Boss | G | 33227643 |
| Sgt. Herman Wittenberg | AG | 13124383 |

Passengers
| SSgt. Edgar E. Lyon | | 16033854 |
| SSgt. Michael E. Gruzinski | | 33300726 |
| SSgt. Will T. Lee | | 35722767 |
| Cpl. John F. Muller, Jr. | | 12218746 |

Crew #54 – Aircraft #41-28609
| 2nd Lt. Abraham J. Kittredge | P | 0745907 |
| 2nd Lt. Charles E. Schmidt | CP | 0750861 |
| 2nd Lt. Harold E. Smith | N | 0620728 |
| 2nd Lt. Raymond Junkin | B | 0743791 |
| Sgt. Harry L. Harris | E | 39304390 |
| Sgt. Kenneth H. DeSoto | R | 19186251 |
| SSgt. Grady V. Howell, Jr. | AE | 34038420 |
| Sgt. Clinton D. Stackhouse II | AR | 18053801 |
| Pvt. Benjamin J. Ochart | G | 32501763 |
| Sgt. James C. Hussong | AG | 33552734 |

Passengers
| TSgt. James H. Cackett | | 6974418 |
| Sgt. Billy C. Smith | | 37244995 |
| Sgt. William Thorne | | 32465234 |
| Cpl. Anthony Ruscetti | | 31180117 |

Crew #55 – Aircraft #42-7722
| 1st Lt. Thomas J. Keene | P | 0429657 |
| 2nd Lt. James R. Bettcher | CP | 0751083 |
| 2nd Lt. Harold W. Smith | N | 0690687 |
| 2nd Lt. Edwin G. Moran | B | 0748015 |
| Sgt. Grover Bingham | E | 17122093 |
| Cpl. William J. Demetropoulus | R | 16156816 |
| Sgt. Brona D. Bottoms | AE | 18187423 |
| Sgt. Charlie L. Blanton | AR | 39394021 |
| Sgt. Frederick L. Krepser | G | 18226472 |
| Sgt. George S. Sansburn | AG | 39552768 |

Passengers
| Captain Glassel S. Stringfellow | | 0416361 |
| 2nd Lt. Harry J. Oppelt | | 0801356 |

MSgt. Darwin T. Hall 7011025
Cpl. John R. Shea 31110439

Crew #56 – Aircraft #42-64447
1st Lt. Myers Wahnee P 0724574
F/O Stuart K. Barr CP T-61270
2nd Lt. Richard M. Hager N 0687941
2nd Lt. Walter S. Maszewski B 0688674
SSgt. Walter Farmer E 18214291
Sgt. Lawrence R. Reep R 14166172
Sgt. Richard H. Elliott AE 18277023
Sgt. William W. Cordray AR 35420351
Sgt. John C. Copolla G 11111234
Sgt. Edward J. McGraw AG 16057297
Passengers
1st Lt. Lester F. Miller 0793358
TSgt. Manlio A. DePaoli 32398673
SSgt. William L. Camp 38317479
Sgt. John A. Sager 36331018

Crew #57 – Aircraft #42-52100
1st Lt. Elmer H. Hammer, Jr. P 0389298
F/O Morgan Goodpasture CP T-6118
2nd Lt. Morris A. Thomson N 0689691
2nd Lt. Roger E. Cuddeback B 0671293
SSgt. Edwin J. Wingfield E 19175590
Sgt. Thaddeus M. Domzalski R 6899699
Sgt. Arthur R. Krueger AE 19124094
Sgt. Francis A. Farris AR 36633396
Sgt. Melvin P. Rosencranz G 11081549
Sgt. Fred G. Rowe AG 16162393
Passengers
1st Lt. Thomas J. Maye 01695844
TSgt. Alfio A. Fontanella 11065954
Sgt. Roland B. Wheeler 11071493
Pvt. Emilie E. Tayofa 37344783

715th Bombardment Squadron Staff

Major Kenneth D. Squyres
1st Lt. Kenneth C. Doty
1st. Lt. Jack P. Edwards
1st Lt. Andrew J. Hau
1st Lt. Joseph E. Kaiser
1st Lt. Robert R. Thornton
2nd Lt. John J. Baldwin
2nd Lt. John A. Black
2nd Lt. Henry B. Gabrielson
2nd Lt. Howard A. Garaas
MSgt. William G. Cooper
MSgt. Marvin V. Druen
MSgt. Robert E. Futrell
MSgt. Gerald W. Gensinger
MSgt. Paul R. Riendeau

MSgt. Reynaldo C. Valdez
TSgt. Richard E. Chaney
TSgt. John L. Dodgen
TSgt. Kenneth C. Johnson
TSgt. Albert F. Kolb
TSgt. John E. McGuinness
SSgt. Stewart R. Chandler
SSgt. Herschel I. Hargrove
SSgt. Arthur E. Holder
SSgt. Torbio P. Hernandez
SSgt. Charles G. Manrose
SSgt. Eugene M. Moisanen
SSgt. George E. Murphy
SSgt. Max M. Myers
SSgt. Walter M. Rude
SSgt. Raymond R. Slocum
Sgt. Billie B. Bell
Sgt. Harrell Blackeney
Sgt. Alfred E. Candelaria
Sgt. Keann P. Cates
Sgt. James F. Lynch
Sgt. Herbert P. Nigri
Sgt. Angelo Paradise
Sgt. James C. Powers
Sgt. Vernon L. Siegel
Sgt. John L. Sullivan
Sgt. Joe R. Vito
Sgt. James W. Woodul
Sgt. Murray D. Zimney
Cpl. Albert J. Barnabee
Cpl. Ralph H. Bent
Cpl. Ralph E. Bradford, Jr.
Cpl. William I. Davidson
Cpl. Corydon T. Jennings
Cpl. James J. Kveton
Cpl. Dometrio T. Landi
Cpl. Henry L. Pedicone
Cpl. Donald E. Rink
Cpl. Jerome R. Spielberg
Pfc. Daniel O. Burmeister
Pfc. Hubert L. Pierson

Crew #61 – Aircraft #41-28578
2nd Lt. Thomas A. E. Foster P 0741894
2nd Lt. Francis G. Rogers CP 0687768
2nd Lt. Paul V. Dailey N 0736627
2nd Lt. Donald W. Hanslik B 0739094
SSgt. James F. Brandt E 35741985
SSgt. Arthur Angelo R 11100398
Sgt. Charles F. Hipps AE 33253299
Sgt. Chester W. Janeczko AR 36553040
Sgt. Billie C. Pollard G 38425993
Sgt. Umberto F. Ostarello AG 36604398
Passengers

| | | |
|---|---|---|
| SSgt. Clinton C. Harrison | | 18043198 |
| SSgt. Marvin V. Druen | | 35286136 |
| SSgt. Stewart R. Chandler | | 11023960 |
| Sgt. Corydon T. Jennings | | 6939175 |

Crew #62 – Aircraft #41-28601

| | | |
|---|---|---|
| 2nd Lt. Alfred H. Locke | P | 0680460 |
| 2nd Lt. Errol A. Self | CP | 0805988 |
| 2nd Lt. John N. Hortenstine | N | 0673077 |
| 2nd Lt. Arthur C. Delclisur | B | 0688513 |
| Pvt. Virgil H. Carroll | E | 32414596 |
| Sgt. Frank Cappella | R | 32672842 |
| Sgt. Pedro S. Paez | AE | 39255107 |
| SSgt. Dale R. Van Blair | AR | 16076061 |
| SSgt. Albert N. Spadafore | G | 11090413 |
| SSgt. Henry L. Boisclair | AG | 14084140 |

Passengers

| | | |
|---|---|---|
| 2nd Lt. John L. Baldwin | | 0860778 |
| SSgt. George E. Murphy | | 35250626 |
| Cpl. Hubert L. Pierson | | 12101184 |
| Cpl. Daniel O. Burmeister | | 17146367 |

Crew #63 – Aircraft #42-7739

| | | |
|---|---|---|
| 2nd Lt. John R. McCune | P | 0741991 |
| 2nd Lt. Lloyd F. Morse | CP | 0808577 |
| 2nd Lt. Maurice L. Hooks | N | 0672950 |
| 2nd Lt. James I. Misuraca | B | 0685087 |
| SSgt. Woodrow W. Yager | E | 34374567 |
| SSgt. Norrell B. Sawyer, Jr. | R | 38329184 |
| SSgt. Bernis F. Bowers | AE | 38319652 |
| SSgt. Kirk C. Dickson | AR | 33125833 |
| SSgt. Willard D. Cobb | G | 11085357 |
| SSgt. Earl R. Kennedy | AG | 17100533 |

Passengers

| | | |
|---|---|---|
| 1st Lt. Joseph E. Kaiser | | 0710387 |
| TSgt. Charles G. Manrose | | 17030425 |
| Sgt. Joe R. Vito | | 35576242 |
| Cpl. Dometrio T. Landi | | 13086359 |

Crew #64 – Aircraft #42-7754

| | | |
|---|---|---|
| 1st. Lt. Alvin D. Skaggs | P | 0726497 |
| 2nd Lt. Benjamin F. Baer | CP | 0806604 |
| 2nd Lt. Donald C. Todt | N | 0750173 |
| 2nd Lt. Elbert F. Lozes | B | 0671380 |
| MSgt. George Glevanik | E | 33115177 |
| SSgt. Stanley C. Filipowicz | R | 11039297 |
| SSgt. Ray K. Lee | AE | 17014781 |
| SSgt. William E. Jackson, Jr. | AR | 15102452 |
| SSgt. Eugene Gaskins | G | 14082144 |
| SSgt. Francis X. Sheehan | AG | 16110241 |

Passengers

| | | |
|---|---|---|
| Major Kenneth D. Squyres | | 0406033 |
| 1st Lt. Arch C. Doty | | 0913666 |
| TSgt. Kenneth C. Johnson | | 19017049 |
| Sgt. Albert J. Barnabee | | 36566628 |

Crew #65 – Aircraft #42-7709

| | | |
|---|---|---|
| 2nd Lt. Philip J. Chase | P | 0745621 |
| 2nd Lt. Bernard L. Reed | CP | 0750844 |
| 2nd Lt. Harry K. Farrell, Jr. | N | 0688013 |
| 2nd Lt. Roland B. Hallinger | B | 0673156 |
| SSgt. Arthur S. Meyerowitz | E | 32000985 |
| SSgt. Joseph De Frame | R | 11118062 |
| SSgt. Thomas M. McNamara | AE | 16088447 |
| SSgt. William D. Dunham | AR | 34303305 |
| SSgt. Anthony Walesik | G | 33034529 |
| SSgt. Howard R. Peck | AG | 39272980 |

Passengers

| | | |
|---|---|---|
| 1st Lt. Jack P. Edwards | | 0793143 |
| 2nd Lt. John A. Black | | 0806611 |
| 2nd Lt. Henry B. Gabrielson | | 0860698 |
| Sgt. Herbert F. Jones | | 16114162 |

Crew #66 – Aircraft #42-52116

| | | |
|---|---|---|
| 2nd Lt. John R. Tarrant, Jr. | P | 0676820 |
| 2nd Lt. Roland P. Thomason | CP | 0687719 |
| 2nd Lt. Grant W. Collins | N | 0687992 |
| 2nd Lt. Paul W. Markiewicz | B | 0685084 |
| Sgt. Roger E. Otto | E | 17069645 |
| Sgt. Bernard W. Janata | R | 15377089 |
| Sgt. William H. McAdoo | AE | 33256512 |
| Sgt. Louis A. Marcantonioi, Jr. | AR | 11091423 |
| Sgt. Carl W. Loftus | G | 35643496 |
| Sgt. Alfred Salotti | AG | 12206392 |

Passengers

| | | |
|---|---|---|
| MSgt. Robert E. Futrell | | 17032131 |
| MSgt. Reynolds C. Valdez | | 18102186 |
| TSgt. Albert F. Kolb | | 21141941 |
| SSgt. James W. Woodul | | 6939291 |

Crew #67 – Aircraft #42-7758

| | | |
|---|---|---|
| 2nd Lt. Gail A. Sheldon | P | 0805992 |
| 2nd Lt. Marion L. Peek | CP | 0809036 |
| 2nd Lt. John P. Lahart | N | 0687950 |
| 2nd Lt. Irwin R. Larson | B | 0749973 |
| TSgt. Norris M. Christian | E | 18115189 |
| Sgt. John L. McGrath | R | 31237174 |
| Sgt. Paul S. McCray | AE | 35635979 |
| Sgt. George H. Parker | AR | 14074246 |
| Sgt. Arthur C. Koth | G | 37655875 |
| Sgt. Sam E. Batchelor | AG | 32783734 |

Passengers

| | | |
|---|---|---|
| 1st Lt. Andrew J. Hau | | 0660993 |
| TSgt. Richard E. Chaney | | 36533403 |
| TSgt. John L. Dodgen | | 6928354 |
| SSgt. Billie B. Bell | | 13103027 |

*Appendices*

| Crew #68 – Aircraft #41-29234 | | |
|---|---|---|
| 2nd Lt. Charles W. Billings, Jr. | P | 0805886 |
| 2nd Lt. William H. Thomas | CP | 0751490 |
| 2nd Lt. Everard P. Wandell | N | 0687968 |
| F/O Edward E. George | B | T-1527 |
| SSgt. Robert B. Kerrick | E | 19124091 |
| Cpl. Jack R. Callison | R | 34212424 |
| Sgt. Walter R. Johnson | AE | 13075999 |
| Sgt. Albert R. Kohl | AR | 33394522 |
| Sgt. Ulmer E. Bradley, Jr. | G | 34333520 |
| Sgt. John H. Briana | AG | 31294483 |
| Passengers | | |
| 1st Lt. John E. D. Grunow | | 0422081 |
| SSgt. Eugene M. Moisanen | | 39185209 |
| SSgt. Raymond R. Slocum | | 37211151 |
| Sgt. Keenan B. Cates | | 39271325 |

| Crew #69 – Aircraft #42-7713 | | |
|---|---|---|
| 2nd Lt. William F. Ferguson | P | 0681366 |
| 2nd Lt. Walter J. Bulawa | CP | 0808947 |
| 2nd Lt. Charles P. Weisenburgh | N | 0689706 |
| 2nd Lt. Harry Kasbarian | B | 0688359 |
| SSgt. Walter T. Kelly | E | 14040497 |
| Pfc. Wladyslaw R. Gerafin | R | 12171902 |
| Sgt. Henry W. Allen, Jr. | AE | 14187838 |
| Sgt. Norman F. Acido | AR | 19174760 |
| Sgt. Leon A. Ward, Jr. | G | 19203253 |
| Sgt. Bernard D. Prabucki | AG | 32731039 |
| Passengers | | |
| 1st Lt. Robert R. Thornton | | 0730028 |
| MSgt. Paul R. Riendeau | | 11041902 |
| Sgt. Harrell Blakeney | | 34394406 |
| Cpl. Ralph H. Bent | | 34448756 |

| Crew #70 – Aircraft # 41-28580 | | |
|---|---|---|
| 2nd Lt. John R. Bringardner | P | 0737530 |
| 2nd Lt. Frank B. Jordon | CP | 0750608 |
| 2nd Lt. Marvin V. McCormick | N | 0406793 |
| 2nd Lt. Billy Gregory | B | 0688348 |
| Sgt. R. L. Cockrell | E | 14161944 |
| Sgt. George V. Crump | R | 11068765 |
| SSgt. Lawrence H. Ramming | AE | 18057181 |
| Sgt. Warren O. Watson | AR | 38140198 |
| Sgt. Charles A. Goud | G | 32720062 |
| Sgt. Joseph J. Marganski | AG | 34649814 |
| Passengers | | |
| 2nd Lt. Howard A. Garaas | | 0861175 |
| MSgt. Gerald W. Gensinger | | 19016924 |
| Sgt. Donald E. Rink | | 16115013 |
| Cpl. Ralph E. Bradford, Jr. | | 34649814 |

| Crew #71 – Aircraft #42-7717 | | |
|---|---|---|
| 2nd Lt. Harvey E. Broxton | P | 0747201 |
| 2nd Lt. Dwight W. Covell | CP | 0811341 |
| 2nd Lt. Robert F. Fauerback | N | 0687825 |
| 2nd Lt. Clair E. Sharp | B | 0746837 |
| Sgt. Donald V. Birdsall | E | 12171708 |
| Sgt. Joe K. Corziatti | R | 18194485 |
| Sgt. Robert E. Hudson | AE | 16162433 |
| Sgt. Keith C. Tindall | AR | 17127045 |
| Sgt. Irving Elba | G | 32711613 |
| Sgt. Henry Kubinski | AG | 36555495 |
| Passengers | | |
| TSgt. John E. McGuinness | | 35353545 |
| SSgt. Toribio P. Hernandez | | 18117463 |
| SSgt. Arthur E. Holder | | 38049055 |
| Sgt. Murray D. Zimney | | 32415988 |

| Crew #73 – Aircraft #42-64444 | | |
|---|---|---|
| 2nd Lt. William O. Ross | P | 0745186 |
| 2nd Lt. William T. Burkett | CP | 0686889 |
| 2nd Lt. George W. Wenthe | N | 0690850 |
| 2nd Lt. Richard H. Grant | B | 0666073 |
| SSgt. Arthur F. Mied | E | 16000220 |
| Sgt. Harold L. Kindt | R | 13094588 |
| Sgt. Norman C. Benson | AE | 37176337 |
| Sgt. Eddie J. Guidry | AR | 38263722 |
| Sgt. Charles Susine, Jr. | G | 32722296 |
| Sgt. Jack M. Garrett | AG | 6931747 |
| Passengers | | |
| SSgt. Herschel L. Hargrove | | 19052983 |
| SSgt. Angelo M. Paradise | | 12131025 |
| Sgt. James F. Lynch | | 33318548 |
| Sgt. Thomas P. Nigri | | 12164850 |

| Crew #74 – Aircraft #41-29235 | | |
|---|---|---|
| 2nd Lt. Jack Swayze | P | 0745215 |
| 2nd Lt. Marshall T. McRae | CP | 0678891 |
| 2nd Lt. Gary Young, Jr. | N | 0740686 |
| 2nd Lt. Arthur B. Rayburn | B | 0751843 |
| Sgt. Harry H. Gottlieb | E | 13151960 |
| Sgt. James H. Parker | R | 13119675 |
| Pvt. Ernest L. Wright | AE | 38055990 |
| Sgt. Warren J. Johnson | AR | 19178635 |
| SSgt. Angelo A. Valenzano | G | 12129290 |
| SSgt. William A. Keeler | AG | 35326310 |
| Passengers | | |
| TSgt. Walter M. Rude | | 20944697 |
| TSgt. John L. Sullivan | | 31142763 |
| Sgt. James C. Powers | | 36446898 |
| Sgt. Vernon L. Siegel | | 32409076 |

| Crew #75 – Aircraft #41-28589 | | |
|---|---|---|
| 2nd Lt. Jack L. Black | P | 0422068 |
| 2nd Lt. Joseph Pomfret | CP | 0810214 |
| 2nd Lt. Peter A. Mermert, Jr. | N | 0690696 |
| 2nd Lt. Robert P. Burkartsmeier | B | 0688578 |
| Sgt. Charles C. Nissan | E | 32589610 |

| | | |
|---|---|---|
| Cpl. Eugene J. Dworaczyk | R | 18201999 |
| Sgt. Michael J. Curran | AE | 37319744 |
| Sgt. Wilfred F. Haschle | AR | 37266508 |
| Sgt. Harold Benvenutti | G | 16168965 |
| Sgt. Richard L. Campbell | AG | 39273785 |

Passengers

| | | |
|---|---|---|
| MSgt. William G. Casper | | 20367101 |
| Sgt. William I. Davidson | | 33432354 |
| Cpl. Henry L. Pedicone | | 12185410 |
| Cpl. Jerone R. Spielberg | | 32691072 |

Crew #77 – Aircraft #42-7764

| | | |
|---|---|---|
| 1st Lt. Graham Guyton | P | 0390541 |
| 2nd Lt. Thomas R. Allen, Jr. | CP | 0807659 |
| 2nd Lt. Richard M. Wheelock | N | 0687970 |
| 2nd Lt. Pickney W. Seiders | B | 0751847 |
| SSgt. Woodrow Callahan | E | 6964020 |
| Sgt. Oscar L. Brown | R | 14085761 |
| Sgt. Dennis C. Quinlan | AE | 32556595 |
| Sgt. Kenneth Ciolek | AR | 34685047 |
| Sgt. Ralph K. Brannon | G | 36150354 |
| Sgt. Josh K. Askew | AG | 34563344 |

Passengers

| | | |
|---|---|---|
| Col. James McK. Thompson | | 017992 |
| Captain John S. Laws | | 0415499 |
| Major James R. Patterson | | 0472768 |
| 1st Lt. Minor L. Morgan | | 0660224 |

This is not a complete list of all replacement crews; only those noted in the 448th BG's monthly summaries and official orders are listed. Some of the names are hard to read on the microfilm and ???'s are used to denote an unreadable name.

Date unknown – Crew #28

| | | |
|---|---|---|
| 1st Lt. Leonard V. Thornton | P | 0675887 |
| 2nd Lt. Ronald S. Wildey | CP | 0749373 |
| 2nd Lt. Bernard E. Smith | B | 0685863 |
| 2nd Lt. William J. Regan | N | 0685731 |
| TSgt. William J Nicholson | | 32410427 |
| SSgt. Kenneth W. Kohrback | | 37239893 |
| TSgt. Emerson D. Miller | | 39253429 |
| SSgt. Jack D. Hess | | 12030990 |
| SSgt. Sol J. Schatz | | 32558661 |
| SSgt. Carmen R. Valentino | | 32558661 |

Date unknown – Crew #31

| | | |
|---|---|---|
| Lt. Robert L. Lehman | P | 0750630 |
| Lt. Lorenz G. Johnson | CP | 0752416 |
| Lt. Phillip Baskin | N | 0687806 |
| Lt. Searl J. Collins | B | 0749611 |
| TSgt. Luther D. Rummage | | 14196490 |
| TSgt. Thomas R. Lee | | 18171694 |
| SSgt. Demo D. Sarentos | | 12140760 |

| | | |
|---|---|---|
| Raymond W. Wood | | 19179737 |
| SSgt. Harry D. Quillen | | 12012673 |
| SSgt. Harry W. Kleber | | 33035628 |

Date unknown – Crew #33

| | | |
|---|---|---|
| Lt. Richard L. Henderson | P | 0751323 |
| Lt. Thomas E, Skeffington | CP | 0751968 |
| Lt. John E. Vernor | N | 0689701 |
| Lt. Joseph E. Chester | B | 0688175 |
| SSgt. George L. Mason | | 15065214 |
| SSgt. Henry K. Bonfield | | 13151955 |
| SSgt. Oscar L. Brown | | 14085781 |
| SSgt. Robert H. Hale | | 17154143 |
| SSgt. Robert C. Strobel | | 36613343 |
| SSgt. Opel J. Powell | | 35578003 |

Date unknown – Crew #34

| | | |
|---|---|---|
| Lt. Edman L. Chapman | P | 0687163 |
| F/O Julius L. Engdahl | CP | T61166 |
| Lt. Cyril F. Coverley | N | 01011425 |
| Lt. Robert E. Feron | B | 0752562 |
| SSgt. Dwayne H. Aurand | | 36622424 |
| SSgt. Robert W. Funk | | 32471042 |
| SSgt. William A. Gautney | | 14182365 |
| SSgt. Denzil F. Stumbo | | 15339513 |
| SSgt. Albert J. Bishop | | 36565365 |
| SSgt. Kenneth L. Hall | | 37497934 |

Date unknown – Crew #36

| | | |
|---|---|---|
| Lt. Kenneth E. Weaver | P | 0743145 |
| Lt. Philip B. Thompson | CP | 0725377 |
| Lt Robert M. Wood | N | 0750261 |
| Lt. Doyle W. Tucker | B | 0679398 |
| SSgt. Thad C. Eatherley | | 18166101 |
| Sgt. William J. Perry | | 18136732 |
| SSgt. Herbert H. Graham | | 33250071 |
| Sgt. William S. Byers | | 33244148 |
| Pvt. Harvey L. Dickey | | 18226840 |
| Sgt. James R. Tipton | | 15382085 |

Replacement Crews – December, 1943

24 December 1943 to 714th Squadron

| | |
|---|---|
| 2Lt. William F. Bonner | 0679034 |
| 2Lt. John D. Richmond | 0805983 |
| 2Lt. Robert E. Johnston | 0685506 |
| 2Lt. Thomsa (nmi) Kavala | 0684166 |
| SSgt. Jean A. Beaulieu | 11091530 |
| SSgt. Carl R. Ehret | 35341404 |
| Sgt. Stanley Zabrowski | 16151199 |
| Sgt. Vito J. Bonovitch | 35517626 |
| Sgt. Warren G. Mounsey | 19145767 |
| Sgt. William A. Venslyke | 13060783 |

24 December 1943 to 714th Squadron
2Lt. Russel O. Reindal        P    0743092
2Lt. Bernard H. Mattson       CP   0805950
2Lt. Wilber R. Phillips       N    0801502
2Lt. Jesse (nmi) Gardner      B    067818
SSgt. Boyd L. Hatzell              37233152
SSgt. Louis F. Donoso              33367790
Sgt. Paul M. Freeze                35667923
Sgt. Elden F. Farrar               11118591
Sgt. Ray S. Waters                 34436677
Sgt. David S. Avila                39849304

Replacement Crews – January, 1944

1 January 1944 To 714th Squadron
2nd Lt. Frank Gibson          P
2nd Lt. Garth Connole         CP
2nd Lt. Paul Schauwacker      N
2nd Lt. Roy Allen             B
SSgt. Larry Putgenter         E
TSgt. Bob Smith               R
Sgt. Chuck Barlow             RW
   (Wounded & replaced by Sgt. Julius Rebeles)
Sgt. Frank Benjamin           TG
Sgt. Richard Collins          BG
Sgt. John Daley               LW

7 January 1944 to 715th Squadron
2nd Lt. Ridd J. Solomon       P    0670344
2nd Lt. Pierre L. Delcambre   CP   0806310
2nd Lt. Jay Pace              B    0673931
2nd Lt. Frank Hamouz          N    0676479
SSgt. John Allen                   13026655
SSgt. Robert C. McMahon            17055897
Sgt. Robert H. Norman              39406577
Sgt. John J. Shreve                36530979
Sgt. Lloyd Jackson                 37223205
Sgt. Robert F. Hendricks           16047117

7 January 1944 to 715th Squadron
2nd Lt. Leonard I. Kronheim        0533872
2nd Lt. Harold J. Kreichbaum       0683028
2nd Lt. Donald R. Silverstrom      0736735
2nd Lt. Jack R. Nehrich            0679204
SSgt. John J. Kreyer Jr.           12180074
SSgt. Louis G. Keim                37263325
Sgt. Stanley W. Mazeika            31173936
Sgt. Albert R. Cavalier            13044384
Sgt. Robert D. Crudele             15374093
Sgt. Charles E. Brown              16076433

7 January 1944 to 448th (Squadron unknown)
2Lt. Eward A. Markewicz       P    0677543
2Lt. Donald (nmi) Keegstra         0679887

2Lt. Arthur E. Zander              0805378
2Lt. Richard L. Nardi              0681559
SSgt. Henry J. Opper               12125979
SSgt. David T. Marsh               33300005
Sgt. Robert F. Holland             12171282
Sgt. Emmet F. Fallert              16144615
Sgt. James B. Whyte                12157865
Sgt. Daniel B. Dougherty           12164002

7 January 1944 to 448th (Squadron unknown)
1Lt. Stanley R. Johnson       P    0728879
2Lt. Roland N. Trent          CP   0806748
2Lt. Raymond R. Moder         N    0669370
2Lt. George W. Purcell        B    0667348
SSgt. Jack N. Walker               12185531
SSgt. Clyde W. Hatley              34037345
SSgt. Charles B. Failla            12036815
Sgt. Robert E. Miller              13134915
Sgt. Earl F. Weigolt               37323484
Sgt. Joe E. Jiron                  18046962

7 January 1944 to 448th (Squadron unknown)
2Lt. Joseph G. Liebich        P    0742435
2Lt. William F. Slater             0685239
2Lt. John A. Spiers                0683919
2Lt. Gerald P. Tallent             0679693
SSgt. Ruffins S. Loegering         17135197
SSgt. Howard (nmi) Hayes           17088811
Sgt. Arthur M. Dupuy               11068724
Sgt. Robert C. Marsh               16093962
Sgt. William A. Stewart            35481731
Sgt. Louis H. Silvestry            31258545

28 January 1944 to 712th Squadron
2nd Lt. Vincent E. Liedka     P    0795781
2nd Lt. Zachariah F. Jackson  CP   0681101
2nd Lt. Ronald L. McAllister  N    0683842
2nd Lt. Charles A. Oliver          0686201
SSgt. Duncan K. Thomson            37439809
TSgt. William L. McCauley          33274219
SSgt. Walton E. Gaskins            34463828
Sgt. James A. Garelock             13118837
Sgt. Paul L. Russel                32269037
Sgt. Lee L. Bryant                 34388723

28 January 1944 to 712th Squadron
2nd Lt Robert F. MacKenzie    P    0743045
2nd Lt. Ernest J. Dellia      CP   0671501
2nd Lt. James J. Kenney       B    0743030
F/O Leonard J. Levine         N    T122062
SSgt. Henry (nmi) Kunstler         32612248
SSgt. Albert C. Johnson Jr.        33344383
Sgt. Lyle W. Johnson               37236945
Sgt. Clarence G. Thompson          16079278

| | | | |
|---|---|---|---|
| Sgt. Leo Amell | | 12079325 | |
| Sgt. Felix Slojk | | 32571404 | |

**28 January 1944 to 713th Squadron**

| | | | |
|---|---|---|---|
| 2nd Lt. Max E. Turpin | P | 0675871 |
| 2nd Lt. Eldon H. Gheck | CP | 0805910 |
| 2nd Lt. Jack (nmi) Boykoff | N | 0750177 |
| 2nd Lt. Robert W. Adams | B | 0654122 |
| SSgt. William F. Hallman | | 14142190 |
| SSgt. Clyde A. Burnette | | 14141825 |
| SSgt. Thomas Culpepper | | 14180050 |
| Sgt. George A. Daneau | | 11021582 |
| Sgt. Donald G. Elder | | 15333504 |
| Sgt. William E. Phillips | | 15089281 |

**28 January 1944 to 715th Squadron**

| | | | |
|---|---|---|---|
| 2nd Lt. Stanley C. Cooper | P | 0533917 |
| 2nd Lt. Wesley V. Helvey | CP | 0805916 |
| 2nd Lt. Michael J. Shonesky | N | 0673821 |
| 2nd Lt. Charles E. Faller | B | 0684145 |
| SSgt. Thomas W. Patterson | | 38338554 |
| SSgt. John A. Tasko | | 35351914 |
| Sgt. Frank J. Demaine | | 12189633 |
| Sgt. Albert (nmi) Di Gioia | | 32365095 |
| Sgt. Julius H. Dothage | | 37402124 |
| Sgt. Joseph F. Nickerson | | 35572323 |

**Replacement Crews – February 1944**

**11 February 1944 to 713th Squadron**

| | | |
|---|---|---|
| Lt. Ronald C. Warnock | | 0731707 |
| Lt. Laurence R. Wenzel | | |
| Lt. Sam Bullard | | 0692604 |
| Lt. Frank L. French | | 0889147 |
| SSgt. Delbert B. Coulter | | 38274253 |
| SSgt. John J. Feyti | | 20248174 |
| SSgt. Eugene M. Harmon | | 18168335 |
| SSgt. Glen C. Haun | | 34509553 |
| SSgt. George F. Kerr | | 13155651 |
| SSgt. William F. Morgan | | 15324525 |

**Date unknown – (flew first mission March 1944)**

| | | | |
|---|---|---|---|
| Lt. Robert F. Pinkus | P | 0661557 |
| Lt. Newton I. Willis | CP | 0691074 |
| Lt. Robert F. Fauerbach | N | 0687325 |
| Lt. Edward P. Ward, Jr. | B | 0751689 |
| TSgt. Joe Raphael | | 18025037 |
| SSgt. Samuel P. Scott | | 15336067 |
| Sgt. Frederick A. Clark | | 11040634 |
| Sgt. Charles F. Miller, Jr. | | 37337367 |
| Sgt. Joseph Morroney | | 32722750 |
| Sgt. Mitchell M. Ramonas | | 31270987 |

**Date unknown – (flew first mission in March or April 1944)**

| | | | |
|---|---|---|---|
| Lt. Melvin Alspaugh | P | |
| Lt. Richard Waters | CP | |
| Lt. William Carlson | N | |
| Lt. William Edwards | B | |
| TSgt. Donald Holter | | |
| TSgt. Jack Anderson | | |
| SSgt. Ray Chartier | | |
| SSgt. Charles Adams | | |
| SSgt. Owen Worsman | | |
| SSgt. Harold Barney | | |

**Date unknown (early Spring 1944)**

| | | | |
|---|---|---|---|
| 1st Lt. Harrison C. Mellor | P | | 0428403 |
| 2nd Lt. Douglas J. Eames | CP | | 0800071 |
| 2nd Lt. Frank H. Jacobson | N | | 0692016 |
| 2nd Lt. Marvin T. Goff | B | | 0688343 |
| SSgt. William A. Warren | E | | 18028026 |
| Sgt. Walter T. Bresslor | AE | | 13146208 |
| SSgt. Francis C. Marx | RO | | 32533995 |
| Sgt. Nelson A. Branch | AG | | 20702227 |
| Sgt. Ira R. Allen | AG | | 39908551 |
| Sgt. Mike Little | AG | | 33195801 |

**Date unknown (early Spring 1944)**

| | | | |
|---|---|---|---|
| 2nd Lt. Robert E. Krieger | P | | 0737599 |
| 2nd Lt. John T. Fuller | CP | | 0755542 |
| 2nd Lt. Edward M. Cotter | N | | 0691959 |
| 2nd Lt. John J. Gallo | B | | ???????? |
| SSgt. Harvey A. Penkara | E | | 19000975 |
| SSgt. Keith R. McFarland | AE | | 19116513 |
| SSgt. Glenn L. Bolling | RO | | 17061783 |
| Sgt. Dayton R. Hier | AG | | 15200934 |
| Sgt. Charles E. Logua | AG | | 31016205 |
| Sgt. Lionel E. Vanderman | AG | | 17059738 |

**Date unknown (early Spring 1944)**

| | | | |
|---|---|---|---|
| Lt. Eugene V. Pulcipher | P | | 0750685 |
| Lt. Elmer P. Meier | CP | | 0815366 |
| Lt. William Carcelli | B | | 0692476 |
| Lt. George S. Fahr | N | | 0697831 |
| SSgt. James R. Hardin | | 17077107 | |
| SSgt. Chester J. Romanosky | | 13056696 | |
| Sgt. William H. Durant | | 35623918 | |
| Sgt. William S. Davis | | 14135368 | |
| Sgt. Maynard H. Young | | 31152846 | |
| Sgt. Carl E. Spellman | | 37368208 | |

**Date unknown (early Spring 1944)**

| | | | |
|---|---|---|---|
| Lt. Cherry C. Pitts | P | | 0679115 |
| Lt. John L. Merkling | CP | | 0683045 |
| Lt. Clifford R. Schrom | B | | 0679664 |
| Lt. Charles W. Wilder, Jr. | N | | 0683885 |
| TSgt. Arthur Angelo | | 11100398 | |

| | | | |
|---|---|---|---|
| TSgt. Ernest W. Robinson, Jr. | | | 14055978 |
| TSgt. Samuel Werner | | | 12156211 |
| SSgt. Thomas H. Harwood | | | 34374445 |
| SSgt. Harry F. Vetter | | | 16088095 |
| SSgt. Stanley L. Wilson | | | 39550446 |

Replacement Crews – April 1944

10 April 1944 to 712th Squadron

| | | | |
|---|---|---|---|
| 2nd Lt. William R. Hayes | P | | 0807436 |
| 2nd Lt. F. Schrammel | CP | | 0817282 |
| 2nd Lt. Stephen P. Tiffany | B | | 0698240 |
| 2nd Lt. Calvin J. Ellis | N | | 0689455 |
| SSgt. Richard Anderson | | | 31144511 |
| SSgt. Leon L. Watson | | | 35347140 |
| Sgt. Atlee (nmi) Arola | | | 32729518 |
| Sgt. James H. Johnson | | | 18135249 |
| Sgt. Gabriel A Latsko | | | 33298420 |
| Sgt. Paul W. Sorensen | | | 39568993 |

10 April 1944 to 712th Squadron

| | | | |
|---|---|---|---|
| 2nd Lt. Thomas M. Plese | P | | 0807044 |
| 2nd Lt. Lawrence E. Anderson | CP | | 0806031 |
| 2nd Lt. Richard P. Tustin | B | | 0814449 |
| F/O Lt. Alfred B. Tallman, Jr. | N | | T-1575 |
| SSgt. Roy E. Herndon | | | 17122516 |
| SSgt. Joseph E. Holmes | | | 11117911 |
| Sgt. Roland M. Cheyne | | | 36400251 |
| Sgt. John S. Davis | | | 13100224 |
| Sgt. John B. Murphy, Jr. | | | 11036674 |
| Sgt. Dale B. Stensrud | | | 17155439 |

10 April 1944 to 713th Squadron

| | | | |
|---|---|---|---|
| 2nd Lt. Raymond Peterson | P | | 0808471 |
| 2nd Lt. Joseph H. Wells | CP | | 0695115 |
| 2nd Lt. Barry S. Brook | B | | 0694323 |
| F/O Kenneth A. Moulten | N | | T-122066 |
| SSgt. John R. Myers | | | 15196732 |
| Sgt. Obert T. Bjorseth | | | 37175175 |
| Sgt. John (nmi) Gedz | | | 16041662 |
| SSgt. Denver L. Putman | | | 37324697 |
| Sgt. Frank E. Carlson | | | 20759430 |
| Sgt. Louis A. Mauduit | | | 18171029 |

10 April 1944 to 714th Squadron

| | | | |
|---|---|---|---|
| 2nd Lt. William W. Blanck | P | | 0808340 |
| 2nd Lt. Dee L. Johnson | CP | | 0684740 |
| 2nd Lt. William Carlson | B | | 0814246 |
| 2nd Lt. Joseph (nmi) Risovich | N | | 0754852 |
| SSgt. James A. Chambers | | | 1416246 |
| SSgt. Joseph E. Harney, Jr. | | | 11111620 |
| Sgt. John B. Barker | | | 18215974 |
| Sgt. John L. Burns | | | 37380177 |

| | | | |
|---|---|---|---|
| Sgt. Stanley Pasternak | | | 32326895 |
| Sgt. James R. Tune | | | 38428344 |

10 April 1944 to 715th Squadron

| | | | |
|---|---|---|---|
| 2nd Lt. John W. Cathey | P | | 0806373 |
| 2nd Lt. Arthur J. Brisson | CP | | 0751983 |
| 2nd Lt. Joseph J. Kwederis | B | | 0814346 |
| F/O Carl M. Carlson | N | | T-1686 |
| TSgt. Culmer H. Darby | | | 20304517 |
| SSgt. Clifton W. Linnell | | | 11118117 |
| SSgt. Russell E. Howle | | | 15097532 |
| Sgt. Jack (nmi) Arluck | | | 33426161 |
| Sgt. Anthony J. Novelli | | | 12158891 |
| Sgt. Arnold J. Wetzel | | | 32466922 |

10 April 1944 to 715th Squadron

| | | | |
|---|---|---|---|
| 2nd Lt. Roy A. Fischer | P | | 0747385 |
| 2nd Lt. Victor F. Hoff | CP | | 0751329 |
| 2nd Lt. Robert F. Poole | B | | 0695694 |
| 2nd Lt. John H. Williams | N | | 0741431 |
| SSgt. Trevor V. Chatfield | | | 14067521 |
| SSgt. Leroy S. Dausman | | | 36414751 |
| SSgt. Pietro A. Romano | | | 11130608 |
| Sgt. Ray L. Goudy | | | 16160271 |
| Sgt. Maurice (nmi) Oliver | | | 18117473 |
| Sgt. John C. Olney | | | 38370120 |

Replacement Crews – May 1944

2 May 1944 to 713th Squadron

| | | | |
|---|---|---|---|
| 1st Lt. Charles F. Mills | P | | 0789734 |
| 1st Lt. John L. Guthrie | CP | | 0561170 |
| 2nd Lt. Orlin L. Munns | B | | 0702171 |
| 2nd Lt. John B. Shields | BN | | 0694507 |
| SSgt. Robert G. Cutrone | | | 32737209 |
| SSgt. Dale A. Van Vorce | | | 15070615 |
| Sgt. Thomas M. Alston | | | 19100844 |
| Sgt. Elmer L. Morgan | | | 37500623 |
| Sgt. John C. Wemmert | | | 36814033 |
| Sgt. Emil P. Zahnow | | | 37165866 |

6 May 1944 to 714th Squadron

| | | | |
|---|---|---|---|
| 2nd Lt. Thomas K. Foster | P | | 0744267 |
| 2nd Lt. Robert G. Silver, Jr. | P | | 0751045 |
| 2nd Lt. Robert C. Phillips | CP | | 0697507 |
| 2nd Lt. Donald R. Allen | B | | 0703697 |
| 2nd Lt. Nunci J. Piucci | N | | 0695524 |
| SSgt. Warren W. Duncan | | | 15103389 |
| SSgt. Benjamin C. Kirschner | | | 39405684 |
| Sgt. John R. Foss | | | 34707354 |
| Sgt. Edward H. Owen | | | 16116533 |
| Sgt. Leonard M. Siegel | | | 13142404 |
| Sgt. Robert B. Spruill | | | 14080837 |

6 May 1944 to 715th Squadron
2nd Lt. Ulrich W. Tschanz          P    0805538
2nd Lt. James R. Pruitt            CP   0816575
2nd Lt. Lionel Greenberg           B    0703242
2nd Lt. Charles W. Ponhorwood      N    0744212
SSgt. Arthur O. Archambau               31173522
SSgt. Robert H. Kuck                    16144168
Sgt. Hulick H. Barney                   39857402
Sgt. Arlin W. Turner                    18140043
Sgt. Frederick A. Schoonmaker           32734039
Sgt. Jap R. Wilson, Jr.                 18076005

6 May 1944 to 715th Squadron
2nd Lt. Jack L. Mercer             P    0810916
2nd Lt. John D. Masters            CP   0758712
2nd Lt. John E. Neel               B    0702172
2nd Lt. Warren G. Phillippi        N    0695523
SSgt. Allan M. Johnson                  39408979
SSgt. William G. Rekart                 35546615
Sgt. Thomas Mistretta                   30207255
Sgt. Thomas C. Murphy                   37559557
Sgt. Charles E. Nelson                  32606827
Sgt. Francis A. Oltman                  32000266

10 May 1944 to 712 Squadron
2nd Lt. Raymond A. Wermeyer        P    0749370
2nd Lt. Jerold I. Grosscup         CP   0685117
2nd Lt. Mathew L. Crovitz          B    0686231
2nd Lt. Ralph A. Hagerty           N    0692489
Sgt. Alvin W. Shaw                      38002227
SSgt. Herman L. Caruso                  32491419
Sgt. Sigmund Borowicz                   16149171
SSgt. Floyd R. Myers                    13110172
Sgt. George L. Shadi                    32712151
Sgt. Edward F. Jones                    31219006

13 May 1944 to 713th Squadron
2nd Lt. John R.B. Swartzel         P    0755785
2nd Lt. Paul F. Dwyer              CP   0700712
2nd Lt. Francis E. Azevedo         B    0761022
2nd Lt. William L. Conglaton       N    0702017
SSgt. Kenneth E. Buck                   19095637
Sgt. John J. Cornwall                   33099397
Sgt. William A. Hamnert                 37554286
Sgt. Joseph E. Bernier                  31260146
Sgt. Harold C. Coorsey                  37551120
Sgt. Harvey C. May                      18089684

13 May 1944 to 714th Squadron
2nd Lt. Michael Kuchwara           P    0751155
2nd Lt. Robert P. Nimmo            CP   0761826
2nd Lt. Morel H. Papa              B    0698906
2nd Lt. Robert H. Cooper           N    0705951

SSgt. Calvin C. Crumbley                34686327
SSgt. Richard J. Subay                  35535596
Sgt. Lawrence J. Kennedy                31242459
Sgt. Jones J. Hyman, Jr.                14109533
Sgt. Harold J. Romar                    32861806

13 May 1944 to 715th Squadron
2nd Lt. John A. White              P    0128586
2nd Lt. Richard T. Looms           CP   0700634
2nd Lt. D.R. Crandall              B    0761046
2nd Lt. Robert C. Knapp            N    0678368
SSgt. Victor L. Cieslewicz              32829960
SSgt. Frank A. Paladino                 32816209
Sgt. Clyde H. Bush                      14190404
Sgt. Eugene J. Eck, Jr.                 33504264
Sgt. Walter L. Shipley                  33576194
Sgt. Stanley J. Wasielewski             35052631

13 May 1944 to 715th Squadron
2nd Lt. Henry S. Hasiak            P    0702043
2nd Lt. William A. Warke           CP   0693082
2nd Lt. Robert T. White            B    0684924
2nd Lt. Edgar C. Scanlon, Jr.      N    0701312
SSgt. Jerry M. Parsons                  16035535
SSgt. Robert J. Cashin                  37553432
Sgt. Millard L. Brown                   15171462
Sgt. Robert B. Thompson                 17066011
Sgt. Norman H. Johnson                  36740128
Sgt. Gordon W. Wright                   12072510

13 May 1944 to 712th Squadron
1st Lt. Wade H. Williford          P    0796932
2nd Lt. Frederick A. Burns         CP   0690361
2nd Lt. Henry N. Vandersten        B    0816007
2nd Lt. Robert M. Wagoner          N    0747685
SSgt. Charles D. Dennis                 33296402
SSgt. William E. Larecy                 18191917
SSgt. Jack (nmi) Matau                  35373000
SSgt. Elmer E. Hinnenkamp               37243033
SSgt. Eugene A. Lloyd                   13170475
SSgt. Ray H. Wingate                    15354833

14 May 1944 to 713th Squadron
2nd Lt. Alexander J. Shogan        P    0812674
2nd Lt. Joseph R. Dowalo           CP   0761906
2nd Lt. George E. Farschman        B    0700720
2nd Lt. Edgar N. Clyde             N    0701566
SSgt. Edward J. Brunetti                37401441
SSgt. Vincent Majewski                  12098906
Sgt. Robert A. Berman                   16089526
Sgt. Leslie H. Douglas                  38371608
Sgt. Riley L. Golden                    34764141
Sgt. Merril N. Smith                    39909895

14 May 1944 to 714th Squadron
2nd Lt. Ralph T. Welsh            P     0755811
2nd Lt. Alonzo A. Bacon           CP    0760225
2nd Lt. Frank J. Erbacher         B     0700713
2nd Lt. Phillip M. Goplen         N     0699069
SSgt. Isadore A. Buechner               36810940
SSgt. Alfred R. Carrington              13116923
Sgt. Charles T. Berrier, Jr.            18014682
Sgt. Kenneth R. Snyder                  33564734
Sgt. Robert K. Snyder                   33564733
Sgt. Vincent K. Torfin                  37549075

23 May 1944 to 712th Squadron
1st Lt. Wayne L. Carmead          P     0725143
2nd Lt. James A. Burt             CP    0701891
2nd Lt. Glenn R. Mack             B     0712625
2nd Lt. Joseph L. Borsh           N     0697765
SSgt. Frank (nmi) Patico                12163953
SSgt. Ellis W. Register                 39279329
Sgt. Roland F. Alling                   32041217
Sgt. John M. Garrity                    13078399
Sgt. Spencer B. Tisenby                 33450852
Sgt. Jack (nmi) Yeaman                  35094550

23 May 1944 to 714th Squadron
1st Lt. Donald A. Briola          P     0730071
2nd Lt. Thornton M. Brown         CP    0701898
2nd Lt. Walter K. Kurk            B     0712602
2nd Lt. James P. Bonnell          N     0697763
SSgt. Mathew M. Gill                    11101269
SSgt. Raymond L. Swinehart              36725295
Sgt. Howard O. Giles                    32916714
Sgt. Carl M. Johnson                    37662603
Sgt. Russell F. Kelly                   37601472
Sgt. William E. Wallace                 36417002

24 May 1944 to 714th Squadron
1st Lt. William H. Gibson         P     0437393
2nd Lt. Arthur M. Coleman         CP    0700864
F/O John K. Elam                  N     T-123927
1st Lt. Lawrence W. Nichols, Jr.  B     0668746
SSgt. John P. Gallagher                 32748370
Sgt. Earl W. Lathrop                    37513289
Sgt. Clarence H. Landon                 17127399
Sgt. Leonard C. Lundgren                39569989
Sgt. Gordon C. Wendling                 37554062
Sgt. Marion J. Barbre                   13033490

25 May 1944 to 713th Squadron
2nd Lt. Edward K. Schultz, Jr.    P     0693070
2nd Lt. Aaron E. Caplan           CP    0703006
2nd Lt. Vaiden U. Dozier          B     0690385
2nd Lt. Michael J. Kentosh        N     0701428
SSgt. Francis S. Dorman                 39121768

SSgt. Ralph F. Hannah                   34601186
Sgt. Frank J. Bernard                   15087102
Sgt. Alfred C. Dupuis                   31269193
Sgt. Hervey N. Whitfield                38370714
Sgt. Rudolph J. Ujcic                   33427246

25 May 1944 to 714th Squadron
2nd Lt. Peter D. MacVean          P     0886043
2nd Lt. John E. Hurley            CP    0706182
2nd Lt. John Savich               B     0712676
2nd Lt. Lawrence E. Carney, Jr.   N     0697988
SSgt. Charles E. Messerli               37470822
SSgt. Harry G. Pace, Jr.                16059707
Sgt. Marshall L. Adamson                37553875
Sgt. John J. Ruelle                     36451286
Sgt. Robert W. Slack                    39912946
Sgt. Leo Williams                       38450998

26 May 1944 to 713th Squadron
2nd Lt. Louis U. Weitzel          P     0886046
2nd Lt. Norman R. Doughty         CP    0163732
2nd Lt. Vincent A. Scarpino       B     0712677
2nd Lt. Allen D. Cassady          N     0697990
SSgt. Donald M. Langland                39280955
SSgt. Leo F. Czekuc                     16086691
Sgt. Fornie G. Burley, Jr.              38467657
Sgt. Jack W. Clair                      39404665
Sgt. Jay L. Hendrickson                 37552207
Sgt. Charles Mathison                   34802625

26 May 1944 to 715th Squadron
2nd Lt. Andrew N. Panicci         P     0809034
2nd Lt. Miles B. Drawhorn         CP    0701989
2nd Lt. John H. Schlicher, Jr.    B     0712679
2nd Lt. Duane G. Christensen      N     0697767
SSgt. Howard A. Lindstrom, Jr.          11096394
SSgt. James S. Bourne, Jr.              20449914
Sgt. Walter E. Anderson                 37553220
Sgt. Lawrence R. Chandler               36420424
Sgt. Walter L. Maddox                   19143330
Sgt. Leland C. Wright                   35606442

27 May 1944 to 712th Squadron
2nd Lt. Clive J. Howell           P     0812266
2nd Lt. Victor D. Dothcek         CP    0697993
2nd Lt. Arthur B. Majestic        B     0701940
F/O Lt. Robert J. Branizza        N     T-125180
SSgt. Herschel O. Hamblin               13062783
SSgt. Bertil S. Johnson                 35099377
Sgt. George J. Grubisa                  32916135
Sgt. James L. Vajgl                     35527301
Sgt. Alexander Istvanovich              37551499
Sgt. Sammie D. Vinson                   14177911

27 May 1944 to 712th Squadron
| | | |
|---|---|---|
| 2nd Lt. Kenneth D. Miller | P | 0816344 |
| 2nd Lt. Donald E. Ericson | CP | 0820732 |
| 2nd Lt. Gilbert S. Newman | B | 0707312 |
| 2nd Lt. Charles A. Albrecht | N | 0706767 |
| SSgt. Carl C. Gillespie, Jr. | | 35618326 |
| SSgt. Jesus A. Balderrama | | 18097389 |
| Sgt. Robert J. Allaire | | 31321177 |
| Sgt. Charles J. Balkash | | 12030926 |
| Sgt. Donald C. Butzer | | 33168929 |
| Sgt. Chester H. Anderson | | 19124405 |

28 May 1944 to 712th Squadron
| | | |
|---|---|---|
| 2nd Lt. Glenn F. Jones | P | 0745107 |
| 2nd Lt. Lloyd W. Kilmer | CP | 0699677 |
| 2nd Lt. James R. Cooper | B | 0712780 |
| 2nd Lt. Donald S. Powell | N | 07015080 |
| SSgt. Wilton L. Tawwater | | 38369444 |
| SSgt. Lester C. Smith, Jr. | | 34705391 |
| Sgt. James B. Baker | | 38453284 |
| Sgt. Harry P. Barker | | 17076386 |
| Sgt. Olaf Bratland | | 37549500 |
| Sgt. Oscar E. Clayton | | 33628616 |

28 May 1944 to 713th Squadron
| | | |
|---|---|---|
| Capt. Alfred C. Fox | P | 01699253 |
| 1st Lt. Louis Cepelak | CP | 01172408 |
| 2nd Lt. Leo L. Lovel | B | 0708267 |
| 2nd Lt. Dudley T. Hall | N | 0684964 |
| SSgt. Martin A. Wiencek | | 36635105 |
| SSgt. Harold Goltz | | 36725357 |
| Sgt. Sherman Hammon | | 39462177 |
| Sgt. William G. McCabe | | 38467568 |
| Sgt. Neal H. Pauley | | 19133990 |
| Sgt. Verne S. Franklin | | 37515321 |

28 May 1944 to 714th Squadron
| | | |
|---|---|---|
| 2nd Lt. Gaylord E. Felton | P | 0691250 |
| 2nd Lt. Vito R. Scorrano | CP | 0702339 |
| 2nd Lt. Raymond W. Bearden | B | 0708390 |
| 2nd Lt. John T. Polashek | N | 0699053 |
| SSgt. William P. Brown | | 15057778 |
| SSgt. Joseph F. Hollwood, Jr. | | 11103701 |
| Sgt. Maynard D. Davis, Jr. | | 20441422 |
| Sgt. Raymond W. Duran | | 32863803 |
| Sgt. William D. George | | 14142421 |
| Sgt. Gerald D. Yoquelet | | 35550361 |

28 May 1944 to 715th Squadron
| | | |
|---|---|---|
| 2nd Lt. Leroy A. Dunston | P | 0693793 |
| 2nd Lt. Charles F. Gavalek | CP | 0708441 |
| 2nd Lt. Robert D. Kitchingman | B | 0699376 |
| F/O Walter Woodward | N | T-123256 |
| SSgt. Earl W. Bernard | | 16014444 |
| SSgt. Emil A. Glos, Jr. | | 39279338 |
| Sgt. Joseph I. Berg | | 36809886 |
| Sgt. Tony Burciaga | | 38415479 |
| Sgt. Blaine O. Bonewell | | 38563284 |
| Sgt. Julian Proctor | | 34651641 |

Replacement Crews – June, 1944

June 2 1944 to 713th Squadron
| | | |
|---|---|---|
| 2nd Lt. Leland L. Beckman | P | 0693915 |
| 2nd Lt. Charles A. Yant | CP | 0819243 |
| 2nd Lt. William F. Gavenda | B | 0706926 |
| 2nd Lt. Morris F. Epps | N | 0698536 |
| SSgt. Avery L. Knight | | 20951889 |
| SSgt. Robert R. Reed | | 31257194 |
| Sgt. Dowey Conn | | 15119160 |
| Sgt. George C. Copeland | | 18108467 |
| Sgt. Michael J. Eannone | | 32891699 |
| Sgt. Dan A. Wais | | 35788495 |

2 June 1944 to 713th Squadron
| | | |
|---|---|---|
| 2nd Lt. George W. Wilson | P | 0692843 |
| 2nd Lt. William Beck | CP | 0817378 |
| 2nd Lt. Robert J. Gann | B | 0707040 |
| 2nd Lt. Stash J. Fridye | N | 0698543 |
| SSgt. Emil J. Lukas | | 16077719 |
| SSgt. Edward J. McNulty | | 12121713 |
| Sgt. Richard J. Liedahl | | 37558286 |
| Sgt. Albert J. McKinnon, Jr. | | 38465380 |
| Sgt. Robert A. Rudolph | | 12148861 |
| Sgt. Joseph B. Starek | | 33423281 |

4 June 1944 to 712th Squadron
| | | |
|---|---|---|
| 2nd Lt. Steward W. Felker | P | 0813491 |
| 2nd Lt. Walter J. Johnson | CP | 0702441 |
| 2nd Lt. James F. Beaver | B | 0760987 |
| 2nd Lt. Robert G. Edwards | N | 0699063 |
| SSgt. Wilbur L. Riffle | | 33416298 |
| SSgt. James J. Donovan | | 33054081 |
| Sgt. William M. Henderson | | 39573130 |
| Sgt. Robert D. Pollock | | 37507377 |
| Sgt. Fred W. Waxler | | 16170709 |
| Cpl Andrew W. Provencher | | 31321195 |

4 June 1944 to 715th Squadron
| | | |
|---|---|---|
| 2nd Lt. Leroy Conner | P | 0690934 |
| 2nd Lt. Rex H. George | CP | 0708445 |
| 2nd Lt. Thaddeus L. Grochowski | B | 0702270 |
| 2nd Lt. Alton L. Kraft | N | 0702055 |
| SSgt. Albert R. Sabo | | 15112917 |
| SSgt. John A. Shawkey | | 33410285 |
| Sgt. John H. Bretthauer | | 32864850 |
| Sgt. Henry H. Mazer | | 11019077 |
| Sgt. David P. Patterson | | 19178400 |
| Sgt. Donald L. Wright | | 35094415 |

5 June 1944 to 712th Squadron (This must be a reassignment to the 712th Squadron. Note that this crew arrived earlier on May 13th, 1944 and was assigned to the 715th Squadron. Officers & men were listed in a different rotation.)

| | | | |
|---|---|---|---|
| 2nd Lt. William A. Warke | P | 0693082 |
| 2nd Lt. Robert T. White | CP | 0684924 |
| 2nd Lt. Edgar C. Scanlon, Jr. | B | 0701312 |
| 2nd Lt. Henry S. Hasiak | N | 0702043 |
| TSgt. Robert J. Cashin | | 37553432 |
| TSgt. Jerry M. Parsons | | 16035535 |
| SSgt. Robert W. Thompson | | 17066011 |
| SSgt. Norman H. Johnson | | 36740128 |
| SSgt. Gordon W. Wright | | 12072510 |
| SSgt. Millard L. Brown | | 15171462 |

5 June 1944 to 712th Squadron

| | | | |
|---|---|---|---|
| 2nd Lt. George W. Booth | P | 0815061 |
| 2nd Lt. Richard W. Davies | CP | 0717575 |
| 2nd Lt. Somon G. Pilson | B | 0821353 |
| 2nd Lt. Bryandt S. Wilson | N | 0698625 |
| SSgt. George M. Hansen | | 36440522 |
| SSgt. Robert D. Long | | 39281985 |
| Sgt. Norman H. Cohen | | 11054347 |
| Sgt. Stephen V. Lawnicki | | 36649419 |
| Sgt. William N. Crisler, Jr. | | 18177131 |
| Sgt. Floyd D. Leverett | | 38449723 |

5 June 1944 to 714th Squadron

| | | | |
|---|---|---|---|
| 2nd Lt. Mauro Dellaselva | P | 0816706 |
| 2nd Lt. Jerome Israel | CP | 0712572 |
| 2nd Lt. Cyrus J. Alexander | B | 0704866 |
| 2nd Lt. Milton V. Bates | N | 0703423 |
| SSgt. Leo H. Mays | | 33532342 |
| SSgt. Leo J. Lorenz, Jr. | | 16083793 |
| Sgt. Raymond R. Baldridge | | 38415285 |
| Sgt. James M. Childers | | 36343748 |
| Sgt. Gerald M. Brooks | | 36458478 |
| Sgt. Raymond L. Wolhaupter | | 37563766 |

6 June 1944 to 715th Squadron

| | | | |
|---|---|---|---|
| 2nd Lt. Wilmer E. Goad | P | 0684313 |
| 2nd Lt. Donald L. Farrar | CP | 0692542 |
| 2nd Lt. Robert C. Hagan | B | 0694668 |
| 2nd Lt. Matt C. Reynolds | N | 0690822 |
| SSgt. Jack W. Dougan | | 33411369 |
| SSgt. James C. Flowe | | 34606522 |
| SSgt. William D. Crist | | 17166279 |
| SSgt. Anthony A. Raschi | | 11105581 |
| SSgt. Samuel Sherkin | | 32619909 |
| Sgt. Glenn K. Copeland | | 17128185 |

7 June 1944 to 712th Squadron

| | | | |
|---|---|---|---|
| 2nd Lt. Aldrich A. Drahos | P | 0695733 |
| 2nd Lt. Harry Schwartz | CP | 0820841 |
| 2nd Lt. Robert E. Lee | B | 0712609 |
| F/O James R. Horst | N | T-124037 |
| SSgt. John H. Copeland | | 35544943 |
| SSgt. Lorene R. Alexander | | 34771241 |
| Sgt. Wilbur M. Hedblade | | 39573636 |
| Sgt. James L. Houston, Jr. | | 14135586 |
| Sgt. Arthur H. Peterson | | 32393682 |

7 June 1944 to 714th Squadron

| | | | |
|---|---|---|---|
| 2nd Lt. Billie C. Blanton | P | 0663471 |
| 2nd Lt. George O. Brown | CP | 0701892 |
| 2nd Lt. George E. Klein | B | 0712592 |
| 2nd Lt. Fred Berkard | N | 0697984 |
| SSgt. Adrian J. Denbroeder | | 31242442 |
| SSgt. Robert P. Lawson | | 39477278 |
| Sgt. Armer L. McMann III | | 33506938 |
| Sgt. Paul E. Sherlock | | 35790464 |
| Sgt. Salvatore J. Sparacio | | 32639951 |
| Sgt. Bernard Stelzer | | 12127655 |

7 June 1944 to 713th Squadron

| | | | |
|---|---|---|---|
| 2nd Lt. Charles E. Fouche | P | 0695786 |
| 2nd Lt. Grenville K. Baker | CP | 0819192 |
| 2nd Lt. Samuel E. Martin | B | 0712072 |
| 2nd Lt. Alan N. Houghton | N | 0704204 |
| SSgt. Eugene E. Mortimer | | 15121484 |
| SSgt. Harry A. Striker, Jr. | | ????????? |
| Sgt. Charles B. Lackner | | 16132383 |
| Sgt. Walter Dukas | | 36741204 |
| Sgt. Albert Marquez | | 16187438 |
| Sgt. William O. Watts | | 6396841 |

7 June 1944 to 715th Squadron

| | | | |
|---|---|---|---|
| 2nd Lt. Marshall L. Johnson | P | 0693924 |
| 2nd Lt. Victor W. Caswell | CP | 0703403 |
| 2nd Lt. Robert E. Mortensen | B | 0818727 |
| F/O Thomas L. Symonds | N | T-123654 |
| SSgt. Jesse M. Stephen | | 15020112 |
| Sgt. Mas W. Barker | | 37237540 |
| Sgt. Isidora Luna | | 38366920 |
| Sgt. James McKibban | | 19185110 |
| Sgt. Bill McClennen | | 13124001 |
| Sgt. Robert B. Morgan | | 18139966 |

10 June 1944 to 712th Squadron

| | | | |
|---|---|---|---|
| 2nd Lt. William C. Dogger | P | 0697624 |
| 2nd Lt. Harvey L. Luke, Jr. | CP | 0712862 |
| 2nd Lt. Joseph Remitz | B | 0705429 |
| 2nd Lt. Raymond E. Sumrell | N | 0703535 |
| SSgt. Wallace Kaplan | | 32417924 |
| SSgt. Richard B. Smith | | 35542432 |
| Sgt. Arthur H. Keeling | | 36422811 |
| Sgt. Clarence W. Kronbetter | | 36459122 |
| Sgt. Ben Maness, Jr. | | 34800896 |
| Sgt. John G. Shia | | 11065201 |

10 June 1944 to 712th Squadron
| | | |
|---|---|---|
| 2nd Lt. Carl A. Eggert | P | 0692864 |
| 2nd Lt. Augustine J. Adomanis | CP | 0820690 |
| 2nd Lt. John W. Cone | B | 0713134 |
| 2nd Lt. Robert E. Hosse | N | 0704203 |
| SSgt. Harry N. Harris | | 16146257 |
| SSgt. Edward P. Schichtel | | 32131077 |
| Sgt. Robert Lloyd | | 33430722 |
| Sgt. Donald F. Mach | | 36459074 |
| Sgt. Harley D. Plante | | 31299299 |
| Sgt. John S. Sexton | | 34138838 |

10 June 1944 to 713th Squadron
| | | |
|---|---|---|
| 1st Lt. Alwyn F. Palmerton | P | 0693942 |
| 2nd Lt. Frederick L. Butler | CP | 0699506 |
| 2nd Lt. John H. McBroom | B | 0817843 |
| 2nd Lt. Roy D. Thompson | N | 0712703 |
| SSgt. Harold Shapiro | | 39277650 |
| Sgt. James E. Wells | | 14170895 |
| Sgt. George H. Bonner | | 31315653 |
| Sgt. Gordon L. Lowe | | 34721324 |
| Sgt. Walter Modjeska | | 35552764 |
| Sgt. Prentiss H. Price | | 34613906 |

10 June 1944 to 714th Squadron
| | | |
|---|---|---|
| 2nd Lt. John J. Milliken | P | 0816799 |
| 2nd Lt. James B. Shafter | | 0820080 |
| 2nd Lt. Stanley Friedman | | 0708437 |
| F/O William C. Barnes | | T123242 |
| SSgt. Leonard J. White | | 20368143 |
| SSgt. Walker L. Lawson | | 33429366 |
| Sgt. Raymand T. Golembiewski | | 36809243 |
| Sgt. Carrol C. Kendall | | 33722172 |
| Sgt. Lionel V. Lozon | | 36589152 |

17 June 1944 to 714th Squadron
| | | |
|---|---|---|
| 2nd Lt. Edward J. Malone | P | 0663538 |
| 2nd Lt. Theodore A. Sengerman | CP | 0823724 |
| 2nd Lt. Wilfred L. Kimball | B | 0708206 |
| 2nd Lt. William F. Kerner | N | 0709923 |
| SSgt. Carl V. Hoppe | | 36562772 |
| SSgt. William P. Lantz | | 3503614 |
| Sgt. Webb M. Floyd | | 20407423 |
| Sgt. Frank D. Hobbs | | 31374364 |
| Sgt. Walter F. Krueger | | 39619583 |
| Pvt. Richard W. Beteau | | 31265832 |

17 June 1944 to 714th Squadron
| | | |
|---|---|---|
| 2nd Lt. Michael M. Senkewitz | P | ??????? |
| 2nd Lt. Milton Halpern | CP | ??????? |
| F/O Michel M. Herbst | B | T-124758 |
| 2nd Lt. Leon Shapiro | N | 0708277 |
| SSgt. James P. Bennett | | 328???81 |
| SSgt. ????? D. Connell | | 17083219 |
| Sgt. Martin M. Bingham | | 32929444 |
| Sgt. Kenneth F. Bradfield | | 37528316 |
| Sgt. Ray B. Crepes | | 16169420 |
| Sgt. Darwin Miller | | 36867001 |

17 June 1944 to 715th Squadron
| | | |
|---|---|---|
| 2nd Lt. Cecil R. Frensko | P | 0696355 |
| 2nd Lt. Odus E. Jones | CP | 0819405 |
| 2nd Lt. Richard B. Kimball | B | 0712132 |
| 2nd Lt. Thomas C. Breckson | N | 0708?46 |
| SSgt. Stanley S. Tabel | | 3131?899 |
| SSgt. Floyd K. Gardner | | 36386669 |
| Sgt. John J. Hadamik | | 36650832 |
| Sgt. Bernard A. Kelley | | 17120834 |
| Sgt. John G. Legner | | 35788445 |
| Sgt. Antonio A. Sciachitano | | 17155164 |

21 June 1944 to 712th Squadron
| | | |
|---|---|---|
| Capt. Gordon R. Koons | P | 0428507 |
| 2nd Lt. Edward E. Hanson | CP | 0820006 |
| 2nd Lt. Ralph C. Dimick | B | 0713007 |
| 2nd Lt. Charles W. Bonner | N | 0703429 |
| SSgt. Robert M. Avery | | 19055485 |
| SSgt. Elven O. Coleman | | 14142308 |
| Sgt. Edward E. McGinnis | | 39208597 |
| Sgt. James C. Popp | | 33675524 |
| Sgt. John L. Meeker | | 38223332 |
| Sgt. Enoch C. Slack | | 34607237 |

21 June 1944 to 713th Squadron
| | | |
|---|---|---|
| Capt. Andrew J. Andersen, Jr. | P | 0391093 |
| 2nd Lt. John W. Allen, Jr. | CP | 0763828 |
| 2nd Lt. William G. Newlon | B | 0712419 |
| 2nd Lt. Vernon L. Gilmore | N | 0698883 |
| SSgt. Mathew W. Spahn | | 36216344 |
| SSgt. John H. Reis | | 15113353 |
| Sgt. Otto R. Palumbo | | 32840613 |
| Sgt. Robert C. Schweitzer | | 33682852 |
| Sgt. Robert F. Kalous | | 32880507 |
| Sgt. Charles F. Donati, Jr. | | 32787054 |

21 June 1944 to 714th Squadron
| | | |
|---|---|---|
| 2nd Lt. Lloyd H. Haddock | P | 0695739 |
| 2nd Lt. Robert J. Rentschler | CP | 0819414 |
| 2nd Lt. Crystal Lang | B | 0712606 |
| 2nd Lt. Pat Farris | N | 0706827 |
| SSgt. Franklin Holtmeier | | 33171371 |
| SSgt. Richard M. Kennedy | | 12096831 |
| Sgt. Ernesto Concepcion | | 32886753 |
| Sgt. Lester E. Seabaugh | | 37519903 |
| Sgt. William V. Pyke | | 19161730 |
| Sgt. Everett W. Marah | | 37620884 |

22 June 1944 to 714th Squadron
1st Lt. Harold C. Turpin           P     0807541
2nd Lt. Charles W. Wilson          CP    0704863
2nd Lt. Robert J. Nystrom          B     0715412
2nd Lt. Theodore J. Conway, Jr.    N     0703454
Sgt. Thomas A. Loughlin, Jr.             36851964
SSgt. Keith D. Wann                      39117819
SSgt. Walter D. Bookamer                 13109114
Sgt. James S. Smidy                      33667892
Sgt. Joseph Skledar, Jr.                 33431803
Sgt. Leo A. VanWorst                     32813059

22 June 1944 to 714th Squadron
1st Lt. James H. Brownell          P     0392071
2nd Lt. Roy M. Johnston            CP    0704941
2nd Lt. Lawrence R. Hastings       B     0713177
2nd Lt. John E. Cumming            N     0703461
SSgt. Stanley E. Womack                  39279099
SSgt. Beurall K. Binns                   38452541
Sgt. Walter Johnson, Jr.                 37497081
Sgt. Robert B. Kress                     35632387
Sgt. Lee A. Parsons                      35218774
Sgt. Clarence M. Smith                   34607622

22 June 1944 to 715th Squadron
1st Lt. Marcus S. Horton           P     0885996
2nd Lt. Floyd D. Mahl              CP    0712415
2nd Lt. Erwin J. Kaidy             B     0764139
2nd Lt. Milton Q. Alber            N     0685414
SSgt. Kenneth W. Prieb                   17131824
SSgt. Lynn H. Satterfied                 34726926
Sgt. Robert E. Morel                     35800893
Sgt. Robert L. Reeves                    19136151
Sgt. Harold E. Fowler                    34677771
Sgt. Fred A. Ozbirn                      39039218

27 June 1944 to 712th Squadron
2nd Lt. James E. Hande             P     0700924
2nd Lt. James W. Kelly             CP    0767562
2nd Lt. Stanley Milberg            B     0711458
F/O Thomas S. Dolan                N     T-2515
SSgt. Ernest J. Atchley                  14123141
SSgt. Arthur N. Weisz                    15374533
Sgt. James R. Martin                     14190906
Cpl. Frank Gavura                        35892692
Cpl. John C. Guisto                      32894262
Cpl. Walter L. Buhr                      37538204

27 June 1944 to 712th Squadron
2nd Lt. Harry W. Kraus             P     0761129
2nd Lt. Edward T. Luszcz           CP    0768150
2nd Lt. Lawrence J. Wolfe          B     0708605
2nd Lt. Norval S. Hovey            N     0766270
Sgt. John Q. Adams                       36813515

Sgt. Kuell Hinson                        34729617
Sgt. Earl R. Kennedy                     38439451
Cpl. Earl J. Estes                       35892315
Cpl. Warren W. Fankhauser                35759849
Cpl. William K. Perry                    38419774

27 June 1944 to 713th Squadron
2nd Lt. William C. Beall           P     0761852
2nd Lt. Harry E. Betts             CP    0767977
2nd Lt. Harvey C. Baker            B     0709463
2nd Lt. Charles E. Langton, Jr.    N     0766294
SSgt. Otto K. Smith                      18202005
Sgt. John C. Bruno                       32605097
Sgt. John M. Jenkins, Jr.                34602346
Cpl. George W. Dickinson                 20225687
Cpl. Carl E. Himes                       33757446
Cpl. Peter J. Murray                     32456666

27 June 1944 to 714th Squadron
2nd Lt. James J. Schierbrock       P     0699776
2nd Lt. Charles W. Meining         CP    0767599
2nd Lt. Willis E. Cobb             B     0711341
2nd Lt. Glen M. Larsen             N     0766295
SSgt. Francis P. Horan                   19089823
SSgt. Robert E. Simmons                  11082391
Sgt. George D. Jeffries                  15377838
Sgt. Joseph T. Marek                     39557546
Sgt. Verlon A. Pallmer                   38507597
Cpl. William W. Rowe                     38352599

28 June 1944 to 713th Squadron
F/O Elliott J. Sidey               P     T-1882
2nd Lt. John P. Deren              CP    0712088
F/O Melvin Krisel                  B     T-12349
2nd Lt. William E. Sallade II      N     0716961
SSgt. John S. Thomson                    13112482
SSgt. Louis A. Owens                     35094874
SSgt. Ralph E. Mull                      33236631
Sgt. Ernest R. Easterling                14130912
Sgt. John M. MacDonald                   39618727
Sgt. Michael Perkowski                   32826132

28 June 1944 to 713th Squadron
2nd Lt. Dale E. Grubb              P     0697412
2nd Lt. Edwin W. Carnahan          CP    0705227
2nd Lt. Bernard Epstein            B     0712535
2nd Lt. Norman G. Marks            N     0716922
SSgt. Frank S. Thomas                    34800965
SSgt. John E. Everett                    34608727
Sgt. John O. Barnes                      37282755
Sgt. John R. Etherington                 32487582
Sgt. Clare W. Hubbard                    36459200
Sgt. Eusebio Rodriguez, Jr.              18197677

28 June 1944 to 714th Squadron
| | | |
|---|---|---|
| 2nd Lt. Huntington S. Gruening | P | 0760729 |
| 2nd Lt. Thomas C. Dahlgren | CP | 0768472 |
| 2nd Lt. John A. Bolton | B | 0710843 |
| 2nd Lt. William B. Norman | N | 0765341 |
| Sgt. Frank C. Dow | | 14102360 |
| SSgt. Lester F. Haughton | | 16111367 |
| Sgt. Ray L. Syrles | | 38405523 |
| Cpl. George E. Hyde, Jr. | | 37538367 |
| Cpl. Frank J. Wilcheck | | 37620869 |
| Pfc. Edwin B. Hungerford | | 37611178 |

28 June 1944 to 715th Squadron
| | | |
|---|---|---|
| 2nd Lt. Richard M. Moody | P | 0761256 |
| 2nd Lt. Douglas P. Pederson | CP | 0768209 |
| 2nd Lt. Jack Glicksman | B | 0709343 |
| 2nd Lt. Robert M. Thompson | N | 0765768 |
| SSgt. William R. Gamble | | 33675510 |
| Sgt. Robert A. Bacon | | 33453497 |
| Sgt. Raymond R. Grambau | | 16134775 |
| Sgt. Joseph J. Hannon, Jr. | | 20262532 |
| Cpl. William J. Maynard | | 35873288 |
| Cpl. Harold E. McBurney, Jr. | | 37675906 |

28 June 1944 to 712th Squadron
| | | |
|---|---|---|
| 2nd Lt. Orville L. Daenzer | P | 0700396 |
| 2nd Lt. Paul L. Grossinger | CP | 0825622 |
| 2nd Lt. John C. Morris | B | 0712886 |
| 2nd Lt. Robert S. Wheeler | N | 0717004 |
| SSgt. Charles F. Sparenberg | | 13075941 |
| SSgt. Hugh K. Burleigh | | 13021606 |
| Sgt. Clifford Blalock | | 36539045 |
| Sgt. John A. Czarnowski | | 31089542 |
| Sgt. Ned W. Thomas | | 33255948 |

28 June 1944 to 712th Squadron
| | | |
|---|---|---|
| 2nd Lt. Harold A. Piper | P | 0700535 |
| 2nd Lt. Edwin E. Peckmann | CP | 0764133 |
| 2nd Lt. Daniel F. Mangin | B | 0713037 |
| 2nd Lt. Maurice L. Ashkikaz | N | 0716820 |
| SSgt. Angelo W. Percacciolo | | 32867227 |
| SSgt. Arthur F. Decker | | 36480121 |
| Sgt. Robert C. Baker | | 39212940 |
| Sgt. David N. Cassell | | 33655802 |
| Sgt. Gilbert F. Morris, Jr. | | 38344789 |
| Sgt. Albert I. Schletter | | 32934891 |

28 June 1944 to 713th Squadron
| | | |
|---|---|---|
| 2nd Lt. Wallace C. Score | P | 0819983 |
| 2nd Lt. Leo F. McGeough | CP | 0706542 |
| 2nd Lt. Michael R. Carestio | B | 0706631 |
| 2nd Lt. Ray B. Bremer | N | 0696631 |
| SSgt. William J. Harkins | | 33589393 |
| SSgt. Harry Whitfield, Jr. | | 32725008 |
| Sgt. Curtis O. Brown | | 37085123 |
| Sgt. Edward L. Kellams | | 33540211 |
| Sgt. Daniel A. Paris | | 36867333 |

28 June 1944 to 714th Squadron
| | | |
|---|---|---|
| 2nd Lt. Bennie F. Adams | P | 0818546 |
| 2nd Lt. Robert J. Byrne | CP | 0716828 |
| 2nd Lt. Frank ? Webb | B | 0717311 |
| F/O Richard E. Winslow | N | T-2373 |
| SSgt. Eugene F. Norman | | 20761171 |
| SSgt. Clyde A. Randall | | 39561345 |
| Sgt. Lawrence R. Berger, Jr. | | 13158242 |
| Sgt. Delmar E. Knight | | 37477496 |
| Sgt. Leonard R. Pannell | | 37534218 |
| Sgt. Kenneth O. Ryhal | | 35056220 |

Replacement Crews – July 1944

2 July 1944 to 712th Squadron
| | | |
|---|---|---|
| 1st Lt. William W. Snavely | P | 026257 |
| 2nd Lt. Marion T. Sebastian | CP | 0689169 |
| F/O William R. Morris | N | T-2103 |
| SSgt. Frederick C. Aldrich | | 11084306 |
| SSgt. Lawrence W. Barham | | 39199288 |
| Sgt. Marvin W. Hicks | | 34765397 |
| Sgt. Robert A. Grabowski | | 16136791 |
| Sgt. Thomas A. Logue | | 34506317 |
| Sgt. Frank L. Parkinson, Jr. | | 38370437 |

2 July 1944 to 712th Squadron
| | | |
|---|---|---|
| 2nd Lt. William W. Gilbert | P | 0686239 |
| 2nd Lt. Anson F. Barton | CP | 0821143 |
| 2nd Lt. Willard D. Powers | N | 0691911 |
| SSgt. Edward F. Daum | | 13038479 |
| SSgt. Doriel S. Gilbert | | 39905264 |
| Sgt. William R. Fisher, Jr. | | 17129136 |
| Sgt. Robert R. Jendrusiak | | 32189572 |
| Cpl. Herbert B. Kemp | | 34396777 |
| Cpl. Victor R. Sardue | | 12054350 |

2 July 1944 to 714th Squadron
| | | |
|---|---|---|
| Capt. Walter W. Dillon | P | 024832 |
| 2nd Lt. John E. Harmer | CP | 0817431 |
| 1st Lt. Charles W. Parish | N | 0734495 |
| TSgt. Filmore G. Layman | | 18058079 |
| TSgt. Edward J Bornheimer | | 36125498 |
| SSgt. Harry C. Barney | | 11101925 |
| Sgt. Morton H. Kessler | | 12203638 |
| Sgt. Erwin S. Kostick | | 36285883 |
| Sgt. George E. Nugent | | 37356752 |

2 July 1944 to 715th Squadron
| | | |
|---|---|---|
| 2nd Lt. Douglas O. Morse | P | 0686267 |
| 2nd Lt. John J. Sullivan | CP | 0696718 |

| | | |
|---|---|---|
| 2nd Lt. Raymond F. Knight | N | 0744532 |
| SSgt. Clarence J. Russi | | 35599476 |
| SSgt. Truman K. Anderson | | 18167649 |
| Sgt. Arlie H. von Tersch | | 39452865 |
| Sgt. Stanley Dubee | | 12147992 |
| Sgt. Floyd D. Eckrosh | | 17097350 |
| Sgt. Robert J. Frolli | | 19042400 |

4 July 1944 to 714th Squadron

| | | |
|---|---|---|
| 2nd Lt. Thomas F. Mulligan | P | 0691620 |
| 2nd Lt. Richard H. Wright | CP | 0687730 |
| 2nd Lt. Edwin F. Hewitt | N | 0813700 |
| SSgt. George C. Contois | | 11085347 |
| SSgt. Steven M. Alexander | | 18045347 |
| Sgt. James M. Fisher | | 39905167 |
| Sgt. Hugh L. Manchester | | 32852631 |
| Sgt. Harold V. P. Shultis | | 31285894 |
| Sgt. Lloyd D. Reid | | 31303765 |

4 July 1944 to 713th Squadron

| | | |
|---|---|---|
| 2nd Lt. John D. Sutton | P | 0812683 |
| F/O Henry J. Radziswicz | CP | T-123686 |
| 2nd Lt. Irving Manin | N | 0805948 |
| SSgt. Marvin R. Biazzo | | 32452657 |
| SSgt. Vincent Cautero, Jr. | | 12159538 |
| Sgt. John J. Hattersley | | 32910112 |
| Sgt. Chester W. Hartley | | 37614749 |
| Sgt. Warren A. Jeffries | | 17144034 |
| Sgt. Edward E. Skuba | | 11043905 |

4 July 1944 to 712th Squadron

| | | |
|---|---|---|
| 2nd Lt. Harold S. Spicer | P | 0815255 |
| 2nd Lt. Paul S. Elder | CP | 0819506 |
| 2nd Lt. Frank J. Plaushin | N | 0700763 |
| SSgt. Jerome Stuart | | 32863671 |
| Sgt. Archie A. Brajkovich | | 37666901 |
| Sgt. Calvin C. Burroughs | | 38517521 |
| Sgt. Richard L. Markham | | 13117459 |
| Sgt. Frank J. Kromer | | 33690091 |
| SSgt. Earl L. Anderson | | 35151070 |

4 July 1944 to 712th Squadron

| | | |
|---|---|---|
| 2nd Lt. Roy E. Stahl | P | 0695782 |
| F/O Sylvester Krol | CP | T-125521 |
| 2nd Lt. Leo M. Conner | N | 0768005 |
| 2nd Lt. Warren W. Hoster, Jr. | B | 0703143 |
| SSgt. Larry L. Bush | | 18050809 |
| SSgt. Verlyn L. Colby | | 38396157 |
| Sgt. Leon M. Renier | | 37539361 |
| Sgt. Emmett R. Wallace | | 38451228 |
| Sgt. Harold G. Von Needa, Jr. | | 33504399 |
| Pvt. Carl R. Lane | | 19125608 |

4 July 1944 to 714th Squadron

| | | |
|---|---|---|
| 2nd Lt. William T. Hensey, Jr. | P | 0699416 |
| 2nd Lt. Kenneth E. Ensign | CP | 0768043 |
| F/O Jack S. Comer | N | T-124501 |
| 2nd Lt. Clarence H. Mellor | B | 0766323 |
| Sgt. William A. Lillard | | 34724435 |
| MSgt. Walton B. Kelly | | 18002434 |
| Sgt. George L. Allen | | 34580133 |
| Cpl. Larry L. Archambault | | 35913851 |
| Cpl. Charles E. Grant | | 34763091 |
| Cpl. John J. Thompson | | 32147899 |

4 July 1944 to 712th Squadron

| | | |
|---|---|---|
| 2nd Lt. Rudy H. Johnson | P | 0699423 |
| 2nd Lt. William I. Hall | CP | 0768529 |
| 2nd Lt. Frederick W. Schaefer, Jr. | N | 0710389 |
| 2nd Lt. Theodore E. Matson | B | 0766320 |
| Sgt. John R. Dunkel, Jr. | | 13030915 |
| Sgt. Charles R. Gambino | | 34800374 |
| Cpl. Parry L. Alexander | | 39709129 |
| Cpl. Earl L. Arnett | | 19122288 |
| Cpl. Albert M. Brown | | 38498077 |
| Cpl. Frank S. Brown | | 39709160 |

15 July 1944 to 713th Squadron

| | | |
|---|---|---|
| 2nd Lt. Andrew T. Panchura | P | 0821068 |
| 2nd Lt. Charles M. Epes, Jr. | CP | 0827794 |
| 2nd Lt. Edward L. Burnetta | N | 0712994 |
| 2nd Lt. Garland L. Purvis | B | 0768894 |
| Sgt. Charles M. Lighty | | 19093022 |
| Sgt. Richard D. Baker | | 6565398 |
| Sgt. Carlos M. Dreyfus | | 12146077 |
| Sgt. William J. McCollum | | 33831280 |
| Sgt. Lloyd H. Searle | | 11017107 |
| Sgt. Robert R. Shrode | | 39277959 |

15 July 1944 to 713th Squadron

| | | |
|---|---|---|
| 2nd Lt. Edmond G. Postemsky | P | 0821075 |
| 2nd Lt. Clifford B. Unwin | CP | 0827089 |
| 2nd Lt. Wyllys B. Jennings | N | 0715414 |
| 2nd Lt. James T. Bell | B | 0769089 |
| Sgt. Alessandro L. Falconi | | 31231533 |
| SSgt. LeRoy J. Canning | | 13151559 |
| Sgt. LeRoy J. Brizzolara | | 32771334 |
| Cpl. Adley V. Benoit | | 38486715 |
| Cpl. Charles C. Payne | | 34737456 |
| Cpl. Julien Polge | | 32937895 |

15 July 1944 to 715th Squadron

| | | |
|---|---|---|
| 2nd Lt. Carl H. Holt | P | 0821282 |
| 2nd Lt. John F. Cushman | CP | 0827787 |
| 2nd Lt. Gordon L. Britt | N | 0712372 |
| 2nd Lt. James E. Fields | B | 0769111 |
| Sgt. James F. Kiely | | 11069461 |

| | | | |
|---|---|---|---|
| SSgt. Williams P. Jones | | | 14024677 |
| SSgt. Romagene Tiner | | | 38175134 |
| Sgt. Bertram Charnow | | | 32693364 |
| Cpl. Michael J. Hill | | | 33703706 |
| Cpl. Robert H. Kessler | | | 12204555 |
| | | | |
| **15 July 1944 to 715th Squadron** | | | |
| 2nd Lt. Harry G. Allen, Jr. | P | | 0820694 |
| 2nd Lt. Donald L. Allen | CP | | 0823215 |
| 2nd Lt. Harry O. Wolfe, Jr. | N | | 0715414 |
| 2nd Lt. Donald E. Burke | B | | 0769089 |
| Cpl. Harold Freedman | | | 31309882 |
| SSgt. Albert R. Pizzoli | | | 13030192 |
| Sgt. Charles H. Carn | | | 32800347 |
| Sgt. Jesse L. Shugars | | | 36300412 |
| Cpl. Clifford T. Cashbit | | | 6711225 |
| Cpl. Edward H. Cristello | | | 13171893 |
| | | | |
| **19 July 1944 to 712th Squadron** | | | |
| F/O Albert J. Lewis | P | | T-123500 |
| 2nd Lt. John E. Briggs | CP | | 0713345 |
| 2nd Lt. William D. Fitch, Jr. | N | | 0717419 |
| 2nd Lt. Robert F. Rikard | B | | 0706903 |
| Sgt. William F. Smith | | | 33568893 |
| SSgt. Thomas E. Martin | | | 6925869 |
| Sgt. Clinton W. Engledow | | | 39555929 |
| Sgt. Cloyd G. Jordon | | | 12083990 |
| Sgt. Louis C. Schlorman | | | 35622664 |
| Cpl. Maurice M. Finberg | | | 37573487 |
| | | | |
| **19 July 1944 to 712th Squadron** | | | |
| 2nd Lt. Donald G. Ginevan | P | | 0817832 |
| 2nd Lt. Billy R. Greenwade | CP | | 0820260 |
| 2nd Lt. Albert F. Duval | B | | 0716390 |
| 2nd Lt. Charles F. Swanson | N | | 0717118 |
| Sgt. Robert J. Jones | | | 12093034 |
| SSgt. John V. Kary | | | 19071394 |
| Sgt. Robert H. Castell-Blanch | | | 39120014 |
| Sgt. Ned A. Chidester | | | 17014458 |
| Sgt. John H. Mitchell | | | 32910987 |
| Pvt. Charles D. Jones | | | 19169839 |
| | | | |
| **19 July 1944 to 714th Squadron** | | | |
| F/O Hosea E. Matthaes | P | | T-123558 |
| 2nd Lt. Thomas E. Miller | CP | | 0825942 |
| 2nd L.t Aurel H. Muntean | B | | 0717476 |
| 2nd Lt. Martin W. Richards | N | | 0706902 |
| SSgt. Selwyn Kaplan | | | 12088705 |
| Sgt. Vincent J. Haley | | | 37356759 |
| Sgt. Eugene M. Petagine | | | 32712083 |
| Cpl. Virgil C. Combs, Jr. | | | 37539520 |
| Cpl. Robert T. Davis | | | 37622762 |
| Cpl. Frank DeCola | | | 13131566 |

| | | | |
|---|---|---|---|
| **19 July 1944 to 715th Squadron** | | | |
| F/O Dodson B. Graybeal | P | | T-123734 |
| F/O William G. Payne | CP | | T-124156 |
| 2nd Lt. Walter A. Ford | B | | 0717420 |
| F/O Wesley W. Palmer | N | | T-2225 |
| Sgt. Donald C. Renkel | | | 32767448 |
| SSgt. Joe Dovico | | | 37654273 |
| Cpl. John C. Adams | | | 39037615 |
| Cpl. Richard J. Byrne | | | 33733437 |
| Cpl. William J. Hartman | | | 33795867 |
| Cpl. Douglas E. Walker | | | 38507353 |
| | | | |
| **26 July 1944 to 712th Squadron** | | | |
| 2nd Lt. Ernest G. Brock | P | | 0696423 |
| 2nd Lt. Paul E. Cooper | CP | | 0760790 |
| 2nd Lt. Paul Horvath | N | | 0712730 |
| F/O Edward H. Kiefer | B | | T-2521 |
| Sgt. William R. Frees | | | 39169984 |
| Sgt. Ernest Jerrell | | | 35702077 |
| Cpl. Edward R. Hess | | | 12122431 |
| Cpl. Eugene E. Loose | | | 17122186 |
| Cpl. Clyde C. White, Jr. | | | 39323460 |
| SSgt. Russell C. Miller, Jr. | | | 33432255 |
| | | | |
| **26 July 1944 to 713th Squadron** | | | |
| 2nd Lt. Sherman F. Furey, Jr. | P | | 0699838 |
| 2nd Lt. William J. Dickinson | CP | | 0768372 |
| 2nd Lt. Clifford Linder | B | | 0712617 |
| 2nd Lt. Rex B. Olson | N | | 0766344 |
| Sgt. Charles Granato | | | 38458037 |
| Sgt. Philip J. Donovan | | | 31259007 |
| Sgt. Florence T. McCarthy | | | 31308816 |
| Cpl. Colston H. Browne | | | 32957421 |
| Cpl. Walter C. Hoke | | | 33507193 |
| Cpl. Stephen L. Kott | | | 32872778 |
| | | | |
| **26 July 1944 to 714th Squadron** | | | |
| 2nd Lt. Bernard Hansen | P | | 0699815 |
| 2nd Lt. Earl W. Aldrich | CP | | 0767961 |
| 2nd Lt. William C. Richardson | B | | 0713226 |
| 2nd L.t Enrico P. Maggenti | N | | 0766513 |
| Sgt. Curtis C. Drouillard | | | 36570067 |
| SSgt. Kenneth T. Rariden, Jr. | | | 16149505 |
| Sgt. Lee R. Cosby | | | 18135077 |
| Sgt. Jacob A. Womack | | | 38508815 |
| Cpl. John Birkhead | | | 35727385 |
| Cpl. Linn C. Garrison | | | 37484571 |
| | | | |
| **26 July 1944 to 715th Squadron** | | | |
| 2nd Lt. Parmely T. Ferrie | P | | 0701915 |
| F/O Theodore E. Bujalski | CP | | T-2725 |
| 2nd Lt. Michael S. Onderick | B | | 0712652 |
| 2nd Lt. Harold R. Ingebrigtsen | N | | 0766288 |
| Sgt. James W. Riley | | | 31111543 |

## Appendices

| | | |
|---|---|---|
| SSgt. Ernest J. Hudgens | | 38413048 |
| Sgt. Leonard H. Campbell | | 18007499 |
| Sgt. George F. Mahar | | 32025159 |
| Sgt. Eugene L. Pointer | | 20759244 |
| Cpl. Charles C. Wolfe | | 20819988 |

30 July 1944 to 715th Squadron

| | | |
|---|---|---|
| Capt. James J. Allen | P | 048090 |
| 2nd Lt. Albert A. Pempek | CP | 0699905 |
| F/O Ralph F. O'Neil | B | T123131 |
| SSgt. James A. Lepold | | 16105957 |
| SSgt. Doanld Zeldin | | 33582669 |
| Sgt. Joseph P. Hadzima | | 35528870 |
| Sgt. Frank J. McGrady | | 32413339 |
| Sgt. Miles G. Seamans | | 31099065 |
| Sgt. John E. Wilson | | 33675465 |

31 July 1944 to 712th Squadron

| | | |
|---|---|---|
| 2nd Lt. William H. Wilhelmi | P | 0705193 |
| F/O John E. Lariviere | CP | T-125458 |
| 2nd Lt. Billy J. Baker | B | 0723267 |
| 2nd Lt. John W. Bice | N | 0722136 |
| Cpl. John G. Vahle, Jr. | | 17071832 |
| Cpl. Lewis T. Kidston | | 31236198 |
| Sgt. Maxwell W. Mackenzie | | 39192491 |
| Cpl. Joseph Longo | | 36818513 |
| Cpl. Jack J. O'Donnel | | 15333914 |
| Cpl. Julius Sakovics | | 13153800 |

31 July 1944 to 713th Squadron

| | | |
|---|---|---|
| 2nd Lt. Gordon F. Hillman | P | 0753285 |
| 2nd Lt. Alden J. Hershiser | CP | 0768545 |
| 2nd Lt. Herbert E. MacNeil | B | 0712626 |
| 2nd Lt. Donald G. Ziebell | N | 0717014 |
| SSgt. Benson F. Quisenberry | | 37220343 |
| SSgt. Leonard R. Saunders | | 38400777 |
| Sgt. Walter G. Cheslock, Jr. | | 33603748 |
| Sgt. Anthony J. Dachille | | 33795339 |
| Sgt. Frank G. Robertson | | 34713843 |
| Sgt. Stanley Z. Swiencki | | 33603529 |

31 July 1944 to 713th Squadron

| | | |
|---|---|---|
| 2nd Lt. Allan C. Wight | P | 0705191 |
| 2nd Lt. Murray R. Apfelbaum | CP | 0715930 |
| 2nd Lt. Donald V. Flanders | B | 0715104 |
| 2nd Lt. Donald H. Longley | N | 0772187 |
| SSgt. Arthur M. Harrington | | 32028157 |
| Sgt. Robert G. Kennohan | | 39130128 |
| Cpl. Donald E. Corson | | 18097867 |
| Cpl. Benjamin F. Edwards, Jr. | | 34637549 |
| Cpl. James A. Hauersperger | | 36743743 |
| Cpl. Philip G. Zapp | | 15113968 |

31 July 1944 to 714th Squadron

| | | |
|---|---|---|
| 2nd Lt. Sidney R. Williamson | P | 0705469 |
| 2nd Lt. Howard O. Sandbeck | CP | 0715930 |
| 2nd Lt. Miles S. Baldwin | N | 0723555 |
| 2nd Lt. Raymond T. Binkley | B | 0706675 |
| Sgt. William J. Arnone | | 19180982 |
| Sgt. Lawrence N. Keeran | | 18166757 |
| Cpl. Edwin C. Brannan | | 34829295 |
| Cpl. Donald W. Clapp, Jr. | | 31190423 |
| Cpl. Dante J. Macario | | 33727569 |
| Cpl. Charles A. Spapperi | | 36693347 |

Replacement Crews– August 1944

3 August 1944 to 712th Squadron

| | | |
|---|---|---|
| 2nd Lt. Herbert H. Jonson | P | 0764033 |
| 2nd Lt. Lawrence H. Daniels | CP | 0772305 |
| 2nd Lt. Alexander J. Prieske | B | 0713105 |
| 2nd Lt. John D. Caldwell | N | 0769003 |
| Cpl. Edward V. Langowski | | 16089625 |
| Cpl. George A. Constable | | 33262058 |
| Cpl. William M. Craigmile | | 16171891 |
| Cpl. Luttia W. Rehar | | 33759444 |
| Cpl. William C. Price | | 39920430 |
| Cpl. John R. Rainwater | | 38350831 |

3 August 1944 to 713th Squadron

| | | |
|---|---|---|
| 2nd Lt. John A. Jordon | P | 0764691 |
| 2nd Lt. Norman S. Harris | CP | 0809951 |
| 2nd Lt. Dwain Butler | B | 0713118 |
| 2nd Lt. Harold Cherry | N | 0768937 |
| Cpl. Charles A. Mainini | | 39694607 |
| Cpl. Joseph Hollowatch | | 32558010 |
| SSgt. Earl Fannin | | 15017821 |
| Cpl. John McConnell | | 32866523 |
| Cpl. Walter G. Rush | | 39017617 |
| Cpl. Bill B. Rysor | | 38589094 |

3 August 1944 to 714th Squadron

| | | |
|---|---|---|
| 2nd Lt. Wilbur C. Bryson | P | 0764111 |
| 2nd Lt. William L. Horrell | CP | 0772380 |
| 2nd Lt. Harry D. Freivogel | B | 0713163 |
| 2nd Lt. Bernard L. Parsons | N | 0769060 |
| Cpl. Henry A. Remsburg | | 35157582 |
| Cpl. Robert R. Brady | | 33626111 |
| Cpl. Gasper W. Interrante | | 33832534 |
| Cpl. Donald J. Kamler | | 32820201 |
| Pvt. Francis B. Neumann | | 17114026 |
| Cpl. Robert C. Weber | | 35216675 |

3 August 1944 to 715th Squadron

| | | |
|---|---|---|
| 2nd Lt. Howard E. Doane | P | 0763970 |
| 2nd Lt. Earl H. Pattee | CP | 0772473 |
| 2nd Lt. Eugene Forman | B | 0715014 |

| | | | |
|---|---|---|---|
| 2nd Lt. John S. Moll, Jr. | N | 0769055 | |
| Cpl. Zigmund Ozimkowski | | 36565548 | |
| Cpl. Luke G. Eresnahan | | 42030668 | |
| Cpl. Melvin F. Foss | | 19022561 | |
| Cpl. Charles J. Jackson | | 15103207 | |
| Cpl. Archie J. Taylor | | 38563189 | |
| Cpl. Robert I. Ussak | | 12086689 | |

**8 August 1944 to 714th Squadron**

| | | | |
|---|---|---|---|
| 2nd Lt. Harold G. Soldan | P | 0701967 | |
| 2nd Lt. Clement L. Maher | CP | 0770700 | |
| 2nd Lt. Burrell E. Weaver | B | 0723217 | |
| 2nd Lt. John B. Wade, Jr. | N | 0772901 | |
| Cpl. James W. McConkie | | 19120091 | |
| Cpl. James C. Alexander | | 33514768 | |
| Cpl. Peter J. Campbell | | 11113767 | |
| Cpl. Dale L. Emlet | | 13122914 | |
| Cpl. Arthur E. Evans | | 35339481 | |
| Cpl. Philip G. Farnsworth | | 31318641 | |

**8 August 1944 to 714th Squadron**

| | | | |
|---|---|---|---|
| 2nd Lt. John M. Buxton | P | 0703311 | |
| 2nd Lt. Howard D. Hinckley | CP | 0771716 | |
| 2nd Lt. Earl T. Brown | B | 0722813 | |
| 2nd Lt. Robert D. Butler | N | 0772901 | |
| Cpl. Leo E. Stephens | | 37558474 | |
| Cpl. Louis D. Babini | | 31420626 | |
| Cpl. John L. Burkhead | | 36770874 | |
| Cpl. Alex Kocheran | | 33414605 | |
| Cpl. Donald M. McCoy | | 35763351 | |
| Cpl. James R. Robinson | | 35756585 | |

**8 August 1944 to 714th Squadron**

| | | | |
|---|---|---|---|
| 2nd Lt. Floyd C. Reynolds | P | 0710233 | |
| 2nd Lt. Curtis H. Hockett | CP | 0699370 | |
| 2nd Lt. Rex L. Furness | N | 0723344 | |
| Cpl. Kenneth W. Eastman | | 16125335 | |
| Cpl. Leon T. Crisp | | 36639569 | |
| Cpl. Calvin H. Ellis | | 36600851 | |
| Cpl. Anthony F. Turk | | 35311311 | |
| Cpl. Shirley L. Wahl | | 36884022 | |
| Pvt. Ralph W. Polhamus | | 16150299 | |

**8 August 1944 to 715th Squadron**

| | | | |
|---|---|---|---|
| 2nd Lt. Francis I. Botkin | P | 0668304 | |
| 2nd Lt. Richard W. Goshorn | CP | 0771623 | |
| 2nd Lt. George H. Pogge | B | 0718974 | |
| 2nd Lt. Thomas N. Mize | N | 0773427 | |
| Cpl. Lawrence M. Wilhelm | | 37604383 | |
| TSgt. Charles A. Shipp | | 18025053 | |
| Cpl. James R. Boatright | | 17130553 | |
| Cpl. Armand J. Dansereau | | 31439980 | |
| Cpl. Kenneth D. Johnson | | 36696966 | |
| Cpl. Harold M. Macauley, Jr. | | 36364958 | |

**9 August 1944 to 712th Squadron**

| | | | |
|---|---|---|---|
| F/O James C. Weaver | P | T-123563 | |
| F/O Anthony A. Kolinski | CP | T-2784 | |
| 2nd Lt. Hugh Ewing, Jr. | B | 0723333 | |
| F/O Paul Hyman | N | T-125702 | |
| Cpl. Charles A. Koon | | 14181097 | |
| Cpl. John F. Curran | | 31423296 | |
| Cpl. Vincent J. DePalma | | 32757453 | |
| Cpl. James P. Kane | | 20382599 | |
| Cpl. Albert R. O'Donnell | | 19047939 | |
| Cpl. Harold H. Brakhage | | 38274268 | |

**9 August 1944 to 713th Squadron**

| | | | |
|---|---|---|---|
| F/O Joseph E. Mlynarczyk | P | T-123724 | |
| F/O Sidney R. Hallman, Jr. | CP | T-3102 | |
| 2nd Lt. James B. Faircloth, Jr. | B | 0723597 | |
| F/O Kenneth C. Goodrich | N | Y-3365 | |
| Cpl. Harold L. Gilmore | | 15131285 | |
| Cpl. Bernard J. Deick | | 37579649 | |
| Cpl. Martin Dolinsky | | 32926985 | |
| Cpl. John S. Knoy | | 35146610 | |
| Cpl. Andrew Kuriatnyk | | 17012914 | |
| Cpl. Kenneth Ryan | | 36460485 | |

**9 August 1944 to 714th Squadron**

| | | | |
|---|---|---|---|
| 2nd Lt. Frank E. Bastian, Jr. | P | 0702217 | |
| 2nd Lt. Donald W. Disbrow | CP | 0715735 | |
| F/O David O. Holst | N | T-125814 | |
| Cpl. William J. Degnan, Jr. | | 16137912 | |
| Cpl. Robert E. Coletti | | 12220736 | |
| Cpl. Paul C. DiGiacomo | | 12206795 | |
| Cpl. Dewey A. Holst | | 37533407 | |
| Cpl. Frederick G. Theobold | | 12089380 | |
| Cpl. William O. Wilbur, Jr. | | 11046908 | |

**9 August 1944 to 715th Squadron**

| | | | |
|---|---|---|---|
| 2nd Lt. David D. Cooper | P | 0719018 | |
| F/O Severyn G. Szudarek | CP | T-123549 | |
| F/O Paul C. Happ | B | T-2146 | |
| 2nd Lt. Herbert M. Lerner | N | 0771943 | |
| Cpl. Merwyn G. Cook | | 18162136 | |
| Cpl. Ralph R. Capps | | 38563914 | |
| Cpl. Isaiah H. Houston, Jr. | | 13063152 | |
| Cpl. Thomas J. Kinsey | | 33709627 | |
| Cpl. William F. Smith | | 36192203 | |
| Cpl. Roy S. Willis | | 36461713 | |

**10 August 1944 to 713th Squadron**

| | | | |
|---|---|---|---|
| 1st Lt. Arthur C. Nelson | P | 0691054 | |
| 1st Lt. Charles R. Bastien | CP | 0702218 | |
| 1st Lt. Lee J. Woods | B | 0703839 | |
| 1st Lt. Robert G. Schultz | N | 0695540 | |
| TSgt. Robert N. Carter | | 39278692 | |
| TSgt. William J. Conroy | | 35585394 | |

SSgt. Donald E. Preston           12174207
SSgt. Robert E. Tracy             16152396
SSgt. Alfred G. Kaiser            38394572
SSgt. Cletus L. Kennedy           16053455

12 August 1944 to 712th Squadron
2nd Lt. William N. Stonebraker   P    0705264
2nd Lt. John R. Richards         CP   0820825
2nd Lt. Francis J. Bergin        B    0723274
2nd Lt. Robert W. Ross           N    0772801
SSgt. Wilbur J. Vogel                 34444049
Sgt. John T. Powers                   37046266
Sgt. Clarence E. Williams             34775842
Cpl. Curtis L. Cagle                  34809077
Cpl. Jesse J. Myers                   38564460
Cpl. Orville D. Stuard                38463954

12 August 1944 to 713th Squadron
2nd Lt. Richard C. Vogel         P    0701980
2nd Lt. Leslie M. Sellers        CP   0713563
2nd Lt. Myron B. Koth            B    0719099
2nd Lt. Warren H. Neville        N    0717091
Sgt. Charles G. Genkinger             35794316
Sgt. Clyde L. Turner, Jr.             14147654
Cpl. James F. Burnett                 36698275
Cpl. John S. Phillips                 17057058
Cpl. Carl G. Stenberg                 31423442

12 August 1944 to 714th Squadron
2nd Lt. John C. Rowe             P    0699740
2nd Lt. Bruce J. Anderson        CP   0771263
2nd Lt. Oscar Rudnick            B    0716766
2nd Lt. Richard H. Best          N    0722263
Cpl. Joseph H. Zonyk                  16118907
Cpl. Martin H. Miller, Jr.            35583733
Cpl. Charles W. Robertson             35217685
Cpl. John Roche                       11048564
Cpl. Robert L. Sammons                35775968
Cpl. Francis E. Scott                 37622104

12 August 1944 to 715th Squadron
2nd Lt. Elvin M. Sheffield       P    0706079
2nd L.t Jack G. Miller           CP   0771095
2nd Lt. John P. Blottie          B    0718483
2nd Lt. George M. Steel, Jr.     N    0716776
Sgt. Harold C. Riepenhoff             15327526
Sgt. Frederick E. Recuparo            32545098
Sgt. George J. Swift                  35588372
Cpl. George E. Crane                  32760452
Cpl. Arthur E. Koch                   31292972
Cpl. Thomas S. Tinney                 14192708

16 August 1944 to 712th Squadron
2nd Lt. Edgar M. Jones           P    0705698

2nd Lt. John P. Zima             CP   0771209
2nd Lt. James C. Powell          B    0719142
F/O John R. Tierney              N    T-1330
Cpl. Kazmer J. Szabo                  35606817
TSgt. Floyd W. Jenkins                17027698
Cpl. Howard J. Casey                  18156360
Cpl. Nelson E. DeVaughan              34727348
Cpl. Romeo G. Valentino               33600679
Cpl. Robert M. Williams               6898122

16 August 1944 to 713th Squadron
2nd Lt. Joseph M. Madden         P    0818715
2nd Lt. Harry D. Gouge           CP   0771384
2nd Lt. Attilio Pasquinelli, Jr. B    0694702
2nd Lt. Ivan A. Brewer           N    0773293
Cpl. Nathan Bernstein                 33739991
Cpl. William O.L. Broberg             33687791
Cpl. Daniel F. Daly, Jr.              32892950
Cpl. Ray F. Gipp                      36836560
Cpl. Ronald L. Kincade                35755768
Cpl. Donald C. O'Connor               35351387

16 August 1944 to 713th Squadron
2nd Lt. William D. Smith         P    0700560
2nd Lt. Leon E. Lyon             CP   0709159
2nd Lt. Lewis R. Hyde            N    0722318
2nd Lt. Michael A. Cocchiola     B    01176285
Sgt. Charles B. Ellis                 31282488
Cpl. Leon G. Farnham                  12094220
SSgt. Earl M. Aspin                   17157076
Cpl. Benjamin W. Johnson, Jr.         38539599
Cpl. James B. Malone                  34829133
Cpl. Frank M.J. Stelmochowski         36657060

16 August 1944 to 714th Squadron
2nd Lt. Ralph J. Camburn         P    0822637
2nd Lt. Dale T. Corder           CP   0711886
2nd Lt. Eugene W. Fichtenkort    N    0718092
F/O Harold H. Dorfman            B    T-126347
Sgt. Blase J. Benziger                12178397
Sgt. Eldon E. Preisel                 16070121
Cpl. Herbert R. Barney, Jr.           31306708
Cpl. Ira M. Welkowitz                 12122677
Cpl. William O. Wheeler               35627714

Replacement Crews – September 1944

7 September 1944 to 712th Squadron
2nd Lt. William C. Holden        P    0709581
2nd Lt. Harold E. Bishop         CP   0718531
2nd Lt. Harvey B. Nachman        N    0722956
2nd Lt. Ross B. West             B    0776841
Cpl. Edward S. Sherman                16079232
Cpl. Horace J. Gardner                35787511

| | |
|---|---|
| Cpl. Patrick S. Raspante | 12205391 |
| Cpl. Clair D. Rowe | 37682574 |
| Cpl. Waldon D. Walls | 33436328 |
| Cpl. David A. Webster | 34771489 |

**7 September 1944 to 712th Squadron**

| | | |
|---|---|---|
| 2nd Lt. Robert W. Westbrook | P | 0705865 |
| 2nd Lt. Bille B. Morrison | CP | 0721908 |
| 2nd Lt. Patrick J. Pariavecchia | N | 0205799 |
| 2nd Lt. Daniel M. Boone | B | 0776605 |
| Cpl. Charles J. Schulz, Jr. | | 32649915 |
| Cpl. Edward J. Bednar | | 33505277 |
| Cpl. Kenneth C. Blodgett | | 14130523 |
| S/Sgt. Howard M. Bullis | | 6908104 |
| Cpl. Leslie D. Haneline | | 38599343 |
| Cpl. Wilbert A. Shander | | 16142213 |

**7 September 1944 to 714th Squadron**

| | | |
|---|---|---|
| 2nd Lt. Robert L. Mains | P | 0680467 |
| 2nd Lt. Allan L. Lake | CP | 0206031 |
| 2nd Lt. John B. Hankin, Jr. | N | 0828688 |
| 2nd Lt. John W. Johnson | B | 0776565 |
| Cpl. Charles E. Cupp, Jr. | | 36854167 |
| Cpl. Harry J. Allen | | 39621301 |
| Cpl. Charles H. Daman | | 39463985 |
| Cpl. Frank S. Merkovich | | 36314180 |
| Sgt. Antonio Munoz, Jr. | | 1800??96 |
| Cpl. Anthony C. Villari | | 35919619 |

**7 September 1944 to 714th Squadron**

| | | |
|---|---|---|
| 2nd Lt. James S. Thomas | P | 0705165 |
| 2nd Lt. Harold C. Hardesty | CP | 0721393 |
| 2nd Lt. Solomon Block | N | 0206046 |
| 2nd Lt. Glenn D. Vanderpool | B | 0776593 |
| Cpl. William E. Bynum | | 14100252 |
| Sgt. Earl W. Horntvedt | | 16111521 |
| Cpl. Roy E. Rudy | | 33238210 |
| Cpl. Larrel C. Scott | | 18193185 |
| Cpl. Denham Ward | | 35879442 |
| Cpl. Hanover Weaver | | 34730275 |

**9 September 1944 to 713th Squadron**

| | | |
|---|---|---|
| 2nd Lt. George E. Franklin, Jr. | P | 0711174 |
| 2nd Lt. Virgil H. Gage | CP | 0709231 |
| 2nd Lt. John R. Nettles | N | 0776776 |
| F/O Jack E. Wright | B | T126383 |
| Cpl. John R. Freaney | | 32930606 |
| Sgt. Harold L. Goettsch | | 37195076 |
| Sgt. John S. Carroll | | 33387406 |
| Sgt. Isadore A. Epstein | | 11100206 |
| Sgt. George E. Letlow, Jr. | | 38544775 |
| Sgt. Joseph R. Rossi | | 35610657 |

**9 September 1944 to 714th Squadron**

| | | |
|---|---|---|
| 2nd Lt. L.C. Barneycastle | P | 0685258 |
| 2nd Lt. Albert S. Broadfoot, Jr. | CP | 0719548 |
| 2nd Lt. Reuben Young, Jr. | N | 0722941 |
| 2nd Lt. Bernard X. Ferrari | B | 0717048 |
| TSgt. James C. Lunt | | 19064266 |
| SSgt. Eathen P. Newcomb | | 15058708 |
| Cpl. Russell B. Alvis | | 13094122 |
| Cpl. Patrick H. George | | 19022087 |
| Cpl. Roland L. Grubbs | | 37488400 |
| Cpl. Aron W. Smith | | 37476724 |

**9 September 1944 to 715th Squadron**

| | | |
|---|---|---|
| 2nd Lt. Joe P. Bowers | P | 0711120 |
| 2nd Lt. Ebonezer J. Murphy | CP | 0772446 |
| 2nd Lt. John A. Smith | N | 0703909 |
| 2nd Lt. Richard F. Drowne | B | 0717044 |
| Cpl. Matthew F. Fourneyron | | 12162834 |
| Sgt. Herbert C. Dennis, Jr. | | 33321540 |
| Cpl. Raymond G. Burke | | 36730017 |
| Cpl. Edwin F. Handzlik | | 36585389 |
| Cpl. Leon E. Kelly | | 34671680 |
| Cpl. Paul D. Levey | | 36654552 |

**9 September 1944 to 715th Squadron**

| | | |
|---|---|---|
| 1st Lt. Downey L. Thomas, Jr. | P | 0725006 |
| 2nd Lt. Reynold R. Peterson | CP | 020557998 |
| 2nd Lt. Wilbur I. Padgett | N | 0829543 |
| F/O David E. Ellis | B | T-3759 |
| Cpl. Melvin H. Free | | 39290455 |
| Sgt. Bobbie E. Carlisle | | 18116352 |
| Cpl. Felix Edwards | | 34731320 |
| Cpl. Jefferson D. Johns | | 34818295 |
| Cpl. Louis Noday | | 35609985 |
| Cpl. Cadis W. Owen | | 38518826 |

**10 September 1944 to 715th Squadron**

| | | |
|---|---|---|
| 2nd Lt. Charles P. Quirk | P | 0822179 |
| 2nd Lt. Edward J. Rutter | CP | 0826005 |
| 2nd Lt. Victor Troese | N | 0558009 |
| 2nd Lt. Alexander J. Walczak | B | 01031776 |
| Sgt. Kenneth V. Olson | | 37557035 |
| Sgt. John C. Lyles | | 34709491 |
| Cpl. Edwin R. Hoover | | 33235826 |
| Cpl. Clyde W. Levan | | 33330267 |
| Cpl. John L. Sharpless | | 13175457 |
| Pvt. Roy E. Hicks | | 36880290 |

**10 September 1944 to 715th Squadron**

| | | |
|---|---|---|
| 2nd Lt. Peter Protich | P | 0822802 |
| 2nd Lt. Harold F Closz | CP | 0828382 |
| 2nd Lt. Leo R. Nikula | N | 0723503 |
| 2nd Lt. James R. Huss | B | 0776898 |
| Cpl. John A. Logan | | 39279056 |

| | | |
|---|---|---|
| Cpl. Michael Molish | | 12062930 |
| Cpl. Jose M. Saenz | | 38533905 |
| Cpl. Wayman C. Snyder | | 14157850 |
| Cpl. Jack N. Sweet | | 32489878 |
| Cpl. Bernard Weiss | | 39264151 |

**16 September 1944 to 712th Squadron**

| | | |
|---|---|---|
| 2nd Lt. Walter W. Shue | P | 0823996 |
| 2nd Lt. Carl E. Martin | CP | 0710984 |
| 2nd Lt. Robert L. Eirich | N | 0723596 |
| 2nd Lt. Charles E. Thompson | B | 0717918 |
| Sgt. Leonard G. Kubelik | | 16080851 |
| Sgt. Ralph W. Lee | | 18124429 |
| Cpl. Thomas R. Elliott | | 35119862 |
| Cpl. Carmi D. Ferguson | | 39333902 |
| Cpl. John E. Meintzer | | 15174501 |
| Cpl. Emil E. Nemec | | 36755454 |

**16 September 1944 to 713th Squadron**

| | | |
|---|---|---|
| 2nd Lt. Charles A. Platt | P | 0704853 |
| 2nd Lt. Glenn W. Doyle | CP | 0709531 |
| 2nd Lt. Harold J. Weeks, Jr. | N | 02058596 |
| 2nd Lt. John W. Snider | B | 0773346 |
| Cpl. Jessie F. Kinsey | | 39574662 |
| Cpl. Marvin L. Davis | | 39701060 |
| Cpl. Pete Blair | | 33419020 |
| Cpl. Liborie W. Papalia | | 32676287 |
| Cpl. Ernest L. Zimmerman | | 38567796 |

**16 September 1944 to 714th Squadron**

| | | |
|---|---|---|
| 2nd Lt. Wesley J. Isaacson | P | 0710948 |
| 2nd Lt. Alfred L. Kopitzki | CP | 0823091 |
| 2nd Lt. Raymond E. Custer | N | 02058425 |
| 2nd Lt. Richard M. Styslo | B | 0776923 |
| Cpl. Robert M. Pittman | | 38391518 |
| Cpl. George T. Loupinas | | 36855654 |
| Cpl. Delvin H. Meyer | | 38558995 |
| Cpl. Richard A. Morties | | 32941623 |
| Cpl. Alfred L. Secor | | 39194577 |
| Cpl. George Suchorsky | | 32909556 |

**16 September 1944 to 715th Squadron**

| | | |
|---|---|---|
| 2nd Lt. Richard J. Hambleton | P | 0772357 |
| 2nd Lt. John R. Walker | CP | 0720970 |
| 2nd Lt. James A. Ennis | N | 0699151 |
| 2nd Lt. William J. Southern | B | 0772891 |
| Cpl. Enno C.M. Krotke | | 19101360 |
| Cpl. Harry S. Hunter, Jr. | | 33693748 |
| Cpl. Thomas D. Johnson | | 18216222 |
| Cpl. John J. Riordan | | 33614084 |
| Cpl. William W. Rousher, Jr. | | 33710491 |
| Cpl. Edward D. Smith | | 33545769 |

**18 September 1944 to 712th Squadron**

| | | |
|---|---|---|
| 2nd Lt. Axel D. Johnson | P | 0705396 |
| 2nd Lt. Vincent Luine | CP | 02056532 |
| 2nd Lt. Omero Menegazzi | N | 0721481 |
| 2nd Lt. Bernard L. Nogues | B | 0772992 |
| Cpl. Bernard E. Coons | | 39858706 |
| Cpl. James R. Delaney | | 16137713 |
| Cpl. Bailey A. McNair | | 34705728 |
| Cpl. Richard P. Miller | | 13159210 |
| Cpl. Robert C. Mount | | 36368248 |
| Cpl. Robert S. Yetter | | 33322965 |

**21 September 1944 to 713th Squadron**

| | | |
|---|---|---|
| 2nd Lt. Stephen H. Hodgson | P | 0800887 |
| 2nd Lt. Edward S. Connell | CP | 0822894 |
| 2nd Lt. Edwin A. Scales | N | 0723442 |
| F/O Albert B. Vanderhoof, Jr. | B | T-126277 |
| Cpl. Robert E. Coker | | 34659229 |
| Cpl. Julian A. Duncan | | 34810727 |
| Cpl. Cecil C. Gwennap | | 35923419 |
| Cpl. Ernest J. Kelley | | 32835857 |
| Cpl. Vincent R. Stakun | | 31392279 |
| Cpl. George Wyda | | 33689697 |

**21 September 1944 to 714th Squadron**

| | | |
|---|---|---|
| 2nd Lt. John M. Ray, Jr. | P | 0741086 |
| 2nd Lt. John M. Pearce | CP | 0811433 |
| 2nd Lt. Arnold W. Rubin | N | 0723175 |
| F/O Ara J. Adams | B | T-126430 |
| Cpl. Hobart F. Chester | | 36420756 |
| Cpl. Peter J. Fager | | 39617321 |
| Cpl. John L. Garrity | | 31416602 |
| Cpl. Francis T. Hildenberger | | 33828036 |
| Cpl. Peter E. Lane | | 34650923 |
| Cpl. Edward H. Webb | | 16134528 |

**Replacement Crew – October 1944**

**31 October 1944 to 712th Squadron**

| | | |
|---|---|---|
| 1st Lt. John J. Caldwell | P | 0437194 |
| 2nd Lt. Richard D. Mace | CP | 0779964 |
| 1st Lt. Michael N. Opacick | N | 01280832 |
| 1st Lt. Delmo M. Pearce | B | 0670230 |
| SSgt. Louis R. Thibert | | 16021388 |
| Cpl. Noble Germany | | 14192853 |
| Cpl. Richard C. Gilchrist | | 11099681 |
| Cpl. Raymond R. Kutchinski | | 32923000 |
| Cpl. Robert S. McLoughlin | | 32315454 |
| Cpl. Godfrey R. Wood | | 11004227 |

**31 October 1944 to 713th Squadron**

| | | |
|---|---|---|
| 2nd Lt. James G. Blank | P | 0567334 |
| 2nd Lt. John K. Huber | CP | 0831942 |
| 2nd Lt. Bernard F. Pargh | N | 02065215 |

| | | |
|---|---|---|
| 2nd Lt. Francis X. Pollio | B | 0505968 |
| Cpl. William T. Harriman | | 17111211 |
| Cpl. Morris L. Kanarek | | 13133595 |
| Cpl. Christopher C. King | | 32747076 |
| Cpl. Derward E. Morrow | | 11043372 |
| Cpl. John A. Woldman, Jr. | | 35398840 |

Replacement Crews – November 1944

6 November 1944 to 715th Squadron
| | | |
|---|---|---|
| 2nd Lt. Elmo A. Solberg | P | 0446778 |
| 2nd Lt. Charles A. Spiesse | | 0689392 |
| F/O Walter S. Deibel | | T128954 |
| 2nd Lt. John S. McHugh | | 0717089 |
| SSgt. Robert G. Perry | | 12035659 |
| Sgt. Roy F. Zieske | | 39614444 |
| Cpl. Henry B. Brandt | | 36782592 |
| Cpl. John E. Conrad, Jr. | | 36836339 |
| Cpl. Clarence G. Gerhart | | 13029453 |
| Cpl. Steve Kachmar | | 33614499 |

7 November 1944 to 712th Squadron
| | | |
|---|---|---|
| 1st. Lt. Irwin W. Ruge | P | 0855352 |
| 2nd Lt. Joseph H. Leroy | CP | 02057328 |
| 2nd Lt. George S. Robertson | N | 02066045 |
| 2nd Lt. Walter T. Foreman | B | 0780246 |
| SSgt. Randall D. Fowler | | 190003642 |
| Cpl. Maurice E. Bordner | | 36685995 |
| Cpl. Myron Revak | | 33611590 |
| Cpl. Laurence W. Scholny | | 36676361 |
| Cpl. John T. Ziino | | 36652218 |

13 November 1944 to 712th Squadron
| | | |
|---|---|---|
| 2nd Lt. Joseph F. Steffan | P | 0719786 |
| 2nd Lt. Henry Y. Edgerton | CP | 02061656 |
| 2nd Lt. Gerald J. Gottlieb | N | 02068377 |
| SSgt. Edward C. Murphy, Jr. | | 35370754 |
| Cpl. Harold Resnikoff | | 18231599 |
| Cpl. Davis E. Denning | | 34870892 |
| Cpl. James H. Gilmore, Jr. | | 14185681 |
| Cpl. Donald A. Payne | | 23178062 |
| Cpl. William Wright, Jr. | | 328?8722 |

13 November 1944 to 712th Squadron
| | | |
|---|---|---|
| 2nd Lt. Irving Smarinsky | P | 0710613 |
| 2nd Lt. Horace B. Lane | CP | 0771451 |
| F/O Sidney V. Peters, Jr. | N | T-129862 |
| F/O Arthur Hoffman | B | T-5541 |
| Cpl. Edward J. Grossman | | 32931086 |
| Cpl. Harold C. Haynes | | 39712761 |
| Cpl. Francis J. Kelly, Jr. | | 11103783 |
| Cpl. Gerard J. Perry | | 31323473 |
| Cpl. Robert Turner, Jr. | | 38519135 |
| Cpl. Elton L. Nichols | | 385186?2 |

13 November 1944 to 712th Squadron
| | | |
|---|---|---|
| 2nd Lt. William L. Voight | P | 0774829 |
| F/O Lt. Fred S. Risinger, Jr. | CP | T-3539 |
| 2nd Lt. Edward J. O'Donoghue | N | 0926916 |
| F/O Willis D. Lonn | B | T5548 |
| Cpl. Dale K. Huson | | 39558319 |
| Cpl. Richard N. Kudukis | | 16102597 |
| Cpl. John J. Noone | | 33908745 |
| Cpl. Wilmer L. Polk | | 38392524 |
| Cpl. Jerry J. Russ | | 36635?08 |
| Cpl. Joseph S. Ulakovich | | 35235649 |

13 November 1944 to 713th Squadron
| | | |
|---|---|---|
| 2nd Lt. Harry R. Mulrain | P | 0828221 |
| 2nd Lt. Thomas F. Murphy | CP | 02068446 |
| 2nd Lt. Joseph O'Connor | N | 0832207 |
| Cpl. Warren R. Dolan | | 36893315 |
| Cpl. William K. Jann | | 42072013 |
| Cpl. Roger E. Leland | | 31281991 |
| Cpl. Jack C. McKay | | 37672915 |
| Cpl. Carey A. Stephens, Jr. | | 34796131 |
| Cpl. James S. Champion, Jr. | | 33713755 |

13 November 1944 to 713th Squadron
| | | |
|---|---|---|
| 2nd Lt. Frederick W. Tod | P | 0776133 |
| F/O Lt. Warren N. Peterson, Jr. | CP | T-2849 |
| 2nd Lt. Howard R. Morton | N | 01055394 |
| 2nd Lt. Herman J.D. Jame | B | 0783073 |
| Cpl. Robert F. Harrison | | 37483658 |
| Cpl. Robert L. Koscki | | 37563258 |
| Cpl. Chester J. Labus | | 33795580 |
| Cpl. Joseph W. Noonan | | 37632374 |
| Cpl. John R. Peterson | | 36483559 |
| Cpl. James R. Turnley | | 33715358 |

13 November 194 to 714th Squadron
| | | |
|---|---|---|
| 2nd Lt. Donald G. Stuhmer | P | 0827990 |
| 2nd Lt. Boardman G. Getsinger, Jr. | CP | 0831691 |
| 2nd Lt. William J. Nugent | N | 02068448 |
| Cpl. Keith B. Mink | | 31450251 |
| Cpl. John B. Klein, Jr. | | 37140182 |
| Cpl. Robert P. McDonald | | 36435101 |
| Cpl. Robert J. Dyer | | 37683078 |
| Cpl. Kenneth T. Ritter | | 33642094 |

14 November 1944 to 712th Squadron
| | | |
|---|---|---|
| 1st Lt. James R. C. Cook | P | 01633141 |
| 1st Lt. Alonzo D. McAllister | CP | 0448096 |
| 2nd Lt. Edwin F. Slowick | N | 0818961 |
| 2nd Lt. Dean E. Peterson | B | 02064312 |
| TSgt. John P. Delaney | | 6289405 |
| Cpl. Garland E. Flinn | | 39472117 |
| Cpl. Franklin A. Halferty | | 35346939 |
| Cpl. Howard W. Ivery | | 35814638 |

| | | |
|---|---|---|
| Cpl. Richard H. McAdams | | 32609119 |
| Cpl. Joseph N. Rodgriquez | | 16083229 |

**15 November 1944 to 713th Squadron**

| | | |
|---|---|---|
| 1st Lt. Howard L. Smith | P | 0675638 |
| 2nd Lt. John A. Harron | CP | 0721400 |
| 2nd Lt. Waldo J. Marolf | N | 02065172 |
| 2nd Lt. Alvin C. Nickerson | B | 02065407 |
| Cpl. Arthur W. Carter, Jr. | | 12178347 |
| Cpl. John J. Dunden, Jr. | | 32937983 |
| Cpl. Harry E. Huster, Jr. | | 13083681 |
| Cpl. David L. Phillips | | 35633197 |
| Cpl. Melvin Schlenoff | | 33902781 |
| Cpl. Mervin E. Schwartz | | 37589627 |

**15 November 1944 to 714th Squadron**

| | | |
|---|---|---|
| 2nd Lt. John J. Opman | P | 0822071 |
| 2nd Lt. Donald A. Ouellette | CP | 0831779 |
| 2nd Lt. Charles J. Ballantine | N | 02065209 |
| Cpl. Henry A. Calika | | 36865856 |
| Cpl. Louis Kaplan | | 12177527 |
| Cpl. John M. Roche | | 36677473 |
| Cpl. Harry J. Steeves | | 11141377 |
| Cpl. Edmund B. Szymczak | | 42027284 |
| Cpl. James F. Wivinis | | 16188906 |

**15 November 1944 to 714th Squadron**

| | | |
|---|---|---|
| 2nd Lt. Harlyn H. Schroeder | P | 0826014 |
| 2nd Lt. Delwin D. Roorda | CP | 02062050 |
| 2nd Lt. Joe F. Castle | N | 01046037 |
| Cpl. Norwood A. D. Adler | | 35909933 |
| Cpl. Raymond M. Bailey | | 35779718 |
| Cpl. Thomas L. Economy | | 32055026 |
| Cpl. Frank C. Lippman | | 42001514 |
| Cpl. Erwin A. Schilling | | 16190123 |
| Cpl. Edward L. Vetterneck | | 36823304 |

**15 November 1944 to 714th Squadron**

| | | |
|---|---|---|
| 2nd Lt. Sylvester J. Peresie | P | 0824201 |
| 2nd Lt. William J. Smith | CP | 0778984 |
| F/O Henning E. Helsing | N | T-129680 |
| F/O Anthony L. Germele | B | T-5453 |
| Cpl. Wayne Sarver | | 19163202 |
| SSgt. Noyle A. Wright | | 19021145 |
| Cpl. William F. Eaton, Jr. | | 35433622 |
| Cpl. Raymond J. Lewis | | 32948304 |
| Pvt. Michele P. Ricciarll, Jr. | | 42101769 |
| Cpl. James O. Yokley | | 14194515 |

**15 November 1944 to 712th Squadron**

| | | |
|---|---|---|
| 2nd Lt. Guilford D. Wikender | P | 0778638 |
| F/O Lex W. Jones | CP | T-128986 |
| F/O Walter D. Fortner | N | T-131825 |
| F/O Ernest R. Belinskas | B | T-5589 |

| | |
|---|---|
| Cpl. David O. Anthony | 36682781 |
| Cpl. William T. Matejka | 36831384 |
| Cpl. Charles W. Peacock | 13151754 |
| Cpl. James E. Riddle | 13189973 |
| Cpl. Charles R. Steele | 33437571 |
| Cpl. Warren F. Wheelock | 13070564 |

**15 November 1944 to 713th Squadron**

| | | |
|---|---|---|
| 2nd Lt. Paul J. Jones | P | 0718659 |
| 2nd Lt. James Mucha | CP | 02059408 |
| 2nd Lt. Herman Engel, Jr. | N | 02065525 |
| Cpl. Charles W. Blaney, Jr. | | 36759119 |
| Cpl. Leonard E. Dailey | | 38540952 |
| Cpl. Edward W. Danecki | | 36826828 |
| Cpl. Alvin J. Stout | | 37703586 |
| Cpl. William J. Wilson | | 34892326 |
| Pvt. Albert J. Dentley | | 34824130 |

**16 November 1944 to 712th Squadron**

| | | |
|---|---|---|
| 2nd Lt. Earl Furnace | P | 0711175 |
| 2nd Lt. Thorpe L. Friar | CP | 0770626 |
| 2nd Lt. Norman W. Kanwisher | N | 02056413 |
| SSgt. Nathan J. Malkin | | 32087779 |
| SSgt. James E. Howell | | 33737992 |
| Sgt. Alexander Yarosky | | 12101126 |
| Sgt. Morris E. Gannon | | 34684506 |
| Sgt. William P. Franks | | 12203781 |
| Sgt. James C. Allison | | 39616772 |

**16 November 1944 to 712th Squadron**

| | | |
|---|---|---|
| 2nd Lt. Joseph B. Brown | P | 0708484 |
| 2nd Lt. Richard C. Seymour | CP | 0772233 |
| 2nd Lt. Orvie O. Casto, Jr. | N | 0722944 |
| SSgt. Phillip Mazzagatti | | 38420922 |
| SSgt. Francis M. Louthan | | ?6294269 |
| Sgt. Jack D. Cowdin | | 18046184 |
| Sgt. Ben E. Vegors, Jr. | | 19142303 |
| Sgt. William R. Kamedish | | 17158055 |
| Sgt. Leroy R. Romig, Jr. | | 33832822 |

**16 November 1944 to 713th Squadron**

| | | |
|---|---|---|
| 2nd Lt. Walter Bobak | P | 0768921 |
| 2nd Lt. Alfred Christ | CP | 0776203 |
| 2nd Lt. Richard C. Wagner | N | 0712953 |
| 2nd Lt. Leslie A. Beaton | B | ??????? |
| SSgt. Richard Mickelson | | 17157786 |
| SSgt. Oranze J. Ruscitti | | 39331840 |
| SSgt. Collis Carlee | | 14184267 |
| Sgt. Morris E. Gannon | | 34684506 |
| Sgt. Donald Beck | | 17132532 |
| Sgt. Arthur Myers | | 35875778 |
| Sgt. Franklin Morgan | | 35706996 |
| Sgt. Richard L. Dietrick | | 13144154 |

16 November 1944 to 714th Squadron
| | | |
|---|---|---|
| Capt. Edward M. Wall | P | 0795106 |
| 1st Lt. Willis H. Young | CP | 0776203 |
| 1st Lt. Charles F. Reeves | PN | 0712359 |
| 1st Lt. Herman Salyer | DRN | 0702965 |
| 1st Lt. Norman Segal | B | 0668794 |
| TSgt. Ansel J. Gladish | | 16028322 |
| TSgt. Robert M. Carlton | | 11129862 |
| Sgt. Victor A. Jensen | | 17028486 |
| SSgt. Clarence S. Scollard | | 3745?625 |
| SSgt. Willoughby | | 39084353 |
| SSgt. Harold G. Beams | | 11071294 |

16 November 1944 to 715th Squadron
| | | |
|---|---|---|
| 2nd Lt. Daniel R. Durbin, Jr. | P | 0704903 |
| 2nd Lt. Thomas L. McQueid | CP | 0886697 |
| 2nd Lt. Donald G. Leetch | N | 0712854 |
| SSgt. James H. Count | | 42001035 |
| SSgt. Joy C. Christensen | | 19115992 |
| Sgt. James E. McCown | | 14200127 |
| Sgt. Joseph C. Svaton | | 12175709 |
| Sgt. Harry F. McCurdy | | 12159282 |
| Sgt. Alfonso C. Gessonius | | 39272064 |

17 November 1944 to 712th Squadron
| | | |
|---|---|---|
| 2nd Lt. James J. Shafter | P | 0771554 |
| 2nd Lt. John R. Paxson | CP | 02057389 |
| 2nd Lt. Neal W. Pettit | N | 0206618 |
| SSgt. Walter W. Petrovich | | 35601205 |
| Cpl. Virgil F. Beall | | 38076333 |
| Cpl. Daniel G. Graham, Jr. | | 16088972 |
| Cpl. William L. Kaiser | | 39577186 |
| Cpl. Taylor L. Tarkington | | 18177172 |
| Cpl. Anderson C. Wright | | 33836022 |

17 November 1944 to 713th Squadron
| | | |
|---|---|---|
| 2nd Lt. Kenneth A. Wheeler | P | 07??679 |
| 2nd Lt. William B. Wiveol | CP | 0775691 |
| 2nd Lt. Cecil L. Pullen | N | 02065825 |
| 2nd Lt. Robert H. Piccolo | B | 0787890 |
| Cpl. Robert Drummond | | 32953248 |
| Cpl. Robert S. Messner | | 12228588 |
| Cpl. Paul V. Oskowski | | 33920308 |
| Cpl. Edward W. Pinner | | 16078736 |
| Cpl. Russel R. Sage | | 35704763 |
| Cpl. Edward Wagner | | 33920010 |

17 November 1944 to 714th Squadron
| | | |
|---|---|---|
| 2nd Lt. Stanley L. Winter | P | 0776175 |
| 2nd Lt. Howard A. Courtnery | CP | 0719572 |
| F/O Roland T. Hauver | N | T-129678 |
| F/O Leo J. Dymerski | B | T-5395 |
| Cpl. Robert D. Bosworth | | 32936851 |
| Cpl. Walden L. Gibbs | | 38350835 |
| Cpl. Gomber D. Hess | | 33919766 |
| Cpl. Paul J. Misera | | 33919908 |
| Cpl. Linwood H. Peaslee | | ?1323568 |
| Cpl. George J. Weinberger | | 39206663 |

17 November 1944 to 715th Squadron
| | | |
|---|---|---|
| 2nd Lt. Albert B. Sanders, Jr. | P | 0721222 |
| 2nd Lt. Wallace K. Grimes | CP | 02059504 |
| 2nd Lt. Joseph L. Nathan | N | 02066023 |
| SSgt. Gilbert A. Matthias | | 36247431 |
| Cpl. Irvin F. Alvey | | 36685927 |
| Cpl. Robert F. Eldridge | | 17168478 |
| Cpl. Warren E. Lutin | | 36868144 |
| Cpl. John P. Royski | | 422020409 |
| Cpl. Pat J. Terrarova | | ?2238842 |

17 November 1944 to 715th Squadron
| | | |
|---|---|---|
| 2nd Lt. Courtland C. Crandall | P | 0825?81 |
| 2nd Lt. Allan J. Carey, Jr. | CP | 0829162 |
| 2nd Lt. Marshall K. Dan | N | 02066023 |
| Sgt. Lewis W. Miller | | 33158948 |
| Cpl. Joseph M. Guthrie | | 13201139 |
| Cpl. Marcus K. Jorgensen | | 36685028 |
| Cpl. John J. Madden | | 32440676 |
| Cpl. Milton M. Olson | | 36647360 |
| Cpl. Oscar W. Olson, Jr. | | 15134212 |

17 November 1944 to 715th Squadron
| | | |
|---|---|---|
| 2nd Lt. Maurice Holmen | P | 0771029 |
| 2nd Lt. Clayton J. Berg | CP | 0777561 |
| 2nd Lt. Harry H. Davis | N | 0723589 |
| SSgt. Carl O. Hightower | | 18031704 |
| Sgt. Bernard T. Fusco | | 20250323 |
| Cpl. Michael D. Castine | | 14190720 |
| Cpl. Robert B. Karoff, Jr. | | 32874036 |
| Cpl. Neil M. Luhrs | | 36873022 |
| Cpl. James F. Cummings, Jr. | | 15394578 |

17 November 1944 to 715th Squadron
| | | |
|---|---|---|
| 2nd Lt. Henry E. Mielke | P | 0775475 |
| 2nd Lt. Harlan R. Edson | CP | 02065971 |
| 2nd Lt. Raymond J. Kincaid | N | 0779157 |
| Cpl. Laurence A. Harris | | 36685434 |
| Cpl. Floyd E. Hudson | | 34916461 |
| Cpl. Floyd L. Johnston | | 35558872 |
| Sgt. Jerome A. Korte | | 36477450 |
| Cpl. Harry A. McClure, Jr. | | 17168640 |
| Cpl. Carl A. Newpher | | 33681761 |

17 November 1944 to 715th Squadron
| | | |
|---|---|---|
| 2nd Lt. Paul G. O'Neil | P | 0771029 |
| 2nd Lt. Theodore R. Zimmerman | CP | 02058344 |
| 2nd Lt. Robert J. Wilkins | N | 02066088 |
| Sgt. Angelo A. Adelizzi | | 16036369 |

| Cpl. Donald Brittingham | | 13141526 | 2nd Lt. John Potgeter | CP | 0777808 |
| Cpl. Dean E. Coolman | | 16159504 | F/O Jules Klingsberg | N | T-132959 |
| Cpl. John R. Gerber | | 35225207 | 1st Lt. Frank P. Law | B | 01296394 |
| Cpl. Harold L. Lang | | 39704462 | Cpl. William K. Krebs | | 37704414 |
| Cpl. Myron Rosenthal | | 36869665 | Cpl. Elbert I. Moore | | 34732029 |
| | | | Cpl. Jerry L. Obermiller | | 37603717 |

**17 November 1944 to 715th Squadron**

Cpl. Donald J. O'Rourke — 36866359
2nd Lt. David G. Anderson — P — 0720103
Cpl. Charles P. Swindler — 33544297
2nd Lt. Quinton B. McLay — CP — 02058848
Cpl. Elmer C. Witty — 36832293
2nd Lt. ????? A. Hammond — N — 02065546
Cpl. William B. Davis — 33541273

**18 December 1944 to 714th Squadron**

Cpl. Theodore E. Dyson — 11104482
2nd Lt. Albert C. Hardies, Jr. — P — 0829471
Cpl. James W. Harrison — 20926606
F/O Hugh R. Jones — T128271
Cpl. Joseph E. Jendzeisyk — 12240678
F/O Stanley Misroch — T132975
Cpl. Raymond W. Kubik — 36882355
F/O Larry R. Ohlson — T7062
Cpl. Joseph M. Szeliga — 36015701
Cpl. Bursley C. Ferguson — 6879680

Cpl. Edmund E. Fraind — 39556122

**Replacement Crews – December 1944**

Cpl. Frederick E. Hoffman — 36699292
Cpl. Thomas G. Osborne, Jr. — 16140239

**11 December 1944 to 715th Squadron**

Cpl. Frederick A. Patrick — 35543725
2nd Lt. Knute P. Stalland — P — 0721875
Cpl. Dwight C. Phillips — 38510723
2nd Lt. Theodore Warner, Jr. — CP — 0928014

**18 Dcember 1944 to 714th Squadron**

2nd Lt. John M. Heard — N — 02069015
Cpl. Oryn M. Blashe — 36841206
2nd Lt. Vincent B. Hoyer — P — 0829481
Cpl. Bobbie C. Glass — 38478645
F/O Strother D.P. Ysinger — T132989
Cpl. Frank E. Grogan, Jr. — 14124065
2nd Lt. James A. Woods III — 02059?46
Cpl. John M. Kropp — 36819181
2nd Lt. Edward J. Ondrasik — 0782827
Cpl. Dale W. Overy — 35296070
SSgt. Lee T. Lain, Jr. — 18014396
Cpl. Joseph F. Parks — 35296070
SSgt. Lucas J. Warian — 6913013

Cpl. Milton A. Nichols — 33541302

**13 December 1944 to 714th Squadron**

Cpl. Donald L. Reedy — 35884363
2nd Lt. Francis R. Piliere — P — 0829290
Cpl. Robert A. Sorich — 36637032
2nd Lt. Raymond E. Gale — CP — 02063602
Cpl. Joseph Szopo — 33710144
2nd Lt. Joseph F. Chaput — N — 02068949
Sgt. William M. Garrett, Jr. — 38414271

**18 December 1944 to 714th Squadron**

Cpl. Ronald F. Burke — 11114531
2nd Lt. Hershel Hausman — P — 0833101
Cpl. Milton Greenfield — 33791521
2nd Lt. Willis P. Risdon — 02062115
Cpl. Charles H. J. Nigrin — 33734022
2nd Lt. Donald E. Bodiker — 02069922
Cpl. Nicholas W. Porcaro — 33828756
F/O John W. Keller — T6901
Cpl. Norman R. Veenstra — 31410192
Sgt. Earl H. Case — 35034551

Sgt. Marshall L. Kisch — 37269858

**13 December 1944 to 715th Squadron**

Cpl. Lester S. McGown — 37560171
2nd Lt. Frederick J. Hahner — P — 0829199
Cpl. Louis Sharu, Jr. — 36854293
F/O John Kelley — CP — T-129030
Cpl. Harry J. Tamplain — 38502206
2nd Lt. John F. Cascio — N — 20268947
Cpl. Chalres O. Thompson, Jr. — 38568170
Sgt. Robert O. Felt — 13040574
Cpl. Harold W. Hudspeth — 36677535

**18 December 1944 to 714th Squadron**

Cpl. Joseph A. Hutchinson — 29924714
2nd Lt. Thomas B. Horton — P — 0829480
Cpl. Wallace D. Laufer — 36768931
2nd Lt. Edward A. Schreiber — 01?17686
Cpl. Edward H. McCuen — 18167550
F/O Ralph A. Nicholas, Jr. — T133232
Cpl. Bernard M. O'Leary, Jr. — 12205415
F/O Stanley F. Jurkanis — T??543

Sgt. Rufus R. Nickell — 33131295

**17 December 1944 to 713th Squadron**

Cpl. Thomas G. Brehm — 33690646
2nd Lt. William A. Hammes — P — 0829469
Cpl. Troy C. Jones, Jr. — 14189133

| | | |
|---|---|---|
| Cpl. Harry A. Kearney, Jr. | | 1216??08 |
| Cpl. Kelly M. Smith | | 20710785 |
| Cpl. Clement E. Staley | | 13187417 |

**18 December 1944 to 715th Squadron**

| | | |
|---|---|---|
| 2nd Lt. James E. Guynes, Jr. | P | 0831159 |
| 2nd Lt. James M. Larson | | 02061983 |
| 1st Lt. Carrol A. Michels | | 01844474 |
| F/O Aram G. Kadehjian | | T132946 |
| Cpl. James M. Hammond, Jr. | | 34816783 |
| Cpl. Lloyd E. Newton | | 18110354 |
| Cpl. Joseph H. Mull, Jr. | | 14174063 |
| Cpl. Edmund A. Schierenbeck | | 37547985 |
| Cpl. John E, Schilling | | 36841583 |
| Cpl. Donald F. Schleicher | | 34903798 |

**18 December 1944 to 715th Squadron**

| | | |
|---|---|---|
| 2nd Lt. Sam Hailey | P | 0831163 |
| 2nd Lt. Jack McDaniel | | 0721827 |
| 2nd Lt. Edward E. Bowman | | 02069927 |
| 2nd Lt. Kendall K. Kimberlin | | 0785408 |
| Cpl. Edward P. French | | 39918501 |
| Cpl. William C. Houk | | 33707696 |
| Cpl. John W. Hottinger | | 33806343 |
| Cpl. Richard L. Oatis | | 35345?10 |
| Cpl. John H. O'Neill | | 11801463 |
| Cpl. William E. Schmidt | | 17112418 |

**20 December 1944 to 714th Squadron**

| | | |
|---|---|---|
| 2nd Lt. Kay L. Flinders | P | 0721683 |
| 2nd Lt. Lurty M. Reid | CP | 0831800 |
| 2nd Lt. John J. McNamee | N | 02068430 |
| Cpl. Nick Anast | | 36696595 |
| Cpl. Henry F. Devine | | 16056638 |
| Cpl. Earl B. Jordan | | 34782003 |
| Cpl. William B. Meharg | | 34793089 |
| Cpl. Edwin S. Mitchell | | 33735195 |
| Cpl. William E. Myers | | 1310?591 |

**20 December 1944 to 715th Squadron**

| | | |
|---|---|---|
| 2nd Lt. Robert L. Stewart | P | 0721683 |
| 2nd Lt. Harold E. Daniels | CP | 0778657 |
| F/O Charles A. Bales | N | T-126387 |
| F/O Charles M. Shumaker, Jr. | B | T-5625 |
| Cpl. William H. Hadley | | 15304043 |
| Cpl. Henry A. Holkenbrink | | 36693927 |
| Cpl. Bernard P. Martin | | 32810002 |
| Cpl. Victor L. Miller | | 12209561 |
| Cpl. Robert V. Ray | | 36764240 |
| Pvt. Frederick Z. Conley | | 32948839 |

**20 December 1944 to 715th Squadron**

| | | |
|---|---|---|
| 2nd Lt. Raymond E. Bunday | P | 07?1341 |
| 2nd Lt. George H. Helm | CP | 0674042 |
| 2nd Lt. Neal P. Schumacher | N | 02070152 |
| 2nd Lt. Leslie N. Taliferro | B | 0785438 |
| Cpl. Clarence H. Bales | | 3772?311 |
| Cpl. Charles V. Clark | | 35233954 |
| Cpl. Glenn D. Hill | | 17147109 |
| Cpl. Paul J. Hurton | | 31427378 |
| Cpl. James P. Mathews, Jr. | | 34829996 |
| Cpl. Harry D. Repp, Jr. | | 13141053 |

**20 December 1944 to 715th Squadron**

| | | |
|---|---|---|
| 2nd Lt. Robert A. Paeschke | P | 0776007 |
| 2nd Lt. Hugh S. Holburn | CP | 0828623 |
| 2nd Lt. Harold W. Onstad | N | 02068452 |
| Sgt. Paul N. Boyerl | | 16132172 |
| Cpl. Richmond H. Dugger, Jr. | | 33645037 |
| Cpl. Robert E. George | | 38534143 |
| Cpl. Frank P. Rinaldi | | 15375752 |
| Cpl. Herbert D. Smith | | 33543226 |
| Cpl. Reno A. Tonegate | | 39403855 |

**Replacement Crews – January 1945**

**17 January 1945 to 713th Squadron**

| | | |
|---|---|---|
| 2nd Lt. Forrest E. McCready | P | 0721826 |
| 2nd Lt. Harold O. Pittenger | CP | 02071826 |
| 2nd Lt. Eddie O. McLaughlin, Jr. | N | 0928913 |
| 2nd Lt. Earl S. Patterson | B | 02069712 |
| Cpl. Claud E. Lamoy, Jr. | | 34848201 |
| Cpl. Pat H. Cochran | | 37706247 |
| Cpl. Arthur J. Helganz | | 36574014 |
| Cpl. Merle L. Law | | 35172764 |
| Cpl. Darwin D. Dague | | 36594095 |
| Cpl. Eugene T. Short | | 36837020 |

**20 January 1945 to 714th Squadron**

| | | |
|---|---|---|
| 1st Lt. Samuel H. Moseley | P | 0727645 |
| 2nd Lt. Charles H. Herring | CP | 0719070 |
| F/O Forrest F. Hauser | N | T-127938 |
| Cpl. George J. Jacobs | | 15119406 |
| Cpl. Sidney Friedman | | 36760306 |
| Cpl. John F. McCarthy | | 32845101 |
| Cpl. Robert C. Rhinard | | 16153786 |
| Cpl. Theodore C. Mower | | 33708165 |
| Cpl. George A. Watkins | | 33812489 |

**20 January 1945 to 713th Squadron**

| | | |
|---|---|---|
| 2nd Lt. Paul E. Homan | P | 02059519 |
| 2nd Lt. Francis Schlansky | | 0823708 |
| F/O Basil J. St. Dennis | | T-133883 |
| F/O Aurelius S. Hinds II | | T-7174 |
| Cpl. Louis T. Camden | | 35295848 |
| Cpl. Leonard A. Fiser | | 37678743 |
| Cpl. Brooks Garner | | 14128585 |
| Cpl. Robert A. Pope | | 11067320 |

| | | | | | | |
|---|---|---|---|---|---|---|
| Cpl. Claude B. Snowden, Jr. | | | 34796531 | F/O Harold W. Goodman | N | T-134557 |
| Cpl. August R. Treuting | | | 38493344 | F/O Douglas V. Buffinton | B | T-134393 |
| | | | | Cpl. Gilbert R. Schenks | | 38473375 |
| 20 January 1945 to 715th Squadron | | | | Cpl. Bruno J. Murski | | 18061283 |
| 2nd Lt. Hugh McFarland | | | 02059558 | Cpl. Kenneth R. Knowles | | 14185007 |
| 2nd Lt. Alexander M. Durant, Jr. | | | 0834771 | Cpl. Leon P. Stone | | 3851334? |
| F/O John B. Young | | | T-134801 | Cpl. John N. Geratey | | 16173370 |
| 2nd Lt. Donald C. McCleary | | | 01314163 | Cpl. Daniel W. Taylor | | 33646051 |
| Sgt. Girthel R. Morrison | | | 16065555 | | | |
| Sgt. Herman L. Waters | | | 34587603 | 11 February 1945 to 715th Squadron | | |
| Cpl. Gordon F. Briquelet | | | 36326978 | 2nd Lt. Douglas W. Torrance | P | 0830752 |
| Cpl. Donald R. Kiehn | | | 35898639 | F/O Lt. Alexander A. Calomeni | CP | T-65763 |
| Cpl. Elmer J. Mroczek | | | 35065303 | 2nd Lt. Frederick T. McKinley | N | 02074504 |
| Cpl. Fred Yule | | | 31425712 | Cpl. Richard G. Brede | | 39461036 |
| | | | | Cpl. Harold M. Burt | | 36880261 |
| Replacement Crews – February 1945 | | | | Cpl. Lawrence J. Caruso | | 42009565 |
| | | | | Cpl. James E. Gleason, Jr. | | 31302882 |
| 2 February 1945 to 713th Squadron | | | | Cpl. Edward Paretti | | 32974580 |
| 2nd Lt. George R. Onufer | P | | 0720343 | Cpl. Hairman M. Merrill | | 34598186 |
| 2nd Lt. Carl T. Wiley, Jr. | CP | | 02058322 | | | |
| F/O Harry S. Wells | N | | T-133897 | 11 February 1945 to 715th Squadron | | |
| Cpl. James O. Attaway | | | 14185093 | 2nd Lt. Horace R. Rigel | P | 0834016 |
| Cpl. Irving H. Horn | | | 12219597 | F/O Earl B. Saxe | CP | T-130599 |
| Cpl. Frederick D. Neilsen | | | 33682680 | 2nd Lt. John R. Williams | N | 02072937 |
| Cpl. Edward J. Parciak | | | 31379608 | Cpl. Vernon E. Burkowski | | 33734815 |
| Cpl. Robert F. Schreier | | | 17071898 | Cpl. Franklin H. Hastings | | 33389960 |
| Cpl. Charles E. Smith | | | 13128493 | Cpl. Louis J. Ladas | | 31205305 |
| | | | | Cpl. William Schneider | | 36761014 |
| 2 February 1945 to 713th Squadron | | | | Pvt. Francis J. Chelland | | 13056160 |
| 2nd Lt. Neil R. McCluhan | P | | 02059556 | Pvt. Lucian A. Whipple, Jr. | | 14045409 |
| 2nd Lt. Elias L. King | CP | | 02059938 | | | |
| 2nd Lt. John K. Zeigler | N | | 02071774 | 11 February 1945 to 713th Squadron | | |
| 2nd Lt. John C. Carabello | B | | 0785567 | 2nd Lt. Karl W. Augenstein | P | 0834347 |
| Cpl. John Berardelli | | | 1308705? | F/O Herman Decktor | CP | T-65766 |
| Cpl. Donald S. Clark | | | 1532739? | F/O John F. Sharpe, Jr. | N | T-134593 |
| Cpl. Wade R. Dodds | | | 3514797? | Cpl. Salvatore C. DeRosa | | 32987179 |
| Cpl. Kirby L. Lyle | | | 3468498? | Cpl. Norman R. Dunphe | | 31369563 |
| Cpl. Chris Snow, Jr. | | | 3694707? | Cpl. John F. Gant | | 17136302 |
| Cpl. Frederick W. Wichman | | | 3529400? | Cpl. William A. Poland | | 6947419? |
| | | | | Cpl. Wilbur Semelvers | | 35216214 |
| 2 February 1945 to 714th Squadron | | | | Cpl. Robert J. Konkol | | 33609314 |
| 2nd Lt. Richard H. Page | P | | 0719717 | | | |
| 2nd Lt. Frederick D. Smit | CP | | 0928938 | 15 February 1945 to 713th Squadron | | |
| F/O Richard I. Wooderson | N | | T-133901 | 2nd Lt. Frederick E. Clarke, Jr. | P | 02061880 |
| Cpl. Robert F. McClatchey | | | 36758004 | F/O David G. Coomer | CP | T-65686 |
| Cpl. Edmund J. Misbach, Jr. | | | 1109547? | 2nd Lt. Urbain L. Doyle | N | 02074678 |
| Cpl. Herbert P. Neville | | | 3140761? | Cpl. Frank B. Barrett | | 14184648 |
| Cpl. William H. Peel | | | 3462495? | Cpl. Harry D. Mitchell III | | 13140966 |
| Cpl. John Snyder, Jr. | | | 3144044? | Cpl. William J. Roberts | | 31390075 |
| Cpl. Stanley W. Thatcher | | | 1312673? | Cpl. Dalfino T. Sarina | | 39417289 |
| | | | | Cpl. Robert H. Steeves | | 31405559 |
| 5 February 1945 to 713th Squadron | | | | Cpl. George D. Moreno | | 39710307 |
| 2nd Lt. Edward V. Anderson | P | | 02058100 | | | |
| F/O John W. Stanford | CP | | T-647211 | | | |

15 February 1945 to 714th Squadron
| 2nd Lt. Gilbert N. Davis | P | 02064877 |
| F/O Donald E. Fletcher | CP | T-6569 |
| 2nd Lt. Edward M. Furman | N | 01011479 |
| Cpl. Seymour Kaufman | | 32790620 |
| Cpl. Dorman Robertson, Jr. | | 38603342 |
| Cpl. Paul A. Sarica | | 33695897 |
| Cpl. Fred Stephens | | 36879277 |
| Cpl. Clarence R. Winters | | 33837338 |
| Cpl. Alfred W. Morin | | 11102998 |

22 February 1945 to 714th Squadron
| 2nd Lt. Stanley E. Guiney | P | 02062098 |
| 2nd Lt. John R. Bohannan | C | 0837440 |
| 2nd Lt. Edward R. Casey | N | 0274665 |
| Cpl. Robert H. Corregan | | 12089866 |
| Cpl. Howard P. Martin | | 33344841 |
| Cpl. Nick Mosora | | 35098758 |
| Cpl. Stanley L. Trebbs | | 36758034 |
| Cpl. William J. Yuhas | | 33709727 |
| Cpl. Francis G. McCarthy | | 31310306 |

22 February 1945 to 715th Squadron
| 2nd Lt. Albert W. Halfhill, Jr. | | 02061489 |
| 2nd Lt. Leo T. Fleisch | | 0837596 |
| F/O Craig F. Dinsbier | | T-137315 |
| TSgt. James W. Mayfield | | 6938444 |
| Sgt. Joseph F. Macone | | 11056977 |
| Cpl. George F. Lange | | 11138664 |
| Cpl. Francis A. Marrocco | | 31369915 |
| Cpl. Frank M. Zelenitz | | 35216736 |
| Cpl. George Zuniga | | 18090196 |

27 February 1945 to 713th Squadron
| 2nd Lt. Donald M. Ford | P | 02058765 |
| 2nd Lt. Ralph J. Radowick | CP | 0930445 |
| 2nd Lt. Clement R. Grosso | N | 02074401 |
| 1st Lt. William M. Jones | B | 01285766 |
| Cpl. Van W. Fowers | | 16197961 |
| Cpl. Harold B. Heyler | | 36881409 |
| Cpl. Norman C. Poorman | | 38686629 |
| Cpl. Victor F. Roys | | 33730924 |
| Cpl. Charles L. Wingo | | 33904910 |
| Cpl. Edward R. Mikiua | | 33608921 |

27 February 1945 to 713th Squadron
| 2nd Lt. Elmer M. Homelvig | P | 02050191 |
| 2nd Lt. Delbert M. Gablock | CP | 02063603 |
| 2nd Lt. Ferris W. Kennedy | N | 0930408 |
| F/O Victor Q. Smith | B | T-5715 |
| Cpl. John R. Cray | | 37358202 |
| Cpl. Charles E. Nelms | | 37627536 |
| Cpl. Cleatus G. Stone | | 36742647 |
| Cpl. Walton J. Tombari | | 31381182 |
| Cpl. Dale T. Wreisner | | 17145062 |

| Cpl. Emory G. Repass | | 36657226 |

27 February 1945 to 713th Squadron
| 2nd Lt. Forrest F. Anderson | P | 0832341 |
| 2nd Lt. Arthur R. Seat, Jr. | CP | 02062385 |
| 2nd Lt. Frank W. Leonard | N | 02074477 |
| F/O Jerome Brown | B | T-6830 |
| Cpl. Elberon G. Andrews | | 35912737 |
| Cpl. Benjamin S. Daniel | | 16138345 |
| Cpl. Douglas J. Fowler | | 34765092 |
| Cpl. Charles E. Schmucker | | 37706944 |
| Cpl. John W. Wideman | | 37704363 |
| Cpl. Harry Hutchinson | | 35221450 |

27 February 1945 to 714th Squadron
| 2nd Lt. Harry S. Constable | P | 0806223 |
| 2nd Lt. Clayton H. Johnson | CP | 02061962 |
| 2nd Lt. Joseph J. McConnell, Jr. | N | 02074493 |
| F/O John J. Northrup | B | T-5773 |
| Cpl. Calvin J. Barnett | | 16130439 |
| Cpl. William J. Davis, Jr. | | 37722410 |
| Cpl. Byron A. Thomas | | 39925056 |
| Cpl. Milton Ulanoff | | 12129063 |
| Cpl. Bernard J. Varney, Jr. | | 31339387 |
| Cpl. Thomas G. Treadwell | | 14140979 |

27 February 1945 to 714th Squadron
| 2nd Lt. William E. Dupree | P | 0832118 |
| 2nd Lt. Harold Major, Jr. | CP | 02067316 |
| 2nd Lt. Charles B. McDonald | N | 02074495 |
| F/O Harry E. Floyd, Jr. | B | T-5696 |
| Sgt. Peter A. Tell | | 32075346 |
| Cpl. John D. DeLang | | 17067161 |
| Cpl. Paul D. Hester | | 18228003 |
| Cpl. Walter A. Scheel | | 35836693 |
| Pvt. William A. Oiler | | 35436498 |
| Cpl. Richard E. Sprenkle | | 3351255? |

27 February 1945 to 714th Squadron
| 2nd Lt. Paul E. Westrick | P | 02062099 |
| 2nd Lt. Everett R. Pickering | CP | 02067346 |
| 2nd Lt. Leon H. Martin | N | 02074515 |
| 2nd Lt. Thomas V. Scott | B | 0785138 |
| Sgt. William J. Grize | | 31124923 |
| Cpl. Donald J. Eidson | | 20953835 |
| Cpl. Roderic H. Landreth | | 19215638 |
| Cpl. Stuart D. Van Deventer | | 3858893? |
| Pvt. Herman Watts | | 1508761? |
| Cpl. John J. Benyo | | 3360843? |

Replacement Crews – March 1945

17 March 1945 to 713th Squadron
| F/O James F. Wagner | P | T-64147 |
| F/O Jerome F. Wassman | CP | T-65186 |

| | | | | | | |
|---|---|---|---|---|---|---|
| F/O John W. Allan | N | T-135845 | 29 March 1945 to 714th Squadron | | | |
| Sgt. Linwood O. Patten | | 33062946 | 2nd Lt. John C. McCoy | P | 02057337 | |
| Cpl. Robert B. Bailey | | 33904874 | F/O Bernard J. Banas | CP | T-65103 | |
| Cpl. Robert D. Catone | | 12229951 | F/O Clarence H. Leimer | N | T-135784 | |
| Cpl. Gerald L. Goble | | 13133481 | F/O Saverio J. Juliano | B | T-134418 | |
| Cpl. Charles P. Holbrook | | 11104593 | Cpl. Bronislaw J. Kardys | | 31252333 | |
| Cpl. Harrison G. Newcomb, Jr. | | 34540299 | Cpl. Angelo J. Leonetti | | 13125368 | |
| | | | Cpl. Frank J. Matula | | 12100312 | |
| 17 March 1945 to 713th Squadron | | | Cpl. Abe L. Morgan | | 34738387 | |
| 2nd Lt. David L. Davis | P | 0835454 | Cpl. Irria J. Peterson | | 38174375 | |
| 2nd Lt. John E. Morrison | CP | 0837510 | Cpl. James E. Smith | | 34634691 | |
| 2nd Lt. David M. Chambers, Jr. | N | 02077198 | | | | |
| 2nd Lt. William R. Fulton | B | 02071953 | 29 March 1945 to 715th Squadron | | | |
| Sgt. George S. Van Horn | | 33364717 | 2nd Lt. Edward S. Lytle | P | 02061999 | |
| Cpl. Johnny R. Clary | | 18210123 | 2nd Lt. Joseph Schweitzer | CP | 0783844 | |
| Cpl. William J. Hawthorne, Jr. | | 31336558 | 2nd Lt. Walter J. David | N | 02076574 | |
| Cpl. Harold W. Poland | | 32266456 | Cpl. Donald T. Fox | | 13113363 | |
| Cpl. James I. Sessums | | 14150844 | Cpl. James A. Lamb | | 35884404 | |
| Cpl. Wayne J. Roache | | 39918707 | Cpl. Stanley Dressler | | 33260322 | |
| | | | Cpl. James L. Hicks, Jr. | | 33658449 | |
| 17 March 1945 to 715th Squadron | | | Cpl. Philbert N. Weber | | 17132130 | |
| 2nd Lt. John F. Walker | P | 0821814 | | | | |
| 2nd Lt. Stephen Blazek, Jr. | CP | 02071938 | 29 March 1945 to 715th Squadron | | | |
| F/O Donald E. Francisco, Jr. | N | T-66102 | 2nd Lt. Walter J. Peters | P | 02061755 | |
| F/O Frank Barilla | B | T-137618 | 2nd Lt. William P. Leibensperger | CP | 02062232 | |
| SSgt. Edward L. James | | 36425125 | 2nd Lt. William P. Leftwich | | 0928110 | |
| Sgt. John P. O'Conner | | 11062033 | Cpl. Dewey R. Brosey | | 35417282 | |
| Cpl. Simon E. Bukovitz | | 33429594 | Cpl. Alfonso A. Coronado | | 39421226 | |
| Cpl. Joseph T. Hurley | | 31372110 | Cpl. Thomas W. Koopman, Jr. | | 32803329 | |
| Cpl. Leonard A. Poritz | | 34787460 | Cpl. Donald R. Morrison | | 19142756 | |
| Cpl. Richard S. Shely, Jr. | | 15119576 | Cpl. William J. Wheeler, Jr. | | 34903986 | |
| | | | Cpl. Louis F. Thronson | | 36834799 | |
| 29 March 1945 to 713th Squadron | | | | | | |
| 2nd Lt. James K. McFerren | P | 02058231 | 29 March 1945 to 715th Squadron | | | |
| F/O Alfred M. Bettman | CP | T-65109 | 2nd Lt. Odis O. Willett | P | 0928360 | |
| 2nd Lt. Ralph E. Williams | N | 02072939 | 2nd Lt. George C. Bothwell, Jr. | CP | 0930903 | |
| Cpl. Walter D. Grindle, Jr. | | 34946134 | F/O Adelbert W. Zeitlow | N | T-137479 | |
| Cpl. Edward V. Ladas | | 31378761 | Sgt. William H. Bixler | | 33111866 | |
| Cpl. Verl D. Moore | | 13188308 | Sgt. Robert E. Youngquist | | 20625440 | |
| Cpl. William C. Morrow | | 18193279 | Cpl. Eugene F. Dunn | | 13142661 | |
| Cpl. Lyle W. Peebles | | 33712182 | Cpl. Delmar L. Fouts | | 17070006 | |
| Cpl. Dean G. Shumaker | | 15127506 | Cpl. Vincent P. Pronesti | | 33797807 | |
| | | | Cpl. Frederick F. Shook | | 13188211 | |
| 29 March 1945 to 714th Squadron | | | | | | |
| 2nd Lt. John W. Trostle | P | 0834062 | 30 March 1945 to 715th Squadron | | | |
| 2nd Lt. Joe H. Davis | CP | 0836899 | 2nd Lt. Warren C. Howard | P | 0718644 | |
| 2nd Lt. Harold W. Harkey | N | 02075368 | 2nd Lt. John F. Moran | CP | 0783808 | |
| Cpl. Thomas M. Duke | | 13142682 | F/O Francis H. Wander | N | T-137476 | |
| Cpl. Burton R. Johnson | | 39334101 | Cpl. John G. Brough | | 39924939 | |
| Cpl. Wynton B. Hudson | | 18098130 | Cpl. Willis C. Conkle | | 35919380 | |
| Cpl. Robert M. Martin | | 14193914 | Cpl. Karl G. Kersh | | 38545998 | |
| Cpl. Adelard O. Soucy | | 31366556 | Cpl. Victor L. Patti | | 17136617 | |
| Cpl. Francis B. Higgins | | 34817643 | Cpl. Miles C. Taylor | | 18242673 | |
| | | | Cpl. Charles S. Cobb | | 14190321 | |

Replacement Crews – April 1945

2 April 1945 to 713th Squadron
| | | |
|---|---|---|
| 1st Lt. Harry H. Mortimore | P | 01012827 |
| F/O Melville W. Cave | CP | T-129249 |
| F/O Leonard E. Roecker, Jr. | N | T-7267 |
| 2nd Lt. Luther C. Shelton, Jr. | B | 0832712 |
| Sgt. Lyle R. Stalnaker | | 6661496 |
| Cpl. Carl W. Anderson, Jr. | | 39465799 |
| Cpl. Dean A. Brinkerhoff | | 39858488 |
| Cpl. Edward E. Carson II | | 35069576 |
| Cpl. Clinton Frankenfield | | 39218669 |
| Cpl. Harry B. Puckett | | 37732855 |

2 April 1945 to 713th Squadron
| | | |
|---|---|---|
| 2nd Lt. John Thomas, Jr. | P | 0835218 |
| F/O Lawrence Supienko | CP | T-63533 |
| 2nd Lt. Ansel W. Stork | N | 02069150 |
| F/O Joseph A. Benjamin | B | T-8114 |
| Cpl. Joseph A. Felipe | | 39140140 |
| Cpl. Ted M. Harris | | 39727606 |
| Cpl. Raymond E. Neuse | | 18233046 |
| Cpl. William W. Oakey | | 36838835 |
| Cpl. Joseph P. Pilla | | 37627405 |
| Pvt. James E. King | | 34801078 |

2 April 1945 to 714th Squadron
| | | |
|---|---|---|
| 2nd Lt. George H. Roos, Jr. | P | 0833590 |
| 2nd Lt. Shirley W. John | CP | 02070846 |
| 2nd Lt. Peter L. Hollod | N | 02073125 |
| 2nd Lt. William A. Denten | B | 0786902 |
| Cpl. William L. LeFore | | 39473173 |
| Cpl. Walter R. Budrey | | 11138060 |
| Cpl. Lee E. Cale | | 37728044 |
| Cpl. Woodrow J. Dawson | | 14191354 |
| Cpl. Clyde W. Holder, Jr. | | 34923610 |
| Cpl. Stephen P. L???? | | 13141847 |

2 April 1945 to 715th Squadron
| | | |
|---|---|---|
| 2nd Lt. William O. Whetsell | P | 0835306 |
| F/O James Beadling | CP | T-133671 |
| 2nd Lt. Lawrence C. Taylor | N | 02069157 |
| 2nd Lt. James E. Muenker | B | 0785206 |
| Cpl. Roy H. Bassler | | 31328988 |
| Cpl. Paul B. Davis | | 14159877 |
| Cpl. William T. Hall, Jr. | | 18242578 |
| Cpl. Robert S. Peterson | | 11139116 |
| Cpl. Robert W. Putney | | 36645347 |
| Cpl. William A. Stone, Jr. | | 34657359 |

2 April 1945 to 715th Squadron
| | | |
|---|---|---|
| 2nd Lt. William A. Winburn III | P | 0833237 |
| 2nd Lt. John A. Brugeman | CP | 0930698 |
| 2nd Lt. Bernard S. Vagnoni | N | 020691?? |
| F/O Lawrence W. Tolj, Jr. | B | T-8294 |
| Cpl. Terrence F. Coyle, Jr. | | 42031581 |
| Cpl. Frank R. Delgrosso | | 33731246 |
| Cpl. John F. Durbin | | 35833754 |
| Cpl. Russell L. Jewett | | 39918807 |
| Cpl. Roy Richburg, Jr. | | 18242685 |
| Cpl. Franklin D. Stevens | | 39616979 |

17 April 1945 to 715th Squadron
| | | |
|---|---|---|
| Capt Daniel H. LaPointe | P | 0727029 |
| 2nd Lt. Robert J. Hertell | CP | 01014677 |
| 2nd Lt. George H. Thomas | N | 02076780 |
| F/O Vincent Donato | B | T-9126 |
| SSgt. James Thonis | | 6147780 |
| Sgt. Edward G. Knertz | | 20253672 |
| Cpl. Jack J. Coats | | 36854728 |
| Cpl. Robert E. Greenwell | | 14160706 |
| Cpl. Clay D. McCraw | | 38608644 |
| Cpl. Laverne C. Stricker | | 16133091 |

17 April 1945 to 715th Squadron
| | | |
|---|---|---|
| 2nd Lt. Joseph H. Andres | P | 02061462 |
| 2nd Lt. Raymond W. Peterson | CP | 02071040 |
| 2nd Lt. Baruyr A. Poladian | N | 01052521 |
| 2nd Lt. Eugene Stetz | B | 0788108 |
| Sgt. Bernard V. Wright | | 6874699 |
| Cpl. James E. Helmuth | | 38563671 |
| Cpl. Robert B. Loy | | 16032276 |
| Cpl. Daniel J. Marek | | 38421916 |
| Cpl. Vernon E. Moore, Jr. | | 34508607 |
| Cpl. John A. Waynick | | 14200281 |

19 April 1945 to 714th Squadron
| | | |
|---|---|---|
| 2nd Lt. Leroy D. Sentor | P | 0828794 |
| 2nd Lt. Robert H. Grimes | CP | 02057280 |
| 2nd Lt. William H. Byrd | N | 02065953 |
| TSgt. Ervin D. Sengstock | | 36828946 |
| TSgt. James H. Triplet | | 17159733 |
| SSgt. Coleman E. King | | 39295906 |
| SSgt. Arthur J. Rekart | | 37612440 |
| SSgt. Leo M. Schade | | 37624028 |
| SSgt. Albert M. Kaplan | | 12100300 |
| SSgt. John D. O'Leary | | 12064281 |

19 April 1945 to 715th Squadron
| | | |
|---|---|---|
| 1st Lt. Robert E. Langenfeld | P | 0720526 |
| 2nd Lt. George W. Warner | CP | 0776156 |
| 2nd Lt. Charles W. Parker | N | 0783148 |
| TSgt. Robert A. Noel | | 38478377 |
| TSgt. Glen M. Hotz | | 15130720 |
| SSgt. James E. Davenport | | 16189182 |
| SSgt. William J. Potthoff | | 32984688 |
| SSgt. Robert J. Krause | | 36840120 |
| SSgt. Philip G. Smith | | 12176765 |
| SSgt. Edward P. Speers | | 13072554 |

## Appendix 4
# LEADERSHIP

Group Commanders
Col. James McK. Thomson        25 May 1943 - 1 April 1944              KIA
Col. Gerry L. Mason            3 April 1944 - 14 November 1944
Col. Charles B. Westover       14 November 1944 - 27 May 1945
Lt. Col. Lester F. Miller      27 May 1945 – return

Group Air Executive Officers
Lt. Col. Hubert S. Judy, Jr.   October 1943 - December 1944
Lt. Col. Byron B. Webb         December 1944 - May 1945
Lt. Col. Lester F. Miller      May 1945 – return

Group Ground Executive Officer
Lt. Col. Karl Elver            25 May 1943 - ???
Lt. Col. James R. Patterson    ??? – 20 February 1945
Maj. William H. Searles        20 February 1945 – return

Group Operations Officers
Maj. Ronald V. Kramer          May 1943 - February 1944
Maj. Robert W. Spence          February 1944 - August 1944
Maj. William G. Blum           August 1944 - September 1944            POW
Lt. Col. Lester F. Miller      September 1944 - October 1944
Lt. Col. Byron B. Webb         October 1944 - December 1944
Lt. Col. John Grable           December 1944 - April 1945
Lt. Col. Lester F. Miller      April 1945 - May 1945
Maj. Leroy A. Smith            May 1945 – return

712th Bomb Squadron Commanders
Maj. Robert Campbell           25 May 1943 – 20 March 1944             KIA
Maj. Ronald V. Kramer          20 March 1944 - 9 May 1944              POW
Maj Lester F. Miller           9 May 1944 – September 1944
Maj. Walter Stroud             September 1944 – October 1944
Maj. Gordon R, Koons           Ocotber 1944 – return

713th Bomb Squadron Commanders
Maj. Chester Hackett           May 1943 - December 1943
Lt. Col. Heber H. Thompson     December 1943 – 12 June 1945            KIA

714th Bomb Squadron Commanders
Maj. Glassel Stringfellow      May 1943 – 26 October 1944
Maj. Emil G. Beaudry           26 October 1944 – return

715th Bomb Squadron Commanders
Maj. Ken Squyres               May 1943 - 5 January 1944               KIA
Maj. Jack Edwards              4 January 1944 - 18 March 1944          Interned
Maj. Chester Hackett           18 March 1944 - 20 June 1944            POW
Maj. James Conrad              20 June 1944 – 11 July 1944             Interned
Maj. John Grable, Jr.          11 July 1944 – October 1944
Maj. Walter C. Stroud          October 1944 – return

*Appendix 5*
# MISSIONS

| Date | Target | 8th AF Mission | 448th Mission |
|---|---|---|---|
| 16 Dec 43 | Bremen (canceled) | 156 | |
| 22 Dec 43 | Osnabruk | 161 | 1 |
| 24 Dec 43 | Labroye, France | 164 | 2 |
| 30 Dec 43 | Ludwigshafen | 169 | 3 |
| 31 Dec 43 | LaRochelle, France | 171 | 4 |
| 4 Jan 44 | Kiel | 174 | 5 |
| 5 Jan 44 | Kiel | 176 | 6 |
| 11 Jan 44 | Meppen & Zundberg | 182 | 7 |
| 14 Jan 44 | St. Pierre-d'Jongurer, France | 183 | 8 |
| 21 Jan 44 | Labroye, France | 187 | 9 |
| 24 Jan 44 | Recalled | 191 | |
| 29 Jan 44 | Frankfurt | 198 | 10 |
| 30 Jan 44 | Brunswick | 200 | 11 |
| 3 Feb 44 | Recalled | 206 | |
| 4 Feb 44 | Frankfurt | 208 | 12 |
| 5 Feb 44 | Tours, France | 210 | 13 |
| 6 Feb 44 | St. Pol, France (Recalled) | 212 | 14 |
| 10 Feb 44 | Rijen, Holland | 216 | 15 |
| 11 Feb 44 | St. Pol, France | 218 | 16 |
| 13 Feb 44 | St. Pol, France | 221 | 17 |
| 20 Feb 44 | Gotha | 226 | 18 |
| 21 Feb 44 | Hesepe | 228 | 19 |
| 22 Feb 44 | Enschede, Holland | 230 | 20 |
| 24 Feb 44 | Gotha | 233 | 21 |
| 25 Feb 44 | Furth | 235 | 22 |
| 28 Feb 44 | Escalles-Buchy, France | 237 | 23 |
| 2 Mar 44 | Frankfurt (Recalled) | 244 | 24 |
| 3 Mar 44 | Heligoland (Recalled) | 246 | 25 |
| 5 Mar 44 | Bergerac, France | 248 | 26 |
| 6 Mar 44 | Berlin | 250 | 27 |
| 8 Mar 44 | Berlin/Erkner | 252 | 28 |
| 9 Mar 44 | Nienburg | 253 | 29 |
| 13 Mar 44 | St. Pol, France (Recalled) | 257 | 30 |
| 16 Mar 44 | Friedrichshafen | 262 | 31 |
| 18 Mar 44 | Friedrichshafen | 264 | 32 |
| 20 Mar 44 | Frankfurt | 269 | 33 |
| 22 Mar 44 | Berlin | 273 | 34 |
| 23 Mar 44 | Munster | 275 | 35 |
| 24 Mar 44 | Nancy, France | 277 | 36 |
| 26 Mar 44 | Moyenneville, France | 280 | 37 |
| 28 Mar 44 | Kille, France (Ijmuiden, Holland) | 283 | |
| 29 Mar 44 | Watten, France | 284 | 38 |
| 1 Apr 44 | Pforzheim/Ludwigshafen | 287 | 39 |
| 6 Apr 44 | Watten, France | 290 | 40 |
| 8 Apr 44 | Brunswick | 291 | 41 |
| 9 Apr 44 | Tutow | 293 | 42 |
| 10 Apr 44 | Bourges, France | 295 | 43 |

| Date | Target | | |
|---|---|---|---|
| 11 Apr 44 | Bernburg | 298 | 44 |
| 12 Apr 44 | Oschersleben (Recalled) | 300 | 45 |
| 13 Apr 44 | Lauffen | 301 | 46 |
| 18 Apr 44 | Watten, France | 306 | 48 |
| 18 Apr 44 | Rathenow | 306 | 47 |
| 19 Apr 44 | Watten, France | 308 | 49 |
| 20 Apr 44 | Bonnieres, France | 309 | 50 |
| 21 Apr 44 | Brux (Recalled) | | |
| 22 Apr 44 | Hamm | 311 | 51 |
| 24 Apr 44 | Gablingen | 315 | 52 |
| 25 Apr 44 | Mannheim (Recalled) | 317 | 53 |
| 26 Apr 44 | Paderborn (Recalled) | 319 | 54 |
| 27 Apr 44 | Wizernes, France | 322 | 55 |
| 27 Apr 44 | Blainville-sur-l'Eau, France | 323 | 56 |
| 29 Apr 44 | Berlin | 327 | 57 |
| 1 May 44 | Pas de Calais, France (Recalled) | 332 | |
| 1 May 44 | Brussels, Belgium | 333 | 58 |
| 4 May 44 | Recalled | 338 | |
| 6 May 44 | Siracourt, France | 340 | 59 |
| 7 May 44 | Munster | 342 | 60 |
| 8 May 44 | Brusnwick | 344 | 61 |
| 9 May 44 | Liege, Belgium | 347 | 62 |
| 10 May 44 | Recalled | | |
| 11 May 44 | Mulhouse, France | 351 | 63 |
| 12 May 44 | Bohlen | 353 | 64 |
| 13 May 44 | Tutow | 355 | 65 |
| 19 May 44 | Brunswick | 358 | 66 |
| 20 May 44 | Reims, France (Recalled) | 359 | |
| 21 May 44 | Siracourt, France (Recalled) | 360 | |
| 22 May 44 | Siracourt, France | 361 | 67 |
| 23 May 44 | Orleans/Bricy, France | 364 | 68 |
| 24 May 44 | Orly, France | 367 | 69 |
| 25 May 44 | Mulhouse, France | 370 | 70 |
| 25 May 44 | Fecamp, France | 370 | 71 |
| 27 May 44 | Konz/Karthaus | 373 | 72 |
| 28 May 44 | Zeitz | 376 | 73 |
| 29 May 44 | Tutow | 379 | 74 |
| 30 May 44 | Rotenburg | 380 | 75 |
| 31 May 44 | Woippy, France (Recalled) | 382 | 76 |
| 2 Jun 44 | Beauvoir, France | 384 | 77 |
| 3 Jun 44 | Stella Plage, France | 388 | 78 |
| 4 Jun 44 | Sangatte, France | 389 | 79 |
| 5 Jun 44 | Sangatte, France | 392 | 80 |
| 6 Jun 44 | Calin/Pointe et Raz de-la-Percee, France | 394 | 81 |
| 6 Jun 44 | Vierville, France | 394 | |
| 6 Jun 44 | Coutances, France | 394 | 82 |
| 6 Jun 44 | Caen, France | 395 | 83 |
| 6 Jun 44 | Coutances, France | 395 | 84 |
| 7 Jun 44 | Conches/Vascaeuil, France | 397 | 85 |
| 8 Jun 44 | Orleans, France | 400 | 86 |
| 10 Jun 44 | Evereux/Fauville, France | 403 | 87 |
| 12 Jun 44 | Ploermel, France | 407 | 88 |
| 12 Jun 44 | Conches, France | 407 | 89 |
| 14 Jun 44 | Orleans, France | 412 | 90 |

| Date | Target | | |
|---|---|---|---|
| 15 Jun 44 | Cinq-la-Pile/Le Port Boulet, France | 414 | 91 |
| 16 Jun 44 | St. Omer/Renescure, France | 416 | 92 |
| 17 Jun 44 | LeMans, France | 419 | 93 |
| 18 Jun 44 | Hamburg | 421 | 94 |
| 18 Jun 44 | Watten, France | 421 | 95 |
| 19 Jun 44 | Haute Cote, France | 423 | 96 |
| 20 Jun 44 | Siracourt, France | 425 | 98 |
| 20 Jun 44 | Politz | 425 | 97 |
| 21 Jun 44 | Berlin | 428 | 99 |
| 21 Jun 44 | Siracourt, France | 429 | 100 |
| 22 Jun 44 | Guyancourt, France | 431 | 101 |
| 24 Jun 44 | Melun, France (Recalled) | 437 | 102 |
| 24 Jun 44 | Haute Cote, France | 438 | 103 |
| 25 Jun 44 | Bretigny, France | 442 | 104 |
| 27 Jun 44 | Creil, France | 443 | 105 |
| 28 Jun 44 | Saarbrucken | 445 | 106 |
| 29 Jun 44 | Bernburg | 447 | 107 |
| 2 Jul 44 | Fiefs, France | 450 | 108 |
| 6 Jul 44 | Sully-sur-Loire, France | 455 | 109 |
| 8 Jul 44 | Rilly la Montogne, France (Recalled) | 460 | 110 |
| 11 Jul 44 | Munich | 466 | 111 |
| 12 Jul 44 | Munich | 469 | 112 |
| 13 Jul 44 | Saarbrucken | 471 | 113 |
| 16 Jul 44 | Saarbrucken | 476 | 114 |
| 17 Jul 44 | St. Sylvester, France | 478 | 115 |
| 18 Jul 44 | Grentheville, France | 481 | 116 |
| 19 Jul 44 | Koblenz | 482 | 117 |
| 20 Jul 44 | Schmalkalden | 484 | 118 |
| 21 Jul 44 | Munich | 486 | 119 |
| 23 Jul 44 | Laon/Athies, France | 490 | 120 |
| 24 Jul 44 | Moutreuil, France | 492 | 121 |
| 24 Jul 44 | Moutreuil, France | 492 | 122 |
| 29 Jul 44 | Bremen | 503 | 123 |
| 31 Jul 44 | Ludwigshafen | 507 | 124 |
| 1 Aug 44 | Villaroche, France | 508 | 125 |
| 2 Aug 44 | St. Dizier, France | 510 | 126 |
| 3 Aug 44 | Douai, France | 512 | 127 |
| 4 Aug 44 | Rostock | 514 | 128 |
| 5 Aug 44 | Fellersleben | 519 | 129 |
| 6 Aug 44 | Hemmingstadt | 524 | 130 |
| 7 Aug 44 | Brussels | 527 | 131 |
| 8 Aug 44 | Rouen, France | 530 | 132 |
| 10 Aug 44 | Pacy-sur-Armancon, France | 537 | 133 |
| 11 Aug 44 | St. Florentin, France | 541 | 134 |
| 12 Aug 44 | Laon/Courran, France | 545 | 135 |
| 13 Aug 44 | Rouen, France | 548 | 136 |
| 14 Aug 44 | Dijon, France | 552 | 137 |
| 15 Aug 44 | Plantlunne | 554 | 138 |
| 16 Aug 44 | Magdeburg | 556 | 139 |
| 18 Aug 44 | Laneuveville, France | 562 | 140 |
| 24 Aug 44 | Brunswick | 568 | 141 |
| 25 Aug 44 | Rostock | 570 | 142 |
| 26 Aug 44 | Ludwigshafen | 576 | 143 |
| 27 Aug 44 | Oranienburg (Recalled) | 584 | 144 |

## Appendices

|  | Truckin Mission, Orleans, France | | |
|---|---|---|---|
|  | Truckin Mission, Orleans, France | | |
|  | Truckin Mission, Orleans, France | | |
|  | Truckin Mission, Orleans, France | | |
|  | Truckin Mission, Orleans, France | | |
| 9 Sep 44 | Gustavsburg | 614 | 145 |
| 11 Sep 44 | Magdeburg | 623 | 146 |
| 12 Sep 44 | Hemmingstadt | 626 | 147 |
| 13 Sep 44 | Ulm (Recalled) | 628 | 148 |
| 18 Sep 44 | Groesbeek, Holland | 639 | 149 |
| 21 Sep 44 | Koblenz | 644 | 150 |
| 22 Sep 44 | Kassel | 645 | 151 |
| 25 Sep 44 | Koblenz | 647 | 152 |
| 26 Sep 44 | Hamm | 648 | 153 |
| 27 Sep 44 | Kassel | 650 | 154 |
| 28 Sep 44 | Kassel | 652 | 155 |
| 30 Sep 44 | Hamm | 655 | 156 |
| 2 Oct 44 | Hamm | 659 | 157 |
| 3 Oct 44 | Pforzheim | 662 | 158 |
| 6 Oct 44 | Harburg | 667 | 159 |
| 7 Oct 44 | Clausthal-Zellerfeld | 669 | 160 |
| 9 Oct 44 | Koblenz | 670 | 161 |
| 14 Oct 44 | Cologne | 677 | 162 |
| 15 Oct 44 | Cologne | 678 | 163 |
| 17 Oct 44 | Cologne | 681 | 164 |
| 19 Oct 44 | Mainz | 683 | 165 |
| 22 Oct 44 | Hamm | 685 | 166 |
| 25 Oct 44 | Neumunster | 688 | 167 |
| 26 Oct 44 | Bottrop | 689 | 168 |
| 30 Oct 44 | Hamburg | 693 | 169 |
| 2 Nov 44 | Bielefeld | 698 | 170 |
| 4 Nov 44 | Geisenkirchen | 700 | 171 |
| 5 Nov 44 | Karlsruhe | 702 | 172 |
| 6 Nov 44 | Minden | 704 | 173 |
| 8 Nov 44 | Rheine | 705 | 174 |
| 9 Nov 44 | Verny, France | 707 | 175 |
| 10 Nov 44 | Hanau | 709 | 176 |
| 16 Nov 44 | Eschweiler | 715 | 177 |
| 21 Nov 44 | Hamburg | 720 | 178 |
| 25 Nov 44 | Bingen | 723 | 179 |
| 26 Nov 44 | Bielefeld | 725 | 180 |
| 27 Nov 44 | Offenburg | 727 | 181 |
| 29 Nov 44 | Bielefeld | 729 | 182 |
| 30 Nov 44 | Neunkirchen | 731 | 183 |
| 4 Dec 44 | Koblenz | 736 | 184 |
| 6 Dec 44 | Minden | 741 | 185 |
| 11 Dec 44 | Hanau | 746 | 186 |
| 12 Dec 44 | Aschaffenberg | 748 | 187 |
| 19 Dec 44 | Ehrang | 756 | 188 |
| 24 Dec 44 | Euskirchen | 760 | 189 |
| 25 Dec 44 | Budesheim | 761 | 190 |
| 28 Dec 44 | Kaiserslautern | 766 | 191 |
| 30 Dec 44 | Mechernich | 770 | 192 |
| 31 Dec 44 | Remagen | 772 | 193 |

| | | | |
|---|---|---|---|
| 2 Jan 45 | Neuwied | 776 | 194 |
| 3 Jan 45 | Neunkirchen | 778 | 195 |
| 5 Jan 45 | Pirmasens | 781 | 196 |
| 6 Jan 45 | Koblenz | 783 | 197 |
| 7 Jan 45 | Achern | 785 | 198 |
| 10 Jan 45 | Wewelwer | 789 | 199 |
| 13 Jan 45 | Worms | 791 | 200 |
| 14 Jan 45 | Hallendorf | 792 | 201 |
| 15 Jan 45 | Kilchberg | 794 | 202 |
| 16 Jan 45 | Dresden | 796 | 203 |
| 28 Jan 45 | Dortmund | 809 | 204 |
| 29 Jan 45 | Munster | 811 | 205 |
| 31 Jan 45 | Brunswick (Recalled) | 813 | 206 |
| 3 Feb 45 | Magdeburg | 817 | 207 |
| 6 Feb 45 | Magdeburg | 821 | 208 |
| 9 Feb 45 | Magdeburg | 824 | 209 |
| 11 Feb 45 | Dulmen | 827 | 210 |
| 14 Feb 45 | Magdeburg | 830 | 211 |
| 15 Feb 45 | Magdeburg | 832 | 212 |
| 16 Feb 45 | Osnabruk | 833 | 213 |
| 19 Feb 45 | Siegen | 835 | 214 |
| 21 Feb 45 | Nurnberg | 839 | 215 |
| 22 Feb 45 | Kreinsen | 841 | 216 |
| 23 Feb 45 | Osnabruk | 843 | 217 |
| 24 Feb 45 | Misberg | 845 | 218 |
| 25 Feb 45 | Aschaffeburg | 847 | 219 |
| 26 Feb 45 | Berlin | 849 | 220 |
| 27 Feb 45 | Halle | 851 | 221 |
| 28 Feb 45 | Meschede | 854 | 222 |
| 1 Mar 45 | Augsburg | 857 | 223 |
| 2 Mar 45 | Magdeburg | 859 | 224 |
| 3 Mar 45 | Magdeburg | 861 | 225 |
| 4 Mar 45 | Stuttgart | 863 | 226 |
| 5 Mar 45 | Harburg | 865 | 227 |
| 8 Mar 45 | Dillenburg | 872 | 228 |
| 9 Mar 45 | Rheine | 875 | 229 |
| 10 Mar 45 | Paderborn | 877 | 230 |
| 11 Mar 45 | Kiel | 881 | 231 |
| 12 Mar 45 | Swinemunde | 883 | 232 |
| 14 Mar 45 | Gutersloh | 886 | 233 |
| 15 Mar 45 | Zossen | 889 | 234 |
| 17 Mar 45 | Hanover | 892 | 235 |
| 18 Mar 45 | Berlin | 894 | 236 |
| 19 Mar 45 | Baumenheim | 896 | 237 |
| 20 Mar 45 | Hemmingstadt | 898 | 238 |
| 21 Mar 45 | Ahlhorn | 901 | 239 |
| 21 Mar 45 | Mulheim | 904 | 240 |
| 22 Mar 45 | Kitzingen | 906 | 241 |
| 23 Mar 45 | Munster | 908 | 242 |
| 24 Mar 45 | Wesel | 911 | 243 |
| 24 Mar 45 | Stormede | 911 | 244 |
| 25 Mar 45 | Buchen | 913 | 245 |
| 30 Mar 45 | Wilhemshaven | 918 | 246 |
| 31 Mar 45 | Brunswick | 920 | 247 |

| | | | |
|---|---|---|---|
| 2 Apr 45 | Alhborg, Denmark (Recalled) | 922 | |
| 4 Apr 45 | Wesendorf | 926 | 248 |
| 5 Apr 45 | Bayreuth | 928 | 249 |
| 6 Apr 45 | Halle | 930 | 250 |
| 7 Apr 45 | Duneberg | 931 | 251 |
| 8 Apr 45 | Roth | 932 | 252 |
| 9 Apr 45 | Landsberg | 935 | 253 |
| 10 Apr 45 | Rechlin | 938 | 254 |
| 11 Apr 45 | Regensburg | 941 | 255 |
| 14 Apr 45 | Coubre Point, France | 948 | 256 |
| 15 Apr 45 | Royan, France | 951 | 257 |
| 16 Apr 45 | Landshut | 954 | 258 |
| 18 Apr 45 | Passau | 959 | 259 |
| 20 Apr 45 | Muhldorf | 962 | 260 |
| 21 Apr 45 | Salzburg, Austria (Recalled) | 963 | 261 |
| 25 Apr 45 | Salzburg, Austria | 968 | 262 |

*Appendix 6*
# STATISTICAL SUMMARY

Missions: 262
Men missing in action: 875 (included POWs, internees)
Men killed in action: 85
Men wounded or died of wounds: 119
Aircraft attacking target: 5,824
Bombs dropped: 15,286.1 tons
Supplies dropped: 139.8 tons
Aircraft missing in action: 98
Aircraft abandoned on continent: 17
Aircraft lost to salvage: 31
Total operational losses: 146
    712th BS: 35
    713th BS: 28
    714th BS: 50
    715th BS: 33
Enemy aircraft destroyed: 47
    712th BS: 10
    713th BS: 4
    714th BS: 15
    715th BS: 18
Operational flying time: 42,419:35 hours
Non-operational flying time: 14,213:53 hours

Source is 448th Bombardment Group (H) Statistical Summary dated 30 May 1945.

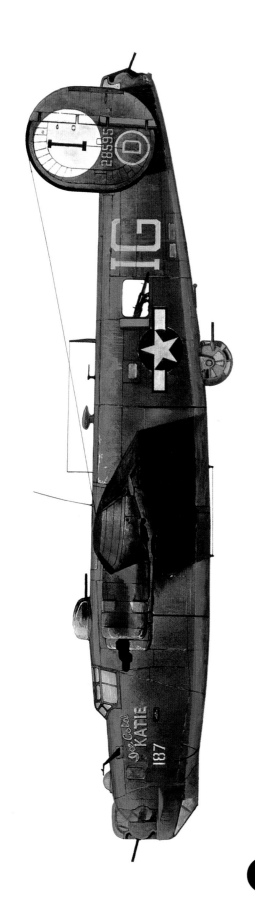
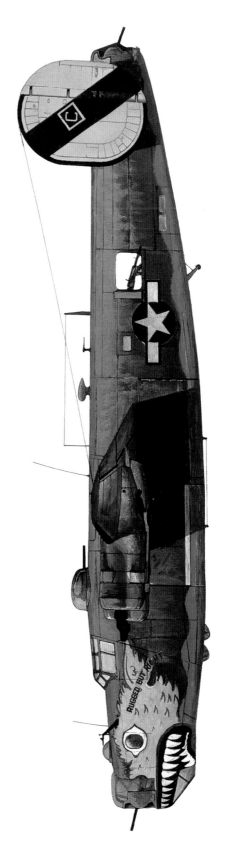

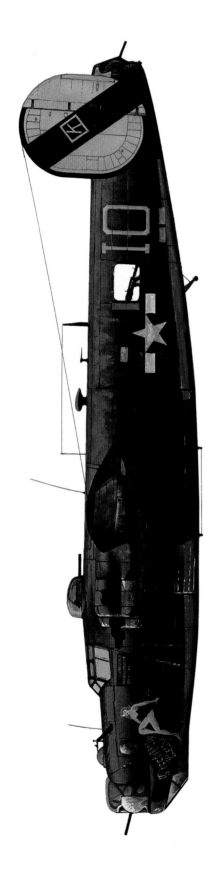

*The 448th Bomb Group Aircraft in Profile*

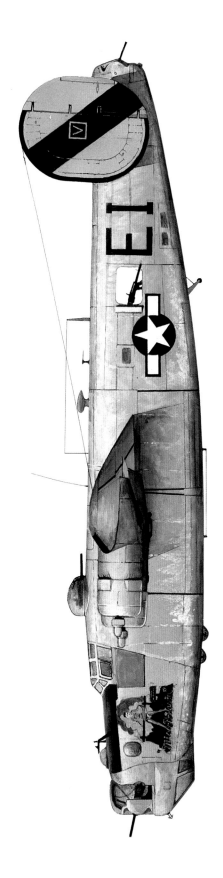

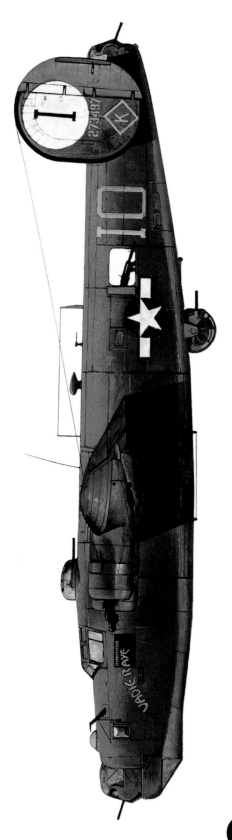
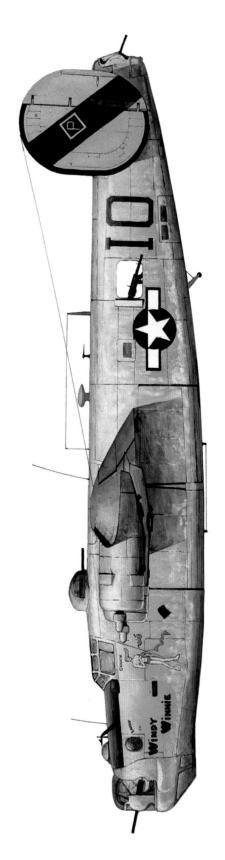

*The 448th Bomb Group Aircraft in Profile*

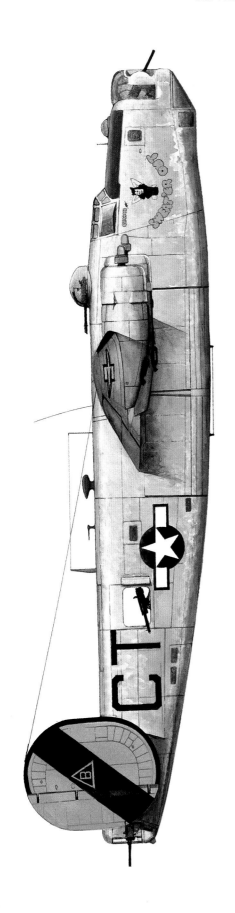

9

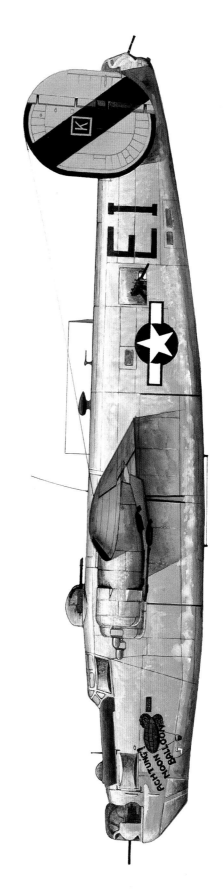

10

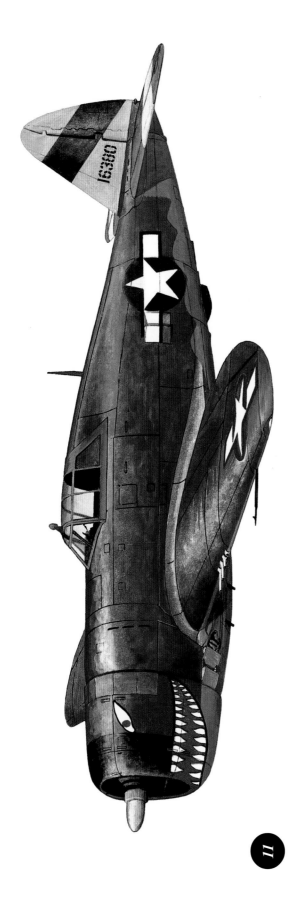
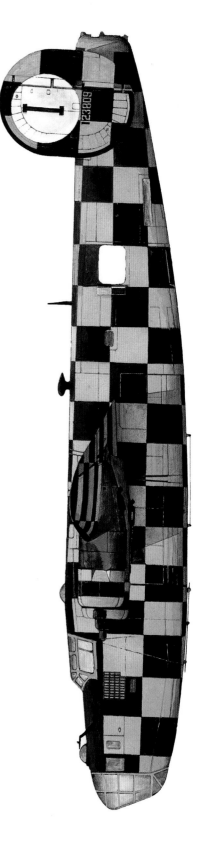

# PROFILE DESCRIPTIONS

1. ICE COLD KATIE. B-24H-1-DT. 41-29595. 713th Bomb Squadron. This Douglas-built B-24 was one of the Group's original aircraft brought from the U.S. to England in November 1943 by Lt. James Bell and crew. It displays the early olive drab paint scheme and tail markings. It completed twenty-seven missions before it was involved in a crash at the end of the runway with three other B-24s during the night of 22 April 1944.

2. RUGGED BUT RIGHT. B-24H-20-FO. 42-94953. 715th Bomb Squadron. Originally named MISANTHROPE this Ford-built aircraft was transferred to the Group in early 1945. Still in overall olive drab, it carried the later version tail markings. Notice the fuselage codes are not evident on this aircraft. It is probable the IO letters were added to the fuselage but it flew for some time without the markings. It completed at least twenty-three missions with the Group until the end of the war when it returned to the U.S.

3. PICCIDILLY LILLY. B-24H-20-CF. 42-50341. 715th Bomb Squadron. Commonly referred to as PICCADILLY LILLY, the actual spelling on the nose art is with an "i." This Consolidated-built Liberator was shot down on its 106th mission on 24 March 1945. The olive drab paint scheme attests to its age as the new arrivals to Group were silver or natural metal finish. The geometric shape on the tail indicates the squadron while the letter is the individual call letter of that aircraft.

4. SONIA. B-24H-25-FO. 42-95270. 713th Bomb Squadron. This aircraft first flew with the 712th BS before it was transferred to the 713th BS. This depiction shows the aircraft sporting the fuselage code of the 712th BS, CT, and the tail markings of the 713th BS, the circle. It flew in this configuration until the fuselage codes were eventually changed to match the tail markings. It flew eighty-three missions before the end of the war when it returned to the U.S.

5. MISS B-HAV'N. B-24J-1-FO. 42-95620. 714th Bomb Squadron. This aircraft was manufactured by Ford and saw service with three of the Group's four squadrons, 713th BS, 712th BS and finally the 714th BS. Nose art on the opposite side of the aircraft showed the same character riding a bomb. It was an original lead aircraft (non-radar) for the Group and accumulated sixty missions before it was lost to jet fighters on 4 April 1945.

6. FRISCO FRISKY. B-24J-1-DT. 42-51246. 713th Bomb Squadron. This Douglas-built aircraft arrived at Seething during the early summer of 1944 flying its first mission with the Group on 8 June 1944 and its last on 24 March 1945. It completed over 96 missions before it returned to the U.S. after the war.

7. VADIE RAYE. B-24J-50-CO. 42-73497. 715th Bomb Squadron. A Block 50 J model produced by Consolidated, it arrived at Seething on 19 January 1944. Among the first J models in Europe, it flew nineteen missions with the Group before intruder aircraft heavily damaged it during the night of 22 April 1944. After landing the aircraft exploded and burned.

8. WINDY WINNIE. B-24J-65-CF. 44-10599. 715th Bomb Squadron. Although depicted with the 715th Bomb Squadron, this J model also served with the 712th BS and the 714th BS. This aircraft first flew with the Group on 11 September 1944 and flew fifteen missions before it crash landed near Verdun, France on 28 January 1944. It was subsequently salvaged on 26 March 1945 after damage was determined to be too extensive for repairs.

9. SWEAT'ER OUT. B-24L-10-FO. 44-49593. 712th Bomb Squadron. This aircraft, produced by Ford, was one of the few L models flown by the Group. The L model was an attempt by manufacturers to speed up production as well as lighten the aircraft. One of the features was a lightweight tail turret shown in this example. This particular aircraft served as a lead aircraft flying eight missions in March and April 1945 before returning to the U.S. at war's end.

10. ACHTUNG! NOON BALLOON. B-24M-5-FO. 44-50540. 714th Bomb Squadron. This Ford-built M model was the last model received by the Group. The Group received only a handful of these M models before the war ended. This aircraft flew its first mission with the Group on 31 March 1945 and completed twelve missions before it returned to the U.S. after the war.

11. P-47C-5-RE. 41-6380. This unnamed "hack" aircraft was inherited by the Group from unknown origins to monitor formation assembly. The Group Air Executive Officer, Lt. Col. Hubert Judy, used this aircraft extensively to aid in assembly and give instructions to the formation before they departed the English coast on combat missions. This same aircraft also sported a slightly different paint scheme earlier than the livery depicted. It consisted of a yellow band on the engine cowl without the shark's mouth.

12. YOU CAWN'T MISS IT. B-24D-5-CO. 41-23809. Before arriving at Seething in January 1944, this aircraft was HELLSADROPPIN II of the 93rd BG. After it was declared war weary, the 448th BG repainted it and used it as an assembly ship. Commonly called CHECKERBOARD by the crews, it flew numerous assembly missions and even was listed on combat formation sheets on 3, 5, and 11 June 1944. The airframe eventually ran out of hours and was salvaged in early January 1945.

# INDEX

Abott, Thomas, 128
Adams, Ara, 220
Ahlstrom, Lennart, 215
Ahrendt, Carl, 113, 185
Allen, George, 72, 171
Allen, Tom, 50, 125
Alspaugh, Melvin, 90, 101, 126, 127
Ambrosini, Henry, 97
Anast, Nick, 186
Anderson, Chester, 131
Anderson, David, 193
Anderson, Ed, 200, 215
Anderson, Forrest, 221
Anderson, Harald, 215
Anderson, Lawrence, 97
Angelo, Arthur, 38
Apple, Thomas, 17, 22, 23, 74, 92
Aquinta, Carl, 164
Arluck, John, 97
Arnold, Seul, 16, 32
Ash, Robert, 75
Augustine, Karl, 202
Ausfresser, Seymour, 25, 27, 89, 128
Ayrest, Robert, 24, 27, 59

Back, 152
Baer, Ben, 29 ,41, 56, 144
Bailey, Ray, 194, 216
Bailey, William, 112, 113, 114
Baker, Grenville, 146
Balzer, Wally, 34, 41, 42, 53, 150, 153, 178
Barak, Jack, 42
Baranofsky, Stanley, 24, 80
Barilla, Frank, 229
Barkett, Thomas, 205
Barnett, Kenneth, 68
Barneycastle, L. C., 178
Barr, Stuart, 75, 76
Bastian, Frank, 143
Baukus, William, 77
Beanland, Harold, 225
Beaver, James, 127
Beck, Donald, 194

Beckman, Leland, 114
Bernard, Fred, 123
Bettcher, James, 116
Billings, Charles (Chuck), 24, 29
Bingham, Grover, 116
Binkley, Raymond, 186
Birdsall, Donald, 63
Birkhead, John, 178
Black, Jack, 80
Blalock, Clifford, 171
Blaney, Chuck, 214
Blank, James, 190, 226, 231
Blanton, Billie, 123, 124
Blanton, Charlie, 57, 104, 116
Block, Solomon, 207
Blum, William, 20, 23, 24, 25, 40, 58, 75, 86, 89, 128, 148
Boartfield, Ralph, 194
Boatright, James, 143
Bobak, Walter, 194
Boll, Raymond, 128
Bollschweiler, Wallace, 16
Bonner, Bill, 51
Booth, George, 126
Borsch, Joseph, 126
Botkin, Francis, 142
Bottoms, B. D., 18, 24, 41, 43, 57, 58, 111, 116, 151
Boula, Frank, 28
Bowers, Joe, 171
Bramhall, Aaron, 72
Brant, James, 43
Bretthauer, Johnny, 152
Bricker, James, 14, 74
Bringardner, John, 68, 72
Broadfoot, Albert, 186, 192, 205
Brock, Gordon, 187, 190, 202, 204, 217, 221
Broila, Donald, 134
Brown, 128
Brown, Charles, 36, 37
Brown, James, 73
Brown, Ralph, 100
Brown, William, 68

Broxton, Harvey, 29, 63, 85
Bullard, Sam, 66
Bunday, Raymond, 186, 189
Bunde, Herb, 80
Burchett, Milton, 187
Burke, Ronald, 211
Burzenski, Stephen, 39
Bushek, Joseph, 128
Buxton, John, 140

Caballero, Manuel, 97
Cagle, V. B., 16
Caldwell, John, 231
Calomeni, Alexander, 225
Campbell, Lawson, 43, 44, 67, 69
Campbell, Robert, 74
Cannon, Alfred, 126
Carlson, Carl, 97
Carney, Lawrence, 121
Carroll, Robert, 61, 74
Caruso, Larry, 225
Casey, Ralph, 97
Castell-Blanche, Robert, 135
Casto, Orvie, 182
Cates, Aubrey, 128
Cathey, John, 97
Chaney, Dick, 188
Chapman, Ed, 45, 86
Charette, Albert (Albee), 29, 87
Chartiers, Ray, 90
Chase, Phillip, 43
Chesser, 190
Chu, Ed, 192, 193, 197, 199, 205, 212, 217, 218, 219, 220, 221, 223
Cieslewicz, Victor, 119
Clark, Donald, 211
Clark, James, 97
Cleary, James, 16
Clift, Donald, 50
Cline, Clair, 63
Cobb, Willard, 70
Cochran, Pat, 187
Cohee, Raymond, 93, 97

*Index*

Cohen, Simon, 80
Coleman, Donald, 25
Colletti, Robert, 143
Conn, Billy, 145
Conn, Dewey, 114
Conner, Leroy, 115, 117, 122, 123, 126, 126, 128, 131, 133, 134, 136, 141, 152, 153, 155, 156, 157, 162, 166, 167
Conrad, James, 124
Cooksey, George, 113
Cooper, Stanley, 63
Copeland, George, 114
Covell, Dwight, 29, 63
Cruikshank, Frank, 32, 189
Crumbley, Calvin, 125
Cupp, Charles, 220
Curtis, James, 50
Cushman, John, 190
Custor, Ray, 188
Cuthbert, William, 88

Dague, Darwin, 187
Daley, John, 71, 72
Daneau, George, 97
Davis, Roy, 87
Davy, Mildred, 31
DeFillipino, Nick, 164
DeFranze, Jospeh, 43
Degnan, William, 143
Delay, Harold, 67
Delcambre, Pierre, 29
Della Selva, Mauro, 136
Demetropolous, William, 116
Dempsey, Jay, 128
Denbroeder, Adrian, 124
Dennis, Herbert, 171
Deren, John, 137
DeSoto, Kenneth, 43
Dickey, Harvey, 80
DiFrancisco, Donald, 225
DiGiacoma, Phil, 143
DiLorenzo, Albert, 15, 25, 28, 93, 97
Disborn, Donald, 143
Dogger, William, 121
Dogger, William, 129
Donovan, James, 122
Dorfman, Harold, 207, 209, 210
Drahos, Aldrich, 117
Drawhorn, Miles, 143
Drouin, Ed, 47
Dunham, William, 43
Dunn, Reginald, 16, 22
DuPont, George, 12, 13, 16, 19, 31, 41, 89, 93, 109, 188, 229
Durbin, Daniel, 193
Durley, E. P. (Earle), 16, 44, 83, 85
Dutka, John, 80
Dworaczyk, Eugene, 81

Eannone, Michael, 114
East, Garland, 116
Easterling, Earnest, 137
Edman, Lawrence, 67
Edwards, Jack, 51, 55, 74
Edwards, William, 75

Eggert, Carl, 114
Eggert, Carl, 152, 168
Elba, Irving, 63
Elder, Donald, 97
Elliot, Richard, 76
Ellis, Calvin, 210, 212, 219
Elver, Karl, 16, 32
Engdahl, Leroy, 44, 57, 102, 104, 105
Engel, Herman, 214
Everett, Benjamin, 33, 42, 230

Fager, Peter, 220
Ferguson, William, 32, 50
Ferrie, Parmley, 138
Feyti, John, 66
Fields, James, 182
Filipowicz, Stanley, 12, 14, 43, 60, 68, 78, 88, 91, 110
Fischer, Leslie, 113
Fitzgerald, Obed, 122
Flannery, Tom, 30
Flinder, Kay, 229, 186, 221
Ford, Donald, 229
Foss, John, 104
Foster, Thomas, 38, 43
Fowler, Harold, 124
Fox, Alfred, 126
Fox, Edward, 43
Fox, Roland, 118, 119
Franklin, George, 183
French, Frank, 66
Fuller, Michael, 112

Gaffney, George, 232
Gaines, Samuel, 32
Gaither, Clifford, 109
Gamble, William, 125, 155, 164, 167, 168, 169, 171
Garceau, Wayne, 121
Gardner, Robert, 137
Garland, Walter, 128
Garrett, William, 211
Garrison, Linn, 178
Garrity, Jack, 220
Gaskins, Gene, 16, 41, 56, 60, 61, 90, 101
Gelling, Ray, 43
Genarlsky, Frank, 167
Germany, Noble, 229, 232
Ghormley, Jack, 67
Gianakos, Nikolas, 16
Gibson, Frank, 91, 107
Gilbert, William, 141
Ginevan, Donald, 135
Givens, John, 171
Glass, Bobbie, 214
Gleason, James, 225
Glevanik, George, 12, 13, 14, 1516, 41, 56, 60, 61, 70, 74, 84, 88, 90, 99, 101
Glicksman, Jack, 125
Goldenberg, Royal, 76
Goodpasture, Morgan, 66
Goshorn, Richard, 143
Gottlieb, Gerald, 213
Graham, Daniel, 210
Greene, George, 191

Grubb, Dale, 125
Gruening, Huntington, 136
Grunow, John, 24, 27, 32, 44, 72
Guynes, James, 199, 202
Guyton, Graham, 50

Haas, Jerome, 128
Hackett, Chester, 14, 38, 87, 115
Hallinger, Roland, 34
Halpern, Milt, 136
Halvorson, Haari, 97
Hamby, Jesse, 80
Hammer, Elmer, 18, 25, 29, 45, 190
Hammes, William, 227
Haney, Edward, 34
Hanslik, Don, 43
Harper, Bob, 30
Harrell, Clarence, 16
Harrison, Paul, 76, 85
Harrison, Robert, 97
Hastings, Franklin, 224
Hau, Andy, 12, 14
Hausman, Hershel, 183, 184, 187, 196, 203, 207, 208, 216, 222, 223
Haynes, Bordie, 93
Heidrich, Herb, 164
Helganz, Arthur, 187
Helvey, Wesley, 63
Henderson, Rich, 23
Hensey, William, 171
Hershiser, Alden, 220
Hess, Edward, 137
Hewitt, Edwin, 124
Hicks, Marvin, 123, 125, 128, 129, 135, 138, 140, 154, 158, 163
Hillman, Gordon, 172
Hillman, Gordon, 184
Hoey, Patrick, 231
Hoffman, Arthur, 203
Holden, William, 215
Holmberg, Harry, 229
Holst, David, 143
Holst, Dewey, 143
Holt, Carl, 168, 182, 190
Homan, Paul, 192, 193, 195, 198, 199, 217, 223, 226, 227
Homelvig, Elmer, 212
Hood, James, 72
Hooks, Maurice, 20, 22
Horton, Marcus, 124, 175, 203
Howard, Orland, 97
Howell, Arthur, 159
Howell, Cleve, 118
Howell, Grady, 93
Hudgens, Ernest, 135
Hudson, Robert, 63
Hughey, Edward, 38
Hutton, Charles, 128

Ingalls, Maynard, 12
Ingelsby, Jim, 164
Irish, Gail, 188
Irons, W. R., 151
Irwin, W. E., 151
Isgrig, Benjamin, 113

Israel, Edward, 113
Issacson, Wesley, 188

Jackson, Bill, 90
Jackson, William, 88
James, Freddie, 188
James, Herman, 215
Janeczko, Chester, 43
Janson, Martin, 34
Jarol, Seymour, 121
Jasura, Peter, 152
Johnson, Ben, 158
Johnson, Walter, 91, 92
Jones, Charles, 135
Jones, Glenn, 122
Jones, Johnny, 93
Jones, Paul, 193, 214
Jonson, Herbert, 142
Jordan, Earl, 186, 205
Jordan, Max, 44
Judy, Hubert, 17, 22, 32, 72, 73, 81, 84, 85, 149

Kain, Jeese, 113
Kaiser, Joseph, 20, 22, 101, 167
Kaiser, William, 210
Kaplan, Selwyn, 205
Kapnick, Arnold, 34
Kearney, Jerry, 210
Keene, Thomas, 18, 29, 43, 77, 111, 115, 116, 117
Kepner, William, 111, 177
Kerniss, Fred, 105, 109, 110, 113, 120, 121, 127, 130, 134, 153, 175, 177, 181, 187, 188, 190
Kerrick, Bob, 93
Kessler, Robert, 128, 129, 133, 138, 140, 141, 147, 150, 152, 149, 158, 163, 166, 167, 172
Key, Carrol, 23
Kimball, Richard, 54
Kinsey, Jesse, 156
Kittredge, Abraham, 43
Klein, George, 124
Knights, Patricia, 31
Knorr, Charles, 24, 29, 80
Kohl, Al, 92
Kovalchick, Vladimir, 113
Kraft, Alton, 167
Kramer, Ronald, 16, 32, 100
Krepser, Frederick, 116, 151
Kubinski, Henry, 63
Kuchwara, Michael, 125
Kurk, Walter, 134
Kushner, John, 126
Kuzminski, Anthony, 88

Labus, Chester, 215
Lamb, James, 188
Lambertson, Robert, 50, 92, 100, 102
Lamoy, Jr., Claud, 187
Lane, Harold, 43
Lanphear, Byron, 38
LaPoint, Bert, 19, 31, 33, 164
Larson, Robert, 124
Law, Merle, 187

Lawnicki, Stephen, 126
Laws, John, 133
Lazarus, Arthur, 34
Lee, Cater, 14, 16, 22, 23, 84, 92, 105
Lee, Ralph, 162
Lee, Ray, 13, 14, 83, 90
Lehman, Robert, 95
Lewis, Albert, 149
Lewis, Robert, 73
Liebich, Joseph, 83
Liedka, Vincent, 83
Litman, Irvin, 28
Lloyd, Robert, 152
Locke, Alfred, 98
Long, Andrew, 100
Lougheed, J. H., 149
Loyd, Ira, 63
Lozes, Elbert, 16, 34, 35, 39, 41, 48, 60, 66, 83, 100, 101
Lyles, John, 171
Lyons, Edgar, 182

Macauley, Harold, 143
MacDonald, John, 171
MacKenzie, Robert, 84
Madden, Joseph, 133
Maggenti, Enrico, 143
Main, Robert, 220
Maino, Alfred, 65
Mains, Robert, 219
Major, Harold, 219
Malone, Edward, 123, 181, 194
Manning, David, 38
Markewicz, Edward, 60
Marroon, Floyd, 54
Marshall, Austin, 16
Martin William, 88
Martin, Robert, 44, 61, 69
Mason, Gerry, 82, 94, 105, 106, 110, 115, 119, 126, 133, 134, 158, 169
Masters, John, 118
Matthaes, Hosea, 134
May, Bert, 82
May, Robert, 231, 232
Maynard, Henry, 97
Maynard, William, 125
Mazzagatti, Phillip, 182
McAllister, Ronald, 60
McBride, Charles, 12, 13, 17, 80
McCluhan, Neil, 210
McConkie, Jim, 179
McCoy, John, 225
McCready, Forrest, 187
McCune, John, 15, 18, 20, 22, 29, 32, 51, 55, 58, 69, 70, 87, 94
McFarland, Hugh, 196, 197, 209
McGinnis, Edward, 126
McHugh, John, 214
McKain, Armor, 124
McKinley, Fredrick, 225
McLaughlin, Eddie, 187
McLaughlin, Newton, 73, 161
McNamara, Thomas, 43
McVean, Peter, 121
Meents, Edward, 67

Meigs, Ralph, 93
Meining, Charles, 175
Mellor, Harrison, 79
Mercer, Jack, 118
Merkling, John, 29
Merkovich Frank, 219
Merrill, Hairman, 225
Meyerowitz, Arthur, 43
Mickelson, Richard, 194
Middleworth, Leroy, 190
Mied, Arthur, 70
Mielke, Henry, 175
Miller, Lester, 85, 113, 116, 170, 198, 230, 231
Miltner, Robert, 74
Misuraca, James, 66
Mobbs, Mildred, 150
Monefeldt, Leonard, 211
Moody, Richard, 125, 141
Moore, Emmett, 50
Moran, Edwin, 57, 116
Morgan, Minor, 82
Morris, William, 125
Morrison, Girthel, 209
Morse, Douglas, 141
Morse, Lloyd, 20
Moseley, Kent, 83
Mucha, Jim, 214
Mulrain, Harry, 186
Munoz, Antonio, 171
Musselman, Everett, 73
Myers, William, 222, 229

Nardi, Richard, 60
Nathan, Joseph, 176
Nelson, Kenneth, 232
Newcomb, Eathan, 178
Nickerson, Joseph, 63
Nimmo, Robert, 117
Nininger, Harold, 97
Nissen, Charles, 81
Nugent, James, 72

O'Brien, Jack, 20, 24, 25, 27, 58, 88, 100, 106, 128
O'Neil, Paul, 184
O'Neil, Ralph, 137
Olhaber, John, 159
Onks, Marvin, 25
Owen, Edward, 104
Owens, Louis, 171

Pace, Artie, 29
Padgett, Guy, 63
Padilla, Vincent, 122, 200
Paeschke, Robert, 190
Palmerton, Elwyn, 120
Panicci, Andrew, 143
Paretti, Edward, 225
Parker, Jack, 23, 26, 93
Parkinson, Ken, 32
Patterson, Earl, 187
Paxson, John, 210, 211, 212
Peek, Marion, 92, 95
Pegher, Jim, 92

## Index

Perez, Sigifredo, 19
Perkowski, Michael, 171
Perry, Gerald, 203
Peterson, Warren, 215
Petrovich, Walter, 210
Pettit, Neal, 208, 210, 212, 219
Phillips, Frank, 75
Phillips, John, 149
Piliere, Francis, 183, 184, 211
Pilley, Ernest, 146
Pinkus, Robert, 77
Pittinger, Harold, 187
Pitts, Cherry, 29, 70, 90, 92
Platt, Charles, 157, 185, 204
Podolsky, Harold, 128
Pokorny, Richard, 232
Pomfret, Joseph, 81
Ponge, William, 97
Popa, Aurel, 125
Postemsky, Edmond, 144
Powell, Furman, 92
Prieb, Kenneth, 124
Protich, Peter, 168
Prouty, Robert, 188
Pulcipher, Eugene, 90

Quirk, John, 171

Rademacher, Rudolf, 214
Ray, John, 212, 220
Ray, Phil, 125
Rebeles, Julius, 77, 84, 87, 89, 96, 103, 137
Reddit, Joseph, 40
Reid, William, 34
Reindal, Russel, 55, 74, 82, 87, 106
Rethal, John, 188
Rhodes, John, 24, 27
Riendeau, Paul "Frenchy", 12
Rigel, Horace, 222
Risinger, Fred, 210
Robicheau, George, 93
Robinson, Ernest, 87, 90
Roffwarg, Eli, 193
Rogers, Francis, 43
Rogers, William, 93, 97
Roorda, Delwin, 195
Roosevelt, Franklin D., 224
Ross, William, 70
Rowe, Clair, 212
Rowe, John, 141, 145, 149, 162, 163, 175, 180, 183, 184, 185, 191, 194, 203, 204
Ruge, Irwin, 187
Runyan, Theodore, 34, 145

Sampson, Bob, 183
Sanders, Albert, 175, 176
Sansburn, George, 116
Sarkovich, Lewis, 12, 26, 62, 65, 95
Schall, 194
Schierbrock, James, 123
Schilling, Erwin, 195
Schleicher, Don, 203
Schlicher, John, 136
Schlund, Karl, 87
Schneider, William, 224

Schroeder, Edward, 28, 58
Schroeder, Harlyn, 179, 186, 194
Schroeder, Henry, 18, 95
Schuman, Donald, 51
Schwartz, Raymond, 160
Schwinn, William, 50, 147, 225
Searles, William, 16, 32, 173
Self, Errol, 98
Shafter, James, 210, 212, 219
Shank, Joseph, 27
Sharp, Claire, 63
Sheehan, Francis, 38, 16, 41, 43, 56, 60, 70, 83, 90
Sheldon, Gail, 55, 77, 87
Sherlock, Paul, 124
Shia, John, 121
Short, Eugene, 187
Sidey, Elliot, 171, 173
Silver, Robert, 104, 107
Skaggs, A. D., 13, 14, 15, 16, 20, 23, 24, 29, 32, 34, 35, 37, 43, 50, 51, 56, 60, 62, 66, 74, 76, 78, 83, 84, 85, 86, 90, 96, 101, 107, 110, 120, 122, 125, 126
Slack, Robert, 121
Slocumb, Ray, 11
Smarinsky, Irving, 202
Smith, Castleton, 74
Smith, Harold, 116
Smith, Leroy, 170, 230
Snavely, William, 123, 125, 128, 133, 135, 148, 150, 154, 156, 158, 166, 184, 191
Snell, Leonard, 60
Snider, John, 156, 185
Snowbarger, Edward, 97
Soldan, Harold, 157, 186
Solomon, Ridd, 29
Sparacio, Salvatore, 124
Spruill, Robert, 104
Spurr, Woody, 148
Squyres, Kenneth, 22
Squyres, Kenneth, 50
Stalland, Knute, 212, 213
Stanford, John, 211, 214, 228
Stebbins, Jack, 188
Steele, Arthur, 13, 18, 24, 26, 27, 88, 128
Steffan, Joseph, 213
Stelzer, Bernard, 124
Stephens, Leo, 140
Stinson, Henry, 224
Stokes, J. P., 32
Stonebraker, William, 144
Stringfellow, Glassel, 104, 106, 112, 120, 121, 170
Stroop, Ray, 25, 28
Suchorsky, George, 188
Suhay, Richard, 125
Sullivan, James, 119
Susino, Charles, 70
Swayze, Jack, 23, 24, 26, 183
Szudarek, Severyn, 169, 175

Tarkington, Taylor, 210, 220
Tarrant, "Moose", 29
Taylor, David, 32
Taylor, Earl, 112

Teague, Alan, 40, 80
Tellman, Alfred, 97
Thalhamer, Richard, 69
Theobold, Frederick, 143
Thompson, James, 11, 12, 32, 35, 50, 78, 79, 80
Thompson, Harriet, 73
Thompson, Heber, 14, 79, 198, 210, 230, 231
Thompson, John, 153
Thompson, Tommy, 14
Thomson, Morris, 18
Thornton, Len, 151
Thornton, Robert, 82
Tod, Fredrick, 212, 215
Todt, Don, 15, 24, 41, 48, 60, 66, 85, 91
Torrance, Douglas, 215, 225
Towles, Raymond, 112
Tozzi, 190
Truman, Harry S., 224
Turner, Jim, 89, 179
Turpin, Harold, 121
Turpin, Max, 97

Unwin, Clifford, 144
Urban, James, 14, 51

Valoc, Russ, 143
Van Blair, Dale, 98
Verran, Thomas, 97
Vetterneck, Edward, 195
Vogel, Richard, 148
Voight, Robert, 45, 74
Voight, William, 182, 207, 209, 210

Wagner, Dick, 194
Wahnee, Myers, 76
Wais, Dan, 115
Wall, Edward, 186
Wandell, Everard, 25, 48
Warke, William, 122
Warner, Theodore, 213
Warnock, Ronald, 66
Warren, William, 79
Wasalik, Anthony, 95
Waters, Ray, 94, 102, 103, 109
Weaver, Kenneth, 80
Weinberger, George, 177
Welkowitz, Ira, 178
Welsh, Ralph, 118
Wenthe, George, 70
Wentzel, 93
Wermert, Peter, 81
Wermeyer, Raymond, 117
Werts, Jack, 133
Westover, Charles, 169, 170, 172, 190, 207, 209, 210, 228, 229, 230
Westphal, Ross, 188
Westrick, Paul, 205
Wheelock, Richard, 50
Whipple, Lucian, 207, 220, 222, 223, 225
White, John, 112, 113, 114, 118, 131
Whitson, William, 214
Whyte, James, 60
Wickham, Dick, 161
Wiegel, 190

Wight, Allen, 184
Wikander, 212
Wilbur, William, 143, 145
Wilder, Charles, 29
Wilhelmi, William, 209, 210
Williams, J. M., 92
Williams, John, 67
Wilson, George, 124, 128
Winter, Bruce, 105
Wisch, Ray, 47

Wolfe, Charles, 138
Wood, Raymond, 194
Wright, 194
Wright, Anderson, 210, 220
Wright, James, 232

Yarnell, Robert, 63
York, Charles, 71
Yuengert, Walter, 50
Yule, Fred, 209

Zabrowski, Stanley, 19, 55, 57, 60, 62, 65, 70, 85, 87, 94, 95, 96, 97, 99, 100, 104, 105
Zeldin, Donald, 134, 137, 139, 142, 144, 147, 166, 167, 168, 169, 182, 208, 209, 220, 221, 223, 225, 226, 227, 228
Zierdt, Kenneth, 113
Zima, John, 142, 144
Zonyk, Joe, 30, 157, 173, 174, 175, 187, 198

*Also from the Publisher*

**The Debden Warbirds: The 4th Fighter Group in World War II.** *Frank E. Speer.* The P-51 Mustangs of the 4th Fighter Group were the first to escort bombers over Berlin during World War II, the first to escort bombers from England to Russia, and at war's end ranked first in the number of enemy aircraft destroyed (over 1,000). This authentic account, gleaned from the Squadron and Tower Diaries, is enhanced by dozens of combat reports and personal accounts from pilots and crews whose day-to-day encounters are faithfully recorded
Size: 8 1/2" x 11" • over 170 b/w and color photos • 224 pp.
ISBN: 0-7643-0725-8 • hard cover • $45.00

**The Eight Ballers: Eyes of the Fifth Air Force - The 8th Photo Reconnaissance Squadron in World War II.** *John Stanaway & Bob Rocker.* Beginning operations in April 1942 with four Lockheed F-4 Lightnings, the 8th Photo Squadron gave the American Army Air Forces its only aerial reconnaissance coverage of the Southwest Pacific during the first part of the war. Over 500 photos, most of which have never been published before. Many new facts are added to the squadron diary that was kept when the 8th operated almost singlehandedly in 1942-43.
Size: 8 1/2" x 11" • over 510 b/w and color photos • 184 pp.
ISBN: 0-7643-0910-2 • hard cover • $39.95

**Wrong Place, Wrong Time: The 305th Bomb Group & the 2nd Schweinfurt Raid.** *George C. Kuhl.* This is true story of the second raid on Schweinfurt, Germany by the Eighth Air Force 1st and 3rd Bombardment Divisions on 14 October 1943. On this day, the Eighth Air Force lost air superiority to the German Luftwaffe in a continuous air battle that lasted over three hours. Many refer to it as the greatest one-day air battle of World War II. Wrong Place, Wrong Time is a study of the 1st Bombardment Division and specifically the 305th Bomb Group on that fateful day.
Size: 6" x 9" • photo section • 248 pp.
ISBN: 0-88740-445-6 • hard cover • $24.95

**The 451st Bomb Group in World War II - A Pictorial History.** *Mike Hill.* This new book is an illustrated history of the 451st Bomb Group in World War II. As part of the USAAF's 15th Air Force, they flew their B-24 Liberators from Castellucia, Italy to heavily defended targets throughout the Mediterranean theater of war. During their combat tour the 451st was awarded three Presidential Unit Citations for their ability to get to the target no matter what the enemy threw at them.
Size: 8 1/2"x11" • over 600 b/w photographs • 160 pp.
ISBN: 0-7643-1287-1 • hard cover • $45.00

**Any Place, Any Time, Any Where: The 1st Air Commandos in World War II.** *R.D. van Wagner.* As an air arsenal which was totally unique in its composition and application, it combined the firepower of P-51A fighters and B-25H bombers with the logistical tentacles of C-47 transports, CG-4A gliders, L-5 and L-1 light planes, and UC-64 bush planes to reach far behind Japanese lines. Unorthodox and eclectic, the 1st Air Commandos serve as a model for conventional and special operations today.
Size: 8 1/2" x 11" • over 200 b/w photographs, maps • 176 pp.
ISBN: 0-7643-0447-X • hard cover • $39.95

**Gabby: A Fighter Pilot's Life.** *Francis Gabreski & Carl Molesworth.* If ever a man has earned his place in the annals of military history, that man is Francis "Gabby" Gabreski. His exploits as a fighter pilot in World War II and Korea are legendary; his rise from humble beginnings to success in military and business careers is inspiring. Gabreski's life is a classic American success story. Now, drawing on his private documents and photographs, Gabby, along with writer Carl Molesworth, tells his thrilling eyewitness story with a candor and a vivid style that should earn this brave pilot a whole new generation of admirers.
Size: 8 1/2" x 11" • over 200 b/w photographs, 8 color aircraft profiles • 176 pp.
ISBN: 0-7643-0442-9 • hard cover • $45.00

**Herky! The Memoirs of a Checker Ace.** *Herschel H. Green.* The dramatic life story of one of the legendary USAAF fighter pilots of World War II who fought across the skies over the Mediterranean and southern Europe in the great aerial campaigns against the Luftwaffe – Herschel H. "Herky" Green. By the time Colonel Green was grounded by orders of higher headquarters, he was the leading ace of the 15th Air Force with eighteen aerial victories.
Size: 8 1/2" x 11" • over 150 b/w photographs • 192 pp.
ISBN: 0-7643-0073-3 • hard cover • $45.00

**Happy Jack's Go Buggy: A Fighter Pilot's Story.** *Jack Ilfrey.* Many new never before published photographs and a special color photo section. Fly with Ilfrey in his P-38 as he and his unit become the first group of American aircraft to fly from the USA to England. Thrill to the stories of aerial combat over North Africa as Ilfrey becomes one of America's first WWII air aces. Marvel at his flying exploits as a member of the 20th Fighter Group/8th Air Force and join him on his evasion story through German occupied France.
Size: 8 1/2" x 11" • over 190 b/w and color photographs, 3 color aircraft profiles • 128 pp.
ISBN: 0-7643-0664-2 • hardcover • $35.00

**The Jolly Rogers: The 90th Bombardment Group in the Southwest Pacific 1942-1944.** A full-color facsimile reprint of the actual Jolly Rogers war book. No book about any Air Corps unit could ever be complete without the laughter, the irritation, the discomfort, the natives ... the backdrop against which the greatest of all human dramas was played. It is with deep reverence and humility that this book is dedicated to the men whom no book could ever honor sufficiently ... the men of the Jolly Rogers who died for their country.
Size: 11" x 8 1/2" • 500+ color photos, illustrations • 112 pp.
ISBN: 0-7643-0258-2 • hard cover • $39.95